Advanced Mathematics: Concepts and Applied Principles

Advanced Mathematics: Concepts and Applied Principles

Editor: Calanthia Wright

NY RESEARCH PRESS
New York

Published by NY Research Press
118-35 Queens Blvd., Suite 400,
Forest Hills, NY 11375, USA
www.nyresearchpress.com

Advanced Mathematics: Concepts and Applied Principles
Edited by Calanthia Wright

International Standard Book Number: 978-1-63238-626-7 (Hardback)

Cataloging-in-Publication Data

Advanced mathematics : concepts and applied principles / edited by Calanthia Wright.
p. cm.
Includes bibliographical references and index.
ISBN 978-1-63238-626-7
1. Mathematics. I. Wright, Calanthia.
QA36 .A38 2019
510--dc23

Contents

Preface

The purpose of the book is to provide a glimpse into the dynamics and to present opinions and studies of some of the scientists engaged in the development of new ideas in the field from very different standpoints. This book will prove useful to students and researchers owing to its high content quality.

Developed as a subject of abstraction and logic to represent and solve real world problems, mathematics has transcended its boundaries into advanced domains of quantum mechanics, string theory, field theory, group theory and complex computation. It uses the principles of arithmetic, advanced algebra, calculus and geometry to develop mathematical formulations for such domains. Mathematical models of game theory, optimization and approximation theories have been used in the fields of economics, chemistry, biology, etc. This book discusses the advanced principles of pure and applied mathematics as well as their applications in modern sciences. While understanding the long-term perspectives of the topics, the book makes an effort in highlighting their impact as a modern tool for the growth of all scientific disciplines. This book is a complete source of reference for researchers, mathematicians and students on the present status of this important field.

At the end, I would like to appreciate all the efforts made by the authors in completing their chapters professionally. I express my deepest gratitude to all of them for contributing to this book by sharing their valuable works. A special thanks to my family and friends for their constant support in this journey.

Editor

1

Multiple Heteroclinic solutions of bilateral difference systems with Laplacian operators

Yuji Liu · Shengping Chen

Abstract Sufficient conditions guaranteeing the existence of three Heteroclinic solutions of a class of bilateral difference systems are established using a fixed point theorem. It is the purpose of this paper to show that the approach to get Heteroclinic solutions of BVPs using multi-fixed-point theorems can be extended to treat the bilateral difference systems with the nonlinear operators $x \to \Delta[p\phi(\Delta x)]$ and $y \to \Delta[q\psi(\Delta y)]$.

Keywords Laplacian operator · Bilateral difference system · Heteroclinic solution · Fixed point theorem

Mathematics Subject Classification 34B10 · 34B15 · 35B10

Introduction

Difference equations appear naturally as analogues and as numerical solutions of differential and delay differential equations having applications in applied digital control [1–3], biology, ecology, economics, physics and so on. Although difference equations are very simple in form, it is extremely difficult to understand thoroughly the behaviors of their solution [25]. In recent years, there have been many papers interested in proving the existence of positive solutions of the boundary value problems (BVPs for short) for the finite difference equations since these BVPs have extensive applications, see the papers [5–17] and the references therein.

Y. Liu (✉) · S. Chen
Department of Mathematics, Guangdong University of Finance and Economics, Guangzhou 510000, People's Republic of China
e-mail: liuyuji888@sohu.com

Contrary to the case of boundary value problems in compact domains, for which a very wide literature has been produced, in the framework of unbounded intervals many questions are still open and the theory presents some critical aspects. One of the main difficulties consists in the lack of good priori estimates and appropriate compact embedding theorems for the usual Sobolev spaces.

Recently, the authors [18–22] studied the existence of solutions of the boundary value problems for infinite difference equations. In [19], the existence of multiple positive solutions of the boundary value problems for second-order discrete equations

$$\begin{cases} \Delta^2 x(n-1) - p\Delta x(n-1) - qx(n-1) + f(n, x(n)) = 0, n \in N, \\ \alpha x(0) - \beta \Delta x(0) = 0, \\ \lim_{n\to+\infty} x(n) = 0 \end{cases} \tag{1.1}$$

was investigated using the cone compression and expansion and fixed point theorems in Frechet spaces with application, where $N = \{0, 1, 2, \ldots\}$ the set of all non-negative integers, $\alpha > 0, \beta > 0$, $p > 0, q > 0$ and f is a continuous function and $\Delta x(n) = x(n+1) - x(n)$.

In paper [22], it was considered the existence of solutions of a class of the infinite time scale boundary value problems. It is easy to see that the results in [22] can be applied to the following BVP for the infinite difference equation

$$\begin{cases} \Delta^2 x(n) + f(n, x(n)) = 0, \quad n \in N, \\ x(0) = 0, \\ x(n) \text{ is bounded.} \end{cases} \tag{1.2}$$

The methods used in [22] are based upon the growth argument and the upper and lower solutions methods.

In [23], motivated by some models arising in hydrodynamics, Rachunek and Rachunkoa studied the second-order non-autonomous difference equation

$$\Delta x(n) = \left(\frac{n}{n+1}\right)^2 (\Delta x(n-1) + h^2 f(x(n))), \quad n \in N,$$

which can be transformed to the following form:

$$\Delta(n^2 \Delta x(n-1)) = h^2 (n+1)^2 f(x(n)), \quad n \in N,$$

where $h > 0$ is a parameter and f is Lipschitz continuous and has three real zeros $L_0 < 0 < L$, conditions for f under which for each sufficiently small $h > 0$ there exists a homoclinic solution of the above equation were presented. The homoclinic solution is a sequence $\{x(n)\}_{n=0}^{\infty}$ satisfying the equation and such that $\{x(n)\}_{n=0}^{\infty}$ is increasing, $x(0) = x(1) \in (L_0, 0)$ and $\lim_{n\to\infty} x(n) = L$.

We note that the difference equations discussed in [19, 22, 23] are those ones defined on $N = \{0, 1, 2, \cdots\}$. The existence of homoclinic solutions for second-order discrete Hamiltonian systems have been studied in [24, 26] by using fountain theorem.

Motivated by above mentioned papers, the purpose of this paper was to investigate the following boundary value problem of the second-order bilateral difference system using a different method

$$\begin{cases} \Delta[p(n)\phi(\Delta x(n))] + f(n, x(n), y(n)) = 0, \quad n \in Z, \\ \Delta[q(n)\psi(\Delta y(n))] + g(n, x(n), y(n)) = 0, \quad n \in Z, \\ \lim\limits_{n\to-\infty} x(n) - \sum\limits_{n=-\infty}^{+\infty} \alpha_n x(n) = 0, \\ \lim\limits_{n\to-\infty} y(n) - \sum\limits_{n=-\infty}^{+\infty} \gamma_n y(n) = 0, \\ \lim\limits_{n\to+\infty} \phi^{-1}(p(n))\Delta x(n) - \sum\limits_{n=-\infty}^{+\infty} \beta_n \Delta x(n) = 0, \\ \lim\limits_{n\to+\infty} \psi^{-1}(q(n))\Delta y(n) - \sum\limits_{n=-\infty}^{+\infty} \delta_n \Delta y(n) = 0, \end{cases} \tag{1.3}$$

where

(a) Z denotes the set of all integers, $\Delta x(n) = x(n+1) - x(n)$,

(b) $p(n), q(n) > 0$ for all $n \in Z$ satisfying

$$\sum_{s=-\infty}^{+\infty} \frac{1}{\phi^{-1}(p(s))} = +\infty, \ \sum_{s=-\infty}^{0} \frac{1}{\phi^{-1}(p(s))} < +\infty,$$
$$\sum_{s=-\infty}^{+\infty} \frac{1}{\psi^{-1}(q(s))} = +\infty, \ \sum_{s=-\infty}^{0} \frac{1}{\psi^{-1}(q(s))} < +\infty,$$

(c) $\alpha_n, \beta_n, \gamma_n, \delta_n \geq 0$ for all $n \in Z$ and satisfy

$$\sum_{n=-\infty}^{+\infty} \alpha_n < 1,$$
$$\sum_{n=-\infty}^{+\infty} \alpha_n \sum_{s=-\infty}^{n-1} \frac{1}{\phi^{-1}(p(s))} < +\infty,$$
$$\sum_{n=-\infty}^{+\infty} \frac{\beta_n}{\phi^{-1}(p(n))} < \frac{1}{\phi^{-1}(1+\beta)} \quad \text{with } \beta > 0,$$

and

$$\sum_{n=-\infty}^{+\infty} \gamma_n < 1,$$
$$\sum_{n=-\infty}^{+\infty} \gamma_n \sum_{s=-\infty}^{n-1} \frac{1}{\psi^{-1}(q(s))} < +\infty,$$
$$\sum_{n=-\infty}^{+\infty} \frac{\delta_n}{\psi^{-1}(q(n))} < \frac{1}{\psi^{-1}(1+\delta)} \quad \text{with } \delta > 0,$$

(d) $f, g : Z \times [0, +\infty) \times [0, +\infty) \to [0, +\infty)$, both f and g are Caratheodory functions (see Definition 1 in "Main results"), and for each $n_0 \in Z, f(n,0,0)^2 + g(n,0,0)^2 \not\equiv 0$ for $n \leq n_0$,

(e) ϕ is defined by $\phi(x) = |x|^{s-2}x$ with $s > 1$, and $\psi(x) = |x|^{t-2}x$ with $t > 1$, their inverse functions are denoted by ϕ^{-1} and ψ^{-1}, respectively.

A pair of bilateral sequences $\{(x(n), y(n))\}$ is called a Heteroclinic solution of BVP (1.3) if $x(n), y(n)$ satisfy all equations in (1.3), $x(n) \geq 0, y(n) \geq 0$ for all $n \in Z$ and either $x(n) > 0$ for all $n \in Z$ or $y(n) > 0$ for all $n \in Z$.

We establish sufficient conditions for the existence of at least three Heteroclinic solutions of BVP (1.3). This paper may be the first one to study the solvability the boundary value problems of bilateral difference systems. The most interesting part in this article is to construct the nonlinear operator and the cone; this constructing method is not found in known papers.

The remainder of this paper is organized as follows: in "Main results", we first give some lemmas, then the main result (Theorem 1 in "Main results") and its proof are presented. An example is given in "An example" to illustrate the main result.

Main results

In this section, we present some background definitions in Banach spaces, state an important three fixed point theorem [4] and prove some technical lemmas. Then the main result is given and proved.

Denote

$$P_n = 1 + \sum_{s=-\infty}^{n-1} \frac{1}{\phi^{-1}(p(s))}, Q_n = 1 + \sum_{s=-\infty}^{n-1} \frac{1}{\psi^{-1}(q(s))}.$$

Definition 1 F is called a Caratheodory function if it satisfies that

$$(x,y) \to F(n, P_n x, Q_n y)$$

is continuous, and for each $r > 0$ there exists a nonnegative bilateral real number sequence $\{\phi_r(n)\}$ with $\sum_{n=-\infty}^{+\infty} \phi_r(n) < +\infty$ such that

$$|F(n, P_n x, Q_n y)| \le \phi_r(n)$$

for all $n \in Z, |x| \le r, |y| \le r$.

As usual, let E be a real Banach space. The nonempty convex closed subset P of E is called a cone in E if $ax \in P$ and $x+y \in P$ for all $x, y \in P$ and $a \ge 0$, and $x \in E$ and $-x \in E$ imply $x = 0$. A map $\varphi : P \to [0, +\infty)$ is a nonnegative continuous concave (or convex) functional map provided φ is nonnegative, continuous and satisfies

$\varphi(tx + (1-t)y) \ge$ (or $\le$) $t\varphi(x) + (1-t)\varphi(y)$ for all $x, y \in P, t \in [0,1]$.

An operator $T; E \to E$ is completely continuous if it is continuous and maps bounded sets into relatively compact sets.

Let E be a real Banach space, P be a cone of E, $\varphi : P \to P$ be a nonnegative convex continuous functional. Denote the sets by

$$P_c = \{x \in P : ||x|| < c\}, \ \overline{P}_c = \{x \in P : ||x|| \le c\}$$

and

$$P(\varphi; b, d) = \{x \in P : \varphi(x) \ge b, ||x|| \le d\}.$$

Lemma 1 Suppose that E is a Banach space and P is a cone of E. Let $T : \overline{P}_c \to \overline{P}_c$ be a completely continuous operator and let φ be a nonnegative continuous concave functional on P. Suppose that there exist $0 < a < b < d \le c$ such that $\varphi(y) \le ||y||$ for all $y \in \overline{P}_c$ and

(C1) $\{y \in P(\varphi; b, d) | \varphi(y) > b\} \ne \emptyset$ and $\varphi(Ty) > b$ for $y \in P(\varphi; b, d)$;

(C2) $||Ty|| < a$ for $||y|| \le a$;

(C3) $\varphi(Ty) > b$ for $y \in P(\varphi; b, c)$ with $||Ty|| > d$.

Then T has at least three fixed points y_1, y_2 and y_3 such that $||y_1|| < a$, $\varphi(y_2) > b$ and $||y_3|| > a$ with $\varphi(y_3) < b$.

Choose

$$X = \left\{ \{x(n) : n \in Z\} : \begin{array}{c} x(n) \in R, n \in Z, \\ \text{there exist the limits} \\ \lim_{n\to+\infty} \frac{x(n)}{P_n}, \ \lim_{n\to-\infty} \frac{x(n)}{P_n} \end{array} \right\}.$$

Define the norm

$$||x||_X = ||x|| = \sup_{n \in Z} \frac{|x(n)|}{P_n}, \quad x \in X.$$

It is easy to see that X is a real Banach space.

Choose

$$Y = \left\{ \{y(n) : n \in Z\} : \begin{array}{c} y(n) \in R, n \in Z, \\ \text{there exist the limits} \\ \lim_{n\to+\infty} \frac{y(n)}{Q_n}, \ \lim_{n\to-\infty} \frac{y(n)}{Q_n} \end{array} \right\}.$$

Define the norm

$$||y||_Y = ||y|| = \sup_{n \in Z} \frac{|y(n)|}{Q_n}, \quad y \in Y.$$

It is easy to see that Y is a real Banach space.

Let $E = X \times Y$ be endowed with the norm

$$||(x,y)|| = \max\{||x||, ||y||\}, \quad (x,y) \in E.$$

Then E is a real Banach space.

Let $h(n) \ge 0$ for every $n \in Z$ be a bilateral sequence with $\sum_{n=-\infty}^{+\infty} h(n)$ converging. Consider the following BVP:

$$\begin{cases} \Delta[p(n)\phi(\Delta x(n))] + h(n) = 0, \quad n \in Z, \\ \lim_{n\to-\infty} x(n) - \sum_{n=-\infty}^{+\infty} \alpha_n x(n) = 0, \\ \lim_{n\to+\infty} \phi^{-1}(p(n))\Delta x(n) - \sum_{n=-\infty}^{+\infty} \beta_n \Delta x(n) = 0. \end{cases} \tag{2.1}$$

Lemma 2 Suppose that (b), (c) and (e) hold. Then $x \in X$ is a solution of BVP (2.1) if and only if

$$x(n) = \frac{1}{1 - \sum_{n=-\infty}^{+\infty} \alpha_n} \sum_{-\infty}^{+\infty} \alpha_n \sum_{s=-\infty}^{n-1} \frac{1}{\phi^{-1}(p(s))} \phi^{-1}\left(A_h + \sum_{t=s}^{+\infty} h(t)\right) + \sum_{s=-\infty}^{n-1} \frac{1}{\phi^{-1}(p(s))} \phi^{-1}\left(A_h + \sum_{t=s}^{+\infty} h(t)\right), \tag{2.2}$$

where $A_h \in \left[0, \frac{1}{\beta}\sum_{s=-\infty}^{+\infty} h(s)\right]$ such that

$$\phi^{-1}(A_h) = \sum_{n=-\infty}^{+\infty} \frac{\beta_n}{\phi^{-1}(p(n))} \phi^{-1}\left(A_h + \sum_{s=n}^{+\infty} h(s)\right). \tag{2.3}$$

Proof **Step 1.** We prove that there a unique $A_h \in \left[0, \frac{1}{\beta}\sum_{s=-\infty}^{+\infty} h(s)\right]$ such that (2.3) holds. In fact, let

$$G(u) = 1 - \sum_{n=-\infty}^{+\infty} \frac{\beta_n}{\phi^{-1}(p(n))}\phi^{-1}\left(1 + \frac{1}{u}\sum_{s=n}^{+\infty} h(s)\right).$$

It is easy to see that G is continuous and increasing on $(-\infty, 0)$ and $(0, +\infty)$ respectively.

One sees from (c) that

$$\lim_{u\to-\infty} G(u) = 1 - \sum_{n=-\infty}^{+\infty} \frac{\beta_n}{\phi^{-1}(p(n))} > 0,$$

$$\lim_{u\to 0^-} G(u) = +\infty,$$

$$\lim_{u\to 0^+} G(u) = -\infty,$$

$$G\left(\frac{1}{\beta}\sum_{s=-\infty}^{+\infty} h(s)\right) \geq 1 - \sum_{n=-\infty}^{+\infty} \frac{\beta_n}{\phi^{-1}(p(n))}\phi^{-1}(1+\beta) \geq 0.$$

Then there is a unique $A_h \in \left[0, \frac{1}{\beta}\sum_{s=-\infty}^{+\infty} h(s)\right]$ such that (2.3) holds.

Step 2. Prove that x satisfies (2.2)–(2.3) if x is a solution of (2.1).

If x is a solution of (2.3), then there exist the limits

$$\lim_{n\to+\infty} \phi^{-1}(p(n))\Delta x(n) = c, \quad \lim_{n\to-\infty} \phi^{-1}(p(n))\Delta x(n)$$

and

$$\phi^{-1}(p(n))\Delta x(n) = \phi^{-1}\left(c + \sum_{s=n}^{+\infty} h(s)\right).$$

So

$$\Delta x(n) = \frac{1}{\phi^{-1}(p(n))}\phi^{-1}\left(c + \sum_{s=n}^{+\infty} h(s)\right).$$

Since $\sum_{n=-\infty}^{0} \frac{1}{\phi^{-1}(p(n))} < +\infty$, then $\lim_{n\to-\infty} x(n) = d \in R$ such that

$$x(n) = d + \sum_{s=-\infty}^{n-1} \frac{1}{\phi^{-1}(p(s))}\phi^{-1}\left(c + \sum_{t=s}^{+\infty} h(t)\right). \quad (2.4)$$

From the boundary conditions in (2.1), we get

$$d = \frac{1}{1 - \sum_{n=-\infty}^{+\infty} \alpha_n} \sum_{n=-\infty}^{+\infty} \alpha_n \sum_{s=-\infty}^{n-1} \frac{1}{\phi^{-1}(p(s))}\phi^{-1}\left(c + \sum_{t=s}^{+\infty} h(t)\right) \quad (2.5)$$

and

$$\phi^{-1}(c) - \sum_{n=-\infty}^{+\infty} \beta_n \phi^{-1}\left(c + \sum_{t=n}^{+\infty} h(t)\right) = 0.$$

By Step 1, we see that $c = A_h$. We now prove that

$$\lim_{n\to+\infty} \frac{x(n)}{1 + \sum_{s=-\infty}^{n-1} \frac{1}{\phi^{-1}(p(s))}} = c. \quad (2.6)$$

In fact, if $c = 0$, then for any $\epsilon > 0$ there exists $H > 0$ such that

$$\phi^{-1}(p(n))|\Delta x(n)| < \frac{\epsilon}{2}, \quad n \geq H.$$

It follows that

$$|x(n)| \leq |x(H)| + \sum_{s=H}^{n-1} |\Delta x(s)| \leq |x(H)| + \frac{\epsilon}{2}\sum_{s=H}^{n-1} \frac{1}{\phi^{-1}(p(s))}, \quad n \geq H.$$

Then

$$\frac{|x(n)|}{1 + \sum_{s=-\infty}^{n-1} \frac{1}{\phi^{-1}(p(s))}} \leq \frac{|x(H)|}{1 + \sum_{s=-\infty}^{n-1} \frac{1}{\phi^{-1}(p(s))}} + \frac{1}{1 + \sum_{s=-\infty}^{n-1} \frac{1}{\phi^{-1}(p(s))}} \frac{\epsilon}{2}\sum_{s=H}^{n-1} \frac{1}{\phi^{-1}(p(s))} < \frac{|x(H)|}{1 + \sum_{s=-\infty}^{n-1} \frac{1}{\phi^{-1}(p(s))}} + \frac{\epsilon}{2}, \quad n \geq H.$$

Since $\sum_{s=-\infty}^{\infty} \frac{1}{\phi^{-1}(p(s))} = \infty$, we can choose $H' > H$ large enough so that

$$\frac{|x(n)|}{1 + \sum_{s=-\infty}^{n-1} \frac{1}{\phi^{-1}(p(s))}} \leq \frac{|x(H)|}{1 + \sum_{s=-\infty}^{n-1} \frac{1}{\phi^{-1}(p(s))}} + \frac{\epsilon}{2} < \epsilon, \quad n \geq H',$$

which implies that

$$\lim_{n\to\infty} \frac{x(n)}{1 + \sum_{s=-\infty}^{n-1} \frac{1}{\phi^{-1}(p(s))}} = 0.$$

If $c \neq 0$, then $\lim_{n\to\infty} \left(\phi^{-1}(p(n))\Delta x(n) - c\right) = 0$. It follows that

$$\lim_{n\to\infty} \phi^{-1}(p(n))\Delta\left[x(n) - c\sum_{s=-\infty}^{n-1} \frac{1}{\phi^{-1}(p(s))}\right] = 0.$$

Then we get similarly that

$$\lim_{n\to\infty} \frac{x(n) - c\sum_{s=-\infty}^{n-1} \frac{1}{\phi^{-1}(p(s))}}{1 + \sum_{s=-\infty}^{n-1} \frac{1}{\phi^{-1}(p(s))}} = 0.$$

Together with $\sum_{s=-\infty}^{\infty} \frac{1}{\phi^{-1}(p(s))} = \infty$, it follows that (2.6) holds. Then $\lim_{n\to-\infty} x(n) = d$, (2.4), (2.5) and (2.6) imply that $x \in X$ and x satisfies (2.2) and (2.3).

Step 3. Prove that $x \in X$ and is a solution of (2.1) if x satisfies (2.2) and (2.3). The proof is simple and is omitted. The proof is complete. □

Let $h(n) \geq 0$ for every $n \in Z$ be a bilateral sequence with $\sum_{n=-\infty}^{+\infty} h(n)$ converging. Consider the following BVP:

$$\begin{cases} \Delta[q(n)\psi(\Delta y(n))] + h(n) = 0, \quad n \in Z, \\ \lim\limits_{n\to-\infty} y(n) - \sum\limits_{n=-\infty}^{+\infty} \gamma_n y(n) = 0, \\ \lim\limits_{n\to+\infty} \psi^{-1}(q(n))\Delta y(n) - \sum\limits_{n=-\infty}^{+\infty} \delta_n \Delta y(n) = 0. \end{cases} \tag{2.7}$$

Lemma 3 Suppose that (b), (c) and (e) hold. Then $y \in Y$ is a solution of BVP (2.7) if and only if

$$y(n) = \frac{1}{1 - \sum\limits_{n=-\infty}^{+\infty} \gamma_n} \sum_{n=-\infty}^{+\infty} \gamma_n \sum_{s=-\infty}^{n-1} \frac{1}{\psi^{-1}(q(s))} \psi^{-1}\left(B_h + \sum_{t=s}^{+\infty} h(t)\right) + \sum_{s=-\infty}^{n-1} \frac{1}{\psi^{-1}(q(s))} \psi^{-1}\left(B_h + \sum_{t=s}^{+\infty} h(t)\right), \tag{2.8}$$

where $B_h \in \left[0, \frac{1}{\delta}\sum_{s=-\infty}^{+\infty} h(s)\right]$ such that

$$\psi^{-1}(B_h) = \sum_{n=-\infty}^{+\infty} \frac{\delta_n}{\psi^{-1}(q(n))} \psi^{-1}\left(B_h + \sum_{s=n}^{+\infty} h(s)\right). \tag{2.9}$$

Proof The proof is similar to that of Lemma 2 and is omitted. □

Let $k_1, k_2 \in Z$ with $k_1 + 2 < k_2$. Denote

$$\mu = \min\left\{\frac{P_{k_1}}{P_{k_2}}, \frac{1}{P_{k_2}}, \frac{1}{\phi^{-1}(p(k_1-1))P_{k_2}}, \frac{Q_{k_1}}{Q_{k_2}}, \frac{1}{Q_{k_2}}, \frac{1}{\psi^{-1}(q(k_1-1))Q_{k_2}}\right\}. \tag{2.10}$$

Choose

$$P = \left\{ (x,y) \in X : \begin{array}{l} x(n) \geq 0, \ y(n) \geq 0 \quad \text{for all } n \in Z, \\ \min\limits_{n\in[k_1,k_2]} \frac{x(n)}{P_n} \geq \mu \sup\limits_{n\in Z} \frac{x(n)}{P_n}, \\ \min\limits_{n\in[k_1,k_2]} \frac{y(n)}{Q_n} \geq \mu \sup\limits_{n\in Z} \frac{y(n)}{Q_n}. \end{array} \right\},$$

It is easy to see that P is a nontrivial cone in X.

Define the nonlinear operator T on P by

$$(T(x,y))(n) = (T_1(x,y)(n), T_2(x,y)(n))$$

with

$$\begin{aligned} T_1(x,y)(n) = {} & \frac{1}{1 - \sum_{n=-\infty}^{+\infty} \alpha_n} \sum_{n=-\infty}^{+\infty} \alpha_n \sum_{s=-\infty}^{n-1} \frac{1}{\phi^{-1}(p(s))} \phi^{-1} \\ & \times \left(A_f + \sum_{t=s}^{+\infty} f(t,x(t),y(t))\right) \\ & + \sum_{s=-\infty}^{n-1} \frac{1}{\phi^{-1}(p(s))} \phi^{-1}\left(A_f + \sum_{t=s}^{+\infty} f(t,x(t),y(t))\right), \\ T_2(x,y)(n) = {} & \frac{1}{1 - \sum_{n=-\infty}^{+\infty} \gamma_n} \sum_{n=-\infty}^{+\infty} \gamma_n \sum_{s=-\infty}^{n-1} \frac{1}{\psi^{-1}(q(s))} \psi^{-1} \\ & \times \left(B_g + \sum_{t=s}^{+\infty} g(t,x(t),y(t))\right) \\ & + \sum_{s=-\infty}^{n-1} \frac{1}{\psi^{-1}(q(s))} \psi^{-1}\left(B_g + \sum_{t=s}^{+\infty} g(t,x(t),y(t))\right), \end{aligned}$$

where $A_f \in \left[0, \frac{1}{\beta} \sum\limits_{s=-\infty}^{+\infty} f(s,x(s),y(s))\right]$ such that

$$\phi^{-1}(A_f) = \sum_{n=-\infty}^{+\infty} \frac{\beta_n}{\phi^{-1}(p(n))} \phi^{-1}\left(A_f + \sum_{s=n}^{+\infty} f(s,x(s),y(s))\right) \tag{2.11}$$

and $B_g \in \left[0, \frac{1}{\delta}\sum_{s=-\infty}^{+\infty} g(s,x(s),y(s))\right]$ such that

$$\psi^{-1}(B_g) = \sum_{n=-\infty}^{+\infty} \frac{\delta_n}{\psi^{-1}(q(n))} \psi^{-1}\left(B_g + \sum_{s=n}^{+\infty} g(s,x(s),y(s))\right). \tag{2.12}$$

Lemma 4 Suppose that (b)–(e) hold. Then $T : P \to P$ is well defined, $(x,y) \in P$ is a positive solution of BVP (1.3) if (x,y) is a fixed point of T, and T is completely continuous.

Proof For $(x,y) \in P$, we know that there exist $r > 0$ such that

$$0 \leq \frac{x(n)}{1 + \sum_{s=-\infty}^{n-1} \frac{1}{\phi^{-1}(p(s))}} = \frac{x(n)}{P_n} \leq r, \quad n \in Z,$$

$$0 \leq \frac{y(n)}{\sum_{s=-\infty}^{n-1} \frac{1}{\psi^{-1}(q(s))} + 1} = \frac{y(n)}{Q_n} \leq r, \quad n \in Z.$$

Since f and g are nonnegative Caratheodory functions, we know that there exists a nonnegative sequence $\phi_n(n)$ with $\sum_{n=-\infty}^{+\infty} \phi_r(n) < +\infty$ such that

$$0 \leq f(n,x(n),y(n)) = f\left(n, P_n\frac{x(n)}{P_n}, Q_n\frac{y(n)}{Q_n}\right) \leq \phi_r(n), \quad n \in Z,$$

$$0 \leq g(n,x(n),y(n)) = g\left(n, P_n\frac{x(n)}{P_n}, Q_n\frac{y(n)}{Q_n}\right) \leq \phi_r(n), \quad n \in Z.$$

By the definitions of T_1 and T_2, we get that

$$T_1(x,y)(n) \geq 0,\ T_2(x,y)(n) \geq 0, \quad n \in Z \tag{2.13}$$

and

$$\begin{cases} \Delta[p(n)\phi(\Delta T_1(x,y)(n))] + f(n,x(n),y(n)) = 0, & n \in Z, \\ \Delta[q(n)\psi(\Delta T_2(x,y)(n))] + g(n,x(n),y(n)) = 0, & n \in Z, \\ \lim_{n\to-\infty} T_1(x,y)(n) - \sum_{n=-\infty}^{+\infty} \alpha_n T_1(x,y)(n) = 0, \\ \lim_{n\to-\infty} T_2(x,y)(n) - \sum_{n=-\infty}^{+\infty} \gamma_n T_2(x,y)(n) = 0, \\ \lim_{n\to+\infty} \phi^{-1}(p(n))\Delta T_1(x,y)(n) - \sum_{n=-\infty}^{+\infty} \beta_n \Delta T_1(x,y)(n) = 0, \\ \lim_{n\to+\infty} \psi^{-1}(q(n))\Delta T_2(x,y)(n) - \sum_{n=-\infty}^{+\infty} \delta_n \Delta T_2(x,y)(n) = 0. \end{cases} \tag{2.14}$$

Since $\Delta[p(n)\phi(\Delta T_1(x,y)(n))] = -f(n,x(n),y(n)) \leq 0$ for all $n \in Z$, we see that $p(n)\phi(\Delta T_1(x,y)(n))$ is decreasing. Then $\phi^{-1}(p(n))\Delta T_1(x,y)(n)$ is decreasing. It is easy to see that

$$\lim_{n\to+\infty} \phi^{-1}(p(n))\Delta T_1(x,y)(n) = \sum_{n=-\infty}^{+\infty} \beta_n \Delta T_1(x,y)(n)$$

$$= \sum_{n=-\infty}^{+\infty} \frac{\beta_n}{\phi^{-1}(p(n))} \phi^{-1}(p(n))\Delta T_1(x,y)(n)$$

$$\geq \sum_{n=-\infty}^{+\infty} \frac{\beta_n}{\phi^{-1}(p(n))} \lim_{n\to+\infty} \phi^{-1}(p(n))\Delta T_1(x,y)(n).$$

Then (c) implies that

$$\lim_{n\to+\infty} \phi^{-1}(p(n))\Delta T_1(x,y)(n) \geq 0.$$

Hence

$$\phi^{-1}(p(n))\Delta T_1(x,y)(n) \geq 0, \quad n \in Z.$$

It follows that $\Delta T_1(x,y)(n) \geq 0$ for all $n \in Z$. So $T_1(x,y)(n)$ is increasing. We consider two cases:

Case 1: there is $n_0 \in Z$ such that

$$\sup_{n\in Z} \frac{T_1(x,y)(n)}{P_n} = \frac{T_1(x,y)(n_0)}{P_{n_0}}.$$

For $n_1, n, n_2 \in Z$ with $n_1 < n < n_2$, Since $\phi^{-1}(p(n))$ $\Delta T_1(x,y)(n)$ is decreasing, we get

$$\phi^{-1}(p(s))\Delta T_1(x,y)(s) \leq \phi^{-1}(p(k))\Delta T_1(x,y)(k)$$

for all $s \geq k$. So there there is λ such that

$$\phi^{-1}(p(s))\Delta T_1(x,y)(s) \leq \lambda \leq \phi^{-1}(p(k))\Delta T_1(x,y)(k),$$
$$s \geq n > k.$$

Then we get

$$(P_n - P_{n_1})\frac{T_1(x,y)(n_2) - T_1(x,y)(n)}{P_{n_2} - P_n}$$

$$= \frac{(P_n - P_{n_1})\sum_{s=n}^{n_2-1} \frac{1}{\phi^{-1}(p(s))} \phi^{-1}(p(s))\Delta T_1(x,y)(s)}{P_{n_2} - P_n}$$

$$= \frac{\sum_{s=n_1}^{n-1} \frac{1}{\phi^{-1}(p(s))} \sum_{s=n}^{n_2-1} \frac{1}{\phi^{-1}(p(s))} \phi^{-1}(p(s))\Delta T_1(x,y)(s)}{\sum_{s=n}^{n_2-1} \frac{1}{\phi^{-1}(p(s))}}$$

$$\leq \frac{\lambda \sum_{s=n_1}^{n-1} \frac{1}{\phi^{-1}(p(s))} \sum_{s=n}^{n_2-1} \frac{1}{\phi^{-1}(p(s))}}{\sum_{s=n}^{n_2-1} \frac{1}{\phi^{-1}(p(s))}}$$

$$= \frac{\lambda \sum_{s=n}^{n_2-1} \frac{1}{\phi^{-1}(p(s))} \sum_{s=n_1}^{n-1} \frac{1}{\phi^{-1}(p(s))}}{\sum_{s=n}^{n_2-1} \frac{1}{\phi^{-1}(p(s))}}$$

$$\leq \frac{\sum_{s=n}^{n_2-1} \frac{1}{\phi^{-1}(p(s))} \sum_{s=n_1}^{n-1} \frac{1}{\phi^{-1}(p(s))} \phi^{-1}(p(s))\Delta T_1(x,y)(s)}{\sum_{s=n}^{n_2-1} \frac{1}{\phi^{-1}(p(s))}}$$

$$= T_1(x,y)(n) - T_1(x,y)(n_1).$$

So

$$(P_n - P_{n_1})\frac{T_1(x,y)(n_2) - T_1(x,y)(n)}{P_{n_2} - P_n}$$
$$+ T_1(x,y)(n_1) - T_1(x,y)(n) \leq 0.$$

It follows that

$$T_1(x,y)(n) \geq \frac{P_{n_2} - P_n}{P_{n_2} - P_{n_1}} T_1(x,y)(n_1) + \frac{P_n - P_{n_1}}{P_{n_2} - P_{n_1}} T_1(x,y)(n_2). \tag{2.15}$$

If $n_0 = k_1$, we get

$$\min_{n\in[k_1,k_2]} \frac{T_1(x,y)(n)}{P_n} \geq \frac{T_1(x,y)(k_1)}{P_{k_2}}$$

$$= \frac{T_1(x,y)(n_0)}{1 + \sum_{s=-\infty}^{n_0-1} \frac{1}{\phi^{-1}(p(s))}} \frac{1 + \sum_{s=-\infty}^{k_1-1} \frac{1}{\phi^{-1}(p(s))}}{1 + \sum_{s=-\infty}^{k_2-1} \frac{1}{\phi^{-1}(p(s))}}$$

$$\geq \frac{P_{k_1}}{P_{k_2}} \sup_{n\in Z} \frac{T_1(x,y)(n)}{1 + \sum_{s=-\infty}^{n-1} \frac{1}{\phi^{-1}(p(s))}}$$

$$\geq \mu \sup_{n\in Z} \frac{T_1(x,y)(n)}{1 + \sum_{s=-\infty}^{n-1} \frac{1}{\phi^{-1}(p(s))}}.$$

If $n_0 > k_1$, choose $n_1 = k_1 - 1$, $n = k_1$ and $n_2 = n_0$, by using (2.15) we have

$$\min_{n\in[k_1,k_2]} \frac{y(n)}{P_n} \geq \frac{T_1(x,y)(k_1)}{P_{k_2}}$$
$$\geq \frac{\frac{P_{n_0}-P_{k_1}}{P_{n_0}-P_{k_1-1}}T_1(x,y)(k_1-1)+\frac{P_{k_1}-P_{k_1-1}}{P_{n_0}-P_{k_1-1}}T_1(x,y)(n_0)}{P_{k_2}}$$
$$\geq \frac{\frac{P_{k_1}-P_{k_1-1}}{P_{n_0}-P_{k_1-1}}T_1(x,y)(n_0)}{P_{k_2}} = \frac{P_{n_0}\frac{P_{k_1}-P_{k_1-1}}{P_{n_0}-P_{k_1-1}}}{P_{k_2}}\frac{T_1(x,y)(n_0)}{P_{n_0}}$$
$$\geq \frac{1}{\phi^{-1}(p(k_1-1))P_{k_2}}\frac{T_1(x,y)(n_0)}{P_{n_0}} \geq \mu \sup_{n\in N_0}\frac{T_1(x,y)(n)}{P_n}.$$

If $n_0 < k_1$, we have

$$\min_{n\in[k_1,k_2]} \frac{T_1(x,y)(n)}{P_n} \geq \frac{T_1(x,y)(k_1)}{P_{k_2}}$$
$$\geq \frac{T_1(x,y)(n_0)}{P_{k_2}} = \frac{P_{n_0}}{P_{k_2}}\frac{T_1(x,y)(n_0)}{P_{n_0}}$$
$$\geq \frac{1}{P_{k_2}}\frac{T_1(x,y)(n_0)}{P_{n_0}} \geq \mu \sup_{n\in N_0}\frac{T_1(x,y)(n)}{P_n}.$$

Case 2: $\sup_{n\in Z}\frac{T_1(x,y)(n)}{P_n} = \lim_{n\to+\infty}\frac{T_1(x,y)(n)}{P_n}$. Choose $n' > k_2$, similarly to Case 1 we can prove that

$$\min_{n\in[k_1,k_2]} \frac{T_1(x,y)(n))}{P_n} \geq \mu\frac{T_1(x,y)(n')}{P_{n'}}.$$

Let $n' \to +\infty$, one sees

$$\min_{n\in[k_1,k_2]} \frac{T_1(x,y)(n))}{P_n} \geq \mu \sup_{n\in N_0}\frac{T_1(x,y)(n)}{P_n}.$$

Case 3: $\sup_{n\in Z}\frac{T_1(x,y)(n)}{P_n} = \lim_{n\to-\infty}\frac{T_1(x,y)(n)}{P_n}$. Choose $n' < k_1$, similarly to Case 1 we can prove that

$$\min_{n\in[k_1,k_2]} \frac{T_1(x,y)(n))}{P_n} \geq \mu\frac{T_1(x,y)(n')}{P_{n'}}.$$

Let $n' \to -\infty$, one sees

$$\min_{n\in[k_1,k_2]} \frac{T_1(x,y)(n))}{P_n} \geq \mu \sup_{n\in N_0}\frac{T_1(x,y)(n)}{P_n}.$$

From Cases 1, 2 and 3, we get

$$\min_{n\in[k_1,k_2]} \frac{T_1(x,y)(n))}{P_n} \geq \mu \sup_{n\in N_0}\frac{T_1(x,y)(n)}{P_n}. \tag{2.16}$$

Similarly we can prove that

$$\min_{n\in[k_1,k_2]} \frac{T_2(x,y)(n))}{Q_n} \geq \mu \sup_{n\in N_0}\frac{T_2(x,y)(n)}{Q_n}. \tag{2.17}$$

From (2.13), (2.16) and (2.17), we know that $T(x,y) \in P$. Thus $T: P \to P$ is well defined.

From (2.14), we get $\Delta T_1(x,y)(n) \geq 0$ for all $n \in Z$. So $T_1(x,y)(n)$ is increasing. Then

$$\lim_{n\to-\infty} T_1(x,y)(n) = \sum_{n=-\infty}^{+\infty}\alpha_n T_1(x,y)(n)$$
$$\geq \sum_{n=-\infty}^{+\infty}\alpha_n \lim_{n\to-\infty} T_1(x,y)(n).$$

It follows that

$$\left(1-\sum_{n=-\infty}^{+\infty}\alpha_n\right)\lim_{n\to-\infty} T_1(x,y)(n) \geq 0.$$

So the assumption (c) implies that $\lim_{n\to-\infty} T_1(x,y)(n) \geq 0$. Similarly, we can prove that $\lim_{n\to-\infty} T_2(x,y)(n) \geq 0$. If there exists n_1, n_2 such that $T_1(x,y)(n_1) = 0$ and $T_2(x,y)(n_2) = 0$, then $T_1(x,y)(n) = T_2(x,y)(n) = 0$ for all $n \leq \min\{n_1, n_2\}$. Hence (2.14) shows us that $f(n,0,0) = 0$ and $g(n,0,0) = 0$ for all $n \leq \min\{n_1,n_2\}$, a contradiction to assumption (d). Hence we know that $(x,y) \in P$ is a positive solution of BVP (1.3) if and only if $(x,y) \in P$ is a fixed point of T.

Now, we prove that T is completely continuous. It suffices to prove that both $T_1 : P \to X$ and $T_2 : P \to Y$ are completely continuous. So we need to prove that both T_1 and T_2 are continuous on P, map bounded subsets into relatively compact sets. We divide the proof into three steps:

Step 1: Prove that both T_1 and T_2 are continuous. For $X_k \in E(k = 0,1,2,\cdots)$ with $X_k \to X_0$ as $k \to +\infty$, we prove that $T(X_k) \to X_0$ as $k \to +\infty$. Suppose that $X_k(n) = (x_k(n), y_k(n))$. Then

$$T_1(X_k)(n) = \frac{\sum_{n=-\infty}^{+\infty}\alpha_n\sum_{s=-\infty}^{n-1}\frac{1}{\phi^{-1}(p(s))}\phi^{-1}\left(A_{fk}+\sum_{t=s}^{+\infty}f(t,x_k(t),y_k(t))\right)}{1-\sum_{n=-\infty}^{+\infty}\alpha_n}$$
$$+\sum_{s=-\infty}^{n-1}\frac{1}{\phi^{-1}(p(s))}\phi^{-1}\left(A_{fk}+\sum_{t=s}^{+\infty}f(t,x_k(t),y_k(t))\right),$$
$$T_2(X_k)(n) = \frac{\sum_{n=-\infty}^{+\infty}\gamma_n\sum_{s=-\infty}^{n-1}\frac{1}{\psi^{-1}(q(s))}\psi^{-1}\left(B_{gk}+\sum_{t=s}^{+\infty}g(t,x_k(t),y_k(t))\right)}{1-\sum_{n=-\infty}^{+\infty}\gamma_n}$$
$$+\sum_{s=-\infty}^{n-1}\frac{1}{\psi^{-1}(q(s))}\psi^{-1}\left(B_{gk}+\sum_{t=s}^{+\infty}g(t,x_k(t),y_k(t))\right),$$

where $A_{fk} \in \left[0, \frac{1}{\beta}\sum_{s=-\infty}^{+\infty}f(s,x_k(s),y_k(s))\right]$ such that

$$\phi^{-1}(A_{fk}) = \sum_{n=-\infty}^{+\infty}\frac{\beta_n\phi^{-1}\left(A_{fk}+\sum_{s=n}^{+\infty}f(s,x_k(s),y_k(s))\right)}{\phi^{-1}(p(n))},$$
$$k = 0,1,2,\ldots \tag{2.18}$$

and $B_{gk} \in \left[0, \frac{1}{\delta}\sum_{s=-\infty}^{+\infty}g(s,x_k(s),y_k(s))\right]$ such that

$$\psi^{-1}(B_{gk}) = \sum_{n=-\infty}^{+\infty} \frac{\delta_n \psi^{-1}\left(B_{gk} + \sum_{s=n}^{+\infty} g(s, x_k(s), y_k(s))\right)}{\psi^{-1}(q(n))},$$
$$k = 0, 1, 2, \ldots. \tag{2.19}$$

We know that there exist $r > 0$ such that

$$0 \leq \frac{x_k(n)}{1 + \sum_{s=-\infty}^{n-1} \frac{1}{\phi^{-1}(p(s))}} = \frac{x_k(n)}{P_n} \leq r, \quad k = 0, 1, 2, \ldots, n \in Z,$$
$$0 \leq \frac{y_k(n)}{\sum_{s=-\infty}^{n-1} \frac{1}{\psi^{-1}(q(s))} + 1} = \frac{y_k(n)}{Q_n} \leq r, \quad k = 0, 1, 2, \ldots, \quad n \in Z.$$

Since f and g are nonnegative Caratheodory functions, we know that there exists a nonnegative sequence $\phi_n(n)$ with $\sum_{n=-\infty}^{+\infty} \phi_r(n) < +\infty$ such that

$$0 \leq f(n, x_k(n), y_k(n)) = f\left(n, P_n \frac{x_k(n)}{P_n}, Q_n \frac{y_k(n)}{Q_n}\right) \leq \phi_r(n), \quad n \in Z,$$
$$0 \leq g(n, x_k(n), y_k(n)) = g\left(n, P_n \frac{x_k(n)}{P_n}, Q_n \frac{y_k(n)}{Q_n}\right) \leq \phi_r(n), \quad n \in Z.$$

We first prove that $A_{fk} \to A_{f0}$ as $k \to +\infty$ and $B_{gk} \to B_{g0}$ as $k \to +\infty$. It is easy to show that

$$0 \leq A_{fk} \leq \frac{1}{\beta} \sum_{s=-\infty}^{+\infty} f(s, x_k(s), y_k(s))$$
$$\leq \frac{1}{\beta} \sum_{s=-\infty}^{+\infty} \phi_r(n), \quad k = 0, 1, 2, \ldots, \quad n \in Z.$$

Without loss of generality, suppose that $A_{fk} \to \overline{A} \neq A_{f0}$. Then there exist two subsequences $A^1_{fk_i}$ and $A^2_{fk_i}$ with $A^1_{fk_i} \to A_1$ and $A^2_{fk_i} \to A_2$ as $i \to +\infty$. From

$$\phi^{-1}(A^j_{fk_i}) = \sum_{n=-\infty}^{+\infty} \frac{\beta_n}{\phi^{-1}(p(n))} \phi^{-1}\left(A^j_{fk_i} + \sum_{s=n}^{+\infty} f(s, x_{k_i}(s), y_{k_i}(s))\right)$$
$$\leq \phi^{-1}\left(\frac{1}{\beta} \sum_{s=-\infty}^{+\infty} \phi_r(n) + \sum_{s=-\infty}^{+\infty} \phi_r(s)\right) \sum_{n=-\infty}^{+\infty} \frac{\beta_n}{\phi^{-1}(p(n))}.$$

Let $i \to +\infty$, we get

$$\phi^{-1}(A_j) = \sum_{n=-\infty}^{+\infty} \frac{\beta_n}{\phi^{-1}(p(n))} \phi^{-1}\left(A_j + \sum_{s=n}^{+\infty} f(s, x_0(s), y_0(s))\right).$$

Together with (2.18), we get $A_1 = A_2 = A_{f0}$. Then $A_{fk} \to A_{f0}$ as $k \to +\infty$. Similarly, we can prove that $B_{gk} \to B_{g0}$ as $k \to +\infty$. These together with the continuous property of f imply that T_1 is continuous at X_0. Similarly, we can prove that T_2 is continuous at X_0. So T is continuous at X_0.

Step 2: For each bounded subset $\Omega \subset P$, prove that $T\Omega$ is bounded. In fact, for each bounded subset $\Omega \subseteq D$, and $(x, y) \in \Omega$. Then there exists $r > 0$ satisfying

$$||(x, y)|| = \max\left\{\sup_{n \in Z} \frac{|x(n)|}{P_n}, \sup_{n \in Z} \frac{|y(n)|}{Q_n}\right\} \leq r.$$

Since f and g are nonnegative Caratheodory functions, we know that there exists a nonnegative sequence $\phi_n(n)$ with $\sum_{n=-\infty}^{+\infty} \phi_r(n) < +\infty$ such that

$$0 \leq f(n, x(n), y(n)) = f\left(n, P_n \frac{x(n)}{P_n}, Q_n \frac{y(n)}{Q_n}\right) \leq \phi_r(n), \quad n \in Z,$$
$$0 \leq g(n, x(n), y(n)) = g\left(n, P_n \frac{x(n)}{P_n}, Q_n \frac{y(n)}{Q_n}\right) \leq \phi_r(n), \quad n \in Z.$$

The method used in Step 1 implies that there exist constants $M > 0$ such that $|A_f| \leq M$ for all $(x, y) \in \Omega$. Then

$$\frac{|T_1(x, y)(n)|}{P_n} = \frac{1}{P_n} \frac{1}{1 - \sum_{n=-\infty}^{+\infty} \alpha_n}$$
$$\times \sum_{n=-\infty}^{+\infty} \alpha_n \sum_{s=-\infty}^{n-1} \frac{1}{\phi^{-1}(p(s))} \phi^{-1}\left(A_f + \sum_{t=s}^{+\infty} f(t, x(t), y(t))\right)$$
$$+ \frac{1}{P_n} \sum_{s=-\infty}^{n-1} \frac{1}{\phi^{-1}(p(s))} \phi^{-1}\left(A_f + \sum_{t=s}^{+\infty} f(t, x(t), y(t))\right)$$
$$\leq \frac{1}{P_n} \frac{1}{1 - \sum_{n=-\infty}^{+\infty} \alpha_n}$$
$$\times \sum_{n=-\infty}^{+\infty} \alpha_n \sum_{s=-\infty}^{n-1} \frac{1}{\phi^{-1}(p(s))} \phi^{-1}\left(M + \sum_{t=-\infty}^{+\infty} \phi_r(t)\right)$$
$$+ \frac{1}{P_n} \sum_{s=-\infty}^{n-1} \frac{1}{\phi^{-1}(p(s))} \phi^{-1}\left(M + \sum_{t=-\infty}^{+\infty} \phi_r(t)\right)$$
$$\leq \frac{1}{1 - \sum_{n=-\infty}^{+\infty} \alpha_n} \phi^{-1}\left(M + \sum_{t=-\infty}^{+\infty} \phi_r(t)\right).$$

Similarly, one has that

$$\frac{|T_2(x, y)(n)|}{Q_n} \leq \frac{1}{1 - \sum_{n=-\infty}^{+\infty} \gamma_n} \psi^{-1}\left(M + \sum_{t=-\infty}^{+\infty} \phi_r(t)\right).$$

It follows that $T\Omega$ is bounded.

Step 3: For each bounded subset $\Omega \subset P$, prove that $T\Omega$ is relatively compact. We need to prove that both $\{T_1(x, y)(n) : (x, y) \in \Omega\}$ and $\{T_2(x, y)(n) : (x, y) \in \Omega\}$ are uniformly equi-convergent as $n \to \pm\infty$. We have

$$\left|\frac{T_1(x,y)(n)}{P_n} - \frac{\sum_{n=-\infty}^{+\infty}\alpha_n \sum_{s=-\infty}^{n-1}\frac{1}{\phi^{-1}(p(s))}\phi^{-1}\left(A_f+\sum_{t=s}^{+\infty}f(t,x(t),y(t))\right)}{1-\sum\limits_{n=-\infty}^{+\infty}\alpha_n}\right|$$
$$\leq \left|\frac{1}{P_n}\frac{1}{1-\sum_{n=-\infty}^{+\infty}\alpha_n}\sum_{n=-\infty}^{+\infty}\alpha_n\sum_{s=-\infty}^{n-1}\frac{1}{\phi^{-1}(p(s))}\phi^{-1}\left(A_f+\sum_{t=s}^{+\infty}f(t,x(t),y(t))\right)\right.$$
$$\left.-\frac{1}{1-\sum_{n=-\infty}^{+\infty}\alpha_n}\sum_{n=-\infty}^{+\infty}\alpha_n\sum_{s=-\infty}^{n-1}\frac{1}{\phi^{-1}(p(s))}\phi^{-1}\left(A_f+\sum_{t=s}^{+\infty}f(t,x(t),y(t))\right)\right|$$
$$+\frac{1}{P_n}\sum_{s=-\infty}^{n-1}\frac{1}{\phi^{-1}(p(s))}\phi^{-1}\left(A_f+\sum_{t=s}^{+\infty}f(t,x(t),y(t))\right)$$
$$\leq \frac{1-P_n}{P_n}\frac{1}{1-\sum_{n=-\infty}^{+\infty}\alpha_n}\sum_{n=-\infty}^{+\infty}\alpha_n\sum_{s=-\infty}^{n-1}\frac{1}{\phi^{-1}(p(s))}\phi^{-1}\left(M+\sum_{t=-\infty}^{+\infty}\phi_r(t)\right)$$
$$+\phi^{-1}\left(M+\sum_{t=-\infty}^{+\infty}\phi_r(t)\right)\sum_{s=-\infty}^{n-1}\frac{1}{\phi^{-1}(p(s))}$$
$$=\phi^{-1}\left(M+\sum_{t=-\infty}^{+\infty}\phi_r(t)\right)\left[\frac{\sum\limits_{n=-\infty}^{+\infty}\alpha_n\sum\limits_{s=-\infty}^{n-1}\frac{1}{\phi^{-1}(p(s))}}{1-\sum_{n=-\infty}^{+\infty}\alpha_n}+1\right]\sum_{s=-\infty}^{n-1}\frac{1}{\phi^{-1}(p(s))}$$
$$\to 0 \text{ uniformly as } n\to\infty.$$

Furthermore, we have

$$\left|\frac{T_1(x,y)(n)}{P_n}-\phi^{-1}(A_f)\right|$$
$$\leq \frac{1}{P_n}\frac{1}{1-\sum\limits_{n=-\infty}^{+\infty}\alpha_n}\sum_{n=-\infty}^{+\infty}\alpha_n\sum_{s=-\infty}^{n-1}\frac{1}{\phi^{-1}(p(s))}\phi^{-1}$$
$$\times\left(A_f+\sum_{t=s}^{+\infty}f(t,x(t),y(t))\right)$$

$$+\left|\frac{1}{P_n}\sum_{s=-\infty}^{n-1}\frac{1}{\phi^{-1}(p(s))}\phi^{-1}\left(A_f+\sum_{t=s}^{+\infty}f(t,x(t),y(t))\right)-\phi^{-1}(A_f)\right|$$
$$\leq\frac{1}{1-\sum\limits_{n=-\infty}^{+\infty}\alpha_n}\sum_{n=-\infty}^{+\infty}\alpha_n\sum_{s=-\infty}^{n-1}\frac{1}{\phi^{-1}(p(s))}\phi^{-1}\left(M+\sum_{t=-\infty}^{+\infty}\phi_r(t)\right)\frac{1}{P_n}$$
$$+\frac{1}{P_n}\sum_{s=-\infty}^{n-1}\frac{|\phi^{-1}(A_f+\sum_{t=s}^{+\infty}f(t,x(t),y(t)))-\phi^{-1}(A_f)|}{\phi^{-1}(p(s))}$$
$$+\phi^{-1}(A_f)\left|\frac{\sum_{s=-\infty}^{n-1}\frac{1}{\phi^{-1}(p(s))}}{P_n}-1\right|$$
$$\leq\frac{1}{1-\sum\limits_{n=-\infty}^{+\infty}\alpha_n}\sum_{n=-\infty}^{+\infty}\alpha_n\sum_{s=-\infty}^{n-1}\frac{1}{\phi^{-1}(p(s))}\phi^{-1}\left(M+\sum_{t=-\infty}^{+\infty}\phi_r(t)\right)\frac{1}{P_n}$$
$$+\frac{1}{P_n}\sum_{s=-\infty}^{n-1}\frac{|\phi^{-1}(A_f+\sum_{t=s}^{+\infty}f(t,x(t),y(t)))-\phi^{-1}(A_f)|}{\phi^{-1}(p(s))}+\phi^{-1}(M)\frac{1}{P_n}.$$

Since $|A_f|\leq M$, $|A_f+\sum_{t=s}^{+\infty}f(t,x(t),y(t))|\leq M$ and ϕ^{-1} is uniformly continuous on $[-M,M]$, then for any $\epsilon>0$ there exists $\sigma>0$ such that $u_1,u_2\in[-M,M]$ and $|u_1-u_2|<\sigma$ imply that $|\phi^{-1}(u_1)-\phi^{-1}(u_2)|<\epsilon$. Since

$$\left|A_f+\sum_{t=s}^{+\infty}f(t,x(t),y(t))-A_f\right|\leq\sum_{t=s}^{+\infty}\phi_r(t)\to 0$$
$$\text{uniformly as } s\to+\infty,$$

then there exists $S>0$ such that

$$\left|A_f+\sum_{t=s}^{+\infty}f(t,x(t),y(t))-A_f\right|<\sigma, s>S,\ (x,y)\in\Omega.$$

So

$$\left|\phi^{-1}\left(A_f+\sum_{t=s}^{+\infty}f(t,x(t),y(t))\right)-\phi^{-1}(A_f)\right|<\epsilon, s>S,$$
$(x,y)\in\Omega$.

It follows that

$$\frac{1}{P_n}\sum_{s=-\infty}^{n-1}\frac{|\phi^{-1}(A_f+\sum_{t=s}^{+\infty}f(t,x(t),y(t)))-\phi^{-1}(A_f)|}{\phi^{-1}(p(s))}$$
$$\leq\frac{1}{P_n}\left[\sum_{s=S+1}^{n-1}\frac{\epsilon}{\phi^{-1}(p(s))}+\sum_{s=-\infty}^{S}\frac{|\phi^{-1}(A_f+\sum_{t=s}^{+\infty}f(t,x(t),y(t)))-\phi^{-1}(A_f)|}{\phi^{-1}(p(s))}\right]$$
$$\leq\frac{1}{P_n}\left[\sum_{s=S+1}^{n-1}\frac{\epsilon}{\phi^{-1}(p(s))}+2\phi^{-1}(M)\sum_{s=-\infty}^{S}\frac{1}{\phi^{-1}(p(s))}\right]$$
$$\leq\epsilon+\frac{2\phi^{-1}(M)\sum_{s=-\infty}^{S}\frac{1}{\phi^{-1}(p(s))}}{P_n}\to 0 \text{ uniformly as } n\to+\infty.$$

Hence

$$\left|\frac{T_1(x,y)(n)}{P_n}-\phi^{-1}(A_f)\right|\to 0 \text{ uniformly as } n\to\infty.$$

One knows that $T_1(\Omega)$ is relatively compact. Similarly we can prove that $T_2(\Omega)$ is relatively compact. Hence $T(\Omega)$ is relatively compact.

From Steps 1, 2 and 3, we know that T is completely continuous. The proof is ended. □

For positive constants a,b,c,d and integers k_1,k_2 with $k_1<k_2$, denote

$$Q=\min\left\{\phi\left(\frac{c(1-\sum_{n=-\infty}^{+\infty}\alpha_n)}{1-\sum_{n=-\infty}^{+\infty}\alpha_n+\sum_{n=-\infty}^{+\infty}\alpha_n\sum_{s=-\infty}^{n-1}\frac{1}{\phi^{-1}(p(s))}}\right)\frac{\delta}{3+3\delta},\right.$$
$$\left.\psi\left(\frac{c(1-\sum_{n=-\infty}^{+\infty}\gamma_n)}{1-\sum_{n=-\infty}^{+\infty}\gamma_n+\sum_{n=-\infty}^{+\infty}\gamma_n\sum_{s=-\infty}^{n-1}\frac{1}{\psi^{-1}(q(s))}}\right)\frac{\beta}{3+3\beta}\right\};$$

$$W=\max\left\{\phi\left(\frac{bP_{k_2}}{\sum_{s=-\infty}^{k_1-1}\frac{1}{\phi^{-1}(p(s))}}\right)\frac{1}{\sum_{t=k_1}^{k_2}\frac{1}{2^{|t|}}},\ \psi\left(\frac{bQ_{k_2}}{\sum_{s=-\infty}^{k_1-1}\frac{1}{\psi^{-1}(q(s))}}\right)\frac{1}{\sum_{t=k_1}^{k_2}\frac{1}{2^{|t|}}}\right\};$$
$$E=\min\left\{\phi\left(\frac{a(1-\sum_{n=-\infty}^{+\infty}\alpha_n)}{1-\sum_{n=-\infty}^{+\infty}\alpha_n+\sum_{n=-\infty}^{+\infty}\alpha_n\sum_{s=-\infty}^{n-1}\frac{1}{\phi^{-1}(p(s))}}\right)\frac{\delta}{3+3\delta},\right.$$
$$\left.\psi\left(\frac{a(1-\sum_{n=-\infty}^{+\infty}\gamma_n)}{1-\sum_{n=-\infty}^{+\infty}\gamma_n+\sum_{n=-\infty}^{+\infty}\gamma_n\sum_{s=-\infty}^{n-1}\frac{1}{\psi^{-1}(q(s))}}\right)\frac{\beta}{3+3\beta}\right\}.$$

Theorem 1 Suppose that (b)–(e) hold. Choose $k_1,k_2\in N$ with $k_1<k_2$. Let μ be defined by (2.10). Furthermore, suppose that there exist $0<a<b<\frac{b}{\mu}<c$ such that

(A1): $f(n,P_n u,Q_n v)\le\frac{Q}{2^{|n|}}$ for all $n\in Z,u,v\in[0,c]$;
$g(n,P_n u,Q_n v)\le\frac{Q}{2^{|n|}}$ for all $n\in Z,u,v\in[0,c]$;

(A2): $f(n,P_n u,Q_n v)\ge\frac{W}{2^{|n|}}$ for all $n\in[k_1,k_2]$, $u,v\in[b,\frac{b}{\mu}]$;
$g(n,P_n u,Q_n v)\ge\frac{W}{2^{|n|}}$ for all $n\in[k_1,k_2]$, $u,v\in[b,\frac{b}{\mu}]$;

(A3): $f(n,P_n u,Q_n v)\le\frac{E}{2^{|n|}}$ for all $n\in Z,u,v\in[0,a]$;
$g(n,P_n u,Q_n v)\le\frac{E}{2^{|n|}}$ for all $n\in Z,u,v\in[0,a]$.

Then BVP (1.3) has at least three positive solutions x_1,x_2,x_3 such that

$$\begin{aligned}
&\sup_{n\in Z}\frac{x_1(n)}{P_n}<a,\ \sup_{n\in Z}\frac{y_1(n)}{Q_n}<a,\\
&\min_{n\in[k_1,k_2]}\frac{x_2(n)}{P_n}>b,\ \min_{n\in[k_1,k_2]}\frac{y_2(n)}{Q_n}>b,\\
&\text{either }\sup_{n\in Z}\frac{x_3(n)}{P_n}>a\text{ or }\sup_{n\in Z}\frac{y_3(n)}{Q_n}>a,\\
&\text{either }\min_{n\in[k_1,k_2]}\frac{x_3(n)}{P_n}<b\text{ or }\min_{n\in[k_1,k_2]}\frac{y_3(n)}{Q_n}<b.
\end{aligned}\tag{2.20}$$

Proof Let E, P and T be defined above. We complete the proof using Lemma 1. Define the functional on $\varphi:P\to R$ by

$$\varphi(x,y)=\min\left\{\min_{n\in[k_1,k_2]}\frac{x(n)}{P_n},\ \min_{n\in[k_1,k_2]}\frac{y(n)}{Q_n}\right\},\ (x,y)\in P.$$

It is easy to see that ϕ is a nonnegative continuous convex functional on the cone P. Choose $d=\frac{b}{\mu}$. Then $0<a<b<d<c$. Now we prove all assumptions in Lemma 1 are satisfied.

(1): Prove that $\varphi(x,y)\le||(x,y)||$ for all $(x,y)\in\overline{P}_c$. It is easy to see that $\varphi(x,y)\le||(x,y)||$ for all $(x,y)\in\overline{P}_c$.

(2): Prove that $T(\overline{P}_c)\subseteq\overline{P}_c$. For $(x,y)\in\overline{P}_c$, we have $||(x,y)||\le c$, then

$$\max\left\{\sup_{n\in Z}\frac{x(n)}{P_n},\ \sup_{n\in Z}\frac{y(n)}{Q_n}\right\}\le c.$$

Then

$$0\le\frac{x(n)}{P_n}\le c,\ 0\le\frac{y(n)}{Q_n}\le c,\ n\in Z.$$

From (A1), we get

$$f(n,x(n),y(n))=f\left(n,P_n\frac{x(n)}{P_n},Q_n\frac{y(n)}{Q_n}\right)\le\frac{Q}{2^{|n|}},n\in Z,$$
$$g(n,x(n),y(n))=g\left(n,P_n\frac{x(n)}{P_n},Q_n\frac{y(n)}{Q_n}\right)\le\frac{Q}{2^{|n|}},n\in Z.$$

So

$$\begin{aligned}
\frac{T_1(x,y)(n)}{P_n}&=\frac{1}{P_n}\frac{1}{1-\sum_{n=-\infty}^{+\infty}\alpha_n}\\
&\times\sum_{n=-\infty}^{+\infty}\alpha_n\sum_{s=-\infty}^{n-1}\frac{1}{\phi^{-1}(p(s))}\phi^{-1}\left(A_f+\sum_{t=s}^{+\infty}f(t,x(t),y(t))\right)\\
&+\frac{1}{P_n}\sum_{s=-\infty}^{n-1}\frac{1}{\phi^{-1}(p(s))}\phi^{-1}\left(A_f+\sum_{t=s}^{+\infty}f(t,x(t),y(t))\right)\\
&\le\frac{1}{P_n}\frac{1}{1-\sum_{n=-\infty}^{+\infty}\alpha_n}\sum_{n=-\infty}^{+\infty}\alpha_n\sum_{s=-\infty}^{n-1}\frac{1}{\phi^{-1}(p(s))}\phi^{-1}\\
&\times\left(\frac{1+\delta}{\delta}\sum_{t=-\infty}^{+\infty}Q2^{-|t|}\right)+\frac{1}{P_n}\sum_{s=-\infty}^{n-1}\frac{1}{\phi^{-1}(p(s))}\phi^{-1}\\
&\times\left(\frac{1+\delta}{\delta}\sum_{t=-\infty}^{+\infty}Q2^{-|t|}\right)\\
&<\left[1+\frac{1}{1-\sum_{n=-\infty}^{+\infty}\alpha_n}\sum_{n=-\infty}^{+\infty}\alpha_n\sum_{s=-\infty}^{n-1}\frac{1}{\phi^{-1}(p(s))}\right]\\
&\phi^{-1}\left(\frac{1+\delta}{\delta}\sum_{t=-\infty}^{+\infty}Q2^{-|t|}\right)\le c.
\end{aligned}$$

Similarly we get

$$\begin{aligned}
\frac{T_2(x,y)(n)}{Q_n}&=\frac{1}{Q_n}\frac{1}{1-\sum_{n=-\infty}^{+\infty}\gamma_n}\\
&\times\sum_{n=-\infty}^{+\infty}\gamma_n\sum_{s=-\infty}^{n-1}\frac{1}{\psi^{-1}(q(s))}\psi^{-1}\left(B_g+\sum_{t=s}^{+\infty}g(t,x(t),y(t))\right)\\
&+\frac{1}{Q_n}\sum_{s=-\infty}^{n-1}\frac{1}{\psi^{-1}(q(s))}\psi^{-1}\left(B_g+\sum_{t=s}^{+\infty}g(t,x(t),y(t))\right)\le c.
\end{aligned}$$

Hence $T(x,y)\in\overline{P}_c$. Then $T(\overline{P}_c)\subseteq\overline{P}_c$.

(3): $\{(x,y)\in P(\varphi;b,d)|\varphi(x,y)>b\}\neq\emptyset$ and $\varphi(T(x,y))>b$ for $(x,y)\in P(\varphi;b,d)$. Since $\frac{b}{\mu}>b$, one sees that $\{(x,y)\in P(\varphi;b,d)$ $|\varphi(x,y)>b\}\neq\emptyset$. For $(x,y)\in P(\varphi;b,d)$, we have

$$\max\left\{\sup_{n\in Z}\frac{x(n)}{P_n},\ \sup_{n\in Z}\frac{y(n)}{Q_n}\right\}\le d=\frac{b}{\mu},$$

and

$$\min\left\{\min_{n\in[k_1,k_2]}\frac{x(n)}{P_n},\ \min_{n\in[k_1,k_2]}\frac{y(n)}{Q_n}\right\}\ge b.$$

Then

$$b\le\frac{x(n)}{P_n},\ \frac{y(n)}{Q_n}\le\frac{b}{\mu},\ n\in[k_1,k_2].$$

It follows from (A2) that

$$f(n,x(n),y(n))=f\left(n,P_n\frac{x(n)}{P_n},Q_n\frac{y(n)}{Q_n}\right)\ge\frac{W}{2^{|n|}},n\in[k_1,k_2],$$
$$g(n,x(n),y(n))=\left(n,P_n\frac{x(n)}{P_n},Q_n\frac{y(n)}{Q_n}\right)\ge\frac{W}{2^{|n|}},n\in[k_1,k_2].$$

Hence

$$\min_{n\in[k_1,k_2]}\frac{T_1(x,y)(n)}{P_n}\geq\frac{T_1(x,y)(k_1)}{P_{k_2}}$$

$$=\frac{1}{P_{k_2}}\frac{1}{1-\sum_{n=-\infty}^{+\infty}\alpha_n}\sum_{n=-\infty}^{+\infty}\alpha_n\sum_{s=-\infty}^{n-1}\frac{1}{\phi^{-1}(p(s))}\phi^{-1}\left(A_f+\sum_{t=s}^{+\infty}f(t,x(t),y(t))\right)$$

$$+\frac{1}{P_{k_2}}\sum_{s=-\infty}^{k_1-1}\frac{1}{\phi^{-1}(p(s))}\phi^{-1}\left(A_f+\sum_{t=s}^{+\infty}f(t,x(t),y(t))\right)$$

$$\geq\frac{1}{P_{k_2}}\sum_{s=-\infty}^{k_1-1}\frac{1}{\phi^{-1}(p(s))}\phi^{-1}\left(\sum_{t=s}^{+\infty}f(t,x(t),y(t))\right)$$

$$>\frac{1}{P_{k_2}}\sum_{s=-\infty}^{k_1-1}\frac{1}{\phi^{-1}(p(s))}\phi^{-1}\left(\sum_{t=k_1}^{k_2}f(t,x(t),y(t))\right)$$

$$\geq\frac{1}{P_{k_2}}\sum_{s=-\infty}^{k_1-1}\frac{1}{\phi^{-1}(p(s))}\phi^{-1}\left(\sum_{t=k_1}^{k_2}\frac{W}{2^{|t|}}\right)\geq b.$$

Similarly, we have

$$\min_{n\in[k_1,k_2]}\frac{T_2(x,y)(n)}{Q_n}>\frac{1}{Q_{k_2}}\sum_{s=-\infty}^{k_1-1}\frac{1}{\psi^{-1}(q(s))}\psi^{-1}\times\left(\sum_{t=k_1}^{k_2}g(t,x(t),y(t))\right)$$

$$\geq\frac{1}{Q_{k_2}}\sum_{s=-\infty}^{k_1-1}\frac{1}{\psi^{-1}(q(s))}\psi^{-1}\left(\sum_{t=k_1}^{k_2}\frac{W}{2^{|t|}}\right)\geq b.$$

Hence $\varphi(T(x,y))>b$ for $(x,y)\in P(\varphi;b,d)$.

(4): $||T(x,y)||<a$ for $||(x,y)||\leq a$. For $||(x,y)||\leq a$, we have

$$\max\left\{\sup_{n\in Z}\frac{x(n)}{P_n},\ \sup_{n\in Z}\frac{y(n)}{Q_n}\right\}\leq a.$$

Then

$$0\leq\frac{x(n)}{P_n}\leq a,\ 0\leq\frac{y(n)}{Q_n}\leq a,\ n\in Z.$$

From (A3), we get

$$f(n,x(n),y(n))=f\left(n,P_n\frac{x(n)}{P_n},Q_n\frac{y(n)}{Q_n}\right)\leq\frac{E}{2^{|n|}},n\in Z,$$

$$g(n,x(n),y(n))=g\left(n,P_n\frac{x(n)}{P_n},Q_n\frac{y(n)}{Q_n}\right)\leq\frac{E}{2^{|n|}},n\in Z.$$

So

$$\frac{T_1(x,y)(n)}{P_n}=\frac{1}{P_n}\frac{1}{1-\sum_{n=-\infty}^{+\infty}\alpha_n}\sum_{n=-\infty}^{+\infty}\alpha_n\sum_{s=-\infty}^{n-1}\frac{1}{\phi^{-1}(p(s))}\phi^{-1}\times\left(A_f+\sum_{t=s}^{+\infty}f(t,x(t),y(t))\right)$$

$$+\frac{1}{P_n}\sum_{s=-\infty}^{n-1}\frac{1}{\phi^{-1}(p(s))}\phi^{-1}\left(A_f+\sum_{t=s}^{+\infty}f(t,x(t),y(t))\right)$$

$$\leq\frac{1}{P_n}\frac{1}{1-\sum\limits_{n=-\infty}^{+\infty}\alpha_n}\sum_{n=-\infty}^{+\infty}\alpha_n\sum_{s=-\infty}^{n-1}\frac{1}{\phi^{-1}(p(s))}\phi^{-1}\times\left(\frac{1+\delta}{\delta}\sum_{t=-\infty}^{+\infty}E2^{-|t|}\right)$$

$$+\frac{1}{P_n}\sum_{s=-\infty}^{n-1}\frac{1}{\phi^{-1}(p(s))}\phi^{-1}\left(\frac{1+\delta}{\delta}\sum_{t=-\infty}^{+\infty}E2^{-|t|}\right)$$

$$<\left[1+\frac{1}{1-\sum_{n=-\infty}^{+\infty}\alpha_n}\sum_{n=-\infty}^{+\infty}\alpha_n\sum_{s=-\infty}^{n-1}\frac{1}{\phi^{-1}(p(s))}\right]\phi^{-1}\times\left(\frac{1+\delta}{\delta}\sum_{t=-\infty}^{+\infty}E2^{-|t|}\right)\leq a.$$

Similarly, we get

$$\frac{T_2(x,y)(n)}{Q_n}=\frac{1}{Q_n}\frac{1}{1-\sum_{n=-\infty}^{+\infty}\gamma_n}\times\sum_{n=-\infty}^{+\infty}\gamma_n\sum_{s=-\infty}^{n-1}\frac{1}{\psi^{-1}(q(s))}\psi^{-1}\left(B_g+\sum_{t=s}^{+\infty}g(t,x(t),y(t))\right)$$

$$+\frac{1}{Q_n}\sum_{s=-\infty}^{n-1}\frac{1}{\psi^{-1}(q(s))}\psi^{-1}\left(B_g+\sum_{t=s}^{+\infty}g(t,x(t),y(t))\right)<a.$$

Hence $||T(x,y)||<a$.

(5): $\varphi(T(x,y))>b$ for $(x,y)\in P(\varphi;b,c)$ with $||T(x,y)||>d$. For $(x,y)\in P(\varphi;b,c)$ with $||T(x,y)||>d$, we have $\varphi(x,y)\leq b$ and $||(x,y)||\leq$. Then

$$\min\left\{\min_{n\in[k_1,k_2]}\frac{x(n)}{P_n},\ \min_{n\in[k_1,k_2]}\frac{y(n)}{Q_n}\right\}\leq b,$$

$$\max\left\{\sup_{n\in Z}\frac{x(n)}{P_n},\ \sup_{n\in Z}\frac{y(n)}{Q_n}\right\}\leq c,$$

and

$$\max\left\{\sup_{n\in Z}\frac{T_1(x,y)(n)}{P_n},\ \sup_{n\in Z}\frac{T_2(x,y)(n)}{Q_n}\right\}>d.$$

So

$$\varphi(T(x,y)) = \min\left\{\min_{n\in[k_1,k_2]}\frac{x(n)}{P_n}, \min_{n\in[k_1,k_2]}\frac{y(n)}{Q_n}\right\}$$
$$\geq \mu \max\left\{\sup_{n\in Z}\frac{T_1(x,y)(n)}{P_n}, \sup_{n\in Z}\frac{T_2(x,y)(n)}{Q_n}\right\} > \mu d = b.$$

Then T has at least three fixed points (x_1, y_1), (x_2, y_2) and (x_3, y_3) such that $||(x_1, y_1)|| < a$, $\psi(x_2, y_2) > b$ and $||(x_3, y_3)|| > a$ with $\psi(x_3, y_3) < b$. Then (x_1, y_1), (x_2, y_2) and (x_3, y_3) satisfy (2.20). The proof is completed. □

An example

In this section, we present an example to illustrate efficiency of Theorem 1.

Example 1 Consider the following boundary value problem of the bilateral difference system:

$$\begin{cases} \Delta[p(n)\Delta x(n)] + f(n, x(n), y(n)) = 0, & n \in Z, \\ \Delta[q(n)\Delta y(n)] + g(n, x(n), y(n)) = 0, & n \in Z, \\ \lim_{n\to-\infty} x(n) = 0, \\ \lim_{n\to-\infty} y(n) = 0, \\ \lim_{n\to+\infty} p(n)\Delta x(n) = 0, \\ \lim_{n\to+\infty} q(n)\Delta y(n) = 0, \end{cases} \tag{3.1}$$

where $p(n) = q(n) = 2^{-n}$, $f, g : Z \times [0, +\infty)^2 \to [0, +\infty)$ are defined by

$$f(n, u, v) = 2^{-|n|}[f_1(2^{-n}u) + f_2(2^{-n}v)]$$
$$g(n, u, v) = 2^{-|n|}[g_1(2^{-n}u) + g_2(2^{-n}v)]$$

with

$$f_1(u) = f_2(u) = 2^{-|n|}\begin{cases} \frac{1}{24}u, & u \in [0, 96], \\ 4 + \frac{235595 - 4}{140 - 96}(u - 96), & u \in [96, 140], \\ 235595, & u \in [140, 3688200], \\ 235595 \times e^{u-3688200}, & u \geq 3688200, \end{cases}$$

$$g_1(u) = g_2(u) = 2^{-|n|}\begin{cases} \frac{1}{24}u, & u \in [0, 96], \\ 4 + \frac{235595 - 4}{140 - 96}(u - 96), & u \in [96, 140], \\ 235595, & u \in [140, 3688200], \\ 235595 \times e^{u-3688200}, & u \geq 3688200. \end{cases}$$

Then (3.1) has at least three positive solutions (x_1, y_1), (x_2, y_2) and (x_3, y_3) satisfying

$$\sup_{n\in Z}\frac{x_1(n)}{2^n} < 96, \quad \sup_{n\in Z}\frac{y_1(n)}{2^n} < 96,$$
$$\min_{n\in[10,12]}\frac{x_2(n)}{2^n} > 140, \quad \min_{n\in[10,12]}\frac{y_2(n)}{2^n} > 140,$$
$$\text{either } \sup_{n\in Z}\frac{x_3(n)}{2^n} > 96 \text{ or } \sup_{n\in Z}\frac{y_3(n)}{2^n} > 96$$
$$\text{either } \min_{n\in[10,12]}\frac{x_3(n)}{2^n} < 140 \text{ or } \min_{n\in[10,12]}\frac{y_3(n)}{2^n} < 140. \tag{3.2}$$

Proof Corresponding to BVP (1.3), $p(n) = q(n) = 2^{-n}$, $\alpha_i = \beta_i = \gamma_i = \delta_i = 0$ for $i = 1, 2, \cdots, n$, $\phi(x) = \psi(x) = x$ with $\phi^{-1}(x) = \psi^{-1}(x) = x$, and

$$\sum_{n=-\infty}^{+\infty}\frac{\beta_n}{\phi^{-1}(p(n))} = 0 < \frac{1}{\phi^{-1}(1+\beta)} \quad \text{with } \beta = 1 > 0,$$
$$\sum_{n=-\infty}^{+\infty}\frac{\delta_n}{\psi^{-1}(q(n))} = 0 < \frac{1}{\psi^{-1}(1+\delta)} \quad \text{with } \delta = 1 > 0.$$

One sees that (b), (c), (d) and (e) hold.

By direct computation, we know that

$$P_n = Q_n = 1 + \sum_{s=-\infty}^{n-1} 2^s = 2^n.$$

Choose the constant $k_1 = 10, k_2 = 12$, $a = 96, b = 140$, $c = 3688200$. It is easy to see that

$$\mu = \min\left\{\frac{P_{k_1}}{P_{k_2}}, \frac{1}{P_{k_2}}, \frac{1}{\phi^{-1}(p(k_1-1))P_{k_2}}, \frac{Q_{k_1}}{Q_{k_2}}, \frac{1}{Q_{k_2}}, \frac{1}{\psi^{-1}(q(k_1-1))Q_{k_2}}\right\} = 2^{-12},$$

$$Q = \min\left\{\phi\left(\frac{c(1-\sum_{n=-\infty}^{+\infty}\alpha_n)}{1-\sum_{n=-\infty}^{+\infty}\alpha_n + \sum_{n=-\infty}^{+\infty}\alpha_n\sum_{s=-\infty}^{n-1}\frac{1}{\phi^{-1}(p(s))}}\right)\frac{\delta}{3+3\delta},\right.$$
$$\left.\psi\left(\frac{c(1-\sum_{n=-\infty}^{+\infty}\gamma_n)}{1-\sum_{n=-\infty}^{+\infty}\gamma_n + \sum_{n=-\infty}^{+\infty}\gamma_n\sum_{s=-\infty}^{n-1}\frac{1}{\psi^{-1}(q(s))}}\right)\frac{\beta}{3+3\beta}\right\} = 614700;$$

$$W = \max\left\{\phi\left(\frac{bP_{k_2}}{\sum_{s=-\infty}^{k_1-1}\frac{1}{\phi^{-1}(p(s))}}\right)\frac{1}{\sum_{t=k_1}^{k_2}\frac{1}{2^{|t|}}}, \psi\left(\frac{bQ_{k_2}}{\sum_{s=-\infty}^{k_1-1}\frac{1}{\psi^{-1}(q(s))}}\right)\frac{1}{\sum_{t=k_1}^{k_2}\frac{1}{2^{|t|}}}\right\} = 10 \times 2^{15};$$

$$E = \min\left\{\phi\left(\frac{a(1-\sum_{n=-\infty}^{+\infty}\alpha_n)}{1-\sum_{n=-\infty}^{+\infty}\alpha_n + \sum_{n=-\infty}^{+\infty}\alpha_n\sum_{s=-\infty}^{n-1}\frac{1}{\phi^{-1}(p(s))}}\right)\frac{\delta}{3+3\delta},\right.$$
$$\left.\psi\left(\frac{a(1-\sum_{n=-\infty}^{+\infty}\gamma_n)}{1-\sum_{n=-\infty}^{+\infty}\gamma_n + \sum_{n=-\infty}^{+\infty}\gamma_n\sum_{s=-\infty}^{n-1}\frac{1}{\psi^{-1}(q(s))}}\right)\frac{\beta}{3+3\beta}\right\} = 16.$$

So

$$f(n, P_n u, Q_n v) = 2^{-|n|}[f_1(u) + f_2(v)],$$
$$g(n, P_n u, Q_n v) = 2^{-|n|}[g_1(u) + g_2(v)].$$

It is easy to check that

(A1): $f(n,P_nu,Q_nv) \le \frac{307530}{2^{|n|}}$ for all $n \in Z, u,v \in [0,3688200]$; $g(n,P_nu,Q_nv) \le \frac{307530}{2^{|n|}}$ for all $n \in Z, u,v \in [0,3688200]$;

(A2): $f(n,P_nu,Q_nv) \ge \frac{5\times 2^{15}}{2^{|n|}}$ for all $n \in [10,12]$, $u,v \in [140,573440]$; $f(n,P_nu,Q_nv) \ge \frac{5\times 2^{15}}{2^{|n|}}$ for all $n \in [10,12]$, $u,v \in [140,573440]$;

(A3): $f(n,P_nu,Q_nv) \le \frac{8}{2^{|n|}}$ for all $n \in Z, u,v \in [0,96]$; $f(n,P_nu,Q_nv) \le \frac{8}{2^{|n|}}$ for all $n \in Z, u,v \in [0,96]$.

Then by Theorem 1, BVP (3.1) has at least three positive solutions $(x_1,y_1),(x_2,y_2)$ and (x_3,y_3) satisfying (3.2). The proof is completed. □

Acknowledgments This work is supported by the Natural Science Foundation of Guangdong province (No: S2011010001900) and the Foundation for High-level talents in Guangdong Higher Education Project.

References

1. Elsayed, E.M., El-Metwally, H.A.: Qualitative Studies of Scalars and Systems of Differene Equations. Lab Lambert Academic Press, Saarbrucken (2012)
2. Elaydi, S.: An Introduction to Difference Equations, 3rd edn. Springer, Berlin (2005)
3. Starr, G.P.: Introduction to applied digital control, 2nd edn. John Wiley and Sons Ltd, New York (2006)
4. Avery, R.I., Peterson, A.C.: Three positive fixed points of nonlinear operators on ordered banach spaces. Comput. Math. Appl. **42**, 313–322 (2001)
5. Agarwal, P.R., O'Regan, D.: Cone compression and expansion and fixed point theorems in Frchet spaces with application. J. Differ. Equ. **171**(2), 412–422 (2001)
6. Agarwal, R.P., O'Regan, D.: Nonlinear Urysohn discrete equations on the infinite interval: a fixed-point approach. Comput. Math. Appl. **42**(3–5), 273–281 (2001)
7. Avery, R.I., Peterson, A.C.: Three positive fixed points of nonlinear operators on ordered Banach spaces. Comput. Math. Appl. **42**(3–5), 313–322 (2001)
8. Cheung, W., Ren, J., Wong, P.J.Y., Zhao, D.: Multiple positive solutions for discrete nonlocal boundary value problems. J. Math. Anal. Appl. **330**(2), 900–915 (2007)
9. Avery, R.I.: A generalization of Leggett-Williams fixed point theorem. Math. Sci. Res. Hot Line **3**(12), 9–14 (1993)
10. Li, Y., Lu, L.: Existence of positive solutions of p-Laplacian difference equations. Appl. Math. Lett. **19**(10), 1019–1023 (2006)
11. Liu, Y., Ge, W.: Twin positive solutions of boundary value problems for finite difference equations with p-Laplacian operator. J. Math. Anal. Appl. **278**(2), 551–561 (2003)
12. Pang, H., Feng, H., Ge, W.: Multiple positive solutions of quasi-linear boundary value problems for finite difference equations. Appl. Math. Comput. **197**(1), 451–456 (2008)
13. Wong, P.J.Y., Xie, L.: Three symmetric solutions of lidstone boundary value problems for difference and partial difference equations. Comput. Math. Appl. **45**(6–9), 1445–1460 (2003)
14. Yu, J., Guo, Z.: On generalized discrete boundary value problems of Emden-Fowler equation. Sci. China (Ser. A Math.) **36**(7), 721–732 (2006)
15. Agarwal, R.P.: Difference Equations and Inequalities: Theory, Methods, and Applications, Second edition, Marcel Dekker Inc, 2000.
16. Ma, R., Raffoul, T.: Positive solutions of three-point nonlinear discrete second order boundary value problem. J. Differ. Equ. Appl. **10**(2), 129–138 (2004)
17. Liu, Y.: Positive solutions of BVPs for finite difference equations with one-dimensional p-Laplacian. Commu. Math. Anal. **4**(1), 58–77 (2008)
18. Agarwal, R.P., O'Regan, D.: Boundary value problems for general discrete systems on infinite intervals. Comput. Math. Appl. **33**(7), 85–99 (1997)
19. Tian, Y., Ge, W.: Multiple positive solutions of boundary value problems for second-order discrete equations on the half-line. J. Differ. Equ. Appl. **12**(2), 191–208 (2006)
20. Agarwal, R.P., O'Regan, D.: Discrete systems on infinite intervals. Comput. Math. Appl. **35**(9), 97–105 (1998)
21. Kanth, A.R., Reddy, Y.: A numerical method for solving two point boundary value problems over infinite intervals. Appl. Math. Comput. **144**(2), 483–494 (2003)
22. Agarwal, R.P., Bohner, M., O'Regan, D.: Time scale boundary value problems on infinite intervals. J. Comput. Appl. Math. **141**(1–2), 27–34 (2002)
23. Rachunek, L., Rachunkoa, I.: Homoclinic solutions of non-autonomous difference equations arising in hydrodynamics. Nonlinear Anal. Real World Appl. **12**, 14–23 (2011)
24. Chen, H., He, Z.: Infinitely many homoclinic solutions for second-order discrete Hamiltonian systems. J. Diff. Equ. Appl. **19**, 1940–1951 (2013)
25. Karpenko, O., Stanzhytskyi, O.: The relation between the existence of bounded solutions of differential equations and the corresponding difference equations. J. Diff. Equ. Appl. **19**, 1967–1982 (2013)
26. Chen, P.: Existence of homoclinic orbits in discrete Hamiltonian systems without Palais-Smale condition. J. Differ. Equ. Appl. **19**, 1981–1994 (2013)

2

Vague ranking of fuzzy numbers

M. Adabitabar Firozja[1] · F. Rezai Balf[1] · S. Firouzian[2]

Abstract In a lot of scientific models in the real world, we confront with comparing fuzzy numbers as decision-making procedures and etc. It will be interest, if we know that, comparison discuss is sometimes ambiguous. Hence, this article focus on ranking fuzzy numbers with protection ambiguity. Our idea for this work is based on this claim that ranking of two fuzzy numbers should be a vague value. However, we utilize the notion of max and min fuzzy simultaneously.

Keywords Fuzzy numbers · Ranking · Vague value

Introduction

In variety of application domains, such as decision making [26], risk assessment [13], linear programming [20], linear systems [12], and artificial intelligence [6], ranking fuzzy numbers are used. This topic has been studied by many researchers. Some researchers employed a distance for ordering of fuzzy numbers such as Abbasbandy and Asady [1], Yao and Wu [25], Allahviranloo and Adabitabar Firozja [4], Deng [21] and Janizade-Haji et al. [14]. Some researchers as [2, 15, 16] presented a defuzzification method for ranking fuzzy numbers. Vincent and Luu in [22] proposed improve their ranking method for fuzzy numbers with integral values. In [7], Deng by using ideal solutions showed a ranking approach. Fortemps and Roubens [11] introduced a ranking method based on area compensation. Some of the other researchers such as Adabitabar firozja et al. [3], Ezzati et al. [9, 10] and Modarres and Sadi-Nezhad [18] proposed a function for ranking. Wang et al. [23] defined the maximal and minimal reference sets and then proposed the ranking method based on deviation degree and relative variation of fuzzy numbers and subsequent Asady in [5] proposed a revised method of ranking LR fuzzy number based on deviation degree with Wang's method. Wang and Luo in [24] presented a ranking approach with positive and negative ideal points. Mahmodi Nejad and Mashinchi [17], introduced ranking fuzzy numbers based on the areas on the left and right sides of fuzzy number. In this paper, we provide a method for calculating the amount of vague value ranking fuzzy numbers.

The paper is organized as follows: The background on fuzzy concepts is presented in Sect. 2. A vague ranking of two fuzzy numbers with its properties is introduced in Sect. 3. Subsequently, in Sect. 4 some examples are presented. Finally, conclusion are drawn in Sect. 5.

Background

There are several definitions of a fuzzy number. In this paper we use the following definition.

Definition 1 [8] A set $\tilde{A}$ is a generalized left right fuzzy numbers (GLRFN) and denoted as $\tilde{A} = (a_1, a_2, a_3, a_4)_{LR}$, if it's membership function satisfies the following:

✉ M. Adabitabar Firozja
m.adabitabar@qaemiau.ac.ir

1 Department of Mathematics Qaemshar Branch, Islamic Azad University, Qaemshahr, Iran

2 Department of Mathematics, Payame Noor University (PNU), Tehran, Iran

$$\mu_{\tilde{A}}(x)=\begin{cases} L(\frac{a_2-x}{a_2-a_1}), & a_1\leq x\leq a_2,\\ 1, & a_2\leq x\leq a_3,\\ R(\frac{x-a_3}{a_4-a_3}), & a_3\leq x\leq a_4,\\ 0, & \text{otherwise}\end{cases} \quad (1)$$

where L and R are strictly decreasing functions defined on [0, 1] and satisfying the conditions:

$$\begin{array}{ll} L(t)=R(t)=1 & \text{if } t\leq 0\\ L(t)=R(t)=0 & \text{if } t\geq 1\end{array}. \quad (2)$$

Remark 1 Trapeziodal fuzzy numbers (TrFN) are special cases of GLRFN with $L(t)=R(t)=1-t$ and we show it as $\tilde{A}=(a_1,a_2,a_3,a_4)$.

Definition 2 A $\alpha-$*level* interval of fuzzy number $\tilde{A}$ is denoted as:

$$[\tilde{A}]^{\alpha}=[A_l(\alpha),A_r(\alpha)]=[a_2-(a_2-a_1)L_A^{-1}(\alpha),\ a_3+(a_4-a_3)R_A^{-1}(\alpha)] \quad (3)$$

Remark 2 Suppose, $\lambda\in R$ then

$$\tilde{A}+\lambda=(a_1+\lambda,a_2+\lambda,a_3+\lambda,a_4+\lambda)_{LR}$$
$$\lambda\tilde{A}=\begin{cases}(\lambda a_1,\lambda a_2,\lambda a_3,\lambda a_4)_{LR} & \lambda\geq 0\\ (\lambda a_4,\lambda a_3,\lambda a_2,\lambda a_1)_{LR} & \lambda<0\end{cases} \quad (4)$$

Definition 3 [3] Let $[A]^{\alpha}=[A_l(\alpha),A_r(\alpha)]$ and $[B]^{\alpha}=[B_l(\alpha),B_r(\alpha)];\alpha\in[0,1]$ be two $\alpha-$cuts of fuzzy numbers. We get

$$[\max\{A,B\}]^{\alpha}=\max\{[A]^{\alpha},[B]^{\alpha}\}=[\max\{A_l(\alpha),B_l(\alpha)\},\ \max\{A_r(\alpha),B_r(\alpha)\}]$$

and

$$[\min\{A,B\}]^{\alpha}=\min\{[A]^{\alpha},[B]^{\alpha}\}=[\min\{A_l(\alpha),B_l(\alpha)\},\ \min\{A_r(\alpha),B_r(\alpha)\}].$$

This definition is showed in Fig. 1.

Definition 4 [3] Let $U=\{u_1,u_2,u_3,...,u_n\}$, a vague set A in U is characterized by a truth-membership function $t_A:U\rightarrow[0,1]$ and a false-membership function $f_A:U\rightarrow[0,1]$, where $t_A(u_i)$ is a lower bound on the grade of membership of u_i derived from the evidence for u_i, $f_A(u_i)$ is a lower bound on the negation of u_i derived from the evidence against u_i, and $t_A(u_i)+f_A(u_i)\leq 1$. The grade of membership of u_i in the vague set A is vague value where bounded by a subinterval $[t_A(u_i),1-f_A(u_i)]$ of [0, 1]. Simply expressed, $A(u_i)=[t_A(u_i),1-f_A(u_i)]$.

For an arbitrary element $a\in[0,1]$, we assume that a is the same as [a, a], namely, $a=[a,a]$. For any $A=[a_1,a_2]$ and $B=[b_1,b_2]$, we can popularize operators such $+$ and $-$ and have $A+B=[a_1+b_1,a_2+b_2]$, $A-B=[a_1-b_2,a_2-b_1]$. Furthermore, we have $A=B\Leftrightarrow a_1=b_1,a_2=b_2$, $A\leq B\Leftrightarrow a_1\leq b_1,a_2\leq b_2$ and $A<B\Leftrightarrow a_1<b_1,a_2<b_2$.

Vague ranking of two fuzzy numbers

Given $\tilde{A},\tilde{B}\in E_{LR}$, are two fuzzy numbers. Regarding to many methods and shortcoming in ranking for fuzzy numbers, it is show that ranking is not deterministic. In other words, we know if $\text{supp}(\tilde{A})\cap\text{supp}(\tilde{B})\neq\phi$ then we can not define a crisp rank for $\tilde{A},\tilde{B}$. Therefore, we claim that ranking of two fuzzy numbers should be a vague value. Some researcher are used the max or min notion for ranking the fuzzy number. But, we utilize the notion of max and min simultaneously. It is trivial maximum of two fuzzy numbers is greater than or equal both of them and minimum of two fuzzy numbers is less than or equal both of them. Therefore, we will present true rate $\tilde{A}\leq\tilde{B}$ as $t_{A\preceq B}$ and false rate $\tilde{A}\leq\tilde{B}$ as $f_{A\preceq B}$ in ranking $\tilde{A}$ and $\tilde{B}$ as follows:

$$\begin{array}{l} t_{A\preceq B}=S\{\{\max\{\tilde{A},\tilde{B}\}\backslash\tilde{A}\}\bigcup\{\min\{\tilde{A},\tilde{B}\}\backslash\tilde{B}\}\}\\ f_{A\preceq B}=S\{\{\max\{\tilde{A},\tilde{B}\}\backslash\tilde{B}\}\bigcup\{\min\{\tilde{A},\tilde{B}\}\backslash\tilde{A}\}\}\end{array} \quad (5)$$

where the signs \ and $S\{.\}$ show subtract of Venn diagram and area, respectively. For this purpose, consider the following Fig. 2.

Where geometrically, $t_{A\preceq B}$ and $f_{A\preceq B}$ defined above is as follows:

$$t_{A\preceq B}=\frac{S(A<B)}{S},\qquad f_{A\preceq B}=\frac{S(B<A)}{S},\qquad S=S(A<B)+S(B<A)+S(A=B) \quad (6)$$

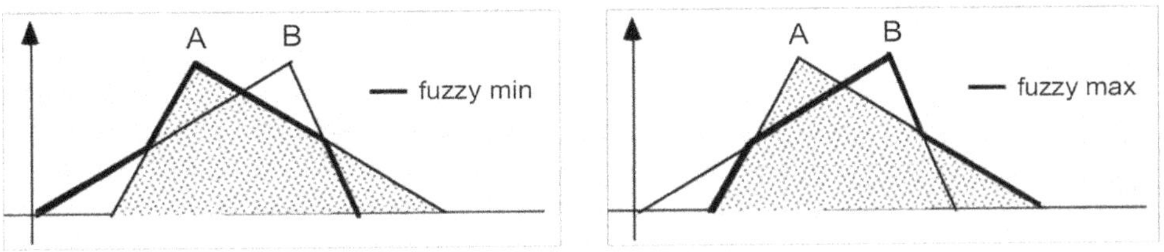

Fig. 1 Fuzzy max and fuzzy min of triangular fuzzy numbers

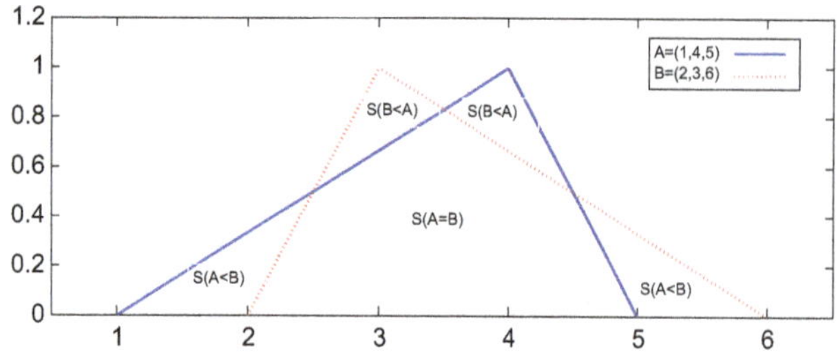

Fig. 2 Comparison of two fuzzy numbers $\tilde{A}, \tilde{B}$

$S(A<B)$ is part of surface B that is located on the right side of A or part of surface A that is located on the left side of B and $S(A=B)$ is common surface of A and B.

Remark 3 For $\tilde{A}, \tilde{B} \in E_{LR}$, $t_{A\preceq B} + f_{A\preceq B} \leq 1$.

Proof

$$t_{A\preceq B} + f_{A\preceq B} = \frac{S(A<B)}{S} + \frac{S(B<A)}{S} = \frac{S(A<B) + S(B<A)}{S(A<B) + S(B<A) + S(A=B)} \leq 1$$

As mentioned in above we used the notion area for ranking. Hence, we introduce two theorems in below. □

Theorem 1 [19] *If points $A(x_1, y_1)$, $B(x_2, y_2)$ and $C(x_3, y_3)$ be arbitrarily coordinates are triangular vertexes in anti-clock wise sense then the area of triangle $\triangle ABC$ is determined as follows:*

$$S_{\triangle ABC} = \frac{1}{2}\left\{\begin{vmatrix} x_1 & y_1 \\ x_2 & y_2 \end{vmatrix} + \begin{vmatrix} x_2 & y_2 \\ x_3 & y_3 \end{vmatrix} + \begin{vmatrix} x_3 & y_3 \\ x_1 & y_1 \end{vmatrix}\right\} \quad (7)$$

Theorem 2 [19] *The area of any regular polygon with $P_j(x_j, y_j)$, $j = 1, \ldots, n$ vertex in anti-clock wise sense is as follows:*

$$S_{P_1P_2\ldots P_n} = \frac{1}{2}\left\{\begin{vmatrix} x_1 & y_1 \\ x_2 & y_2 \end{vmatrix} + \begin{vmatrix} x_2 & y_2 \\ x_3 & y_3 \end{vmatrix} + \cdots + \begin{vmatrix} x_n & y_n \\ x_1 & y_1 \end{vmatrix}\right\} \quad (8)$$

Definition 5 Assume that $\tilde{A}$ and $\tilde{B}$ be two GLRFNs, validity rating of $A \preceq B$ is belong to interval $[t_{A\preceq B}, 1 - f_{A\preceq B}]$ where $t_{A\preceq B}$ minimum accuracy and $1 - f_{A\preceq B}$ maximum accuracy and we show with vague value of rank $\tilde{A} \preceq \tilde{B}$ with $\mathrm{VR}(A \preceq B)$ where

$$\mathrm{VR}(A \preceq B) = [t_{A\preceq B}, 1 - f_{A\preceq B}] \quad (9)$$

Some properties

For $\tilde{A}$ and $\tilde{B} \in E_{LR}$ and $\lambda \in R$:

Proposition 1 *$VR(A \preceq B)$ is a vague value.*

Proof With Remark 1. proof is evident.

Proposition 2 $t_{A\preceq B} = f_{B\preceq A}, f_{A\preceq B} = t_{B\preceq A}$.

Proof With Eq. (6) proof is evident.

Proposition 3 $VR(A \preceq B) = 1 - VR(B \preceq A)$

Proof Regarding to Eq. (9) and Proposition 2. $1 - \mathrm{VR}(B \preceq A) = [1, 1] - [t_{B\preceq A}, 1 - f_{B\preceq A}] = [f_{B\preceq A}, 1 - t_{B\preceq A}] = [t_{A\preceq B}, 1 - f_{A\preceq B}] = \mathrm{VR}(A \preceq B)$

Proposition 4 $VR(\lambda A \preceq \lambda B) = \begin{cases} VR(A \preceq B) & 0 \leq \lambda, \\ VR(B \preceq A) & otherwise. \end{cases}$

Proof Regarding to the Eqs. (9) and (6); if $\lambda \geq 0$ $\mathrm{VR}(\lambda A \preceq \lambda B) = [t_{\lambda A\preceq \lambda B}, 1 - f_{\lambda A\preceq \lambda B}] = [t_{A\preceq B}, 1 - f_{A\preceq B}] = \mathrm{VR}(A \preceq B)$ And if $\lambda < 0$

$\mathrm{VR}(\lambda A \preceq \lambda B) = [t_{\lambda A\preceq \lambda B}, 1 - f_{\lambda A\preceq \lambda B}] = [t_{B\preceq A}, 1 - f_{B\preceq A}] = \mathrm{VR}(B \preceq A)$

Proposition 5 $VR(\lambda + A \preceq \lambda + B) = VR(A \preceq B)$.

Proof Regarding Eqs. (9) and (6)

$$\mathrm{VR}(\lambda + A \preceq \lambda + B) = [t_{\lambda+A\preceq \lambda+B}, 1 - f_{\lambda+A\preceq \lambda+B}]$$

$$= [t_{A\preceq B}, 1 - f_{A\preceq B}] = \mathrm{VR}(A \preceq B).$$

Proposition 6 If $a_4 \leq b_1$ *then* $VR(\tilde{A} \preceq \tilde{B}) = [1, 1]$.

Proof Regarding to Eqs. (9) and (6) proof is evident.

Proposition 7 *If $\tilde{A}$ and $\tilde{B}$ are two GLRFNs then only one of the following relationship is established:*

$VR(A \preceq B) = VR(B \preceq A)$, $VR(A \preceq B) \leq VR(B \preceq A)$ *and* $VR(A \preceq B) \geq VR(B \preceq A)$.

Proof If $t_{A\preceq B} = t_{B\preceq A}$ then with Proposition 2, $f_{A\preceq B} = f_{B\preceq A}$ therefore $\mathrm{VR}(A \preceq B) = \mathrm{VR}(B \preceq A)$. If $t_{A\preceq B} < t_{B\preceq A}$ then with Proposition 2, $f_{A\preceq B} < f_{B\preceq A}$ therefore $\mathrm{VR}(A \preceq B) < \mathrm{VR}(B \preceq A)$. If $t_{A\preceq B} > t_{B\preceq A}$ then with Proposition 2, $f_{A\preceq B} > f_{B\preceq A}$ therefore $\mathrm{VR}(A \preceq B) > \mathrm{VR}(B \preceq A)$.

Definition 6 If $\tilde{A}$ and $\tilde{B}$ are two GLRFNs, validity rating of $A \preceq B$ is belong to interval $\mathrm{VR}(A \preceq B) = [t_{A\preceq B}, 1 - f_{A\preceq B}]$ where $t_{A\preceq B}$ minimum accuracy and $1 - f_{A\preceq B}$ maximum accuracy. With Proposition 9, we define ranking method as follows:

1. If $\mathrm{VR}(A \preceq B) = \mathrm{VR}(B \preceq A) = [0.5, 0.5]$, then can be said $\tilde{A} \approx \tilde{B}$.
2. If $\mathrm{VR}(A \preceq B) = \mathrm{VR}(B \preceq A) = [0, 1]$, then can be said $\tilde{A} = \tilde{B}$.
3. If $\mathrm{VR}(A \preceq B) < \mathrm{VR}(B \preceq A)$ then can be said $\tilde{B} \preceq \tilde{A}$.
4. If $\mathrm{VR}(A \preceq B) = [1, 1]$ or $\mathrm{VR}(B \preceq A) = [0, 0]$ then can be said $\tilde{A} < \tilde{B}$.

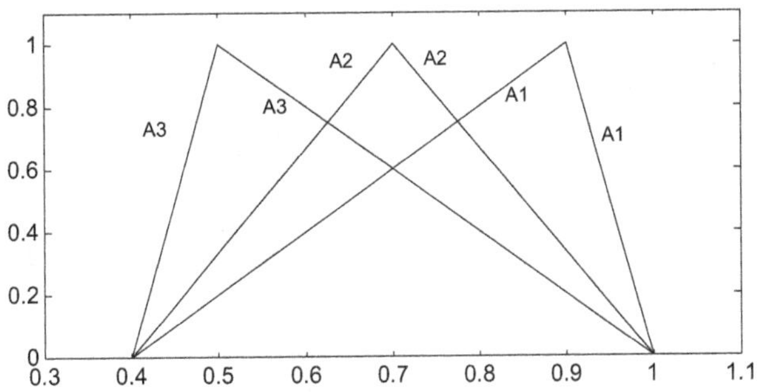

Fig. 3 Set 1-$\{\tilde{A_1}, \tilde{A_2}, \tilde{A_3}\}$

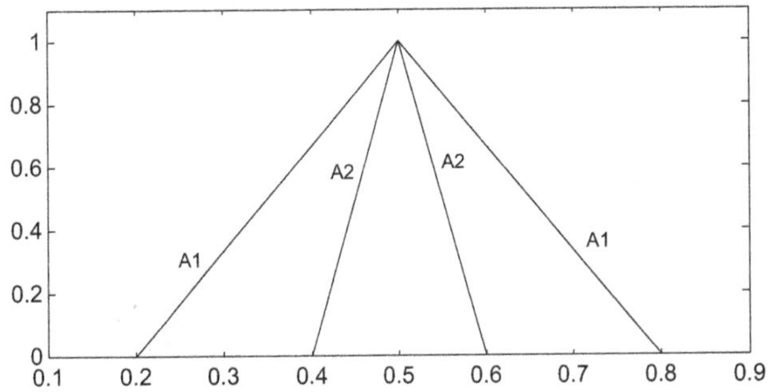

Fig. 4 Set 2-$\{\tilde{A_1}, \tilde{A_2}\}$

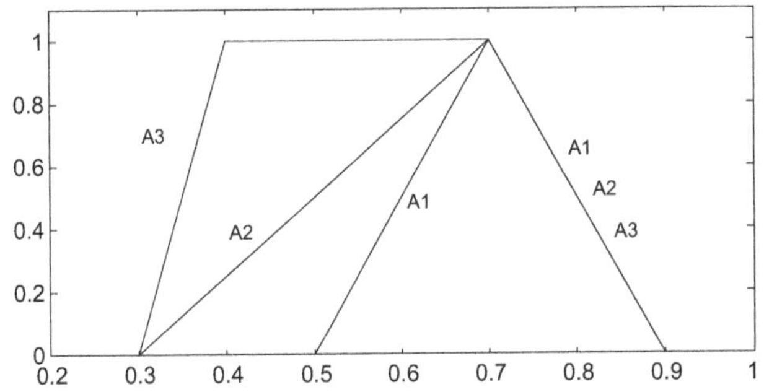

Fig. 5 Set 3-$\{\tilde{A_1}, \tilde{A_2}, \tilde{A_3}\}$

Numerical examples

For description the proposed method some examples constructed as follow [4, 21].

Set 1 $A_1 = (0.4, 0.9, 1)$, $A_2 = (0.4, 0.7, 1)$, $A_3 = (0.4, 0.5, 1)$ where show in Fig. 3.

$\mathrm{VR}(A_2 \leq A_1) = [0.4, 1]$, with Proposition 3 $\mathrm{VR}(A_1 \leq A_2) = [0, 0.6]$ therefore according to Definition 2, $A_1 \succeq A_2$.

$\mathrm{VR}(A_3 \leq A_1) = [0.82, 1]$ and obviously $\mathrm{VR}(A_1 \leq A_3) = [0, 0.18]$ therefore, $A_1 \succeq A_3$.

$\mathrm{VR}(A_3 \leq A_2) = [0.4, 1]$ and it is trivial that $\mathrm{VR}(A_2 \leq A_3) = [0, 0.6]$ therefore, $A_2 \succeq A_3$.

Set 2 $A_1 = (0.2, 0.5, 0.8)$, $A_2 = (0.4, 0.5, 0.6)$, where show in Fig. 4.

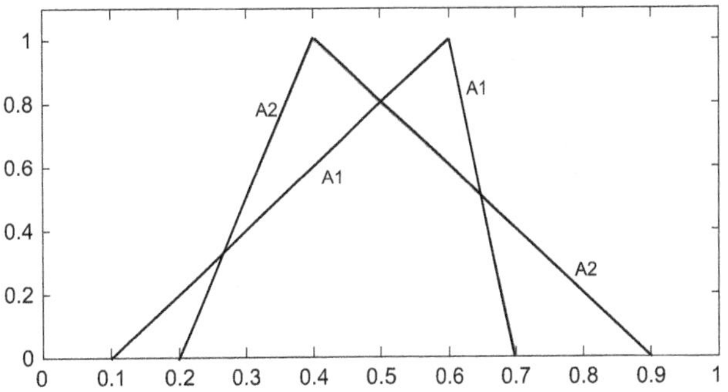

Fig. 6 Set 4-$\{\tilde{A_1}, \tilde{A_2}\}$

$\mathrm{VR}(A_1 \leq A_2) = [0.5, 0.5]$ therefor $\mathrm{VR}(A_2 \leq A_1) = [0.5, 0.5]$ hence $A_1 \approx A_2$.

Set 3 $A_1 = (0.5, 0.7, 0.9)$, $A_2 = (0.3, 0.7, 0.9)$, $A_3 = (0.3, 0.4, 0.7, 0.9)$, where show in Fig. 5.

$\mathrm{VR}(A_3 \leq A_2) = [0.333, 1]$, $\mathrm{VR}(A_3 \leq A_1) = [0.56, 1]$ and $\mathrm{VR}(A_2 \leq A_1) = [0.333, 1]$ as before, it follows that, $A_1 \succeq A_2$, $A_2 \succeq A_3$ and $A_1 \succeq A_3$.

And we consider another example for comparing the current method with

Set 4 $A_1 = (0.1, 0.6, 0.7)$, $A_2 = (0.2, 0.4, 0.9)$ where show in Fig. 6.

$\mathrm{VR}(A_2 \leq A_1) = [0.1931417, 0.83182047]$, that shows $A_2 \succeq A_1$.

Conclusions

In this paper, we showed that ranking of two fuzzy numbers should be a vague value. For this reason, we utilize the notion of max and min simultaneously in order to determining the ambiguity rate in ranking of two fuzzy numbers. It is shown that this approach verifies some properties as stability, transition and complement.

References

1. Abbasbandy, S., Asady, B.: Ranking of Fuzzy numbers by sign distance. Inf. Sci. **176**, 2405–2416 (2006)
2. Shureshjani, R.A., Darehmiraki, M.: A new parametric method for ranking fuzzy numbers. Indag. Math. **24**, 518–529 (2013)
3. Firozja, M.A., Agheli, B., Hosseinzadeh, M.: Ranking function of two LR-fuzzy numbers. J. Intell. Fuzzy Syst. **26**, 1137–1142 (2014)
4. Allahviranloo, T., Firozja, F.M.: Ranking of fuzzy numbers by a new metric. Soft Comput. **14**, 773–782 (2010)
5. Asady, B.: The revised method of ranking LR fuzzy number based on deviation degree. Expert Syst. Appl. **37**, 5056–5060 (2010)
6. Bui, D.T., Bui, Q.T., Nguyen, Q.P., Pradhan, B., Nampak, H., Trinh, P.T.: A hybrid artificial intelligence approach using GIS-

based neural-fuzzy inference system and particle swarm optimization for forest fire susceptibility modeling at a tropical area. Agric. For. Meteorol. **233**, 32–44 (2017)

7. Deng, H.: Comparing and ranking fuzzy numbers using ideal solutions. Appl. Math. Model. **38**, 1638–1646 (2014)
8. Dubios, D., Prade, H.: Fuzzy sets and Systems: Theory and Application. Academic Press, New york (1980)
9. Ezzati, R., Allahviranloo, T., Khezerloo, S., Khezerloo, M.: An approach for ranking of fuzzy numbers. Expert Syst. Appl. **39**, 690–695 (2012)
10. Ezzati, R., Khezerloo, S., Ziari, S.: Application of parametric form for ranking of fuzzy numbers. Iran. J. Fuzzy Syst. **12**, 59–74 (2015)
11. Fortemps, P., Roubens, M.: Ranking and defuzzification method based on area compensation. Fuzzy Sts Syst. **82**, 319–330 (1996)
12. Ghanbari, R.: Solutions of fuzzy LR algebraic linear systems using linear programs. Appl. Math. Model. **39**, 5164–5173 (2015)
13. Gul, M., Guneri, A.F.: A fuzzy multi criteria risk assessment based on decision matrix technique: a case study for aluminum industry. J. Loss Prev. Process Ind. **40**, 89–100 (2016)
14. Janizade-Haji, M., Zare, H.K., Eslamipoor, R., Sepehriar, A.: A developed distance method for ranking generalized fuzzy numbers. Neural Comput. Appl. **25**, 727–731 (2014)
15. Ma, M., Kandel, A., Friedman, M.: A new approach for defuzzification. Fuzzy Sets Syst. **111**, 351–356 (2000)
16. Ma, M., Kandel, A., Friedman, M.: Correction to "A new approach for defuzzification". Fuzzy Sets Syst. **128**, 133–134 (2000)
17. Nejad, A.M., Mashinchi, M.: Ranking fuzzy numbers based on the areas on the left and the right sides of fuzzy number. Comput. Math. Appl. **61**, 431–442 (2011)
18. Modarres, M., Nezhad, S.S.: Ranking fuzzy numbers by preference ratio. Fuzzy Sets Syst. **118**, 429–439 (2001)
19. Balf, F.R.: Ranking efficient units by regular polygon area (RPA) in DEA. Int. J. Ind. Math. **3**, 41–53 (2011)
20. Tan, R.R., Aviso, K.B., Cayamanda, C.D., Chiu, A.S.F., Promentilla, M.A.B., Ubando, A.T., Yu, K.D.S.: A fuzzy linear programming enterprise inputoutput model for optimal crisis operations in industrial complexes. Int. J. Prod. Econ. **181**, 410–418 (2016)
21. Tran, L., Duckstein, L.: Comparison of fuzzy numbers using a fuzzy distance measure. Fuzzy Sets Syst. **130**, 331–341 (2002)
22. Vincent, F.Y., Luu, Q.D.: An improved ranking method for fuzzy numbers with integral values. Appl. Soft Comput. **14**, 603–608 (2014)
23. Wang, Z.X., Liu, Y.J., Fan, Z.P., Feng, B.: Ranking LR fuzzy number based on deviation degree. Inf. Sci. **179**, 2070–2077 (2009)
24. Wang, Y.M., Luo, Y.: Area ranking of fuzzy numbers based on positive and negative ideal point. Comput. Math. Appl. **58**, 1769–1779 (2009)
25. Yao, J.S., Wu, K.: Ranking fuzzy numbers based on decomposition principle and signed distance. Fuzzy Sets Syst. **116**, 275–288 (2000)
26. Zhang, F., Ignatius, J., Limc, C.P., Zhao, Y.: A new method for ranking fuzzy numbers and its application to group decision making. Appl. Math. Model. **38**, 1563–1582 (2014)

3

Some convergence results for nearly asymptotically nonexpansive nonself mappings in CAT(κ) spaces

Aynur Şahin[1] · **Metin Başarır**[1]

Abstract The purpose of this paper is to prove the demiclosedness principle and convergence theorems for nearly asymptotically nonexpansive nonself mappings in CAT(κ) spaces with $\kappa > 0$. Our results extend and improve some recent results announced in the current literature.

Keywords Fixed point · Nearly asymptotically nonexpansive nonself mapping · Demiclosedness principle · Δ-convergence · Strong convergence · CAT(κ) space

Mathematics Subject Classification 47H10 · 54E40

Introduction

Throughout this paper, $\mathbb{N}$ is the set of all positive integers and $\mathbb{R}$ is the set of all real numbers. Let K be a nonempty subset of a metric space (X, d) and $T: K \to X$ be a nonself mapping. Denote by $F(T) = \{x \in K : Tx = x\}$, the set of fixed points of T. A nonself mapping T is said to be nonexpansive if

$$d(Tx, Ty) \leq d(x, y), \quad \forall x, y \in K.$$

A subset K of X is said to be a *retract* of X if there exists a continuous mapping $P: X \to K$ such that $Px = x$ for all $x \in K$. A mapping $P: X \to K$ is said to be a *retraction* if $P^2 = P$. It follows that if P is a retraction, then $Py = y$ for all y in the range of P.

Definition 1 [11] Let K be a nonempty subset of a metric space (X, d) and P be a nonexpansive retraction of X onto K. A nonself mapping $T: K \to X$ is said to be

(i) *Lipschitzian* if for each $n \in \mathbb{N}$, there exists a positive number k_n such that

$$d(T(PT)^{n-1}x, T(PT)^{n-1}y) \leq k_n d(x, y), \quad \forall x, y \in K;$$

(ii) *uniformly L-Lipschitzian* if $k_n = L$ for all $n \in \mathbb{N}$;

(iii) *asymptotically nonexpansive* if $k_n \geq 1$ for all $n \in \mathbb{N}$ with $\lim_{n\to\infty} k_n = 1$.

The class of nearly Lipschitzian nonself mappings is an important generalization of the class of Lipschitzian nonself mappings and was introduced by Khan [19].

Definition 2 [19] Let K be a nonempty subset of a metric space (X, d), P be a nonexpansive retraction of X onto K and fix a sequence $\{a_n\} \subset [0, \infty)$ with $\lim_{n\to\infty} a_n = 0$. A nonself mapping $T: K \to X$ is said to be *nearly Lipschitzian* with respect to $\{a_n\}$ if for each $n \in \mathbb{N}$, there exists a constant $k_n \geq 0$ such that

$$d(T(PT)^{n-1}x, T(PT)^{n-1}y) \leq k_n(d(x, y) + a_n), \quad \forall x, y \in K. \tag{1}$$

The infimum of constants k_n satisfying (1) is denoted by $\eta(T(PT)^{n-1})$ and is called *nearly Lipschitz constant.*

Remark 1 [19] For $n = 1$, the inequality (1) can be written as:

$$d(T(PT)^{1-1}x, T(PT)^{1-1}y) \leq k_1(d(x, y) + a_1),$$

where we have to take a_1 as zero. Thus in this case, we have

---

✉ Aynur Şahin
ayuce@sakarya.edu.tr

Metin Başarır
basarir@sakarya.edu.tr

1 Department of Mathematics, Sakarya University, 54050 Adapazarı, Sakarya, Turkey

$d(T(PT)^{1-1}x, T(PT)^{1-1}y) \leq k_1 d(x,y)$.

Definition 3 [19] A nearly Lipschitzian nonself mapping T with the sequence $\{a_n, \eta(T(PT)^{n-1})\}$ is said to be *nearly asymptotically nonexpansive* if $\eta(T(PT)^{n-1}) \geq 1$ for all $n \in \mathbb{N}$ and $\lim_{n\to\infty} \eta(T(PT)^{n-1}) = 1$.

Agarwal et al. [2] introduced the modified S-iteration process in a Banach space:

$$\begin{cases} x_1 \in K, \\ y_n = (1-\beta_n)x_n + \beta_n T^n x_n, \\ x_{n+1} = (1-\alpha_n)T^n x_n + \alpha_n T^n y_n, \quad n \in \mathbb{N}, \end{cases} \tag{2}$$

where $\{\alpha_n\}$ and $\{\beta_n\}$ are real sequences in [0, 1]. This iteration is independent of those of modified Mann iteration in [31] and modified Ishikawa iteration in [35] and reduces to S-iteration of Agarwal et al. [2] when $T^n = T$ for all $n \in \mathbb{N}$. The convergence of S-iteration for different classes of mappings in different spaces has been studied by many authors (see, e.g., [3–5, 17, 18, 20]).

In this paper, we prove the demiclosedness principle for nearly asymptotically nonexpansive nonself mappings in $CAT(\kappa)$ spaces. Also, we present the strong and Δ-convergence theorems of the modified S-iteration process for mappings of this type in a $CAT(\kappa)$ space. Our results extend and improve the corresponding results of Khan [19], Saluja et al. [30], Khan and Abbas [20] and many other results in this direction.

Preliminaries on CAT(κ) space

For a real number κ, a CAT(κ) space is defined by a geodesic metric space whose geodesic triangle is sufficiently thinner than the corresponding comparison triangle in a model space with the curvature κ. The term 'CAT(κ)' was coined by Gromov [16, p.119] and the initials are in honor of Cartan, Alexandrov and Toponogov, each of whom considered similar conditions in varying degrees of generality. Fixed point theory in CAT(κ) spaces was first studied by Kirk [21, 22]. His works were followed by a series of new works by many authors, mainly focusing on CAT(0) spaces (see, e.g., [1, 9, 10, 12, 13, 15, 23, 24, 27, 29, 30, 33, 34]). Since any CAT(κ) space is a CAT(κ') space for $\kappa' \geq \kappa$ (see [6, p. 165]), all results for a CAT(0) space can immediately be applied to any CAT(κ) space with $\kappa \leq 0$.

Let (*X*, *d*) be a metric space and let $x, y \in X$ with $d(x,y) = l$. A *geodesic path* joining x to y (or, more briefly, a *geodesic* from x to y) is an isometry $c : [0,l] \subset \mathbb{R} \to X$ such that $c(0) = x$ and $c(l) = y$. The image of c is called a *geodesic* (or *metric*) *segment* joining x and y. A geodesic segment joining x and y is not necessarily unique in general. When it is unique, this geodesic segment is denoted by [*x*, *y*]. This means that $z \in [x,y]$ if and only if there exists $\alpha \in [0,1]$ such that $d(x,z) = \alpha d(x,y)$ and $d(y,z) = (1-\alpha)$ $d(x,y)$. In this case, we write $z = (1-\alpha)x \oplus \alpha y$ for simplicity.

The space (*X*, *d*) is said to be a *geodesic space* if every two points of X are joined by a geodesic and X is said to be a *uniquely geodesic* if there is exactly one geodesic joining x to y for each $x, y \in X$. Let $D \in (0,\infty]$. If for every x, $y \in X$ with $d(x,y) < D$, a geodesic from x to y exists, then X is said to be *D-geodesic space*. Moreover, if such a geodesic is unique for each pair of points then X is said to be a *D-uniquely geodesic*. Notice that X is a geodesic space if and only if it is a *D*- geodesic space.

A subset K of X is said to be *convex* if K includes every geodesic segment joining any two of its points. The set K is said to be *bounded* if $\text{diam}(K) = \sup\{d(x,y) : x, y \in K\} < \infty$.

To define a CAT(κ) space, we use the following concept called *model space*. For $\kappa = 0$, the two-dimensional model space $M_\kappa^2 = M_0^2$ is the Euclidean space $\mathbb{R}^2$ with the metric induced from the Euclidean norm. For $\kappa > 0$, M_κ^2 is the two-dimensional sphere $(\frac{1}{\sqrt{\kappa}})\mathbb{S}^2$ whose metric is a length of a minimal great arc joining each of the two points. For $\kappa < 0$, M_κ^2 is the two-dimensional hyperbolic space $(\frac{1}{\sqrt{-\kappa}})\mathbb{H}^2$ with the metric defined by a usual hyperbolic distance.

The diameter of M_κ^2 is denoted by

$$D_\kappa = \begin{cases} \dfrac{\pi}{\sqrt{\kappa}} & \kappa > 0, \\ +\infty & \kappa \leq 0. \end{cases}$$

A *geodesic triangle* $\triangle(x,y,z)$ in a metric space (*X*, *d*) consists of three points *x*, *y*, *z* in *X* (the vertices of $\triangle$) and three geodesic segments between each pair of vertices (the edges of $\triangle$). A *comparison triangle* for the geodesic triangle $\triangle(x,y,z)$ in (*X*, *d*) is a triangle $\overline{\triangle}(\overline{x},\overline{y},\overline{z})$ in M_κ^2 such that

$$d(x,y) = d_{M_\kappa^2}(\overline{x},\overline{y}), d(y,z) = d_{M_\kappa^2}(\overline{y},\overline{z}) \text{ and } d(z,x) = d_{M_\kappa^2}(\overline{z},\overline{x})$$

(see [6, Lemma 2.14]). If $\kappa \leq 0$, then such a comparison triangle always exists in M_κ^2. If $\kappa > 0$, such a comparison triangle exists whenever $d(x,y) + d(y,z) + d(z,x) < 2D_\kappa$. A point $\overline{p} \in [\overline{x},\overline{y}]$ is called a *comparison point* for $p \in [x,y]$ if $d(x,p) = d_{M_\kappa^2}(\overline{x},\overline{p})$.

A geodesic triangle $\triangle(x,y,z)$ in X is said to satisfy the CAT(κ) *inequality* if for any $p, q \in \triangle(x,y,z)$ and for their comparison points $\overline{p}, \overline{q} \in \overline{\triangle}(\overline{x},\overline{y},\overline{z})$, one has

$$d(p,q) \leq d_{M_\kappa^2}(\overline{p},\overline{q}).$$

Now, we are ready to introduce the concept of CAT(κ) space in the following definition taken from [6].

Definition 4

(i) If $\kappa \leq 0$, then a metric space (X, d) is called a CAT(κ) *space* if X is a geodesic space such that all of its geodesic triangles satisfy the CAT(κ) inequality.

(ii) If $\kappa > 0$, then a metric space (X, d) is called a CAT(κ) *space* if it is D_κ-geodesic and any geodesic triangle $\triangle(x, y, z)$ in X with $d(x, y) + d(y, z) + d(z, x) < 2D_\kappa$ satisfies the CAT(κ) inequality.

Notice that in a CAT(0) space (X, d) if $x, y, z \in X$, then the CAT(0) inequality implies

$$\text{(CN)}\ d^2\left(z, \frac{1}{2}x \oplus \frac{1}{2}y\right) \leq \frac{1}{2}d^2(z, x) + \frac{1}{2}d^2(z, y) - \frac{1}{4}d^2(x, y).$$

This is the *(CN) inequality* of Bruhat and Tits [8]. This inequality is extended by Dhompongsa and Panyanak [14] as

$$(\text{CN}^*)d^2(z, (1-\alpha)x \oplus \alpha y) \leq (1-\alpha)d^2(z, x) + \alpha d^2(z, y) - \alpha(1-\alpha)d^2(x, y)$$

for all $\alpha \in [0, 1]$ and $x, y, z \in X$.

Let $R \in (0, 2]$. Recall that a geodesic space (X, d) is said to be *R-convex* (see [26]) if for any three points $x, y, z \in X$, we have

$$d^2(z, (1-\alpha)x \oplus \alpha y) \leq (1-\alpha)d^2(z, x) + \alpha d^2(z, y) - \frac{R}{2}\alpha(1-\alpha)d^2(x, y). \tag{3}$$

It follows from the (CN*) inequality that a CAT(0) space is R-convex for $R = 2$.

The following lemma is a consequence of Proposition 3.1 in [26].

Lemma 1 [27, Lemma 2.3] *Let $\kappa > 0$ and (X, d) be a complete CAT (κ) space with diam $(X) \leq \frac{\pi/2-\epsilon}{\sqrt{\kappa}}$ for some $\epsilon \in (0, \pi/2)$. Then, (X, d) is R-convex for $R = (\pi - 2\epsilon)\tan(\epsilon)$.*

In the sequel, we need the following lemma.

Lemma 2 [6, p. 176] *Let $\kappa > 0$ and (X, d) be a complete CAT (κ) space with diam $(X) \leq \frac{\pi/2-\epsilon}{\sqrt{\kappa}}$ for some $\epsilon \in (0, \pi/2)$. Then*

$$d((1-\alpha)x \oplus \alpha y, z) \leq (1-\alpha)d(x, z) + \alpha d(y, z)$$

for all $x, y, z \in X$ and $\alpha \in [0, 1]$.

We now collect some elementary facts about CAT(κ) spaces. Most of them are proved in the setting of CAT(1) spaces. For completeness, we state the results in a CAT(κ) space with $\kappa > 0$.

Let $\{x_n\}$ be a bounded sequence in a CAT(κ) space X. For $x \in X$, we set $r(x, \{x_n\}) = \limsup_{n\to\infty} d(x, x_n)$. The *asymptotic radius* $r(\{x_n\})$ of $\{x_n\}$ is defined by

$$r(\{x_n\}) = \inf\{r(x, \{x_n\}) : x \in X\}.$$

Further, the *asymptotic center* $A(\{x_n\})$ of $\{x_n\}$ is the set

$$A(\{x_n\}) = \{x \in X : r(x, \{x_n\}) = r(\{x_n\})\}.$$

It is well known [15, Proposition 4.1] that in a CAT(κ) space X with diam$(X) < \frac{\pi}{2\sqrt{\kappa}}$, $A(\{x_n\})$ consists of exactly one point.

Now, we can give the concept of Δ-convergence and collect some of its basic properties.

Definition 5 [23, 25] A sequence $\{x_n\}$ is *Δ-convergent* to $x \in X$ if x is the unique asymptotic center of any subsequence of $\{x_n\}$. In this case, we write $\Delta\text{-}\lim_{n\to\infty} x_n = x$ and call x the Δ-limit of $\{x_n\}$.

Lemma 3 *Let $\kappa > 0$ and (X, d) be a complete CAT (κ) space with diam $(X) \leq \frac{\pi/2-\epsilon}{\sqrt{\kappa}}$ for some $\epsilon \in (0, \pi/2)$. Then, the following statements hold:*

(i) [15, Corollary 4.4] *Every sequence in X has a Δ-convergent subsequence;*

(ii) [15, Proposition 4.5] *If $\{x_n\} \subseteq X$ and $\Delta\text{-}\lim_{n\to\infty} x_n = x$, then $x \in \bigcap_{k=1}^{\infty} \overline{conv}\{x_k, x_{k+1}, \ldots\}$, where $\overline{conv}(A) = \bigcap\{B : B \supseteq A$ and B is closed and convex $\}$.*

By the uniqueness of asymptotic centers, Panyanak [27] obtained the following lemma.

Lemma 4 [27, Lemma 2.7] *Let $\kappa > 0$ and (X, d) be a complete CAT (κ) space with diam $(X) \leq \frac{\pi/2-\epsilon}{\sqrt{\kappa}}$ for some $\epsilon \in (0, \pi/2)$. If $\{x_n\}$ is a sequence in X with $A(\{x_n\}) = \{x\}$ and $\{u_n\}$ is a subsequence of $\{x_n\}$ with $A(\{u_n\}) = \{u\}$ and the sequence $\{d(x_n, u)\}$ converges, then $x = u$.*

The following lemma is crucial in the study of iteration processes in both metric and Banach spaces and it was proved by Qihou [28].

Lemma 5 [28, Lemma 2] *Let $\{a_n\}, \{b_n\}$ and $\{\delta_n\}$ be sequences of non-negative real numbers such that*

$$a_{n+1} \leq (1 + \delta_n)a_n + b_n, \quad \forall n \in \mathbb{N}.$$

If $\sum_{n=1}^{\infty} \delta_n < \infty$ and $\sum_{n=1}^{\infty} b_n < \infty$, then $\lim_{n\to\infty} a_n$ exists.

Demiclosedness principle

It is well known that one of the fundamental and celebrated results in the theory of nonexpansive mappings is Browder's *demiclosedness principle* [7] which states that if K is a nonempty closed convex subset of a uniformly convex

Banach space X and $T : K \to X$ is a nonexpansive mapping, then $I - T$ is demiclosed at 0, that is, for any sequence $\{x_n\}$ in K if $x_n \to x$ weakly and $(I - T)x_n \to 0$ strongly, then $(I - T)x = 0$, where I is the identity mapping of X. Saluja et al. [30] proved the demiclosedness principle for nearly asymptotically nonexpansive self mappings in a CAT(κ) space. Now, we prove the demiclosedness principle for nearly asymptotically nonexpansive nonself mappings in this space.

Theorem 1 *Let $\kappa > 0$ and (X, d) be a complete CAT (κ) space with diam $(X) \le \frac{\pi/2-\epsilon}{\sqrt{\kappa}}$ for some $\epsilon \in (0, \pi/2)$. Let K be a nonempty closed convex subset of X, P be a nonexpansive retraction of X onto K and $T : K \to X$ be a uniformly continuous nearly asymptotically nonexpansive nonself mapping with $F(T) \neq \emptyset$. If $\{x_n\}$ is a sequence in K such that $\lim_{n\to\infty} d(x_n, Tx_n) = 0$ and $\Delta\text{-}\lim_{n\to\infty} x_n = w$, then $w \in K$ and $Tw = w$.*

Proof By Lemma 3, $w \in K$. Now, we define $\Psi(u) = \limsup_{n\to\infty} d(x_n, u)$ for each $u \in K$. Since $\lim_{n\to\infty} d(x_n, Tx_n) = 0$, by induction we can prove that

$$\lim_{n\to\infty} d(x_n, T(PT)^{m-1}x_n) = 0, \quad \forall m \in \mathbb{N}. \tag{4}$$

In fact, it is obvious that the conclusion is true for $m = 1$. Suppose the conclusion holds for m, now we prove that the conclusion is also true for $m + 1$. By the uniform continuity of TP, we have

$$\lim_{n\to\infty} d(T(PT)^{m-1}x_n, T(PT)^m x_n) = 0$$

so that

$$\begin{aligned} d(x_n, T(PT)^m x_n) &\le d(x_n, T(PT)^{m-1}x_n) + d(T(PT)^{m-1}x_n, \\ &\quad \times T(PT)^m x_n) \\ &\to 0 \quad \text{as } n \to \infty. \end{aligned}$$

Equation (4) is proved. This implies that

$$\Psi(u) = \limsup_{n\to\infty} d(T(PT)^{m-1}x_n, u), \quad \text{for each } u \in K \text{ and } m \in \mathbb{N}. \tag{5}$$

In (5), taking $u = T(PT)^{m-1}w$, we have

$$\begin{aligned} \Psi(T(PT)^{m-1}w) &= \limsup_{n\to\infty} d(T(PT)^{m-1}x_n, T(PT)^{m-1}w) \\ &\le \limsup_{n\to\infty} [\eta(T(PT)^{m-1})(d(x_n, w) + a_m)]. \end{aligned}$$

Hence

$$\limsup_{m\to\infty} \Psi(T(PT)^{m-1}w) \le \Psi(w). \tag{6}$$

Furthermore, for any $n, m \in \mathbb{N}$, it follows from the inequality (3) with $\alpha = \frac{1}{2}$,

$$\begin{aligned} d^2\left(x_n, \frac{1}{2}w \oplus \frac{1}{2}T(PT)^{m-1}w\right) &\le \frac{1}{2}d^2(x_n, w) + \frac{1}{2}d^2(x_n, T(PT)^{m-1}w) \\ &\quad - \frac{R}{8}d^2(w, T(PT)^{m-1}w). \end{aligned}$$

Since $\Delta\text{-}\lim_{n\to\infty} x_n = w$, letting $n \to \infty$, we get

$$\begin{aligned} \Psi^2(w) &\le \Psi^2\left(\frac{1}{2}w \oplus \frac{1}{2}T(PT)^{m-1}w\right) \\ &\le \frac{1}{2}\Psi^2(w) + \frac{1}{2}\Psi^2(T(PT)^{m-1}w) - \frac{R}{8}d^2(w, T(PT)^{m-1}w), \end{aligned}$$

which yields that

$$d^2(w, T(PT)^{m-1}w) \le \frac{4}{R}[\Psi^2(T(PT)^{m-1}w) - \Psi^2(w)]. \tag{7}$$

By (6) and (7), we have $\lim_{m\to\infty} d(w, T(PT)^{m-1}w) = 0$. In view of the continuity of TP, we obtain

$$w = \lim_{m\to\infty} T(PT)^m w = \lim_{m\to\infty} TP\left(T(PT)^{m-1}w\right) = TPw = Tw.$$

This completes the proof. □

From Theorem 1, we now derive the following result, yet is new in the literature.

Corollary 1 *Let K be a nonempty bounded closed convex subset of a complete CAT(0) space (X, d), P be a nonexpansive retraction of X onto K and $T : K \to X$ be a uniformly continuous nearly asymptotically nonexpansive nonself mapping. If $\{x_n\}$ is a sequence in K such that $\lim_{n\to\infty} d(x_n, Tx_n) = 0$ and $\Delta\text{-}\lim_{n\to\infty} x_n = w$, then $w \in K$ and $Tw = w$.*

Proof It is well known that every convex subset of a CAT(0) space, equipped with the induced metric, is a CAT(0) space (see [6]). Then, (K, d) is a CAT(0) space and hence it is a CAT(κ) space for all $\kappa > 0$. Notice also that K is R-convex for $R = 2$. Since K is bounded, we can choose $\epsilon \in (0, \pi/2)$ and $\kappa > 0$ so that $\text{diam}(K) \le \frac{\pi/2-\epsilon}{\sqrt{\kappa}}$. The conclusion follows from Theorem 1. □

Convergence theorems of the modified S-iteration process

We start with Δ-convergence of the modified S-iterative sequence for nearly asymptotically nonexpansive nonself mappings in CAT(κ) spaces.

Theorem 2 *Let $\kappa > 0$ and (X, d) be a complete CAT (κ) space with diam $(X) \le \frac{\pi/2-\epsilon}{\sqrt{\kappa}}$ for some $\epsilon \in (0, \pi/2)$. Let*

K be a nonempty closed convex subset of X, P be a nonexpansive retraction of X onto K and $T: K \to X$ be a uniformly continuous nearly asymptotically nonexpansive nonself mapping with the sequence $\{a_n, \eta(T(PT)^{n-1})\}$ such that $\sum_{n=1}^{\infty} a_n < \infty$ and $\sum_{n=1}^{\infty}\left(\eta(T(PT)^{n-1}) - 1\right) < \infty$. Let $\{x_n\}$ be a sequence in K defined by

$$\begin{cases} x_1 \in K, \\ y_n = P((1-\beta_n)x_n \oplus \beta_n T(PT)^{n-1}x_n), \\ x_{n+1} = P((1-\alpha_n)T(PT)^{n-1}x_n \oplus \alpha_n T(PT)^{n-1}y_n), \quad n \in \mathbb{N}, \end{cases} \tag{8}$$

where $\{\alpha_n\}$ and $\{\beta_n\}$ are real sequences in (0, 1) such that $\liminf_{n\to\infty} \alpha_n(1-\alpha_n) > 0$ *and* $\liminf_{n\to\infty} \beta_n(1-\beta_n) > 0$. *If $F(T) \neq \emptyset$, then $\{x_n\}$ is Δ-convergent to a fixed point of T.*

Proof We divide our proof into three steps.

Step 1. First, we prove that

$$\lim_{n\to\infty} d(x_n, p) \text{exists for each } p \in F(T). \tag{9}$$

Let $p \in F(T)$. Since T is a nearly asymptotically nonexpansive nonself mapping, by (8) and Lemma 2, we have

$$\begin{aligned} d(y_n, p) &= d(P((1-\beta_n)x_n \oplus \beta_n T(PT)^{n-1}x_n), p) \\ &\le d((1-\beta_n)x_n \oplus \beta_n T(PT)^{n-1}x_n, p) \\ &\le (1-\beta_n)d(x_n, p) + \beta_n d(T(PT)^{n-1}x_n, p) \\ &\le (1-\beta_n)d(x_n, p) + \beta_n \eta(T(PT)^{n-1}) \\ &\quad (d(x_n, p) + a_n) \\ &\le \eta(T(PT)^{n-1})[(1-\beta_n)d(x_n, p) + \beta_n d(x_n, p)] \\ &\quad + \beta_n \eta(T(PT)^{n-1})a_n \\ &\le \eta(T(PT)^{n-1})d(x_n, p) + \eta(T(PT)^{n-1})a_n. \end{aligned}$$

This implies that

$$\begin{aligned} d(x_{n+1}, p) &= d(P((1-\alpha_n)T(PT)^{n-1}x_n \\ &\quad \oplus \alpha_n T(PT)^{n-1}y_n), p) \\ &\le d((1-\alpha_n)T(PT)^{n-1}x_n \oplus \alpha_n T(PT)^{n-1}y_n, p) \\ &\le (1-\alpha_n)d(T(PT)^{n-1}x_n, p) + \alpha_n d(T(PT)^{n-1}y_n, p) \\ &\le \eta(T(PT)^{n-1})[(1-\alpha_n)(d(x_n, p) + a_n) \\ &\quad + \alpha_n(d(y_n, p) + a_n)] \\ &\le \eta(T(PT)^{n-1})[(1-\alpha_n)d(x_n, p) \\ &\quad + \alpha_n \eta(T(PT)^{n-1})d(x_n, p) \\ &\quad + (1 + \eta(T(PT)^{n-1}))a_n] \\ &\le \left(\eta(T(PT)^{n-1})\right)^2 d(x_n, p) + [\eta(T(PT)^{n-1}) \\ &\quad + \left(\eta(T(PT)^{n-1})\right)^2]a_n \\ &= (1+\sigma_n)d(x_n, p) + \xi_n, \end{aligned} \tag{10}$$

where $\sigma_n = \left(\eta(T(PT)^{n-1})\right)^2 - 1 = (\eta(T(PT)^{n-1}) + 1)(\eta(T(PT)^{n-1}) - 1)$ and $\xi_n = [\eta(T(PT)^{n-1}) + \left(\eta(T(PT)^{n-1})\right)^2]a_n$. Since $\sum_{n=1}^{\infty}\left(\eta(T(PT)^{n-1}) - 1\right) < \infty$ and $\sum_{n=1}^{\infty} a_n < \infty$, it follows that $\sum_{n=1}^{\infty} \sigma_n < \infty$ and $\sum_{n=1}^{\infty} \xi_n < \infty$. Hence, by Lemma 5, we get that $\lim_{n\to\infty} d(x_n, p)$ exists for each $p \in F(T)$.

Step 2. Next, we prove that

$$\lim_{n\to\infty} d(x_n, Tx_n) = 0. \tag{11}$$

Since $\{x_n\}$ is bounded, there exists $R > 0$ such that $\{x_n\}, \{y_n\} \subset B(p, R')$ for all $n \in \mathbb{N}$ with $R' < D_k/2$. In view of (3), we have

$$\begin{aligned} d^2(y_n, p) &\le d^2((1-\beta_n)x_n \oplus \beta_n T(PT)^{n-1}x_n, p) \\ &\le (1-\beta_n)d^2(x_n, p) + \beta_n d^2(T(PT)^{n-1}x_n, p) \\ &\quad - \frac{R}{2}\beta_n(1-\beta_n)d^2(x_n, T(PT)^{n-1}x_n) \\ &\le (1-\beta_n)d^2(x_n, p) + \beta_n[\eta(T(PT)^{n-1}) \\ &\quad (d(x_n, p) + a_n)]^2 \\ &\quad - \frac{R}{2}\beta_n(1-\beta_n)d^2(x_n, T(PT)^{n-1}x_n) \\ &\le (\eta(T(PT)^{n-1}))^2 d^2(x_n, p) + Pa_n \\ &\quad - \frac{R}{2}\beta_n(1-\beta_n)d^2(x_n, T(PT)^{n-1}x_n) \end{aligned} \tag{12}$$

for some $P > 0$. This implies that

$$d^2(y_n, p) \le (\eta(T(PT)^{n-1}))^2 d^2(x_n, p) + Pa_n. \tag{13}$$

From (3) and using (13), we get

$$\begin{aligned} d^2(x_{n+1}, p) &\le d^2((1-\alpha_n)T(PT)^{n-1}x_n \oplus \alpha_n T(PT)^{n-1}y_n, p) \\ &\le (1-\alpha_n)d^2(T(PT)^{n-1}x_n, p) + \alpha_n d^2(T(PT)^{n-1}y_n, p) \\ &\quad - \frac{R}{2}\alpha_n(1-\alpha_n)d^2(T(PT)^{n-1}x_n, T(PT)^{n-1}y_n) \\ &\le (1-\alpha_n)[\eta(T(PT)^{n-1})(d(x_n, p) + a_n)]^2 \\ &\quad + \alpha_n[\eta(T(PT)^{n-1})(d(y_n, p) + a_n)]^2 \\ &\quad - \frac{R}{2}\alpha_n(1-\alpha_n)d^2(T(PT)^{n-1}x_n, T(PT)^{n-1}y_n) \\ &\le (1-\alpha_n)\left(\eta(T(PT)^{n-1})\right)^2 d^2(x_n, p) + Qa_n \\ &\quad + \alpha_n\left(\eta(T(PT)^{n-1})\right)^2\left[\left(\eta(T(PT)^{n-1})\right)^2 d^2(x_n, p) + Pa_n\right] \\ &\quad + La_n - \frac{R}{2}\alpha_n(1-\alpha_n)d^2(T(PT)^{n-1}x_n, T(PT)^{n-1}y_n) \\ &\le \left(\eta(T(PT)^{n-1})\right)^4 d^2(x_n, p) + (Q+M+L)a_n \\ &\quad - \frac{R}{2}\alpha_n(1-\alpha_n)d^2(T(PT)^{n-1}x_n, T(PT)^{n-1}y_n) \end{aligned}$$

$$
\begin{aligned}
&=\left[1+\left(\eta(T(PT)^{n-1})\right)^4-1\right]d^2(x_n,p)+(Q+M+L)a_n\\
&\quad-\frac{R}{2}\alpha_n(1-\alpha_n)d^2(T(PT)^{n-1}x_n,T(PT)^{n-1}y_n)\\
&=[1+(\eta(T(PT)^{n-1})-1)\rho]d^2(x_n,p)+(Q+M+L)a_n\\
&\quad-\frac{R}{2}\alpha_n(1-\alpha_n)d^2(T(PT)^{n-1}x_n,T(PT)^{n-1}y_n)
\end{aligned}
$$

for some $Q,M,L,\rho>0$. This inequality yields that

$$
\begin{aligned}
&\frac{R}{2}\alpha_n(1-\alpha_n)d^2(T(PT)^{n-1}x_n,T(PT)^{n-1}y_n)\\
&\quad\leq d^2(x_n,p)-d^2(x_{n+1},p)+(\eta(T(PT)^{n-1})-1)\\
&\qquad\rho d^2(x_n,p)+(Q+M+L)a_n.
\end{aligned}
$$

Since $\sum_{n=1}^{\infty} a_n<\infty$, $\sum_{n=1}^{\infty}\left(\eta(T(PT)^{n-1})-1\right)<\infty$ and $d(x_n,p)<R'$, we obtain

$$
\sum_{n=1}^{\infty}\alpha_n(1-\alpha_n)d^2(T(PT)^{n-1}x_n,T(PT)^{n-1}y_n)<\infty.
$$

Hence by the fact that $\liminf_{n\to\infty}\alpha_n(1-\alpha_n)>0$, we get

$$
\lim_{n\to\infty} d(T(PT)^{n-1}x_n,T(PT)^{n-1}y_n)=0. \tag{14}
$$

Now, consider (12), we have

$$
\begin{aligned}
d^2(y_n,p)&\leq[1+((\eta(T(PT)^{n-1}))^2-1)]d^2(x_n,p)+Pa_n\\
&\quad-\frac{R}{2}\beta_n(1-\beta_n)d^2(x_n,T(PT)^{n-1}x_n)\\
&\leq[1+(\eta(T(PT)^{n-1}-1)\mu]d^2(x_n,p)+Pa_n\\
&\quad-\frac{R}{2}\beta_n(1-\beta_n)d^2(x_n,T(PT)^{n-1}x_n)
\end{aligned}
$$

for some $\mu>0$. This inequality yields that

$$
\begin{aligned}
&\frac{R}{2}\beta_n(1-\beta_n)d^2(x_n,T(PT)^{n-1}x_n)\\
&\leq d^2(x_n,p)-d^2(y_n,p)+(\eta(T(PT)^{n-1})-1)\mu d^2(x_n,p)+Pa_n.
\end{aligned}
$$

Since $\sum_{n=1}^{\infty}a_n<\infty$, $\sum_{n=1}^{\infty}\left(\eta(T(PT)^{n-1})-1\right)<\infty$, $d(x_n,p)<R'$ and $d(y_n,p)<R'$, we obtain

$$
\sum_{n=1}^{\infty}\beta_n(1-\beta_n)d^2(x_n,T(PT)^{n-1}x_n)<\infty.
$$

Hence by the fact that $\liminf_{n\to\infty}\beta_n(1-\beta_n)>0$, we have

$$
\lim_{n\to\infty} d(x_n,T(PT)^{n-1}x_n)=0. \tag{15}
$$

Now using (15), we get

$$
\begin{aligned}
d(x_n,y_n)&\leq d(x_n,(1-\beta_n)x_n\oplus\beta_nT(PT)^{n-1}x_n)\\
&\leq\beta_n d(T(PT)^{n-1}x_n,x_n)\\
&\to0 \text{ as } n\to\infty.
\end{aligned}
$$

Also, we observe that

$$
\begin{aligned}
d(x_{n+1},x_n)&\leq d((1-\alpha_n)T(PT)^{n-1}x_n\oplus\alpha_nT(PT)^{n-1}y_n,x_n)\\
&\leq(1-\alpha_n)d(T(PT)^{n-1}x_n,x_n)+\alpha_nd(T(PT)^{n-1}y_n,x_n)\\
&\leq(1-\alpha_n)d(T(PT)^{n-1}x_n,x_n)\\
&\quad+\alpha_n[d(T(PT)^{n-1}y_n,T(PT)^{n-1}x_n)+d(T(PT)^{n-1}x_n,x_n)]\\
&=d(T(PT)^{n-1}x_n,x_n)+\alpha_nd(T(PT)^{n-1}y_n,T(PT)^{n-1}x_n)\\
&\to0 \text{ as } n\to\infty.
\end{aligned}
\tag{16}
$$

Therefore, we obtain

$$
\begin{aligned}
d(x_{n+1},y_n)&\leq d(x_{n+1},x_n)+d(x_n,y_n)\\
&\to0 \text{ as } n\to\infty.
\end{aligned}
\tag{17}
$$

Furthermore, since

$$
\begin{aligned}
d(x_{n+1},T(PT)^{n-1}y_n)&\leq d(x_{n+1},x_n)+d(x_n,T(PT)^{n-1}x_n)\\
&\quad+d(T(PT)^{n-1}x_n,T(PT)^{n-1}y_n),
\end{aligned}
$$

using (14)–(16), we have

$$
\lim_{n\to\infty} d(x_{n+1},T(PT)^{n-1}y_n)=0. \tag{18}
$$

Since every nearly asymptotically nonexpansive mapping is nearly Lipschitzian, then we get

$$
\begin{aligned}
d(x_n,Tx_n)&\leq d(x_n,T(PT)^{n-1}x_n)+d(T(PT)^{n-1}x_n,T(PT)^{n-1}y_{n-1})\\
&\quad+d(T(PT)^{n-1}y_{n-1},Tx_n)\\
&=d(x_n,T(PT)^{n-1}x_n)+d(T(PT)^{n-1}x_n,T(PT)^{n-1}y_{n-1})\\
&\quad+d(T(PT)^{1-1}(PT)^{n-1}y_{n-1},T(PT)^{1-1}x_n)\\
&\leq d(x_n,T(PT)^{n-1}x_n)+d(T(PT)^{n-1}x_n,T(PT)^{n-1}y_{n-1})\\
&\quad+k_1d((PT)^{n-1}y_{n-1},x_n)\\
&\leq d(x_n,T(PT)^{n-1}x_n)+\eta(T(PT)^{n-1})[d(x_n,y_{n-1})+a_n]\\
&\quad+k_1d((PT)^{n-1}y_{n-1},x_n)\\
&=d(x_n,T(PT)^{n-1}x_n)+\eta(T(PT)^{n-1})[d(x_n,y_{n-1})+a_n]\\
&\quad+k_1d(PT(PT)^{n-1}y_{n-1},Px_n)\\
&\leq d(x_n,T(PT)^{n-1}x_n)+\eta(T(PT)^{n-1})[d(x_n,y_{n-1})+a_n]\\
&\quad+k_1d(T(PT)^{n-2}y_{n-1},x_n).
\end{aligned}
$$

Hence (15), (17) and (18) imply that $\lim_{n\to\infty} d(x_n,Tx_n)=0$.

Step 3. Now, we prove that $\{x_n\}$ is Δ-convergent to a fixed point of T.

Let $\omega_W(x_n)=\cup A(\{u_n\})$, where the union is taken over all subsequences $\{u_n\}$ of $\{x_n\}$. First, we show that $\omega_W(x_n)\subseteq F(T)$. Let $u\in\omega_W(x_n)$. Then, there exists a subsequence $\{u_n\}$ of $\{x_n\}$ such that $A(\{u_n\})=\{u\}$. By Lemma 3, there exists a subsequence $\{v_n\}$ of $\{u_n\}$ such that $\Delta\text{-}\lim_{n\to\infty}v_n=v\in K$. Also by (11), we have $\lim_{n\to\infty}d(v_n,Tv_n)=0$. It follows from Theorem 1 that $v\in F(T)$. Moreover, by (9), $\lim_{n\to\infty}d(x_n,v)$ exists. Thus,

$u = v$ by Lemma 4. This implies that $\omega_W(x_n) \subseteq F(T)$. Next, we show that $\omega_W(x_n)$ consists of exactly one point. Let $\{u_n\}$ be a subsequence of $\{x_n\}$ with $A(\{u_n\}) = \{u\}$ and let $A(\{x_n\}) = \{x\}$. Since $u \in \omega_W(x_n) \subseteq F(T)$, fsrom (9) $\lim_{n\to\infty} d(x_n, u)$ exists. Again by Lemma 4, $x = u$. Thus, $\omega_W(x_n) = \{x\}$. This means that $\{x_n\}$ is Δ-convergent to a fixed point of T. The proof is completed. □

Next, we discuss the strong convergence of the iterative sequence $\{x_n\}$ defined by (8) for nearly asymptotically nonexpansive nonself mappings in a CAT(κ) space.

Theorem 3 *Let X, K, P, T and $\{x_n\}$ be the same as in Theorem 2. Then, $\{x_n\}$ converges strongly to a fixed point of T if and only if $\liminf_{n\to\infty} d(x_n, F(T)) = 0$ where $d(x, F(T)) = \inf\{d(x,p) : p \in F(T)\}$.*

Proof If $\{x_n\}$ converges to $p \in F(T)$, then $\lim_{n\to\infty} d(x_n, p) = 0$. Since $0 \le d(x_n, F(T)) \le d(x_n, p)$, we have $\liminf_{n\to\infty} d(x_n, F(T)) = 0$.

Conversely, suppose that $\liminf_{n\to\infty} d(x_n, F(T)) = 0$. It follows from (9) that $\lim_{n\to\infty} d(x_n, F(T))$ exists. Thus by hypothesis $\lim_{n\to\infty} d(x_n, F(T)) = 0$. Next, we show that $\{x_n\}$ is a Cauchy sequence. In fact, it follows from (10) that for any $p \in F(T)$

$$d(x_{n+1}, p) \le (1 + \sigma_n) d(x_n, p) + \xi_n, \quad \forall n \in \mathbb{N},$$

where $\sum_{n=1}^{\infty} \sigma_n < \infty$ and $\sum_{n=1}^{\infty} \xi_n < \infty$. Hence for any positive integers n, m, we have

$$\begin{aligned} d(x_{n+m}, x_n) &\le d(x_{n+m}, p) + d(p, x_n) \\ &\le (1 + \sigma_{n+m-1}) d(x_{n+m-1}, p) + \xi_{n+m-1} + d(x_n, p). \end{aligned}$$

Since for each $x \ge 0, 1 + x \le e^x$, we have

$$\begin{aligned} d(x_{n+m}, x_n) &\le e^{\sigma_{n+m-1}} d(x_{n+m-1}, p) + \xi_{n+m-1} + d(x_n, p) \\ &\le e^{\sigma_{n+m-1}+\sigma_{n+m-2}} d(x_{n+m-2}, p) + e^{\sigma_{n+m-1}} \xi_{n+m-2} \\ &\quad + \xi_{n+m-1} + d(x_n, p) \\ &\le \ldots \\ &\le e^{\sum_{i=n}^{n+m-1} \sigma_i} d(x_n, p) + e^{\sum_{i=n+1}^{n+m-1} \sigma_i} \xi_n + e^{\sum_{i=n+2}^{n+m-2} \sigma_i} \xi_{n+1} \\ &\quad + \ldots + e^{\sigma_{n+m-1}} \xi_{n+m-2} + \xi_{n+m-1} + d(x_n, p) \\ &\le (1 + N) d(x_n, p) + N \sum_{i=n}^{n+m-1} \xi_i, \end{aligned}$$

where $N = e^{\sum_{i=1}^{\infty} \sigma_i} < \infty$. Therefore, we have

$$d(x_{n+m}, x_n) \le (1 + N) d(x_n, F(T)) + N \sum_{i=n}^{n+m-1} \xi_i \to 0 \text{ as } n, m \to \infty.$$

This shows that $\{x_n\}$ is a Cauchy sequence in K. Since K is a closed subset in a complete CAT(κ) space X, it is complete. We can assume that $\{x_n\}$ converges strongly to some point $p^\star \in K$. As T is continuous, so $F(T)$ is closed subset in K. Since $\lim_{n\to\infty} d(x_n, F(T)) = 0$, we obtain $p^\star \in F(T)$. This completes the proof. □

Remark 2 In Theorem 3, the condition $\liminf_{n\to\infty} d(x_n, F(T)) = 0$ may be replaced with $\limsup_{n\to\infty} d(x_n, F(T)) = 0$.

Recall that a mapping T from a subset K of a metric space (X, d) into itself is *semi-compact* if every bounded sequence $\{x_n\} \subset K$ satisfying $d(x_n, Tx_n) \to 0$ as $n \to \infty$ has a strongly convergent subsequence.

Senter and Dotson [32, p.375] introduced the concept of Condition (I) as follows.

A nonself mapping $T : K \to X$ with $F(T) \neq \emptyset$ is said to satisfy the *Condition (I)* if there exists a non-decreasing function $f : [0, \infty) \to [0, \infty)$ with $f(0) = 0$ and $f(r) > 0$ for all $r \in (0, \infty)$ such that

$$d(x, Tx) \ge f(d(x, F(T))) \quad \text{for all} \quad x \in K.$$

Using the above definitions, we obtain the following strong convergence theorem.

Theorem 4 *Let X, K, P, T and $\{x_n\}$ be the same as in Theorem 2.*

(i) *If T is semi-compact, then $\{x_n\}$ converges strongly to a fixed point of T.*
(ii) *If T satisfies Condition (I), then $\{x_n\}$ converges strongly to a fixed point of T.*

Proof

(i) It follows from (9) that $\{x_n\}$ is a bounded sequence in K. Also, by (11), we have $\lim_{n\to\infty} d(x_n, Tx_n) = 0$. Then, by the semi-compactness of T, there exists a subsequence $\{x_{n_k}\} \subset \{x_n\}$ such that $\{x_{n_k}\}$ converges strongly to some point $p \in K$. Moreover, by the uniform continuity of T, we have

$$d(p, Tp) = \lim_{k\to\infty} d(x_{n_k}, Tx_{n_k}) = 0.$$

This implies that $p \in F(T)$. Again, by (9), $\lim_{n\to\infty} d(x_n, p)$ exists. Hence p is the strong limit of the sequence $\{x_n\}$. As a result, $\{x_n\}$ converges strongly to a fixed point p of T.

(ii) By virtue of (9), $\lim_{n\to\infty} d(x_n, F(T))$ exists. Further, by Condition (I) and (11), we have

$$\lim_{n\to\infty} f(d(x_n, F(T))) \le \lim_{n\to\infty} d(x_n, Tx_n) = 0.$$

That is, $\lim_{n\to\infty} f(d(x_n, F(T))) = 0$. Since f is a non-decreasing function satisfying $f(0) = 0$ and $f(r) > 0$ for all $r \in (0, \infty)$, it follows that $\lim_{n\to\infty}$

$d(x_n, F(T)) = 0$. Now, Theorem 3 implies that $\{x_n\}$ converges strongly to a point p in $F(T)$.

□

Remark 3

(i) Theorem 1 extends Theorem 3.2 of Saluja et al. [30] from a nearly asymptotically nonexpansive self mapping to a nearly asymptotically nonexpansive nonself mapping.

(ii) Theorem 2 extends Theorem 1 of Khan [19] from a uniformly convex Banach space to a CAT(κ) space considered in this paper.

(iii) Our results extend the corresponding results of Khan and Abbas [20] to the case of a more general class of nonexpansive mappings from a CAT(0) space to a CAT(κ) space considered in this paper.

References

1. Abkar, A., Eslamian, M.: Common fixed point results in CAT(0) spaces. Nonlinear Anal. **74**, 1835–1840 (2011)
2. Agarwal, R.P., O'Regan, D., Sahu, D.R.: Iterative construction of fixed points of nearly asymptotically nonexpansive mappings. J. Nonlinear Convex Anal. **8**(1), 61–79 (2007)
3. Akbulut, S., Gündüz, B.: Strong and Δ-convergence of a faster iteration process in hyperbolic space. Commun. Korean Math. Soc. **30**(3), 209–219 (2015)
4. Başarır, M., Şahin, A.: On the strong and Δ-convergence of new multi-step and S-iteration processes in a CAT(0) space. J. Inequal. Appl. **2013**, 482 (2013)
5. Başarır, M., Şahin, A.: On the strong and Δ-convergence of S-iteration process for generalized nonexpansive mappings on CAT(0) space. Thail. J. Math. **12**(3), 549–559 (2014)
6. Bridson, M., Haefliger, A.: Metric Spaces of Non-Positive Curvature. Springer, Berlin (1999)
7. Browder, F.E.: Semicontractive and semiaccretive nonlinear mappings in Banach spaces. Bull. Am. Math. Soc. **74**, 660–665 (1968)
8. Bruhat, F., Tits, J.: Groupes réductifs sur un corps local. I. Données radicielles valuées. Publ. Math. Inst. Hautes Études Sci. **41**, 5–251 (1972)
9. Chang, S.S., Wang, L., Joseph Lee, H.W., Chan, C.K., Yang, L.: Demiclosed principle and Δ-convergence theorems for total asymptotically nonexpansive mappings in CAT(0) spaces. Appl. Math. Comput. **219**(5), 2611–2617 (2012)
10. Chaoha, P., Phon-on, A.: A note on fixed point sets in CAT(0) spaces. J. Math. Anal. Appl. **320**, 983–987 (2006)
11. Chidume, C.E., Ofoedu, E.U., Zegeye, H.: Strong and weak convergence theorems for asymptotically nonexpansive mappings. J. Math. Anal. Appl. **280**, 364–374 (2003)
12. Cho, Y.J., Ćirić, L., Wang, S.: Convergence theorems for nonexpansive semigroups in CAT(0) spaces. Nonlinear Anal. **74**, 6050–6059 (2011)
13. Dhompongsa, S., Kaewkhao, A., Panyanak, B.: Lim's theorems for multivalued mappings in CAT(0) spaces. J. Math. Anal. Appl. **312**, 478–487 (2005)
14. Dhompongsa, S., Panyanak, B.: On Δ-convergence theorems in CAT(0) spaces. Comput. Math. Appl. **56**, 2572–2579 (2008)
15. Espinola, R., Fernandez-Leon, A.: CAT(κ)-spaces, weak convergence and fixed points. J. Math. Anal. Appl. **353**, 410–427 (2009)
16. Gromov, M.: Hyperbolic groups. In: Gersten, S.M. (ed.) Essays in Group Theory, MSRI Publ. vol. 8, pp. 75–263. Springer (1987)
17. Gündüz, B., Akbulut, S.: On weak and strong convergence theorems for a finite family of nonself I-asymptotically nonexpansive mappings. Math. Morav. **19**(2), 49–64 (2015)
18. Gündüz, B., Khan, S.H., Akbulut, S.: On convergence of an implicit iterative algorithm for nonself asymptotically nonexpansive mappings. Hacet. J. Math. Stat. **43**(3), 399–411 (2014)
19. Khan, S.H.: Weak convergence for nonself nearly asymptotically nonexpansive mappings by iterations. Demonstr. Math. XLVI **I**(2), 371–381 (2014)
20. Khan, S.H., Abbas, M.: Strong and Δ-convergence of some iterative schemes in CAT(0) spaces. Comput. Math. Appl. **61**(1), 109–116 (2011)
21. Kirk, W.A.: Geodesic geometry and fixed point theory. Seminar of Mathematical Analysis (Malaga/Seville, 2002/2003), Colecc. Abierta, vol. 64, pp. 195–225. Univ. Sevilla Secr. Publ, Seville (2003)
22. Kirk, W.A.: Geodesic geometry and fixed point theory II. In: International Conference on Fixed Point Theory and Applications, pp. 113–142, Yokohama Publishers, Yokohama (2004)
23. Kirk, W.A., Panyanak, B.: A concept of convergence in geodesic spaces. Nonlinear Anal. **68**, 3689–3696 (2008)
24. Leustean, L.: A quadratic rate of asymptotic regularity for CAT(0)-spaces. J. Math. Anal. Appl. **325**, 386–399 (2007)
25. Lim, T.C.: Remarks on some fixed point theorems. Proc. Am. Math. Soc. **60**, 179–182 (1976)
26. Ohta, S.: Convexities of metric spaces. Geom. Dedic. **125**, 225–250 (2007)
27. Panyanak, B.: On total asymptotically nonexpansive mappings in CAT(κ) spaces. J. Inequal. Appl. **2014**, 336 (2014)
28. Qihou, L.: Iterative sequences for asymptotically quasi-nonexpansive mappings with error member. J. Math. Anal. Appl. **259**, 18–24 (2001)
29. Saejung, S.: Halpern's iteration in CAT(0) spaces. Fixed Point Theory Appl. **2010**, Article ID 471781 (2010)
30. Saluja, G.S., Postolache, M., Kurdi, A.: Convergence of three-step iterations for nearly asymptotically nonexpansive mappings in CAT(κ) spaces. J. Inequal. Appl. **2015**, 156 (2015)
31. Schu, J.: Weak and strong convergence to fixed points of asymptotically nonexpansive mappings. Bull. Austral. Math. Soc. **43**, 153–159 (1991)
32. Senter, H.F., Dotson, W.G.: Approximating fixed points of nonexpansive mappings. Proc. Am. Math. Soc. **44**, 375–380 (1974)
33. Shahzad, N., Markin, J.: Invariant approximations for commuting mappings in CAT(0) and hyperconvex spaces. J. Math. Anal. Appl. **337**, 1457–1464 (2008)
34. Şahin, A., Başarır, M.: On the strong convergence of a modified S-iteration process for asymptotically quasi-nonexpansive mappings in a CAT(0) space. Fixed Point Theory Appl. **2013**, 12 (2013)
35. Tan, K.K., Xu, H.K.: Fixed point iteration process for asymptotically nonexpansive mappings. Proc. Am. Math. Soc. **122**(3), 733–739 (1994)

4

Construction of multiscaling functions using the inverse representation theorem of matrix polynomials

M. Mubeen[1] · V. Narayanan[1]

Abstract Wavelet analysis deals with finding a suitable basis for the class of L^2 functions. Symmetric basis functions are very useful in various applications. In the case of all wavelets other than the famous Haar wavelet, the simultaneous inclusion of compact supportedness, orthogonality and symmetricity is not possible. Theory of multiwavelets assumes significance since it offers orthogonal, compact frames without losing symmetry. We can also construct symmetric, compactly supported and pseudo-biorthogonal bases which are also possible only in the case of multiwavelets. The properties of a multiwavelet directly depends on the corresponding multiscaling function. A multiscaling function is characterized by a unique symbol function, which is a matrix polynomial in complex exponential. A matrix polynomial can be constructed from its spectral data. Our aim is to find the necessary as well as sufficient conditions a spectral data must satisfy so that the corresponding matrix polynomial is the symbol function of a compactly supported, symmetric multiscaling function $\Phi(x)$. We will construct such a multiscaling function $\Phi(x)$ and its dual $\tilde{\Phi}(x)$ such that the functions $\Phi(x)$ and $\tilde{\Phi}(x)$ form a pair of pseudo-biorthogonal multiscaling functions.

✉ M. Mubeen
mubeen_p100041ma@nitc.ac.in

V. Narayanan
vna@nitc.ac.in

[1] Department of Mathematics, National Institute of Technology Calicut, Calicut, India

Keywords Matrix polynomial · Multiscaling function · Jordan pair · Symmetry

Introduction

Wavelet bases can be constructed using the notion of multiresolution analysis (MRA). In order to generate an MRA, we need to find a function vector $\Phi = (\phi_i)_{i=1}^n$, $\phi_i : \mathbb{R} \to \mathbb{C}$ which generates an MRA. A function vector Φ generates an MRA if it is L^2 stable, compactly supported and satisfies the multiscaling equation

$$\Phi(x) = \sqrt{2}\sum_{k=0}^{l} H_k \Phi(2x - k), \tag{1.1}$$

where $l \in \mathbb{N}$ and $H_k \in \mathbb{C}^{n\times n}$. A function vector Φ which satisfies the multiscaling equation (1.1) is called a multiscaling function or refinable function. To generate an MRA, one needs to find a function vector Φ which satisfies Eq. (1.1). To find such a solution vector, we usually switch over to the frequency domain where Eq. (1.1) becomes

$$\hat{\Phi}(\xi) = H(\xi/2)\hat{\Phi}(\xi/2), \tag{1.2}$$

where

$$H(\xi) = \frac{1}{\sqrt{2}}\sum_{k=0}^{l} H_k e^{-ik\xi}, \tag{1.3}$$

which is called a symbol function or a mask function. The existence of a solution to the multiscaling equation is determined by the nature of the corresponding symbol function. Moreover, the properties of this solution are determined by the nature of the symbol function. In fact, the symbol function $H(\xi)$ is a matrix polynomial in complex exponential. Each matrix polynomial $H(\xi)$ possesses a spectral pair or Jordan pair (*X*, *T*), where *X* is a matrix containing the generalized eigenvectors of $H(\xi)$ and *T* is a block diagonal matrix where each block is a Jordan matrix corresponding to the eigenvalues of $H(\xi)$. Given the pair (*X*, *T*), we can construct a matrix polynomial having (*X*, *T*) as its spectral data. We have to find the properties of a spectral data so that the corresponding matrix polynomial is the symbol function of a compactly supported, symmetric multiscaling function $\Phi(x)$. Also, we have to find a dual multiscaling function $\tilde{\Phi}(x)$ so that the functions $\Phi(x)$ and $\tilde{\Phi}(x)$ form a pair of pseudo-biorthogonal multiscaling functions.

Preliminaries

Let

$$L(\lambda) = \sum_{k=0}^{l} A_k\lambda^k, A_k \in \mathbb{C}^{n\times n}, \lambda \in \mathbb{C} \tag{2.1}$$

be a matrix polynomial of degree *l*. A complex number λ_0 is said to be an eigenvalue of $L(\lambda)$ if Det $L(\lambda_0) = 0$. Then there exists a nonzero vector $x_0 \in \mathbb{C}^n$ such that $L(\lambda_0)x_0 = 0$ and x_0 is called the eigenvector of $L(\lambda)$ corresponding to the eigenvalue λ_0.

Definition 2.1 [1] The chain of vectors $x_0, x_1 \ldots x_k \in \mathbb{C}^n$, $x_0 \neq 0$, is a Jordan chain of length k+1 of the matrix polynomial $L(\lambda)$ if

$$\sum_{p=0}^{i} \frac{L^p(\lambda_0)}{p!} x_{i-p} = 0, \quad i = 0, 1, 2 \ldots k, \tag{2.2}$$

where $L^p(\lambda_0)$ is the pth derivative of $L(\lambda)$ at λ_0.

This is a generalization of the usual notion of a Jordan chain of a square matrix. Let $T \in \mathbb{C}^{nl\times nl}$ and *T* be a block diagonal matrix where each block is a Jordan matrix corresponding to an eigenvalue of $L(\lambda)$, also let $X \in \mathbb{C}^{n\times nl}$ and column vectors of *X* are precisely the Jordan chains of $L(\lambda)$ corresponding to the eigenvalues of $L(\lambda)$. The Jordan chains appear in *X* in the order the corresponding eigenvalues appear in *T*. Then the pair (*X*, *T*) is said to be a Jordan pair. Now we will give the definition of a decomposable pair.

Definition 2.2 [1] A pair of matrices

$$X = [X_1\ X_2] \text{ and } T = \begin{pmatrix} T_1 & 0 \\ 0 & T_2 \end{pmatrix},$$

where $X_1 \in \mathbb{C}^{n\times m}$, $X_2 \in \mathbb{C}^{n\times(nl-m)}$ and $T_1 \in \mathbb{C}^{m\times m}$, $T_2 \in \mathbb{C}^{(nl-m)\times(nl-m)}$ with $0 \leq m \leq nl$ is called a decomposable pair of degree *l* if the matrix

$$S_{l-1} = Col[X_1T_1^i\ X_2T_2^{l-1-i}]_{i=0}^{l-1}$$

is nonsingular. A pair (*X*, *T*) satisfying this property is called a decomposable pair of the regular $n \times n$ matrix polynomial $L(\lambda) = \sum_{i=0}^{l} A_i\lambda^i$ if

$$\sum_{i=0}^{l} A_iX_1T_1^i = 0, \ \sum_{i=0}^{l} A_iX_2T_2^{l-i} = 0. \tag{2.3}$$

Given a decomposable pair (*X*, *T*), we can construct a matrix polynomial $L(\lambda)$ having (*X*, *T*) as its decomposable pair using the inverse representation theorem of matrix polynomials which is stated as follows.

Theorem 2.1 [1] *Let* $(X,T) = ([X_1\ X_2], T_1 \oplus T_2)$ *be a decomposable pair of degree l, and let* $S_{l-2} = Col[X_1T_1^i\ X_2T_2^{l-2-i}]_{i=0}^{l-2}$. *Then, for every* $n \times nl$ *matrix V such that the matrix* $\begin{pmatrix} S_{l-2} \\ V \end{pmatrix}$ *is nonsingular, the matrix polynomial*

$$L(\lambda) = V(I-P)((I \oplus T_2)\lambda - (T_1 \oplus I))(U_0 + U_1\lambda + U_2\lambda^2 + \cdots + U_{l-1}\lambda^{l-1}), \tag{2.4}$$

where

$$P = (I \oplus T_2)[Col(X_1T_1^i\ X_2T_2^{l-1-i})_{i=0}^{l-1}]^{-1}\begin{pmatrix} I \\ 0 \end{pmatrix} S_{l-2} \tag{2.5}$$

and

$$[U_0\ U_1\ U_2 \ldots U_{l-1}] = [Col(X_1T_1^i\ X_2T_2^{l-1-i})_{i=0}^{l-1}]^{-1} \tag{2.6}$$

has (X,T) as its decomposable pair.

If (*X*, *T*) is a Jordan pair of a matrix polynomial $L(\lambda)$, then it is a decomposable pair of $L(\lambda)$ [1]. We can construct a matrix polynomial for a given Jordan pair (*X*, *T*) using the inverse representation theorem. A sufficient condition on a Jordan pair (*X*, *T*) so that the corresponding matrix polynomial is the symbol function of a compactly supported multiscaling function has been derived by us in [2] and is as follows.

Theorem 2.2 [2] *Let (X,T) =* $([X_1\ X_2], T_1 \oplus T_2)$ *be a Jordan pair such that the* $nl \times nl$ *matrix* $(I \oplus T_2) - (T_1 \oplus I)$ *is of full rank. Then there exists a symbol function* $H(\xi)$ *with Jordan pair (X,T) such that the corresponding multiscaling equation* (1.1) *has a solution vector* Φ *such that* $\hat{\Phi}$ *is continuous at 0 with* $\hat{\Phi}(0) \neq 0$.

Thus, by choosing a Jordan pair (*X*, *T*) such that $(I \oplus T_2) - (T_1 \oplus I)$ is of full rank, we can form a multiscaling function Φ. Now our aim is to find the additional conditions on (*X*, *T*) so that this multiscaling function is symmetric also. A multiscaling function vector Φ is symmetric if its each component function is symmetric about some point.

Definition 2.3 [3] The refinable function vector $\Phi = (\phi_i)_{i=1}^n$ is symmetric if each component function $\phi_i, 1 \leq i \leq n$ is symmetric about some point $a_i \in \mathbb{R}$. That is

$$\phi_i(a_i + x) = \phi_i(a_i - x) \, \forall x \in \mathbb{R}, 1 \leq i \leq n. \tag{2.7}$$

The symmetricity of Φ is closely related to the properties of the associated symbol function $H(\xi)$. A sufficient property of $H(\xi)$ for Φ to be symmetric is given by the following Lemma.

Lemma 2.1 [3] *If the symbol $H(\xi)$ satisfies*

$$H(\xi) = A(2\xi)H(-\xi)A(\xi)^{-1}, \tag{2.8}$$

where

$$A(\xi) = \begin{pmatrix} \pm e^{-2ia\xi} & & & & \\ & \pm e^{-2ia\xi} & & & \\ & & \pm e^{-2ia\xi} & & \\ & & & \cdots & \\ & & & & \pm e^{-2ia\xi} \end{pmatrix},$$

then Φ is symmetric about the point a.

In the next section, we will find the conditions on the Jordan pair (X, T) so that the corresponding multiscaling function vector is symmetric based on these results.

A multiscaling function $\Phi(x)$ is said to be orthogonal if

$$\langle \Phi(x-k), \Phi(x-t) \rangle = \int \Phi(x-k)\Phi(x-t)^* dx = \delta_{kt}I, \quad k, t \in \mathbb{Z}. \tag{2.9}$$

In some situations, we use biorthogonal bases or pseudo biorthogonal bases instead of the orthogonal ones. Sometimes, biorthogonal bases with additional properties act more effectively than orthogonal bases. Two multiscaling functions $\Phi(x)$, $\tilde{\Phi}(x)$ are biorthogonal if

$$\langle \Phi(x-k), \tilde{\Phi}(x-t) \rangle = \int \Phi(x-k)\tilde{\Phi}(x-t)^* dx = \delta_{kt}I. \tag{2.10}$$

We call $\tilde{\Phi}(x)$ the dual of $\Phi(x)$. Let $H(\xi)$ and $F(\xi)$ be the symbol functions corresponding to the multiscaling functions $\Phi(x)$ and $\tilde{\Phi}(x)$ respectively. $\Phi(x)$ and $\tilde{\Phi}(x)$ form a pair of biorthogonal bases if and only if $H(\xi)$ and $F(\xi)$ satisfy the perfect reconstruction formula [4],

$$H(\xi)F^*(\xi) + H(\xi+\pi)F^*(\xi+\pi) = I. \tag{2.11}$$

It may happen that instead of the perfect reconstruction formula, $H(\xi)$ and $F(\xi)$ satisfy the generalized condition of perfect reconstruction [5] given below,

$$H(\xi)F^*(\xi) + H(\xi+\pi)F^*(\xi+\pi) = cI, \ c \neq 1. \tag{2.12}$$

Then $\Phi(x)$ and $\tilde{\Phi}(x)$ form a pair of pseudo-biorthogonal multiscaling functions, i.e,

$$\langle \Phi(x-k), \tilde{\Phi}(x-t) \rangle = \int \Phi(x-k)\tilde{\Phi}(x-t)^* dx = \delta_{kt}cI, \, c \neq 1. \tag{2.13}$$

Taking $z = e^{-i\xi}$, we get Eq. (2.12) as,

$$H(z)F^*(z) + H(-z)F^*(-z) = cI, \ c \neq 1. \tag{2.14}$$

If we perform one analysis step followed by one synthesis step using a biorthogonal basis, we get the initial signal exactly. In the case of pseudo biorthogonal basis, an analysis step followed by the synthesis step will produce the initial signal multiplied by c. We can recover the signal exactly by rescaling by c at each synthesis step [5]. In the case of scalar wavelets, $H(z)$ and $F(z)$ are polynomials in z so that $H(1) = F(1) = 1$ and $H(-1) = F(-1) = 0$ [3]. Hence the case $c \neq 1$ is not possible in the case of scalar wavelets.

Our aim is to construct a symbol function of degree 3 by selecting a suitable Jordan pair so that the corresponding multiscaling function $\Phi(x)$ is symmetric, compactly supported and there exists a dual multiscaling function $\tilde{\Phi}(x)$ so that the pair $\{\Phi(x-k) : k \in Z\}$ and $\{\tilde{\Phi}(x-k) : k \in Z\}$ form a pseudo-biorthogonal pair of bases. The condition on $H(z)$ for pseudo-biorthogonality is given by Eq. (2.14). In this article, we will formulate the condition on $H(z)$ for the symmetricity of the corresponding multiscaling function. Then, we will construct a compactly supported, symmetric multiscaling function $\Phi(x)$. Finally, we will construct the dual multiscaling function $\tilde{\Phi}(x)$ so that $\Phi(x)$ and $\tilde{\Phi}(x)$ form a pseudo-biorthogonal pair of multiscaling functions.

Symmetry

In this section, we will define a symmetric matrix polynomial and will show that a multiscaling function vector Φ is symmetric if the corresponding symbol function $H(\xi)$ is symmetric. We will then derive the necessary as well as sufficient properties a Jordan pair (X, T) must possess so that the corresponding matrix polynomial is symmetric.

Definition 3.1 A matrix polynomial

$L(\lambda) = A_0 + A_1\lambda + A_2\lambda^2 + \cdots + A_l\lambda^l$

is said to be symmetric if $A_0 = A_l, A_1 = A_{l-1} \ldots, A_k = A_{l-k}$.

Lemma 3.1 *If the symbol function*

$$H(\xi) = A_0 + A_1 e^{-i\xi} + A_2 e^{-2i\xi} + \cdots + A_l e^{-il\xi} \tag{3.1}$$

of degree l is symmetric, then the corresponding multiscaling function Φ *is symmetric about the point* $\frac{l}{2}$.

Proof Given that

$H(\xi) = A_0 + A_1 e^{-i\xi} + A_2 e^{-2i\xi} + \cdots + A_l e^{-il\xi}$

is symmetric, i.e. $A_0 = A_l$, $A_1 = A_{l-1} \ldots$, $A_k = A_{l-k}$. Then we can see that,

$$\begin{aligned} H(\xi) &= e^{-il\xi} H(-\xi) \\ &= e^{-2il\xi} e^{il\xi} H(-\xi) \\ &= e^{-2il\xi} I_n H(-\xi) e^{il\xi} I_n. \end{aligned} \tag{3.2}$$

Taking $a = \frac{l}{2}$, we get

$$\begin{aligned} &H(\xi) = e^{-4ia\xi} I_n H(-\xi) e^{2ia\xi} I_n \\ &\Rightarrow H(\xi) = A(2\xi) H(-\xi) A(\xi)^{-1}, \end{aligned} \tag{3.3}$$

where

$$A(\xi) = \begin{pmatrix} e^{-2ia\xi} & & & & \\ & e^{-2ia\xi} & & & \\ & & e^{-2ia\xi} & & \\ & & & \cdots & \\ & & & & e^{-2ia\xi} \end{pmatrix}.$$

i.e. $H(\xi)$ satisfies Eq. (2.8). Hence, by Lemma 2.1 we can assert that the corresponding multiscaling function Φ is symmetric about the point $a = \frac{l}{2}$. □

Our aim is to find the necessary as well as sufficient conditions on a Jordan pair such that the corresponding multiscaling function Φ is symmetric. We have shown that the multiscaling function corresponding to a symmetric symbol function is symmetric. Since the symbol function is a matrix polynomial, our problem changes to finding the properties of Jordan pair of a symmetric matrix polynomial. A crucial necessary property of Jordan pair of a symmetric matrix polynomial is given by the following Lemma.

Lemma 3.2 *If* $L(\lambda)$ *is a symmetric matrix polynomial, then its Jordan pair (X, T) has the property that if* $\lambda_0 \neq 0$ *is an eigenvalue of* $L(\lambda)$ *with eigenvector* x_0*, then* $\frac{1}{\lambda_0}$ *is also an eigenvalue with the same eigenvector* x_0*. If 0 is an eigenvalue of* $L(\lambda)$*, then* $L(\lambda)$ *will have an infinite eigenvalue with the eigenvector that of 0.*

Proof Given that

$L(\lambda) = A_0 + A_1\lambda + A_2\lambda^2 + \cdots + A_l\lambda^l$

is a symmetric matrix polynomial. If 0 is an eigenvalue of $L(\lambda)$, then since $\tilde{L}(\lambda) = \lambda^l L(\frac{1}{\lambda}) = L(\lambda)$, the matrix polynomial $\tilde{L}(\lambda)$ also has an eigenvalue 0. i.e. $L(\lambda)$ has an eigenvalue at infinity (By definition). Now, let $\lambda_0 \neq 0$ is an eigenvalue with eigenvector x_0, then we have

$L(\lambda_0)x_0 = 0.$

Since $L(\lambda)$ is a symmetric matrix polynomial, we have

$$L(\lambda_0) = \lambda_0^l L(\frac{1}{\lambda_0}). \tag{3.4}$$

Then

$$L(\lambda_0)x_0 = \lambda_0^l L(\frac{1}{\lambda_0})x_0 = 0. \tag{3.5}$$

Since $\lambda_0 \neq 0$, we have

$$L(\frac{1}{\lambda_0})x_0 = 0, \tag{3.6}$$

i.e. $\frac{1}{\lambda_0}$ is also an eigenvalue with the same eigenvector x_0. □

Lemma 3.2 states that for a symmetric matrix polynomial, it is necessary that the eigenvalues occur in reciprocals. Our attempt is to construct a symmetric matrix polynomial by selecting a suitable Jordan pair (X, T) with only finite eigenvalues. We will state the sufficient properties of a Jordan pair (X, T) such that the corresponding matrix polynomial is symmetric.

Theorem 3.1 *Let (X,T) be a Jordan pair where* $X \in n \times nl$ *and T is a diagonal matrix of order nl with entries being eigenvalues,* $n \in \mathbb{N}$, $n \geq 2$ *and l is even. T has only nonzero elements neither of which equals 1. Assume that the eigenvalues in T occur in reciprocals with same eigenvectors in X. i.e. if* λ_0 *is an eigenvalue in T with eigenvector* x_0*, then* $\frac{1}{\lambda_0}$ *is also an eigenvalue in T with same eigenvector* x_0*. Then a matrix polynomial with Jordan pair (X,T) is symmetric.*

Proof Given that (X, T) is a Jordan pair where $X \in n \times nl$ and T is a diagonal matrix such that $T \in nl \times nl$. Let λ_i $(i = 1, 2 \cdots \frac{nl}{2})$ are the eigenvalues in T. Since the eigenvalues occur in reciprocals, it follows that $\frac{1}{\lambda_i}$ $(i = 1, 2 \cdots \frac{nl}{2})$ are also eigenvalues. Assume that $L(\lambda) = A_0 + A_1\lambda + A_2\lambda^2 + \cdots + A_l\lambda^l$ is a matrix polynomial with the Jordan pair (X, T). We have to show that $L(\lambda)$ is symmetric, i.e. $A_0 = A_l$, $A_1 = A_{l-1} \ldots$, $A_k = A_{l-k}$ or we have to show that

$L(\lambda) = \lambda^l L(\frac{1}{\lambda}).$

Assume the contrary that $L(\lambda) \neq \lambda^l L(\frac{1}{\lambda})$, or the matrix polynomial $N(\lambda) = L(\lambda) - \lambda^l L(\frac{1}{\lambda}) \neq 0$. The sum of algebraic multiplicities of eigenvalues of a matrix polynomial will be the degree of its Determinant [1]. Since $N(\lambda)$ is a nonzero matrix polynomial of degree l and order n, the sum of algebraic multiplicities of the eigenvalues of $N(\lambda)$ will be less than or equal to nl . Now we will show that if $N(\lambda) \neq 0$, then the total algebraic multiplicity exceeds nl, which is a contradiction.

We claim that λ_i and $\frac{1}{\lambda_i}$ $(i = 1, 2 \cdots \frac{nl}{2})$ are eigenvalues of $N(\lambda)$ with same eigenvectors they had for $L(\lambda)$. To prove this, suppose that λ_i is an eigenvalue of $L(\lambda)$ with eigenvector x_i for some i. Then we have $L(\lambda_i)x_i = 0$ and $L(\frac{1}{\lambda_i})x_i = 0$. Now,

$$
\begin{aligned}
N(\lambda_i)x_i &= (L(\lambda_i) - \lambda_i^l L(\frac{1}{\lambda_i}))x_i \\
&= L(\lambda_i)x_i - \lambda_i^l L(\frac{1}{\lambda_i})x_i = 0,
\end{aligned}
$$

and

$$
\begin{aligned}
N(\frac{1}{\lambda_i})x_i &= (L(\frac{1}{\lambda_i}) - \frac{1}{\lambda_i^l}L(\lambda_i))x_i \\
&= L(\frac{1}{\lambda_i})x_i - \frac{1}{\lambda_i^l}L(\lambda_i)x_i = 0.
\end{aligned}
$$

Thus λ_i and $\frac{1}{\lambda_i}$ are eigenvalues of $N(\lambda)$ for $i = 1, 2 \cdots \frac{nl}{2}$. Thus we get a total of $\frac{nl}{2} + \frac{nl}{2} = nl$ eigenvalues for $N(\lambda)$. Now $N(\lambda)$ is given by

$$
\begin{aligned}
N(\lambda) &= (A_0 - A_l) + (A_1 - A_{l-1})\lambda + (A_2 - A_{l-2})\lambda^2 \\
&\quad + \cdots + (A_{l-2} - A_2)\lambda^{l-2} \\
&\quad + (A_{l-1} - A_1)\lambda^{l-1} + (A_l - A_0)\lambda^l. \qquad (3.7)
\end{aligned}
$$

Then

$$
\begin{aligned}
N(1) &= (A_0 - A_l) + (A_1 - A_{l-1}) + (A_2 - A_{l-2}) \\
&\quad + \cdots + (A_{l-2} - A_2) \\
&\quad + (A_{l-1} - A_1) + (A_l - A_0) = 0.
\end{aligned}
$$

$\Rightarrow N(1)y_i = 0$, for linearly independent eigenvectors y_i, $i = 1, 2 \cdots n$.

i.e. 1 is an eigenvalue of $N(\lambda)$ with algebraic multiplicity n. Then the sum of algebraic multiplicities of eigenvalues of $N(\lambda)$ is at least $nl + n$ (there can be other eigenvalues also), which is not possible since the sum of algebraic multiplicities of all eigenvalues of $N(\lambda)$ should not exceed nl. Hence our assumption that $L(\lambda) \neq \lambda^l L(\frac{1}{\lambda})$ is false. We can conclude that $L(\lambda) = \lambda^l L(\frac{1}{\lambda})$, i.e. the matrix polynomial $L(\lambda)$ is symmetric. □

We can construct symmetric matrix polynomials of even degree using the above result. To construct symmetric matrix polynomials of odd degree, we have to select the Jordan pair (X, T) with minor changes. For any symmetric matrix polynomial $L(\lambda)$ of odd degree, we can easily verify that $L(-1) = 0$. Then, we have $L(-1)p_i = 0$, for linearly independent eigenvectors p_i where $i = 1, 2 \cdots n$. Hence -1 is an eigenvalue of $L(\lambda)$ with multiplicity n. Incorporating this change, we state the preceding result for odd values of l.

Theorem 3.2 *Let (X,T) be a Jordan pair where* $X \in n \times nl$ *and T is a diagonal matrix of order nl,* $n \in \mathbb{N}, n \geq 2$ *and l is odd. T has only nonzero elements neither of which equals 1. Assume that the eigenvalues in T occur in reciprocals with same eigenvectors in X. Also,* -1 *is an eigenvalue in T with multiplicity n. Then a matrix polynomial with Jordan pair (X,T) is symmetric.*

Proof Given that -1 occurs n times in T, then there will be $nl - n$ eigenvalues in T other than -1. Given that they occur in reciprocals, i.e. if λ_i is an eigenvalue in T, then $\frac{1}{\lambda_i}$ is also an eigenvalue in T. Thus we have, for $i = 1, 2 \cdots \frac{nl-n}{2}$, λ_i and its reciprocal $\frac{1}{\lambda_i}$ are eigenvalues in T, and together they constitute $nl - n$ eigenvalues (Here $nl - n$ is always even since l is odd). Now, let $L(\lambda) = A_0 + A_1\lambda + A_2\lambda^2 + \cdots + A_l\lambda^l$ be a matrix polynomial with Jordan pair (X, T), we have to show that

$$
L(\lambda) = \lambda^l L(\frac{1}{\lambda}).
$$

As we did in the last proof, assume the contrary that $L(\lambda) \neq \lambda^l L(\frac{1}{\lambda})$, or the matrix polynomial $N(\lambda) = L(\lambda) - \lambda^l L(\frac{1}{\lambda}) \neq 0$. Since $N(\lambda)$ is a nonzero matrix polynomial of degree l and order n, the sum of algebraic multiplicities of the eigenvalues of $N(\lambda)$ can be maximum nl. We claim that λ_i and $\frac{1}{\lambda_i}$ $(i = 1, 2 \cdots \frac{nl-n}{2})$ are eigenvalues of $N(\lambda)$ with same eigenvectors they had for $L(\lambda)$. To prove this, suppose that λ_i is an eigenvalue of $L(\lambda)$ with eigenvector x_i for some i. Then we have, $L(\lambda_i)x_i = 0$ and $L(\frac{1}{\lambda_i})x_i = 0$. Now,

$$
\begin{aligned}
N(\lambda_i)x_i &= (L(\lambda_i) - \lambda_i^l L(\frac{1}{\lambda_i}))x_i \\
&= L(\lambda_i)x_i - \lambda_i^l L(\frac{1}{\lambda_i})x_i = 0,
\end{aligned}
$$

and

$$
\begin{aligned}
N(\frac{1}{\lambda_i})x_i &= (L(\frac{1}{\lambda_i}) - \frac{1}{\lambda_i^l}L(\lambda_i))x_i \\
&= L(\frac{1}{\lambda_i})x_i - \frac{1}{\lambda_i^l}L(\lambda_i)x_i = 0.
\end{aligned}
$$

i.e. λ_i and $\frac{1}{\lambda_i}$ are eigenvalues of $N(\lambda)$ for $i = 1, 2 \cdots \frac{nl-n}{2}$. Thus, we get a total of $\frac{nl-n}{2} + \frac{nl-n}{2} = nl - n$ eigenvalues for

$N(\lambda)$. Since -1 is an eigenvalue of $L(\lambda)$ with algebraic multiplicity n, $L(-1)p_i = 0$ for linearly independent eigenvectors p_i, $i = 1, 2 \cdots n$. Then we have,

$$N(-1)p_i = (L(-1) - (-1)^l L(\frac{1}{-1}))p_i$$
$$= L(-1)p_i - (-1)^l L(-1)p_i = 0,$$

for $i = 1, 2 \cdots n$.

i.e. -1 is an eigenvalue of $N(\lambda)$ with algebraic multiplicity n. Then, the sum of algebraic multiplicities of eigenvalues of $N(\lambda)$ is at least $nl - n + n = nl$. Now, $N(\lambda)$ is given by

$$N(\lambda) = (A_0 - A_l) + (A_1 - A_{l-1})\lambda + (A_2 - A_{l-2})\lambda^2 + \cdots + (A_{l-2} - A_2)\lambda^{l-2} + (A_{l-1} - A_1)\lambda^{l-1} + (A_l - A_0)\lambda^l.$$

Then

$$N(1) = (A_0 - A_l) + (A_1 - A_{l-1}) + (A_2 - A_{l-2}) + \cdots + (A_{l-2} - A_2) + (A_{l-1} - A_1) + (A_l - A_0) = 0$$

$\Rightarrow N(1)y_i = 0$ for linearly independent eigenvectors y_i, $i = 1, 2 \cdots n$.

i.e. 1 is an eigenvalue of $N(\lambda)$ with algebraic multiplicity n. Then, the sum of algebraic multiplicities of eigenvalues of $N(\lambda)$ is at least $nl + n$ (there can be other eigenvalues also), which is not possible since the sum of algebraic multiplicities of all eigenvalues of $N(\lambda)$ cannot exceed nl. Hence, our assumption that $L(\lambda) \neq \lambda^l L(\frac{1}{\lambda})$ is false. We conclude that $L(\lambda) = \lambda^l L(\frac{1}{\lambda})$, i.e. the matrix polynomial $L(\lambda)$ is symmetric. □

Construction of multiscaling function

We have obtained the properties of the spectral data of a matrix polynomial so that it is a symbol function of a symmetric multiscaling function Φ. Since each entry in this symbol function is a trigonometric polynomial (algebraic polynomial in $z = e^{-i\xi}$), the associated multiscaling function is compactly supported [6]. We will obtain the multiscaling function by employing the cascade algorithm [3]. The cascade algorithm will converge if the multiscaling coefficients satisfy certain properties and if the initial functions are appropriately chosen. Let H_0, H_1, H_2, $H_3 \in \mathbb{C}^{2\times 2}$ be the set of multiscaling coefficients and define the 8×8 matrix D as

$$D = \begin{pmatrix} H_0 & 0 & 0 & 0 \\ H_2 & H_1 & H_0 & 0 \\ 0 & H_3 & H_2 & H_1 \\ 0 & 0 & 0 & H_3 \end{pmatrix}. \quad (4.1)$$

Let D_0 be the 3×3 sub block matrix of D at the top left, D_k is the sub matrix 'k' steps to the left. Then

$$D_0 = \begin{pmatrix} H_0 & 0 & 0 \\ H_2 & H_1 & H_0 \\ 0 & H_3 & H_2 \end{pmatrix} \quad (4.2)$$

and

$$D_1 = \begin{pmatrix} H_1 & H_0 & 0 \\ H_3 & H_2 & H_1 \\ 0 & 0 & H_3 \end{pmatrix}. \quad (4.3)$$

Definition 4.1 [3] The recursion coefficients H_k of a matrix refinement equation with dilation factor m satisfy the sum rules of order p if there exist vectors y_0, $y_1 ... y_{p-1}$ with $y_0 \neq 0$ such that

$$\sum_{t=0}^{n} \binom{n}{t} m^t (-i)^{n-t} y_t D^{n-t} H(\frac{2\pi s}{m}) = \delta_{0s} y_n, s = 0, 1, 2 \ldots m-1, \quad (4.4)$$

for $n = 0 ... p-1$.

Theorem 4.1 [3] *Assume that $H(\xi)$ satisfies the sum rules of order 1, and the joint spectral radius*

$$\rho(D_0|F_1, D_1|F_1 \ldots D_{m-1}|F_1) = \lambda < 1,$$

where F_1 is the orthogonal complement of the common left eigenvector $e^* = (\mu_0^*, \mu_0^* \ldots \mu_0^*)$ of $D_0, D_1 ... D_{m-1}$. Then, the cascade algorithm has a unique solution Φ which is Holder continuous of order α for any $\alpha < -\log_m \lambda$.

A method to find $H(\xi)$ which satisfies the sum rules of order 1 by suitably selecting the Jordan pair (X, T) is given in [2]. Based on that method, we construct the symbol function $H(\xi)$ so that it satisfies the sum rules of order 1. While finding the multiscaling coefficients H_k, we ensure that the value of the joint spectral radius

$$\rho(D_0|F_1, D_1|F_1 \ldots D_{m-1}|F_1)$$

is less than 1. Then by Theorem 4.1, the cascade algorithm converges for the set of multiscaling coefficients H_k. An example of this construction is given as follows. We will start with a Jordan pair (X, T) satisfying the conditions

1. $I - T$ is of full rank (Theorem 2.2)
2. The eigenvalues are nonzero and not equal to 1. They occur in reciprocals with same eigenvectors. Also, -1 is an eigenvalue with multiplicity n (in the following example, $n = 2$) (Theorem 3.2)

Without loss of generality, we take $X \in \mathbb{C}^{2\times 6}$ and $T \in \mathbb{C}^{6\times 6}$ satisfying the above listed conditions, and are given by,

$$X = \begin{pmatrix} 0.2785 & 0.2785 & 0.9575 & 0.9575 & 1.0000 & 0.8003 \\ 0.5469 & 0.5469 & 0.9649 & 0.964 & 0 & 0.1419 \end{pmatrix}$$

and

$$T = \begin{pmatrix} 34 & 0 & 0 & 0 & 0 & 0 \\ 0 & 0.0294 & 0 & 0 & 0 & 0 \\ 0 & 0 & -33 & 0 & 0 & 0 \\ 0 & 0 & 0 & -0.0303 & 0 & 0 \\ 0 & 0 & 0 & 0 & -1 & 0 \\ 0 & 0 & 0 & 0 & 0 & -1 \end{pmatrix}.$$

It can be verified that $I - T$ is of rank 6, then there exists a matrix polynomial with Jordan pair (X, T) which is a symbol function of a multiscaling function vector Φ (Theorem 2.2). Since (X, T) satisfies the conditions in Theorem 3.2, the matrix polynomial $H(\xi)$ must be symmetric. By employing the procedure to find the multiscaling coefficients H_k given in [2], we have obtained the multiscaling coefficients as follows.

$$H_0 = \begin{pmatrix} 0.0647 & -0.0442 \\ 0.0525 & -0.0400 \end{pmatrix},$$

$$H_1 = \begin{pmatrix} 0.6424 & 0.0442 \\ -0.0525 & 0.4642 \end{pmatrix},$$

$$H_2 = \begin{pmatrix} 0.6424 & 0.0442 \\ -0.0525 & 0.4642 \end{pmatrix},$$

$$H_3 = \begin{pmatrix} 0.0647 & -0.0442 \\ 0.0525 & -0.0400 \end{pmatrix}.$$

We get

$$H(0) = \begin{pmatrix} 1 & 0 \\ 0 & 0.6 \end{pmatrix}$$

and

$$H(-1) = \begin{pmatrix} 0 & 0 \\ 0 & 0 \end{pmatrix}.$$

For $y_0 = [1 \quad 0]$, we get $y_0 H(0) = y_0$ and $y_0 H(-1) = [0 \quad 0]$. Thus, H satisfies the sum rules of order 1.

Now, our attempt is to find the multiscaling function vector Φ corresponding to the above multiscaling coefficients. For that, we need to ensure that the conditions in Theorem 4.1 are satisfied. The matrices D, D_0 and D_1 associated to H are given by

$$D = \begin{pmatrix} 0.0915 & -0.0625 & 0 & 0 & 0 & 0 & 0 & 0 \\ 0.0742 & -0.0565 & 0 & 0 & 0 & 0 & 0 & 0 \\ 0.9085 & 0.0625 & 0.9085 & 0.0625 & 0.0915 & -0.0625 & 0 & 0 \\ -0.0742 & 0.6565 & -0.0742 & 0.6565 & 0.0742 & -0.0565 & 0 & 0 \\ 0 & 0 & 0.0915 & -0.0625 & 0.9085 & 0.0625 & 0.9085 & 0.0625 \\ 0 & 0 & 0.0742 & -0.0565 & -0.0742 & 0.6565 & -0.0742 & 0.6565 \\ 0 & 0 & 0 & 0 & 0 & 0 & 0.0915 & -0.0625 \\ 0 & 0 & 0 & 0 & 0 & 0 & 0.0742 & -0.0565 \end{pmatrix},$$

$$D_0 = \begin{pmatrix} 0.0915 & -0.0625 & 0 & 0 & 0 & 0 \\ 0.0742 & -0.0565 & 0 & 0 & 0 & 0 \\ 0.9085 & 0.0625 & 0.9085 & 0.0625 & 0.0915 & -0.0625 \\ -0.0742 & 0.6565 & -0.0742 & 0.6565 & 0.0742 & -0.0565 \\ 0 & 0 & 0.0915 & -0.0625 & 0.9085 & 0.0625 \\ 0 & 0 & 0.0742 & -0.0565 & -0.0742 & 0.6565 \end{pmatrix},$$

$$D_1 = \begin{pmatrix} 0.9085 & 0.0625 & 0.0915 & -0.0625 & 0 & 0 \\ -0.0742 & 0.6565 & 0.0742 & -0.0565 & 0 & 0 \\ 0.0915 & -0.0625 & 0.9085 & 0.0625 & 0.9085 & 0.0625 \\ 0.0742 & -0.0565 & -0.0742 & 0.6565 & -0.0742 & 0.6565 \\ 0 & 0 & 0 & 0 & 0.0915 & -0.0625 \\ 0 & 0 & 0 & 0 & 0.0742 & -0.0565 \end{pmatrix},$$

$\rho(D_0|F_1, D_1|F_1) = 0.7757 < 1,$

where F_1 is the orthogonal complement of the common left eigenvector

$$\begin{pmatrix} 0.5774 \\ 0.0000 \\ 0.5774 \\ 0.0000 \\ 0.5774 \\ 0.0000 \end{pmatrix}$$

of D_0, D_1. Also H satisfies the sum rules of order 1. Thus H satisfies the conditions in Theorem 4.1. The cascade algorithm will converge to a multiscaling function Φ which is Holder continuous if the initial function is chosen as a piecewise linear function that interpolates on the set of integers [3]. The values of Φ^0 at the integers are obtained by finding the 1-eigenvector of the matrix D. Since the symbol function H is a matrix polynomial of degree 3, the support(Φ^0) is contained in [0,3]. We have to find the values of Φ^0 at the points 0,1,2 and 3. For that, find the 1-eigenvector of the matrix D, which is given by

$$\begin{pmatrix} 0.0000 \\ 0.0000 \\ 0.7071 \\ 0.0000 \\ 0.7071 \\ 0.0000 \\ 0.0000 \\ 0.0000 \end{pmatrix} = \begin{pmatrix} \Phi^0(0) \\ \Phi^0(1) \\ \Phi^0(2) \\ \Phi^0(3) \end{pmatrix}.$$

Now choose the initial function Φ^0 as the piecewise linear function that interpolates at the points 0, 1, 2 and 3. Then, the cascade algorithm will converge to the solution Φ (Fig. 1) which is compactly supported in [0,3] and is symmetric about the point 1.5.

The obtained multiscaling function is compactly supported and symmetric. Thus, we are able to construct a compactly supported multiscaling function which is symmetric about the point 1.5.

Construction of pseudo biorthogonal symmetric multiscaling functions

In the previous sections, we constructed a symbol function $H(z)$ and the corresponding multiscaling function $\Phi(x)$ which is symmetric and compactly supported. Now, we have to construct the dual multiscaling function $\tilde{\Phi}(x)$. For that, we have to find the dual symbol $F(z)$ corresponding to $H(z)$ so that the generalized condition of perfect reconstruction (2.14) holds. For a given symbol $H(z)$, a dual symbol $F(z)$ satisfying Eq. (2.14) exists if Determinants of $H(z)$ and $H(-z)$ do not have common roots [5]. Now, the Jordan pair (X, T) that we selected for constructing $H(z)$ is given by,

$$X = \begin{pmatrix} 0.2785 & 0.2785 & 0.9575 & 0.9575 & 1.0000 & 0.8003 \\ 0.5469 & 0.5469 & 0.9649 & 0.964 & 0 & 0.1419 \end{pmatrix}$$

and

$$T = \begin{pmatrix} 34 & 0 & 0 & 0 & 0 & 0 \\ 0 & 0.0294 & 0 & 0 & 0 & 0 \\ 0 & 0 & -33 & 0 & 0 & 0 \\ 0 & 0 & 0 & -0.0303 & 0 & 0 \\ 0 & 0 & 0 & 0 & -1 & 0 \\ 0 & 0 & 0 & 0 & 0 & -1 \end{pmatrix}.$$

The diagonal entries of T are the eigenvalues of the obtained symbol function $H(z)$. Looking at the diagonal entries of T, it is clear that negative of an eigenvalue is not again an eigenvalue. Since eigenvalues of $H(z)$ are precisely the roots of the Determinant of $H(z)$, we can say that negative of a root of Determinant of $H(z)$ is not again its

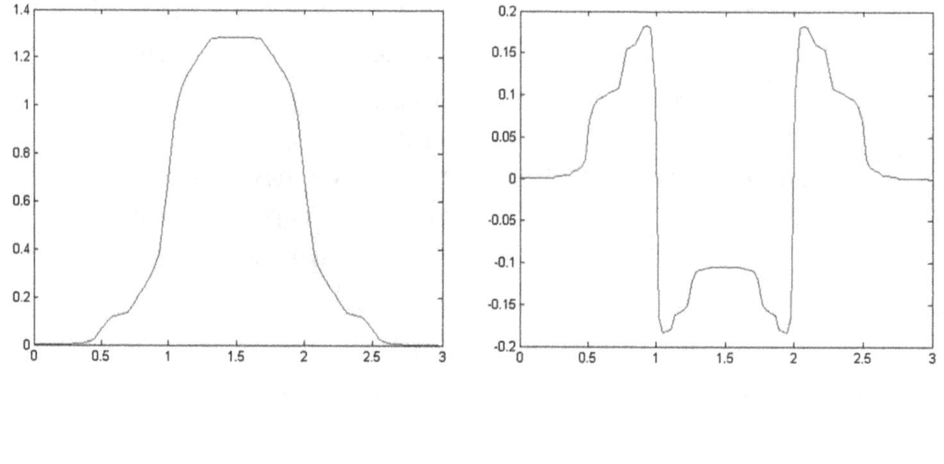

Fig. 1 The two components ϕ_0 and ϕ_1 of the multiscaling function Φ corresponding to the obtained multiscaling coefficients. Here both components ϕ_0 and ϕ_1 are symmetric and are compactly supported in [0,3]

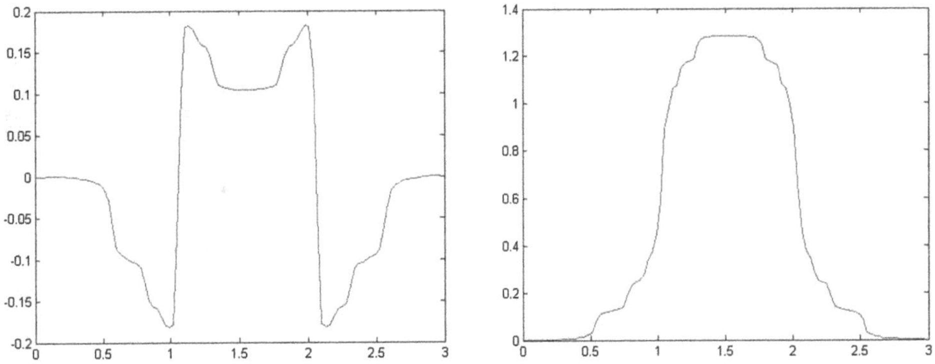

Fig. 2 The two components $\tilde{\phi}_0$ and $\tilde{\phi}_1$ of the dual multiscaling function $\tilde{\Phi}$. Here $\tilde{\phi}_0$ and $\tilde{\phi}_1$ are symmetric and compactly supported

root. In other words, Determinants of $H(z)$ and $H(-z)$ do not have common roots. Thus, there exists a dual symbol $F(z)$ so that the generalized condition of perfect reconstruction is satisfied. Using the cofactor method given in [5], we got the dual symbol function $F(z)$ corresponding to $H(z)$ as

$$F(z) = F_{-3}z^{-3} + F_{-2}z^{-2} + F_{-1}z^{-1} + F_0 + F_1 z + F_2 z^2 + F_3 z^3 + F_4 z^4 + F_5 z^5,$$

where

$$F_{-2} = e^{-4}\begin{bmatrix} -0.4660 & -0.6116 \\ 0.5149 & 0.7537 \end{bmatrix},$$

$$F_{-1} = e^{-3}\begin{bmatrix} 0.5936 & 0.1305 \\ -0.1099 & 0.6629 \end{bmatrix},$$

$$F_0 = \begin{bmatrix} -0.0401 & -0.0525 \\ 0.0442 & 0.0646 \end{bmatrix},$$

$$F_1 = \begin{bmatrix} 0.4638 & 0.0525 \\ -0.0442 & 0.6419 \end{bmatrix}, F_2 = \begin{bmatrix} 0.4638 & 0.0525 \\ -0.0442 & 0.6419 \end{bmatrix},$$

$$F_3 = \begin{bmatrix} -0.0401 & -0.0525 \\ 0.0442 & 0.0646 \end{bmatrix},$$

$$F_4 = e^{-3}\begin{bmatrix} 0.5936 & 0.1305 \\ -0.1099 & 0.6629 \end{bmatrix}, F_5 = e^{-4}\begin{bmatrix} -0.4660 & -0.6116 \\ 0.5149 & 0.7537 \end{bmatrix}.$$

Here $F_5 = F_{-2}$, $F_4 = F_{-1}$, $F_3 = F_0$, $F_2 = F_1$, i.e. The properties of the multiscaling coefficients H_k which enable the symmetricity of $\Phi(x)$ are preserved and hence the dual multiscaling function $\tilde{\Phi}(x)$ will also be symmetric. Since the entries in the matrix coefficients F_{-2}, F_{-1}, F_4, F_5 are very small, the support of $\tilde{\Phi}(x)$ will be almost similar to that of $\Phi(x)$. The components of the dual multiscaling function $\tilde{\Phi}(x)$ are given in Fig. 2.

Thus we have obtained a symmetric and compactly supported dual multiscaling function $\tilde{\Phi}(x)$ so that the functions $\Phi(x)$ and $\tilde{\Phi}(x)$ form a pair of pseudo-biorthogonal multiscaling functions.

Results

1. We defined a symmetric matrix polynomial analogous to symmetric scalar polynomials (Definition 3.1)
2. If the symbol $H(\xi)$ is a symmetric matrix polynomial of degree l, then the corresponding multiscaling function Φ will be symmetric about the point $\frac{l}{2}$ (Lemma 3.1)
3. The eigenvalues of a symmetric matrix polynomial $L(\lambda)$ occur in reciprocals with same eigenvectors, i.e. If λ_0 is an eigenvalue of $L(\lambda)$ with eigenvector x_0, then $\frac{1}{\lambda_0}$ is also an eigenvalue of $L(\lambda)$ with same eigenvector x_0 (Lemma 3.2)
4. Let (X, T) be a Jordan pair where $X \in n \times nl$ and T is a diagonal matrix of order nl, $n \in \mathbb{N}$, $n \geq 2$ and l is even. Assume that T has only nonzero elements neither of which equals 1 and eigenvalues of T occur in reciprocals with same eigenvectors in X. Then a matrix polynomial with Jordan pair (X, T) is symmetric (Theorem 3.1)
5. In the above result, if l is an odd number then also the matrix polynomial is symmetric provided that -1 is an eigenvalue of T (or diagonal entry in T) with algebraic multiplicity n (Theorem 3.2)

Conclusions

We have found the necessary as well as sufficient conditions on a Jordan pair (X, T) such that the corresponding matrix polynomial $H(\xi)$ is symmetric. We selected a Jordan pair satisfying these conditions and constructed a symmetric matrix polynomial $H(\xi)$. Using cascade algorithm, we found the multiscaling function Φ for which the matrix polynomial $H(\xi)$ acts as a symbol function. Since $H(\xi)$ is a symmetric matrix polynomial, we saw that Φ is also symmetric. Finally we constructed the dual multiscaling function $\tilde{\Phi}(x)$ which is symmetric and compactly supported so that the functions $\Phi(x)$ and $\tilde{\Phi}(x)$ form a pair of pseudo-biorthogonal multiscaling functions.

References

1. Gohberg, I., Lancaster, P., Rodman, L.: Matrix polynomials, vol. 58. Siam, Philadelphia, USA (1982)
2. Mubeen, M., Narayanan, V.: Inverse representation theorem for matrix polynomials and multiscaling functions. Fractals, wavelets, and their applications, pp. 319–339. Springer International Publishing, Switzerland (2014)
3. Keinert, F.: Wavelets and multiwavelets. CRC Press, Boca Raton, Florida (2003)
4. Strela, V.: A note on construction of biorthogonal multi-scaling functions. Wavelets, multiwavelets and their applications: AMS special session on wavelets, multiwavelets and their applications, January 1997, San Diego, California, vol. 216, p. 149 (1998)
5. Strela, V., Strang, G.: Pseudo-biorthogonal multiwavelets and finite elements (1997) **(preprint)**
6. Plonka, G., Strela, V.: Construction of multiscaling functions with approximation and symmetry. SIAM J. Math. Anal. **29**(2), 481–510 (1998)

5

Some fixed point theorems for (α, β)-admissible Geraghty type contractive mappings and related results

Sumit Chandok[1]

Abstract In this paper, we introduced the concept of (α, β)-admissible Geraghty type contractive mappings. Sufficient conditions for the existence of a fixed point for such class of generalized nonlinear contractive mappings in metric spaces are provided. As applications, we derive a fixed point theorem for these contractions whenever the space is endowed with a graph. Some interesting consequences of our theorems are also obtained. The proved results generalize and extend various well-known results in the literature. Some examples are illustrated for the usability of the results.

Keywords Fixed point · (α, β)-admissible · Contraction mappings · Metric space

Mathematics Subject Classification 47H10 · 54H25

Introduction and preliminaries

Fixed point theory has gained very large impetus due its wide range of applications in various fields such as engineering, economics, computer science, and many others. It is well known that the contractive-type conditions are very indispensable in the study of fixed point theory and Banach's fixed point theorem [1] for contraction mappings is one of the pivotal results in analysis. This theorem that has been extended and generalized by various authors (see, e.g., [2–15]) and has many applications in mathematics and other related disciplines as well.

In an attempt to generalize the Banach contraction principle, many researchers extended the following result in certain directions.

Theorem 1.1 (see, [8]) *Let* (X, d) *be a complete metric space and* $T : X \to X$ *be a mapping. Assume that there exists a function* $\theta : [0, \infty) \to [0, 1]$ *such that, for any bounded sequence* $\{t_n\}$ *of positive reals,* $\theta(t_n) \to 1$ *implies* $t_n \to 0$ *and* $d(Tx, Ty) \leq \theta(d(x, y))d(x, y)$ *for all* $x, y \in X$. *Then, T has a unique fixed point.*

In 2012, Samet et al. [15] introduced the concepts of α-contractive and α-admissible mappings and established various fixed point theorems for such class of mappings defined on complete metric spaces. Thereafter, the existence of fixed points of α-admissible contractive-type mappings in complete metric spaces has been studied by several researchers (see [2, 11, 13, 14] and references cited therein). In this paper, we introduced the concept of (α, β)-admissible Geraghty type contractive mappings. Sufficient conditions for the existence of a fixed point for such class of generalized nonlinear contractive mappings in metric spaces are provided. As applications, we derive a fixed point theorem for these contractions whenever the space is endowed with a graph. Several other interesting results for cyclic mappings and ordered metric spaces are also deduced. The proved results using the concept of (α, β)-admissible mappings generalize and extend various well-known results in the literature. Some examples are illustrated for the justification of the results.

To start with we give some notations and introduce some definitions which will be used in the sequel.

✉ Sumit Chandok
chansok.s@gmail.com

[1] Department of Mathematics, Khalsa College of Engineering and Technology, Punjab Technical University, Amritsar 143001, India

Definition 1.1 Let X be a non-empty set, $T : X \to X$ and $\alpha, \beta : X \times X \to \mathbb{R}^+$. We say that T is an (α, β)-admissible mapping if $\alpha(x,y) \geq 1$ and $\beta(x,y) \geq 1$ implies $\alpha(Tx, Ty) \geq 1$, and $\beta(Tx, Ty) \geq 1$, for all $x, y \in X$.

Definition 1.2 Let X be a non-empty set, $T : X \to X$ and $\alpha : X \times X \to \mathbb{R}^+$. We say that T is an α-admissible mapping if $\alpha(x,y) \geq 1$ and implies $\alpha(Tx, Ty) \geq 1$, for all $x, y \in X$.

Definition 1.3 Let (X, d) be a metric space, and $\alpha, \beta : X \times X \to [0, \infty)$. X is (α, β)-regular if $\{x_n\}$ is a sequence in X such that $x_n \to x \in X$, $\alpha(x_n, x_{n+1}) \geq 1$, $\beta(x_n, x_{n+1}) \geq 1$, for all n, there exists a subsequence $\{x_{n_k}\}$ of $\{x_n\}$ such that $\alpha(x_{n_k}, x_{n_k+1}) \geq 1$, $\beta(x_{n_k}, x_{n_k+1}) \geq 1$ for all $k \in \mathbb{N}$ and $\alpha(x, Tx) \geq 1$, $\beta(x, Tx) \geq 1$.

Definition 1.4 Let (X, d) be a metric space, and $\alpha : X \times X \to [0, \infty)$. X is a α-regular if $\{x_n\}$ is a sequence in X such that $x_n \to x$, $\alpha(x_n, x_{n+1}) \geq 1$, there exists a subsequence $\{x_{n_k}\}$ of $\{x_n\}$ such that $\alpha(x_{n_k}, x_{n_k+1}) \geq 1$ for all $k \in \mathbb{N}$ and $\alpha(x, Tx) \geq 1$.

Definition 1.5 Let $(X, d, \leq)$ be an ordered metric space, and $\alpha : X \times X \to [0, \infty)$. X is an ordered α-regular if $\{x_n\}$ is a sequence in X such that $x_n \to x$, $\alpha(x_n, x_{n+1}) \geq 1$, and $x_n \leq x_{n+1}$ then there exists a subsequence $\{x_{n_k}\}$ of $\{x_n\}$ such that $\alpha(x_{n_k}, x_{n_k+1}) \geq 1$ and $x_{n_k} \leq x_{n_k+1}$ for all $k \in \mathbb{N}$ and $\alpha(x, Tx) \geq 1$.

Throughout the paper, $F(T)$ denotes the set of fixed points of T.

Main results

We say Θ be a family of functions $\theta : [0, \infty) \to [0, 1)$ such that for any bounded sequence $\{t_n\}$ of positive reals, $\theta(t_n) \to 1$ implies $t_n \to 0$.

We say Ψ be a family of functions $\psi : [0, \infty) \to [0, \infty)$ such that ψ is continuous, strictly increasing and $\psi(0) = 0$.

Definition 2.1 Let (X, d) be a metric space, $T : X \to X$ and $\alpha, \beta : X \times X \to \mathbb{R}^+$. A mapping T is said to be (α, β)-Geraghty type-I rational contractive mapping if there exists a $\theta \in \Theta$, such that for all $x, y \in X$, following condition holds:

$$\alpha(x, Tx)\beta(y, Ty)\psi(d(Tx, Ty)) \leq \theta(\psi(M(x, y)))\psi(M(x, y)), \tag{2.1}$$

where $M(x,y) = \max\{d(x,y), d(x,Tx), d(y,Ty), \frac{d(x,Tx)d(y,Ty)}{1+d(x,y)}, \frac{d(x,Tx)d(y,Ty)}{1+d(Tx,Ty)}\}$ and $\psi \in \Psi$.

Theorem 2.1 *Let (X, d) be a complete metric space, T is self-mapping, $T : X \to X$, and $\alpha, \beta : X \times X \to \mathbb{R}^+$. Suppose that the following conditions are satisfied:*

(i) *T is an (α, β)-admissible mapping;*
(ii) *T is an (α, β)-Geraghty type-I rational contractive mapping;*
(iii) *there exists $x_0 \in X$ such that $\alpha(x_0, Tx_0) \geq 1$ and $\beta(x_0, Tx_0) \geq 1$;*
(iv) *either T is continuous or X is (α, β)-regular.*

Then, T has a fixed point $x \in X$ and $\{T^n x_0\}$ converges to x.

Further, if for all $x, y \in F(T)$, *with $x \neq y$ such that $\alpha(x, Tx) \geq 1$, $\alpha(y, Ty) \geq 1$ and $\beta(x, Tx) \geq 1$, $\beta(y, Ty) \geq 1$, then T has a unique fixed point in X.*

Proof Let $x_0 \in X$ such that $\alpha(x_0, Tx_0) \geq 1$ and $\beta(x_0, Tx_0) \geq 1$. Now, we can construct the sequences $\{x_n\}$ in X by $x_n = T^n x_0 = Tx_{n-1}$, for $n \in \mathbb{N}$.

Moreover, we assume that if $x_{n_0} = x_{n_0+1}$, for some $n_0 \in \mathbb{N}$, then x_{n_0} is a fixed point of T. Consequently, we suppose that $x_n \neq x_{n+1}$ for all $n \in \mathbb{N}$.

Since T is (α, β)-admissible mapping, $\alpha(x_0, Tx_0) = \alpha(x_0, x_1) \geq 1$, $\alpha(Tx_0, Tx_1) = \alpha(x_1, x_2) \geq 1$, $\alpha(Tx_1, Tx_2) = \alpha(x_2, x_3) \geq 1$. Hence, by induction, we get $\alpha(x_n, x_{n+1}) \geq 1$ for all $n \geq 0$.

Similarly, $\beta(x_n, x_{n+1}) \geq 1$ for all $n \geq 0$.

Consider (2.1), we have

$$\begin{aligned}\psi(d(x_{n+1}, x_{n+2})) &= \psi(d(Tx_n, Tx_{n+1})) \\ &\leq \alpha(x_n, Tx_n)\beta(x_{n+1}, Tx_{n+1})\psi(d(Tx_n, Tx_{n+1})) \\ &\leq \theta(\psi(M(x_n, x_{n+1})))\psi(M(x_n, x_{n+1})),\end{aligned}$$

where

$$\begin{aligned}M(x_n, x_{n+1}) &= \max\left\{d(x_n, x_{n+1}), d(x_n, Tx_n), d(x_{n+1}, Tx_{n+1}), \frac{d(x_n, Tx_n)d(x_{n+1}, Tx_{n+1})}{1 + d(x_n, x_{n+1})}, \frac{d(x_n, Tx_n)d(x_{n+1}, Tx_{n+1})}{1 + d(Tx_n, Tx_{n+1})}\right\} \\ &= \max\left\{d(x_n, x_{n+1}), d(x_n, x_{n+1}), d(x_{n+1}, x_{n+2}), \frac{d(x_n, x_{n+1})d(x_{n+1}, x_{n+2})}{1 + d(x_n, x_{n+1})}, \frac{d(x_n, x_{n+1})d(x_{n+1}, x_{n+2})}{1 + d(x_{n+1}, x_{n+2})}\right\} \\ &= \max\{d(x_n, x_{n+1}), d(x_{n+1}, x_{n+2})\}.\end{aligned} \tag{2.2}$$

Now, if $M(x_n, x_{n+1}) = d(x_{n+1}, x_{n+2})$, then

$$\begin{aligned}\psi(d(x_{n+1}, x_{n+2})) &\leq \theta(\psi(M(x_n, x_{n+1})))\psi(M(x_n, x_{n+1})) \\ &\leq \theta(\psi(M(x_n, x_{n+1})))\psi(d(x_{n+1}, x_{n+2})) \\ &< \psi(d(x_{n+1}, x_{n+2})),\end{aligned}$$

which is a contradiction, using the properties of ψ.

Therefore, it implies that $M(x_n, x_{n+1}) = d(x_n, x_{n+1})$ and

$$\begin{aligned}\psi(d(x_{n+1}, x_{n+2})) &\leq \theta(\psi(M(x_n, x_{n+1})))\psi(M(x_n, x_{n+1})) \\ &\leq \theta(\psi(M(x_n, x_{n+1})))\psi(d(x_n, x_{n+1})) \\ &\leq \psi(d(x_n, x_{n+1})).\end{aligned} \tag{2.3}$$

Hence, using the properties of ψ, we conclude that

$$d(x_{n+1},x_{n+2}) \leq d(x_n,x_{n+1}), \quad (2.4)$$

for every $n \in \mathbb{N}$. Therefore, sequence $\{d(x_n,x_{n+1})\}$ is decreasing and for the non-negative decreasing sequence $\{d(x_n,x_{n+1})\}$, there exists some $r \geq 0$, such that

$$\lim_{n\to\infty} d(x_n,x_{n+1}) = r. \quad (2.5)$$

Further from (2.3), it implies that

$$\frac{\psi(d(x_{n+1},x_{n+2}))}{\psi(M(x_n,x_{n+1}))} \leq \theta(\psi(M(x_n,x_{n+1}))) < 1. \quad (2.6)$$

On letting $n \to \infty$ in above inequality, we have $\lim_{n\to\infty} \theta(\psi(M(x_n,x_{n+1}))) = 1$, and $\theta \in \Theta$, $\lim_{n\to\infty} \psi(M(x_n,x_{n+1})) = 0$, which yields that

$$r = \lim_{n\to\infty} d(x_n,x_{n+1}) = 0. \quad (2.7)$$

Now, we will show that $\{x_n\}$ is a Cauchy sequence. Suppose, to the contrary, that $\{x_n\}$ is not a Cauchy sequence.

Then, there exists $\delta > 0$ for which we can find subsequences $\{x_{n_k}\}$ and $\{x_{m_k}\}$ of $\{x_n\}$ with $n_k > m_k > k$ such that

$$d(x_{n_k},x_{m_k}) \geq \delta \quad (2.8)$$

Further, corresponding to m_k, we can choose n_k in such a way that it is the smallest integer with $n_k > m_k$ and satisfying (2.8), we have

$$d(x_{n_k-1},x_{m_k}) < \delta. \quad (2.9)$$

Using triangle inequality, we have

$$0 < \delta = d(x_{n_k},x_{m_k}) \leq d(x_{n_k},x_{n_k-1}) + d(x_{n_k-1},x_{m_k}) < \delta + d(x_{n_k},x_{n_k-1}) \quad (2.10)$$

Letting $k \to \infty$ and using (2.7) and (2.8), we obtain

$$\lim_{k\to\infty} d(x_{n_k},x_{m_k}) = \delta \quad (2.11)$$

Again, using triangle inequality, we have

$$d(x_{n_k},x_{m_k}) \leq d(x_{n_k},x_{n_k-1}) + d(x_{n_k-1},x_{m_k-1}) + d(x_{m_k-1},x_{m_k}),$$

and

$$d(x_{n_k-1},x_{m_k-1}) \leq d(x_{n_k},x_{n_k-1}) + d(x_{n_k},x_{m_k}) + d(x_{m_k-1},x_{m_k}).$$

Therefore,

$$\begin{aligned} d(x_{n_k},x_{m_k}) &\leq d(x_{n_k},x_{n_k-1}) + d(x_{n_k-1},x_{m_k-1}) + d(x_{m_k-1},x_{m_k}) \\ &\leq 2d(x_{n_k},x_{n_k-1}) + d(x_{n_k},x_{m_k}) + 2d(x_{m_k-1},x_{m_k}) \end{aligned} \quad (2.12)$$

Letting $k \to \infty$ in (2.12) and using (2.7), (2.11), we get

$$\lim_{k\to\infty} d(x_{n_k-1},x_{m_k-1}) = \delta. \quad (2.13)$$

Put $x = x_{m_k}$ and $y = x_{n_k}$ in (2.1), we obtain

$$\begin{aligned} \psi(d(Tx_{m_k},Tx_{n_k})) &\leq \alpha(x_{m_k},Tx_{m_k})\beta(x_{n_k},Tx_{n_k})\psi(d(Tx_{m_k},Tx_{n_k})) \\ &\leq \theta(\psi(M(x_{m_k},x_{n_k})))\psi(M(x_{m_k},x_{n_k})), \end{aligned} \quad (2.14)$$

where

$$\begin{aligned} M(x_{m_k},x_{n_k}) &= \max\Big\{ d(x_{m_k},x_{n_k}), d(x_{m_k},Tx_{m_k}), d(x_{n_k},Tx_{n_k}), \\ &\qquad \frac{d(x_{m_k},Tx_{m_k})d(x_{n_k},Tx_{n_k})}{1+d(x_{m_k},x_{n_k})}, \frac{d(x_{m_k},Tx_{m_k})d(x_{n_k},Tx_{n_k})}{1+d(Tx_{m_k},Tx_{n_k})}\Big\} \\ &= \max\Big\{ d(x_{m_k},x_{n_k}), d(x_{m_k},x_{m_k+1}), d(x_{n_k},x_{n_k+1}), \\ &\qquad \frac{d(x_{m_k},x_{m_k+1})d(x_{n_k},x_{n_k+1})}{1+d(x_{m_k},x_{n_k})}, \frac{d(x_{m_k},x_{m_k+1})d(x_{n_k},x_{n_k+1})}{1+d(x_{m_k+1},x_{n_k+1})}\Big\} \end{aligned} \quad (2.15)$$

Therefore, $\psi(d(x_{m_{k+1}},x_{n_{k+1}})) \leq \theta(\psi(M(x_{m_k},x_{n_k})))\psi(M(x_{m_k},x_{n_k}))$. On taking limit $k \to \infty$, we have $\psi(\delta) \leq \lim_{k\to\infty} \theta(\psi(M(x_{m_k},x_{n_k})))\psi(\delta)$, that is $1 \leq \lim_{k\to\infty} \theta(\psi(M(x_{m_k},x_{n_k})))$, which implies that $\lim_{k\to\infty} \theta(\psi(M(x_{m_k},x_{n_k}))) = 1$. Consequently, we obtain $\lim_{k\to\infty} M(x_{m_k},x_{n_k}) = 0$ and hence $\lim_{k\to\infty} d(x_{m_k+1},x_{n_k+1}) = 0$ which is a contradiction.

This shows that $\{x_n\}$ is a Cauchy sequence. Since X is complete, there exists $x^* \in X$ such that $x_n \to x^*$.

First, we suppose that T is continuous. Therefore, we have

$$x^* = \lim_{n\to\infty} x_{n+1} = \lim_{n\to\infty} Tx_n = T\lim_{n\to\infty} x_n = T\,x^*.$$

Now, we suppose that X is (α, β)-regular.

Therefore, there exists a subsequence $\{x_{n_k}\}$ of $\{x_n\}$ such that $\alpha(x_{n_k-1},x_{n_k}) \geq 1$ and $\beta(x_{n_k-1},x_{n_k}) \geq 1$ for all $k \in \mathbb{N}$ and $\alpha(x^*,Tx^*) \geq 1$ and $\beta(x^*,Tx^*) \geq 1$. Now, from inequality (2.1) with $x = x_{n_k}$ and $y = x^*$, we obtain

$$\begin{aligned} \psi(d(x_{n_k+1},Tx^*)) &= \psi(d(Tx_{n_k},Tx^*)) \\ &\leq \alpha(x_{n_k},Tx_{n_k})\beta(x^*,Tx^*)\psi(d(Tx_{n_k},Tx^*))) \\ &\leq \theta(\psi(M(x_{n_k},x^*)))\psi(M(x_{n_k},x^*)), \end{aligned} \quad (2.16)$$

where

$$\begin{aligned} M(x_{n_k},x^*) &= \max\Big\{ d(x_{n_k},x^*), d(x_{n_k},Tx_{n_k}), d(x^*,Tx^*), \\ &\qquad \frac{d(x_{n_k},Tx_{n_k})d(x^*,Tx^*)}{1+d(x_{n_k},x^*)}, \frac{d(x_{n_k},Tx_{n_k})d(x^*,Tx^*)}{1+d(Tx_{n_k},Tx^*)}\Big\} \\ &= \max\Big\{ d(x_{n_k},x^*), d(x_{n_k},x_{n_k+1}), d(x^*,Tx^*),. \\ &\qquad \frac{d(x_{n_k},x_{n_k+1})d(x^*,Tx^*)}{1+d(x_{n_k},x^*)}, \frac{d(x_{n_k},x_{n_k+1})d(x^*,Tx^*)}{1+d(x_{n_k+1},Tx^*)}\Big\}. \end{aligned} \quad (2.17)$$

Therefore,
$\psi(d(x_{n_k+1}, Tx^*)) \leq \theta(\psi(M(x_{n_k}, x^*)))\psi(M(x_{n_k}, x^*))$. On taking limit $k \to \infty$, we have $\psi(d(x^*, Tx^*)) \leq \lim_{k\to\infty} \theta(\psi(M(x_{n_k}, x^*)))\psi(d(x^*, Tx^*))$, that is $1 \leq \lim_{k\to\infty} \theta(\psi(M(x_{n_k}, x^*)))$, which implies that $\lim_{k\to\infty} \theta(\psi(M(x_{n_k}, x^*))) = 1$. Consequently, we obtain $\lim_{k\to\infty} M(x_{n_k}, x^*) = 0$ and hence $d(x^*, Tx^*) = 0$, that is, $x^* = Tx^*$.

Further, suppose that x^* and y^* are two fixed points of T such that $x^* \neq y^*$ and $\alpha(x^*, Tx^*) \geq 1$, $\alpha(y^*, Ty^*) \geq 1$ and $\beta(x^*, Tx^*) \geq 1$, $\beta(y^*, Ty^*) \geq 1$. Now applying (2.1), we have

$$\begin{aligned}\psi(d(x^*, y^*)) &= \psi(d(Tx^*, Ty^*))\\ &\leq \alpha(x^*, Tx^*)\beta(y^*, Ty^*)\psi(d(Tx^*, Ty^*))\\ &\leq \theta(\psi(M(x^*, y^*)))\psi(M(x^*, y^*))\end{aligned}$$

where

$$M(x^*, y^*) = \max\left\{d(x^*, y^*), d(x^*, Tx^*), d(y^*, Ty^*), \frac{d(x^*, Tx^*)d(y^*, Ty^*)}{1 + d(x^*, y^*)}, \frac{d(x^*, Tx^*)d(y^*, Ty^*)}{1 + d(Tx^*, Ty^*)}\right\}.$$

Hence,
$\psi(d(x^*, y^*)) \leq \theta(\psi(M(x^*, y^*)))\psi(d(x^*, y^*)) < \psi(d(x^*, y^*))$, which is a contradiction unless $d(x^*, y^*) = 0$, that is, $x^* = y^*$. Hence, T has a unique fixed point. □

Example 2.1 Let $X = [0, \infty)$ be endowed with the usual metric $d(x, y) = |x - y|$ for all $x, y \in X$ and $T : X \to X$ be defined by

$$Tx = \begin{cases} \dfrac{1 - x^2}{8}, & x \in [0, 1] \\ 9x, & x \in (1, \infty) \end{cases} \tag{2.18}$$

Define also $\alpha, \beta : X \times X \to \mathbb{R}^+$, $\theta : [0, \infty) \to [0, 1)$ and $\psi : [0, \infty) \to [0, \infty)$ as

$$\alpha(x, y) = \begin{cases} 1, & (x, y) \in [0, 1] \\ 0, & \text{otherwise} \end{cases}$$

$$\beta(x, y) = \begin{cases} 1, & (x, y) \in [0, 1] \\ 0, & \text{otherwise} \end{cases}$$

$\theta(t) = \frac{1}{2}$ and $\psi(t) = t$.

Now, we prove that Theorem 2.1 can be applied to T (here, a fixed point is $u = \sqrt{17} - 4$), but Theorem 1.1 cannot be applied to T.

Clearly, (X, d) is a complete metric space. We show that T is an (α, β)-admissible mapping. Let $x, y \in X$, if $\alpha(x, y) \geq 1$ and $\beta(x, y) \geq 1$, then $x, y \in [0, 1]$. On the other hand, for all $x \in [0, 1]$, we have $Tx \leq 1$. It follows that $\alpha(Tx, Ty) \geq 1$ and $\beta(Tx, Ty) \geq 1$. Thus, the assertion holds. In reason of the above arguments, $\alpha(0, T0) \geq 1$ and $\beta(0, T0) \geq 1$.

Now, if $\{x_n\}$ is a sequence in X such that $\alpha(x_n, x_{n+1}) \geq 1$, $\beta(x_n, x_{n+1}) \geq 1$ and $x_n \to x \in X$, for all $n \in \mathbb{N} \cup \{0\}$, then $x_n \subseteq [0, 1]$ and hence $x \in [0, 1]$. This implies that $\alpha(x, Tx) \geq 1$, and $\beta(x, Tx) \geq 1$.

Let $x, y \in [0, 1]$. We get

$$\begin{aligned}\alpha(x, Tx)\beta(y, Ty)\psi(d(Tx, Ty)) = |Tx - Ty| &= \frac{1}{8}|x - y||\\ \times x + y| &\leq \frac{1}{2}|x - y|\\ &= \theta(\psi(d(x, y)))\psi(d(x, y)).\end{aligned}$$

Hence, the given inequality is satisfied.

Otherwise, if $\alpha(x, Tx)\beta(y, Ty) = 0$. Then, $\alpha(x, Tx)\beta(y, Ty)\psi(d(Tx, Ty)) = 0 \leq \theta(\psi(M(x, y)))\psi(M(x, y))$. Therefore, all the conditions of Theorem 2.1 are satisfied and T has a fixed point.

Now, let $x = 3$, $y = 4$. We get

$$d(T3, T4) = 9 > \frac{1}{2} = \frac{1}{2}|3 - 4| = \theta(d(3, 4))d(3, 4).$$

Therefore, Theorem 1.1 does not hold for this example.

Definition 2.2 Let (X, d) be a metric space, $T : X \to X$ and $\alpha, \beta : X \times X \to \mathbb{R}^+$. A mapping T is said to be α, β-Geraghty type-II rational contractive mapping if there exists a $\theta \in \Theta$, such that for all $x, y \in X$, following condition holds:

$$[\psi(d(Tx, Ty)) + l]^{\alpha(x,Tx)\beta(y,Ty)} \leq \theta(\psi(M(x, y)))\psi(M(x, y)) + l, \tag{2.19}$$

where $M(x, y) = \max\{d(x, y), d(x, Tx), d(y, Ty), \frac{d(x,Tx)}{d(y, Ty)1 + d(x, y)}, \frac{d(x,Tx)d(y,Ty)}{1+d(Tx,Ty)}\}$, $\psi \in \Psi$ and $l \geq 1$.

Theorem 2.2 *Let (X, d) be a complete metric space, T is self-mapping, $T : X \to X$, and $\alpha, \beta : X \times X \to \mathbb{R}$. Suppose that the following conditions are satisfied:*

(i) *T is an (α, β)-admissible mapping;*
(ii) *T is an (α, β)-Geraghty type-II rational contractive mapping;*
(iii) *there exists $x_0 \in X$ such that $\alpha(x_0, Tx_0) \geq 1$ and $\beta(x_0, Tx_0) \geq 1$;*
(iv) *either T is continuous or X is (α, β)-regular.*

Then, T has a fixed point $x^* \in X$ and $\{T^n x_0\}$ converges to x^*.

Further, if for all $x, y \in F(T)$, with $x \neq y$ such that $\alpha(x, Tx) \geq 1$, $\alpha(y, Ty) \geq 1$ and $\beta(x, Tx) \geq 1$, $\beta(y, Ty) \geq 1$, then T has a unique fixed point in X.

Proof Let $x_0 \in X$ such that $\alpha(x_0, Tx_0) \geq 1$ and $\beta(x_0, Tx_0) \geq 1$. Now, we can construct the sequences $\{x_n\}$ in X by $x_n = T^n x_0 = Tx_{n-1}$, for $n \in \mathbb{N}$.

Moreover, we assume that if $x_{n_0} = x_{n_0+1}$, for some $n_0 \in \mathbb{N}$, then x_{n_0} is a fixed point of T. Consequently, we suppose that $x_n \neq x_{n+1}$ for all $n \in \mathbb{N}$.

Since T is (α, β)-admissible mapping, $\alpha(x_0, Tx_0) = \alpha(x_0, x_1) \geq 1$, $\alpha(Tx_0, Tx_1) = \alpha(x_1, x_2) \geq 1$, $\alpha(Tx_1, Tx_2) = \alpha(x_2, x_3) \geq 1$. Hence, by induction, we get $\alpha(x_n, x_{n+1}) \geq 1$ for all $n \geq 0$.

Similarly, $\beta(x_n, x_{n+1}) \geq 1$ for all $n \geq 0$.

Consider (2.19), we have

$$\begin{aligned}\psi(d(x_{n+1}, x_{n+2})) + l &= \psi(d(Tx_n, Tx_{n+1})) + l\\ &\leq [\psi(d(Tx_n, Tx_{n+1})) + l]^{\alpha(x_n, Tx_n)\beta(x_{n+1}, Tx_{n+1})}\\ &\leq \theta(\psi(M(x_n, x_{n+1})))\psi(M(x_n, x_{n+1})) + l,\end{aligned}$$

where

$$\begin{aligned}M(x_n, x_{n+1}) &= \max\left\{d(x_n, x_{n+1}), d(x_n, Tx_n), d(x_{n+1}, Tx_{n+1}),\right.\\ &\left.\frac{d(x_n, Tx_n)d(x_{n+1}, Tx_{n+1})}{1 + d(x_n, x_{n+1})}, \frac{d(x_n, Tx_n)d(x_{n+1}, Tx_{n+1})}{1 + d(Tx_n, Tx_{n+1})}\right\}\\ &= \max\left\{d(x_n, x_{n+1}), d(x_n, x_{n+1}), d(x_{n+1}, x_{n+2}),\right.\\ &\left.\frac{d(x_n, x_{n+1})d(x_{n+1}, x_{n+2})}{1 + d(x_n, x_{n+1})}, \frac{d(x_n, x_{n+1})d(x_{n+1}, x_{n+2})}{1 + d(x_{n+1}, x_{n+2})}\right\}\\ &= \max\{d(x_n, x_{n+1}), d(x_{n+1}, x_{n+2})\}.\end{aligned} \tag{2.20}$$

Now, if $M(x_n, x_{n+1}) = d(x_{n+1}, x_{n+2})$, then

$$\begin{aligned}\psi(d(x_{n+1}, x_{n+2})) &\leq \theta(\psi(M(x_n, x_{n+1})))\psi(M(x_n, x_{n+1}))\\ &\leq \theta(\psi(M(x_n, x_{n+1})))\psi(d(x_{n+1}, x_{n+2}))\\ &< \psi(d(x_{n+1}, x_{n+2})),\end{aligned}$$

which is a contradiction, using the properties of ψ.

Therefore, it implies that $M(x_n, x_{n+1}) = d(x_n, x_{n+1})$ and

$$\begin{aligned}\psi(d(x_{n+1}, x_{n+2})) &\leq \theta(\psi(M(x_n, x_{n+1})))\psi(M(x_n, x_{n+1}))\\ &\leq \theta(\psi(M(x_n, x_{n+1})))\psi(d(x_n, x_{n+1}))\\ &\leq \psi(d(x_n, x_{n+1})).\end{aligned} \tag{2.21}$$

Hence, using the properties of ψ, we conclude that

$$d(x_{n+1}, x_{n+2}) \leq d(x_n, x_{n+1}), \tag{2.22}$$

for every $n \in \mathbb{N}$. Therefore, sequence $\{d(x_n, x_{n+1})\}$ is decreasing. On the similar lines as in Theorem 2.1, we can prove that

$$r = \lim_{n\to\infty} d(x_n, x_{n+1}) = 0. \tag{2.23}$$

Now, we will show that $\{x_n\}$ is a Cauchy sequence. Suppose, to the contrary, that $\{x_n\}$ is not a Cauchy sequence.

Then, there exists $\delta > 0$ for which we can find subsequences $\{x_{n_k}\}$ and $\{x_{m_k}\}$ of $\{x_n\}$ with $n_k > m_k > k$ such that

$$d(x_{n_k}, x_{m_k}) \geq \delta. \tag{2.24}$$

Further, corresponding to m_k, we can choose n_k in such a way that it is the smallest integer with $n_k > m_k$ and satisfying (2.24), we have

$$d(x_{n_k-1}, x_{m_k}) < \delta. \tag{2.25}$$

Using triangle inequality, we have

$$\begin{aligned}0 < \delta = d(x_{n_k}, x_{m_k}) &\leq d(x_{n_k}, x_{n_k-1}) + d(x_{n_k-1}, x_{m_k}) < \delta\\ &+ d(x_{n_k}, x_{n_k-1})\end{aligned} \tag{2.26}$$

Letting $k \to \infty$ and using (2.23) and (2.24), we obtain

$$\lim_{k\to\infty} d(x_{n_k}, x_{m_k}) = \delta. \tag{2.27}$$

Again, using triangle inequality, we have

$$d(x_{n_k}, x_{m_k}) \leq d(x_{n_k}, x_{n_k-1}) + d(x_{n_k-1}, x_{m_k-1}) + d(x_{m_k-1}, x_{m_k}),$$

and

$$d(x_{n_k-1}, x_{m_k-1}) \leq d(x_{n_k}, x_{n_k-1}) + d(x_{n_k}, x_{m_k}) + d(x_{m_k-1}, x_{m_k}).$$

Therefore,

$$\begin{aligned}d(x_{n_k}, x_{m_k}) &\leq d(x_{n_k}, x_{n_k-1}) + d(x_{n_k-1}, x_{m_k-1}) + d(x_{m_k-1}, x_{m_k})\\ &\leq 2d(x_{n_k}, x_{n_k-1}) + d(x_{n_k}, x_{m_k}) + 2d(x_{m_k-1}, x_{m_k})\end{aligned} \tag{2.28}$$

Letting $k \to \infty$ in (2.28) and using (2.23), (2.27), we get

$$\lim_{k\to\infty} d(x_{n_k-1}, x_{m_k-1}) = \delta. \tag{2.29}$$

Put $x = x_{m_k}$ and $y = x_{n_k}$ in (2.19), we obtain

$$\begin{aligned}\psi(d(Tx_{m_k}, Tx_{n_k})) + l &\leq [\psi(d(Tx_{m_k}, Tx_{n_k})) + l]^{\alpha(x_{m_k}, Tx_{m_k})\beta(x_{n_k}, Tx_{n_k})}\\ &\leq \theta(\psi(M(x_{m_k}, x_{n_k})))\psi(M(x_{m_k}, x_{n_k})) + l,\end{aligned} \tag{2.30}$$

where

$$\begin{aligned}M(x_{m_k}, x_{n_k}) &= \max\left\{d(x_{m_k}, x_{n_k}), d(x_{m_k}, Tx_{m_k}), d(x_{n_k}, Tx_{n_k}),\right.\\ &\left.\frac{d(x_{m_k}, Tx_{m_k})d(x_{n_k}, Tx_{n_k})}{1 + d(x_{m_k}, x_{n_k})}, \frac{d(x_{m_k}, Tx_{m_k})d(x_{n_k}, Tx_{n_k})}{1 + d(Tx_{m_k}, Tx_{n_k})}\right\}\\ &= \max\left\{d(x_{m_k}, x_{n_k}), d(x_{m_k}, x_{m_k+1}), d(x_{n_k}, x_{n_k+1}),\right.\\ &\left.\frac{d(x_{m_k}, x_{m_k+1})d(x_{n_k}, x_{n_k+1})}{1 + d(x_{m_k}, x_{n_k})}, \frac{d(x_{m_k}, x_{m_k+1})d(x_{n_k}, x_{n_k+1})}{1 + d(x_{m_k+1}, x_{n_k+1})}\right\}.\end{aligned} \tag{2.31}$$

Therefore, $\psi(d(x_{m_{k+1}}, x_{n_{k+1}})) \leq \theta(\psi(M(x_{m_k}, x_{n_k})))\psi(M(x_{m_k}, x_{n_k}))$. On taking limit $k \to \infty$, we have $\psi(\delta) \leq \lim_{k\to\infty} \theta(\psi(M(x_{m_k}, x_{n_k})))\psi(\delta)$, that is $1 \leq \lim_{k\to\infty} \theta(\psi(M(x_{m_k}, x_{n_k})))$, which implies that $\lim_{k\to\infty} \theta(\psi(M(x_{m_k}, x_{n_k}))) = 1$. Consequently, we obtain $\lim_{k\to\infty} M(x_{m_k}, x_{n_k}) = 0$ and hence $\lim_{k\to\infty} d(x_{m_k+1}, x_{n_k+1}) = 0$ which is a contradiction.

This shows that $\{x_n\}$ is a Cauchy sequence. Since X is complete, there exists $x^* \in X$ such that $x_n \to x^*$.

First, we suppose that T is continuous. Therefore, we have

$$x^* = \lim_{n\to\infty} x_{n+1} = \lim_{n\to\infty} Tx_n = T \lim_{n\to\infty} x_n = T\, x^*.$$

Now, we suppose that X is α, β-regular.

Therefore, there exists a subsequence $\{x_{n_k}\}$ of $\{x_n\}$ such that $\alpha(x_{n_k-1}, x_{n_k}) \geq 1$ and $\beta(x_{n_k-1}, x_{n_k}) \geq 1$ for all $k \in \mathbb{N}$ and $\alpha(x^*, Tx^*) \geq 1$ and $\beta(x^*, Tx^*) \geq 1$. Now, from inequality (2.19) with $x = x_{n_k}$ and $y = x^*$, we obtain

$$\begin{aligned}\psi(d(x_{n_k+1}, Tx^*)) + l &= \psi(d(Tx_{n_k}, Tx^*)) + l \\ &\leq [\psi(d(Tx_{n_k}, Tx^*))) + l]^{\alpha(x_{n_k}, Tx_{n_k})\beta(x^*, Tx^*)} \\ &\leq \theta(\psi(M(x_{n_k}, x^*)))\psi(M(x_{n_k}, x^*)) + l,\end{aligned} \tag{2.32}$$

where

$$\begin{aligned}M(x_{n_k}, x^*) &= \max\left\{d(x_{n_k}, x^*), d(x_{n_k}, Tx_{n_k}), d(x^*, Tx^*), \frac{d(x_{n_k}, Tx_{n_k})d(x^*, Tx^*)}{1 + d(x_{n_k}, x^*)}, \frac{d(x_{n_k}, Tx_{n_k})d(x^*, Tx^*)}{1 + d(Tx_{n_k}, Tx^*)}\right\} \\ &= \max\left\{d(x_{n_k}, x^*), d(x_{n_k}, x_{n_k+1}), d(x^*, Tx^*), \frac{d(x_{n_k}, x_{n_k+1})d(x^*, Tx^*)}{1 + d(x_{n_k}, x^*)}, \frac{d(x_{n_k}, x_{n_k+1})d(x^*, Tx^*)}{1 + d(x_{n_k+1}, Tx^*)}\right\}.\end{aligned} \tag{2.33}$$

Therefore, $\psi(d(x_{n_k+1}, Tx^*)) \leq \theta(\psi(M(x_{n_k}, x^*)))\psi(M(x_{n_k}, x^*))$. On taking limit $k \to \infty$, we have $\psi(d(x^*, Tx^*)) \leq \lim_{k\to\infty} \theta(\psi(M(x_{n_k}, x^*)))\psi(d(x^*, Tx^*))$, that is $1 \leq \lim_{k\to\infty} \theta(\psi(M(x_{n_k}, x^*)))$, which implies that $\lim_{k\to\infty} \theta(\psi(M(x_{n_k}, x^*))) = 1$. Consequently, we obtain $\lim_{k\to\infty} M(x_{n_k}, x^*) = 0$ and hence $d(x^*, Tx^*) = 0$, that is, $x^* = Tx^*$.

Further, suppose that x^* and y^* are two fixed points of T such that $x^* \neq y^*$ and $\alpha(x^*, Tx^*) \geq 1$, $\alpha(y^*, Ty^*) \geq 1$ and $\beta(x^*, Tx^*) \geq 1$, $\beta(y^*, Ty^*) \geq 1$. Hence from (2.19), we have

$$\begin{aligned}\psi(d(x^*, y^*)) + l &= \psi(d(Tx^*, Ty^*)) + l \\ &\leq [\psi(d(Tx^*, Ty^*)) + l]^{\alpha(x^*, Tx^*)\beta(y^*, Ty^*)} \\ &\leq \theta(\psi(M(x^*, y^*)))\psi(M(x^*, y^*)) + l\end{aligned}$$

where

$$M(x^*, y^*) = \max\left\{d(x^*, y^*), d(x^*, Tx^*), d(y^*, Ty^*), \frac{d(x^*, Tx^*)d(y^*, Ty^*)}{1 + d(x^*, y^*)}, \frac{d(x^*, Tx^*)d(y^*, Ty^*)}{1 + d(Tx^*, Ty^*)}\right\}.$$

Hence, $\psi(d(x^*, y^*)) \leq \theta(\psi(M(x^*, y^*)))\psi(d(x^*, y^*))$, which implies that $\theta(\psi(M(x^*, y^*))) = 1$. Therefore, $d(x^*, y^*) = 0$, that is, $x^* = y^*$. Hence, T has a unique fixed point. □

Example 2.2 Let $X = [0, \infty)$ be endowed with the usual metric $d(x, y) = |x - y|$ for all $x, y \in X$ and $T : X \to X$ be defined by

$$Tx = \begin{cases} \dfrac{x}{x+1}, & x \in [0, 1] \\ 3x, & x \in (1, \infty) \end{cases} \tag{2.34}$$

Define also $\alpha, \beta : X \times X \to \mathbb{R}^+$, $\theta : [0, \infty) \to [0, 1)$ and $\psi : [0, \infty) \to [0, \infty)$ as

$$\alpha(x, y) = \begin{cases} 1, & (x, y) \in [0, 1] \\ 0, & \text{otherwise} \end{cases}$$

$$\beta(x, y) = \begin{cases} 1, & (x, y) \in [0, 1] \\ 0, & \text{otherwise} \end{cases}$$

$\theta(t) = \frac{1}{1+t}$ and $\psi(t) = t$.

Now, we prove that Theorem 2.2 can be applied to T (here, a fixed point is $u = 0$), but Theorem 1.1 cannot be applied to T.

Clearly, (X, d) is a complete metric space. On the same lines of Example 2.1, we can show that T is an (α, β)-admissible mapping.

Let $x, y \in [0, 1]$. We get

$$\begin{aligned}[\psi(d(Tx, Ty)) + l]^{\alpha(x,Tx)\beta(y,Ty)} &= \psi(d(Tx, Ty)) + l \\ &= |Tx - Ty| + l = \left|\frac{x}{x+1} - \frac{y}{y+1}\right| + l \\ &= \left|\frac{y-x}{(1+x)(1+y)}\right| + l \\ &\leq \left|\frac{y-x}{1+y-x}\right| + l \\ &= \theta(\psi(d(x, y)))\psi(d(x, y)) + l.\end{aligned}$$

Hence, the given inequality is satisfied.

Otherwise, $\alpha(x, Tx)\beta(y, Ty) = 0$ and $[\psi(d(Tx, Ty)) + l]^{\alpha(x,Tx)\beta(y,Ty)} = 1 \leq \theta(\psi(M(x, y)))\psi(M(x, y)) + l$. Therefore, all the conditions of Theorem 2.2 are satisfied and T has a fixed point.

Now, let $x = 3$, $y = 4$. We get

$$d(T3, T4) = 3 > \frac{1}{2} = \frac{1}{1 + |3-4|}|3 - 4| = \theta(d(3, 4))d(3, 4).$$

Therefore, Theorem 1.1 does not hold for this example.

Consequences of the main results

In this section, we discuss some consequences of our main results. First, we prove some fixed point theorems for cyclic mappings in metric and ordered metric spaces. Also, we obtain some other related interesting results in the next section. In the last section, we obtain a fixed point theorem whenever the space is endowed with a graph.

Fixed point results for cyclic mappings in metric and ordered metric spaces

Theorem 3.1 *Let (X, d) be a complete metric space, A and B be two nonempty closed subsets of X . Suppose that $\alpha : X \times X \to [0, \infty)$, and $T : A \cup B \to A \cup B$ be a mapping with $TA \subseteq B$ and $TB \subseteq A$, such that $\alpha(Ty, Tx) \geq 1$ if $\alpha(x, y) \geq 1$, where $x \in A$ and $y \in B$. Further assume that T satisfies any one of the following contractive condition for all $x \in A$ and $y \in B$*

$$\alpha(x, y)\psi(d(Tx, Ty)) \leq \theta(\psi(M(x, y)))\psi(M(x, y));$$

or

$$[\psi(d(Tx, Ty)) + l]^{\alpha(x,y)} \leq \theta(\psi(M(x, y)))\psi(M(x, y)) + l,$$

where $M(x, y) = \max\{d(x, y), d(x, Tx), d(y, Ty), \frac{d(x,Tx)}{d(y, Ty)1 + d(x, y)}, \frac{d(x,Tx)d(y,Ty)}{1+d(Tx,Ty)}\}$, $\psi \in \Psi$ *and* $l \geq 1$.

If there exists $x_0 \in A$ such that $\alpha(x_0, Tx_0) \geq 1$, and either T is continuous or X is α-regular, then T has a fixed point in $A \cap B$.

Proof Let $Y = A \cup B$ and $\beta : Y \times Y \to [0, \infty)$ defined as

$$\beta(x, y) = \begin{cases} 1, & \text{if } x \in A, \quad y \in B, \\ 0, & \text{otherwise} \end{cases}$$

Then, (Y, d) is complete metric space. Now, if $x_0 \in A$ is such that $\alpha(x_0, Tx_0) \geq 1$, then also $\beta(x_0, Tx_0) \geq 1$ and hence all the hypotheses of above Theorems 2.1–2.2 hold with $X = Y$. Consequently, T has a fixed point in $A \cup B$, say z. Since $z \in A$ implies $z = Tz \in B$ and $z \in B$ implies $z = Tz \in A$, then $z \in A \cap B$.

Now, assume that $\{x_n\}$ be a sequence in Y such that $\alpha(x_{2n}, x_{2n+1}) \geq 1$ and $\beta(x_2n, x_{2n+1}) \geq 1$ for all $n \in \mathbb{N} \cup \{0\}$ and $x_n \to x$ as $n \to \infty$, then $x_{2n} \in A$ and $x_{2n+1} \in B$. Since B is closed, then $x \in B$ and hence $\alpha(x_{2n}, x) \geq 1$ and $\beta(x_{2n}, x) \geq 1$. We deduce that all the hypotheses of above Theorems 2.1–2.2 are satisfied with $X = Y$ and hence T has a fixed point. □

Theorem 3.2 *Let $(X, d, \leq)$ be an ordered complete metric space, A and B be two nonempty closed subsets of X. Suppose that $\alpha : X \times X \to [0, \infty)$, and $T : A \cup B \to A \cup B$ be a mapping with $TA \subseteq B$ and $TB \subseteq A$, such that $\alpha(Ty, Tx) \geq 1$ if $\alpha(x, y) \geq 1$, where $x \in A$ and $y \in B$. Further assume that T satisfies any one of the following contractive condition for all $x \in A$ and $y \in B$ with $x \leq y$*

$$\alpha(x, y)\psi(d(Tx, Ty)) \leq \theta(\psi(M(x, y)))\psi(M(x, y));$$

or

$$[\psi(d(Tx, Ty)) + l]^{\alpha(x,y)} \leq \theta(\psi(M(x, y)))\psi(M(x, y)) + l,$$

where $M(x, y) = \max\{d(x, y), d(x, Tx), d(y, Ty), \frac{d(x,Tx)d(y,Ty)}{1 + d(x, y)}, \frac{d(x,Tx)d(y,Ty)}{1+d(Tx,Ty)}\}$, $\psi \in \Psi$ *and* $l \geq 1$.

If there exists $x_0 \in A$ such that $\alpha(x_0, Tx_0) \geq 1$ and $x_0 \leq Tx_0$, and either T is continuous and decreasing or X is an ordered α-regular, then T has a fixed point in $A \cup B$.

Proof Consider the complete metric space (Y, d), where $Y = A \cup B$ and define $\beta : Y \times Y \to [0, \infty)$ defined as

$$\beta(x, y) = \begin{cases} 1, & \text{if } x \in A, \quad y \in B, \text{ with } x \leq y \\ 0, & \text{otherwise} \end{cases}$$

Let $\beta(x, y) \geq 1$ for $x, y \in X$, then $x \in A$ and $y \in B$ with $x \leq y$. It follows that $Tx \in B$ and $Ty \in A$ with $Ty \leq Tx$, since T is decreasing. Therefore, $\beta(Ty, Tx) \geq 1$, that is, T is an (α, β)-admissible mapping. Now, let $\alpha(x_0, Tx_0) \geq 1$ with $x_0 \in A$ and $x_0 \leq Tx_0$. From $x_0 \in A$ we have $Tx_0 \in B$ with $x_0 \leq Tx_0$, that is, $\beta(x_0, Tx_0) \geq 1$. Hence, all the hypotheses of above Theorems 2.1–2.2 hold with $X = Y$. Consequently, T has a fixed point in $A \cup B$, say z. Since $z \in A$ implies $z = Tz \in B$ and $z \in B$ implies $z = Tz \in A$, then $z \in A \cap B$.

Now assume that $\{x_n\}$ be a sequence in Y such that $\alpha(x_{2n}, x_{2n+1}) \geq 1$ and $\beta(x_2n, x_{2n+1}) \geq 1$ for all $n \in \mathbb{N} \cup \{0\}$ and $x_n \to x$ as $n \to \infty$, then $x_{2n} \in A$ and $x_{2n+1} \in B$. Since B is closed, and X is an ordered α-regular, we have $x \in B$ and $x_{2n} \leq x$, hence $\alpha(x_{2n}, x) \geq 1$ and $\beta(x_{2n}, x) \geq 1$. We deduce that all the hypotheses of above Theorems 2.1–2.2 are satisfied with $X = Y$ and hence T has a fixed point. □

Related results

Theorem 3.3 *Let (X, d) be a complete metric space, T is self-mapping, $T : X \to X$, and $\alpha : X \times X \to \mathbb{R}$. Suppose that the following conditions are satisfied:*

(i) *T is an (α)-admissible mapping;*
(ii) *T satisfies any one of following rational contractive mapping;*

$$[\psi(d(Tx, Ty)) + l]^{\alpha(x,Tx)\alpha(y,Ty)} \leq \theta(\psi(M(x, y)))\psi(M(x, y)) + l;$$

or

$$\alpha(x, Tx)\alpha(y, Ty)\psi(d(Tx, Ty)) \leq \theta(\psi(M(x, y)))\psi(M(x, y)),$$

where $M(x, y) = \max\{d(x, y), d(x, Tx), d(y, Ty), \frac{d(x,Tx)}{d(y, Ty)1 + d(x, y)}, \frac{d(x,Tx)d(y,Ty)}{1+d(Tx,Ty)}\}$, $\psi \in \Psi$ *and* $l \geq 1$.

(iii) *there exists $x_0 \in X$ such that $\alpha(x_0, Tx_0) \geq 1$ and $\beta(x_0, Tx_0) \geq 1$;*

(iv) *either T is continuous or X is α-regular. Then, T has a fixed point* $x^* \in X$ *and* $\{T^n x_0\}$ *converges to* x^*. *Further, if For all* $x, y \in F(T)$, *with* $x \neq y$ *such that* $\alpha(x, Tx) \geq 1$, *and* $\alpha(y, Ty) \geq 1$, *then T has a unique fixed point in X.*

Corollary 3.1 [11] *Let (X, d) be a complete metric space and* $T : X \to X$ *be an α-admissible mapping. Assume that there exists a function* $\theta \in \Theta$, *such that*

$$(d(Tx, Ty)+l)^{\alpha(x,Tx)\alpha(y,Ty)} \leq \theta(d(x, y))d(x, y) + l$$

for all $x,\ y\ \in X, l \geq 1$. *Suppose that either*

(a) *T is continuous or*
(b) *if* $\{x_n\}$ *is a sequence in X such that* $x_n \to x$, $\alpha(x_n,\ x_{n+1}) \geq\ 1$ *for all n, then* $\alpha(x,\ Tx) \geq 1$.

If there exists $x_0 \in X$ *such that* $\alpha(x_0,\ Tx_0)\ \geq 1$, *then T has a fixed point.*

Corollary 3.2 [11] *Let (X, d) be a complete metric space and* $T : X \to X$ *be an α-admissible mapping. Assume that there exists a function* $\theta \in \Theta$, *such that*

$$\alpha(x, Tx)\alpha(y, Ty)d(Tx, Ty) \leq \theta(d(x, y))d(x, y)$$

for all $x,\ y\ \in X$. *Suppose that either*

(a) *T is continuous or*
(b) *if* $\{x_n\}$ *is a sequence in X such that* $x_n \to x$, $\alpha(x_n,\ x_{n+1}) \geq\ 1$ *for all n, then* $\alpha(x,\ Tx) \geq 1$.

If there exists $x_0 \in X$ *such that* $\alpha(x_0,\ Tx_0)\ \geq 1$, *then T has a fixed point.*

Theorem 3.4 *Let (X, d) be a complete metric space, T is self-mapping,* $T : X \to X$, *and* $\alpha : X \times X \to \mathbb{R}^+$. *Suppose that the following conditions are satisfied:*

(i) *T is an α-admissible mapping;*
(ii) *T satisfies the following contractive condition*

$$\alpha(x, y)\psi(d(Tx, Ty)) \leq \theta(\psi(M(x, y)))\psi(M(x, y));$$

or

$$[\psi(d(Tx, Ty)) + l]^{\alpha(x,y)} \leq \theta(\psi(M(x, y)))\psi(M(x, y)) + l,$$

where $M(x, y) = \max\{d(x, y), d(x, Tx), d(y, Ty), \frac{d(x,Tx)d(y,Ty)}{1+d(x,y)}, \frac{d(x,Tx)d(y,Ty)}{1+d(Tx,Ty)}\}$, $\psi \in \Psi$ *and* $l \geq 1$.

(iii) *there exists* $x_0 \in X$ *such that* $\alpha(x_0,\ Tx_0)\ \geq 1$;
(iv) *either T is continuous or X is α-regular.*

Then, T has a fixed point $x \in X$ *and* $\{T^n x_0\}$ *converges to x.*

Further, if for all $x, y \in F(T)$, *with* $x \neq y$ *such that* $\alpha(x, Tx) \geq 1$, $\alpha(y, Ty) \geq 1$, *then T has a unique fixed point in X.*

Proof Define mapping $\beta : X \times X \to [0, \infty)$ defined as

$$\beta(x, y) = \begin{cases} 1, & \text{if} \quad x, y \in X, \\ 0, & \text{otherwise.} \end{cases}$$

Now using the above Theorems 2.1–2.2, we get the result. □

Fixed point results for graphic contractions

Consistent with Jachymski [12], let (X, d) be a metric space and Δ denotes the diagonal of the Cartesian product $X \times X$. Consider a directed graph G such that the set $V(G)$ of its vertices coincides with X, and the set $E(G)$ of its edges contains all loops, i.e., $E(G) \supseteq \Delta$. We assume G has no parallel edges, so we can identify G with the pair ($V(G)$, $E(G)$). Moreover, we may treat G as a weighted graph (see [12]) by assigning to each edge the distance between its vertices. If x and y are vertices in a graph G, then a path in G from x to y of length m ($m \in \mathbb{N}$) is a sequence $\{x_i\}_{i=0}^m$ of $m + 1$ vertices such that $x_0 = x, x_m = y$ and $(x_{i-1}, x_i) \in E(G)$ for $i = 1, \ldots, m$. A graph G is connected if there is a path between any two vertices. G is weakly connected if $\tilde{G}$ is connected (see for details [3, 12]). Recently, some results have appeared providing sufficient conditions for a mapping to be a Picard operator if (X, d) is endowed with a graph. The first result in this direction was given by Jachymski [12].

Definition 3.1 [12] Let (X, d) be a metric space endowed with a graph G. We say that a self-mapping $T : X \to X$ is a Banach G-contraction or simply a G-contraction if T preserves the edges of G, that is, $(x, y) \in E(G) \Longrightarrow (Tx, Ty) \in E(G)$ for all $x, y \in X$ and T decreases the weights of the edges of G in the following way:

$\exists \alpha \in (0, 1)$ such that for all $x, y \in X$, $(x, y) \in E(G) \Longrightarrow d(Tx, Ty) \leq \alpha d(x, y)$.

Theorem 3.5 *Let (X, d) be a complete metric space endowed with a graph G, T is a self-mapping,* $T : X \to X$, *satisfying*

$$\psi(d(Tx, Ty)) \leq \theta(\psi(M(x, y)))\psi(M(x, y));$$

or

$$\psi(d(Tx, Ty)) + l \leq \theta(\psi(M(x, y)))\psi(M(x, y)) + l,$$

where $M(x, y) = \max\{d(x, y), d(x, Tx), d(y, Ty), \frac{d(x,Tx)}{}d(y, Ty)1 + d(x, y), \frac{d(x,Tx)d(y,Ty)}{1+d(Tx,Ty)}\}$, $\psi \in \Psi$ *and* $l \geq 1$.

Suppose that the following assertions hold:

(i) *for all* $x, y \in X$, $(x, y) \in E(G)$, *implies* $(Tx, Ty) \in E(G)$;
(ii) $\{x_n\}$ *is a sequence in X such that* $x_n \to x$ *as* $n \to \infty$ *and* $(x_n, Tx_n) \in E(G)$, *then* $(x, Tx) \in E(G)$;
(iii) *there exists* $x_0 \in X$ *such that* $(x_0,\ Tx_0)\ \in E(G)$.

Then, T has a fixed point $x \in X$.

Proof Define mapping $\alpha : X \times X \to [0, \infty)$ defined as

$$\alpha(x,y) = \begin{cases} 1, & \text{if } (x,y) \in G, \\ 0, & \text{otherwise}. \end{cases}$$

Now, we show that T is an α-admissible mapping. Suppose that $\alpha(x,y) \geq 1$. Therefore, we have $(x,y) \in E(G)$. From (i), we get $(Tx, Ty) \in E(G)$. So, $\alpha(Tx, Ty) \geq 1$ and T is an α-admissible mapping. Hence, from the definition of α and inequality, we have

$$\alpha(x,y)\psi(d(Tx,Ty)) \leq \theta(\psi(M(x,y)))\psi(M(x,y));$$

or

$$[\psi(d(Tx,Ty)) + l]^{\alpha(x,y)} \leq \theta(\psi(M(x,y)))\psi(M(x,y)) + l,$$

where $M(x,y) = \max\{d(x,y), d(x,Tx), d(y,Ty), \frac{d(x,Tx)d(y,Ty)}{1+d(x,y)}, \frac{d(x,Tx)d(y,Ty)}{1+d(Tx,Ty)}\}$, $\psi \in \Psi$ and $l \geq 1$.

From (iii) there exists $x_0 \in X$ such that $(x_0, Tx_0) \in E(G)$, $\alpha(x_0, Tx_0) \geq 1$. Let $\{x_n\}$ is a sequence in X such that $x_n \to x$ as $n \to \infty$ and $(x_n, Tx_n) \in E(G)$ for all $n \in \mathbb{N}$, then $\alpha(x_n, Tx_n) \geq 1$. Thus, from (ii) we get, $(x, Tx) \in E(G)$. That is, $\alpha(x, Tx) \geq 1$. Therefore, all conditions of Theorem 3.4 hold true and T has a fixed point. □

Remark 3.1 Using the technique of Samet et al. [15], we can obtain corresponding coupled fixed point results from our Theorems 2.1, 2.2.

Acknowledgments The author is thankful to the learned referees for very careful reading and valuable suggestions.

References

1. Banach, S.: Sur les opéerations dans les ensembles abstraits et leur application aux équations integrales. Fund. Math. **3**, 133–181 (1922)
2. Berzig, M., Rus, M.-D.: Fixed point theorems for α-contractive mappings of Meir–Keeler type and applications. Nonlinear Anal. Model. Control. **19**(2), 178–198 (2014)
3. Bojor, F.: Fixed point theorems for Reich type contraction on metric spaces with a graph. Nonlinear Anal. **75**, 3895–3901 (2012)
4. Chandok, S., Choudhury, B.S., Metiya, N.: Some fixed point results in ordered metric spaces for rational type expressions with auxiliary functions. J. Egypt. Math. Soc. **23**(1), 95–101 (2015). doi:10.1016/j.joems.2014.02.002
5. Chandok, S., Khan, M.S., Abbas, M.: Common fixed point theorems for nonlinear weakly contractive mappings. Ukrainian Math. J. **66**(4), 531–537 (2014)
6. Chandok, S., Khan, M.S., Narang, T.D.: Fixed point theorem in partially ordered metric spaces for generalized contraction mappings. Azerbaijan J. Math. **5**(1), 89–96 (2015)
7. Chandok, S., Postolache, M.: Fixed point theorem for weakly Chatterjea type cyclic contractions. Fixed Point Theory Appl. **2013**, 28 (2013) doi:10.1186/1687-1812-2013-28
8. Geraghty, M.: On contractive mappings. Proc. Am. Math. Soc. **40**, 604–608 (1973)
9. Amini-Harandi, A., Emami, H.: A fixed point theorem for contraction type maps in partially ordered metric spaces and application to ordinary differential equations. Nonlinear Anal. **72**, 2238–2242 (2010)
10. Harjani, J., Sadarangani, K.: Fixed point theorems for weakly contractive mappings in partially ordered sets. Nonlinear Anal. **71**, 3403–3410 (2009)
11. Hussain, N., Karapinar, E., Salimi, P., Akbar, F.: α-admissible mappings and related fixed point theorems. J. Inequal. Appl. **2013**, 114 (2013)
12. Jachymski, J.: Equivalent conditions for generalized contractions on (ordered) metric spaces. Nonlinear Anal. **74**, 768–774 (2011)
13. Karapinar, E.: $\alpha - \psi$-Geraghty contraction type mappings and some related fixed point results. Filomat **28**(1), 37–48 (2014)
14. La Rosa, V., Vetro, P.: Common fixed points for α, ψ, ϕ-contractions in generalized metric spaces. Nonlinear Anal.: Model. Control **19**(1), 43–54 (2014)
15. Samet, B., Vetro, C., Vetro, P.: Fixed point theorem for $\alpha - \psi$ contractive type mappings. Nonlinear Anal. **75**, 2154–2165 (2012)

6

On the Dirichlet's type of Eulerian polynomials

Serkan Araci · Mehmet Acikgoz · Erkan Ağyüz

Abstract In the present paper, we introduce the Eulerian polynomials attached to χ using p-adic q-integral on $\mathbb{Z}_p$. Also, we give some new interesting identities via the generating functions of Dirichlet's type of Eulerian polynomials. In addition, by applying Mellin transformation to the generating function of Dirichlet's type of Eulerian polynomials, we define Eulerian L type function which interpolates Dirichlet's type of Eulerian polynomials at negative integers.

Keywords Eulerian polynomials · p-adic q-integral on $\mathbb{Z}_p$ · Mellin transformation · L function

Mathematics Subject Classification 11S80 · 11B68

Introduction

Recently, Kim et al. have studied on the Eulerian polynomials and derived Witt's formula for the Eulerian polynomials together with the relations between Genocchi, Tangent and Euler numbers. For more on this and related issues, see, e.g., [1]. Looking at the arithmetic works of T. Kim, Y. Simsek, H. M. Srivastava and other related mathematicians, they have introduced many various generating functions for types of the Bernoulli, the Euler, the Genocchi numbers and polynomials and derived some new interesting identities (see [1–28] for a systematic work).

Kim originally defined the p-adic integral on $\mathbb{Z}_p$ based on the q-integers (called p-adic q-integral on $\mathbb{Z}_p$) and showed that this integral is related to non-archimedean combinatorial analysis in mathematical physics such as the functional equation of the q-zeta function, the q-Stirling numbers and q-Mahler theory, and so on. We refer the reader to [4, 5].

We now briefly summarize some properties of the usual Eulerian polynomials:

The Eulerian polynomials $\mathcal{A}_n(x)$ are defined as (known as the generating function of Eulerian poynomials)

$$e^{\mathcal{A}(x)t} = \sum_{n=0}^{\infty} \mathcal{A}_n(x)\frac{t^n}{n!} = \frac{1-x}{e^{t(1-x)}-x} \tag{1}$$

where we have used $\mathcal{A}^n(x) := \mathcal{A}_n(x)$, symbolically. The Eulerian polynomials can be generated via the recurrence relation:

$$(\mathcal{A}(t)+(t-1))^n - t\mathcal{A}_n(t) = \begin{cases} 1-t & \text{if } n=0 \\ 0 & \text{if } n \neq 0, \end{cases} \tag{2}$$

(for details, see [1]).

Suppose that p be a fixed odd prime number. Throughout this paper, we use the following notations. By $\mathbb{Z}_p$, we denote the ring of p-adic rational integers, $\mathbb{Q}$ denotes the field of rational numbers, $\mathbb{Q}_p$ denotes the field of p-adic rational numbers, and $\mathbb{C}_p$ denotes the completion of algebraic closure of $\mathbb{Q}_p$. Let $\mathbb{N}$ be the set of natural numbers and $\mathbb{N}^* = \mathbb{N} \cup \{0\}$.

The normalized p-adic absolute value is defined by

S. Araci (✉)
Department of Economics, Faculty of Economics, Administrative and Social Science, Hasan Kalyoncu University, 27410 Gaziantep, Turkey
e-mail: mtsrkn@hotmail.com

M. Acikgoz · E. Ağyüz
Department of Mathematics, Faculty of Arts and Science, University of Gaziantep, 27310 Gaziantep, Turkey
e-mail: acikgoz@gantep.edu.tr

E. Ağyüz
e-mail: erkanagyuz@hotmail.com

$$|p|_p = \frac{1}{p}.$$

In this paper, we assume $|q-1|_p < 1$ as an indeterminate. Let $UD(\mathbb{Z}_p)$ be the space of uniformly differentiable functions on $\mathbb{Z}_p$. For a positive integer d with $(d,p)=1$, set

$$X = X_m = \lim_{\overleftarrow{m}} \mathbb{Z}/dp^m\mathbb{Z},$$

$$X^* = \bigcup_{\substack{0<a<dp \\ (a,p)=1}} a + dp\mathbb{Z}_p$$

and

$$a + dp^m\mathbb{Z}_p = \{x \in X \mid x \equiv a (\operatorname{mod} dp^m)\},$$

where $a \in \mathbb{Z}$ satisfies the condition $0 \leq a < dp^m$.

Firstly, for introducing fermionic p-adic q-integral, we need some basic information which we state here. A measure on $\mathbb{Z}_p$ with values in a p-adic Banach space B is a continuous linear map

$$f \mapsto \int f(x)\mu = \int_{\mathbb{Z}_p} f(x)\mu(x)$$

from $C^0(\mathbb{Z}_p, \mathbb{C}_p)$, (continuous function on $\mathbb{Z}_p$) to B. We know that the set of locally constant functions from $\mathbb{Z}_p$ to $\mathbb{Q}_p$ is dense in $C^0(\mathbb{Z}_p, \mathbb{C}_p)$ so.

Explicitly, for all $f \in C^0(\mathbb{Z}_p, \mathbb{C}_p)$, the locally constant functions

$$f_n = \sum_{i=0}^{p^m-1} f(i) 1_{i+p^m\mathbb{Z}_p} \to f \text{ in } C^0$$

Now, set $\mu(i+p^m\mathbb{Z}_p) = \int_{\mathbb{Z}_p} 1_{i+p^m\mathbb{Z}_p}\mu$. Then $\int_{\mathbb{Z}_p} f\mu$ is given by the following Riemann sum

$$\int_{\mathbb{Z}_p} f\mu = \lim_{m\to\infty} \sum_{i=0}^{p^m-1} f(i)\mu(i+p^m\mathbb{Z}_p)$$

The following q-Haar measure is defined by Kim in [2] and [4]:

$$\mu_q(a+p^m\mathbb{Z}_p) = \frac{q^a}{[p^m]_q}$$

So, for $f \in UD(\mathbb{Z}_p)$, the p-adic q-integral on $\mathbb{Z}_p$ is defined by Kim as follows:

$$I_q(f) = \int_{\mathbb{Z}_p} f(\eta)d\mu_q(\eta) = \lim_{n\to\infty} \frac{1}{[p^n]_q} \sum_{\eta=0}^{p^n-1} q^\eta f(\eta). \tag{3}$$

The bosonic integral is considered as the bosonic limit $q \to 1$, $I_1(f) = \lim_{q\to 1} I_q(f)$. In [8, 9] and [10], similarly, the p-adic fermionic integration on $\mathbb{Z}_p$ is defined by Kim as follows:

$$I_{-q}(f) = \lim_{q\to -q} I_q(f) = \int_{\mathbb{Z}_p} f(x)d\mu_{-q}(x). \tag{4}$$

By (4), we have the following well-known integral equation:

$$q^n I_{-q}(f_n) + (-1)^{n-1} I_{-q}(f) = [2]_q \sum_{l=0}^{n-1} (-1)^{n-1-l} q^l f(l) \tag{5}$$

Here $f_n(x) := f(x+n)$. By (5), we have the following equalities:

If n odd, then

$$q^n I_{-q}(f_n) + I_{-q}(f) = [2]_q \sum_{l=0}^{n-1} (-1)^l q^l f(l). \tag{6}$$

If n even, then we have

$$I_{-q}(f) - q^n I_{-q}(f_n) = [2]_q \sum_{l=0}^{n-1} (-1)^l q^l f(l). \tag{7}$$

Substituting $n=1$ into (6), we readily see the following

$$qI_{-q}(f_1) + I_{-q}(f) = [2]_q f(0). \tag{8}$$

Replacing q by q^{-1} in (8), we easily derive the following

$$I_{-q^{-1}}(f_1) + qI_{-q^{-1}}(f) = [2]_q f(0). \tag{9}$$

In [1], Kim et al. considered $f(x) = e^{-x(1+q)t}$ in (9) and derived Witt's formula of the Eulerian polynomials as follows:

For $n \in \mathbb{N}^*$,

$$I_{-q^{-1}}(x^n) = \frac{(-1)^n}{(1+q)^n} \mathcal{A}_n(-q). \tag{10}$$

In the next section we will introduce $I_{-q^{-1}}(\chi(x)x^n)$ based on the fermionic p-adic q-integral in the p-adic integer ring which will be known as the Eulerian polynomials attached to χ (or Dirichlet's type of Eulerian polynomials) and we will give some new properties.

On the Dirichlet's type of Eulerian polynomials

Throughout this section, we always make use of d as an odd natural number. Firstly, we consider the following equality using (6):

$$\begin{aligned} &\int_{\mathbb{Z}_p} f(x+d)d\mu_{-q^{-1}}(x) + q^d \int_{\mathbb{Z}_p} f(x)d\mu_{-q^{-1}}(x) \\ &= [2]_q \sum_{0\leq l\leq d-1} (-1)^l q^{d-l+1} f(l). \end{aligned} \tag{11}$$

Let χ be a Dirichlet character of conductor d, which is any multiple of p (=*odd*). Then, substituting $f(x) = \chi(x)e^{-x(1+q)t}$ in (11), we compute as follows:

$$\int_{\mathbb{Z}_p} \chi(x+d)e^{-(x+d)(1+q)t}d\mu_{-q^{-1}}(x) + q^d\int_{\mathbb{Z}_p} \chi(x)e^{-x(1+q)t}d\mu_{-q^{-1}}(x) = [2]_q \sum_{0\leq l\leq d-1} (-1)^l q^{d-l+1}\chi(l)e^{-l(1+q)t}$$

After some applications, we see that

$$\int_{\mathbb{Z}_p} \chi(x)e^{-x(1+q)t}d\mu_{-q^{-1}}(x) = [2]_q \sum_{l=0}^{d-1}(-1)^l q^{d-l+1}\chi(l)\frac{e^{-l(1+q)t}}{e^{-d(1+q)t}+q^d}. \tag{12}$$

Let $\mathcal{F}_q(t \mid \chi) = \sum_{n=0}^{\infty} \mathcal{A}_{n,\chi}(-q)\frac{t^n}{n!}$. Then, we introduce the following definition of generating function of Dirichlet's type of Eulerian polynomials.

Definition 1 For $n \in \mathbb{N}^*$, we define the following:

$$\sum_{n=0}^{\infty} \mathcal{A}_{n,\chi}(-q)\frac{t^n}{n!} = [2]_q \sum_{l=0}^{d-1}(-1)^l q^{d-l+1}\chi(l)\frac{e^{-l(1+q)t}}{e^{-d(1+q)t}+q^d}. \tag{13}$$

By (12) and (13), we state the following theorem which is the Witt's formula for Dirichlet's type of Eulerian polynomials.

Theorem 2.1 The following identity holds true:

$$I_{-q^{-1}}(\chi(x)x^n) = \frac{(-1)^n}{(1+q)^n}\mathcal{A}_{n,\chi}(-q). \tag{14}$$

Using (13), we discover the following applications:

$$\begin{aligned}\sum_{n=0}^{\infty} \mathcal{A}_{n,\chi}(-q)\frac{t^n}{n!} &= [2]_q \sum_{l=0}^{d-1}(-1)^l q^{d-l+1}\chi(l)\frac{e^{-l(1+q)t}}{e^{-d(1+q)t}+q^d}\\ &= [2]_q \sum_{l=0}^{d-1}(-1)^l q^{-l+1}\chi(l)e^{-l(1+q)t}\sum_{m=0}^{\infty}(-1)^m q^{-md}e^{-md(1+q)t}\\ &= q[2]_q \sum_{m=0}^{\infty}\sum_{l=0}^{d-1}(-1)^{l+md}\chi(l+md)q^{-(l+md)}e^{-(l+md)(1+q)t}\\ &= q[2]_q \sum_{m=0}^{\infty}(-1)^m\chi(m)q^{-m}e^{-m(1+q)t}.\end{aligned}$$

Thus, we get the following theorem.

Theorem 2.2 The following

$$\mathcal{F}_q(t \mid \chi) = \sum_{n=0}^{\infty} \mathcal{A}_{n,\chi}(-q)\frac{t^n}{n!} = [2]_q \sum_{m=0}^{\infty}\frac{(-1)^m\chi(m)e^{-m(1+q)t}}{q^{m-1}} \tag{15}$$

is true.

By considering Taylor expansion of $e^{-m(1+q)t}$ in (15), we procure the following theorem.

Theorem 2.3 For $n \in \mathbb{N}$, we have

$$\frac{(-1)^n}{q(1+q)^{n+1}}\mathcal{A}_{n,\chi}(-q) = \sum_{m=1}^{\infty}\frac{(-1)^m\chi(m)m^n}{q^m}. \tag{16}$$

From (14) and (16), we easily derive the following corollary:

Corollary 2.4 For $n \in \mathbb{N}$, we have

$$\lim_{n\to\infty}\sum_{m=1}^{p^n-1}\frac{(-1)^m\chi(m)m^n}{q^{m+1}} = 2\sum_{m=1}^{\infty}\frac{(-1)^m\chi(m)m^n}{q^{m-1}}.$$

We now give distribution formula for Dirichlet's type of Eulerian polynomials using p-adic q-integral on $\mathbb{Z}_p$, as follows:

$$\begin{aligned}\int_{\mathbb{Z}_p}\chi(x)x^n d\mu_{-q^{-1}}(x) &= \lim_{m\to\infty}\frac{1}{[dp^m]_{-q^{-1}}}\sum_{x=0}^{dp^m-1}(-1)^x\chi(x)x^n q^{-x}\\ &= \frac{d^n}{[d]_{-q^{-1}}}\sum_{a=0}^{d-1}(-1)^a\chi(a)q^{-a}\\ &\quad\times\left(\lim_{m\to\infty}\frac{1}{[p^m]_{-q^{-d}}}\sum_{x=0}^{p^m-1}(-1)^x\left(\frac{a}{d}+x\right)^n q^{-dx}\right)\\ &= \frac{d^n}{[d]_{-q^{-1}}}\sum_{a=0}^{d-1}(-1)^a\chi(a)q^{-a}\int_{\mathbb{Z}_p}\left(\frac{a}{d}+x\right)^n d\mu_{-q^{-d}}(x).\end{aligned}$$

Thus, we state the following theorem.

Theorem 2.5 The following identity holds true:

$$\frac{(-1)^n}{(1+q)^n}\mathcal{A}_{n,\chi}(-q) = \frac{d^n}{[d]_{-q^{-1}}}\sum_{a=0}^{d-1}(-1)^a\chi(a)q^{-a} \times \int_{\mathbb{Z}_p}\left(\frac{a}{d}+x\right)^n d\mu_{-q^{-d}}(x). \tag{17}$$

Notice that the Eq. (17) is related to q-Genocchi polynomials with weight zero, $\widetilde{G}_{n,q}(x)$, and q-Euler polynomials with weight zero, $\widetilde{E}_{n,q}(x)$, which is defined by Araci et al. and Kim and Choi in [21] and [11], respectively, as follows:

$$\frac{\widetilde{G}_{n+1,q}(x)}{n+1} = \lim_{m\to\infty}\frac{1}{[p^m]_{-q}}\sum_{y=0}^{p^m-1}(-1)^y(x+y)^n q^y \tag{18}$$

and

$$\widetilde{E}_{n,q}(x) = \int_{\mathbb{Z}_p}(x+y)^n d\mu_{-q}(y). \tag{19}$$

By expressions of (17), (18) and (19), we easily discover the following corollary.

Corollary 2.6 For $n \in \mathbb{N}^*$, we have

$$\frac{(-1)^n}{(1+q)^n}\mathcal{A}_{n,\chi}(-q) = \frac{d^n}{(n+1)[d]_{-q^{-1}}}\sum_{a=0}^{d-1}(-1)^a\chi(a)q^{-a}\widetilde{G}_{n+1,q^{-d}}\left(\frac{a}{d}\right).$$

Moreover,

$$\frac{(-1)^n}{(1+q)^n}\mathcal{A}_{n,\chi}(-q) = \frac{d^n}{[d]_{-q^{-1}}}\sum_{a=0}^{d-1}(-1)^a\chi(a)q^{-a}\widetilde{E}_{n,q^{-d}}\left(\frac{a}{d}\right).$$

On the Eulerian L type function

Our goal in this section is to introduce Eulerian L type function by applying Mellin transformation to the generating function of Dirichlet's type of Eulerian polynomials. By (15), for $s \in \mathbb{C}$, we define the following

$$L_E(s \mid \chi) = \frac{1}{\Gamma(s)}\int_0^\infty t^{s-1}\mathcal{F}_q(t \mid \chi)dt$$

where $\Gamma(s)$ is the Euler Gamma function. It becomes as follows:

$$\begin{aligned} L_E(s \mid \chi) &= q[2]_q\sum_{m=0}^{\infty}(-1)^m\chi(m)q^{-m}\left\{\frac{1}{\Gamma(s)}\int_0^\infty t^{s-1}e^{-m(1+q)t}dt\right\} \\ &= \frac{q}{(1+q)^{s-1}}\sum_{m=1}^{\infty}\frac{(-1)^m\chi(m)}{q^m m^s} \end{aligned}$$

So, we give definition of Eulerian L type function as follows:

Definition 2 For $s \in \mathbb{C}$, then we have

$$L_E(s \mid \chi) = \frac{q}{(1+q)^{s-1}}\sum_{m=1}^{\infty}\frac{(-1)^m\chi(m)}{q^m m^s}. \tag{20}$$

Substituting $s = -n$ into (20) and comparing with the Eq. (16), then, relation between Eulerian L type function and Dirichlet's type of Eulerian polynomials is given by the following theorem.

Theorem 3.1 The following equality holds true:

$$L_E(-n \mid \chi) = \begin{cases} -\mathcal{A}_{n,\chi}(-q) & \text{if n odd,} \\ \mathcal{A}_{n,\chi}(-q) & \text{if n even.} \end{cases}$$

References

1. Kim, D.S., Kim, T., Kim, W.J., Dolgy, D.V.: A note on eulerian polynomials. Abstr. Appl. Anal. **2012** (2012). Art. ID:269640. doi:10.1155/2012/269640
2. Kim, T.: On a q-analogue of the p-adic log gamma functions and related integrals. J. Num. Theory **76**, 320–329 (1999)
3. Kim, T., Rim, S.H.: A note on p-adic Carlitz's q -Bernoulli numbers. Bull. Aust. Math. Soc. **62**, 227–234 (2000)
4. Kim, T.: q-Volkenborn integration. Russ. J. Math. Phys. **19**, 288–299 (2002)
5. Kim, T.: Non-archimedean q-integrals associated with multiple Changhee q-Bernoulli Polynomials. Russ. J. Math. Phys. **10**, 91–98 (2003)
6. Kim, T.: p-adic q-integrals associated with the Changhee-Barnes' q-Bernoulli Polynomials. Integral Transform. Spec. Funct. **15**, 415–420 (2004)
7. Kim, T.: q-Generalized Euler numbers and polynomials. Russ. J. Math. Phys. **13**(3), 293–308 (2006)
8. Kim, T.: Some identities on the q-Euler polynomials of higher order and q-stirling numbers by the fermionic p-adic integral on $\mathbb{Z}_p$. Russ. J. Math. Phys. **16**, 484–491 (2009)
9. Kim, T.: On the q-extension of Euler and Genocchi numbers. J. Math. Anal. Appl. **326**, 1458–1465 (2007)
10. Kim, T.: On the analogs of Euler numbers and polynomials associated with p-adic q-integral on $\mathbb{Z}_p$ at $q = 1$. J. Math. Anal. Appl. **331**, 779–792 (2007)
11. Kim, T., Choi, J.: On the q-Euler numbers and polynomials with weight 0. Abstr. Appl. Anal. **2012** (2012). Art. ID:795304. doi:10.1155/2012/795304
12. Jang, L.C.: The q-analogue of twisted Lerch type Euler Zeta functions. Bull. Korean Math. Soc. **47**(6), 1181–1188 (2010)
13. Jang, L.C., Kurt, V., Simsek, Y., Rim, S.H.: q-analogue of the p-adic twisted l-function. J. Concr. Appl. Math. **6**(2), 169–176 (2008)
14. Ozden, H., Cangul, I.N., Simsek, Y.: Multivariate interpolation functions of higher order q-Euler numbers and their applications. Abstr. Appl. Anal. **2008** (2008). Art. ID:390857. doi:10.1155/2008/390857
15. Cangul, I.N., Ozden, H., Simsek, Y.: A new approach to q -Genocchi numbers and their interpolation functions. Nonlinear Anal. **71**, 793–799 (2009)
16. Cetin, E., Acikgoz, M., Cangul, I.N., Araci, S.: A note on the (h,q)-Zeta type function with weight α. J. Inequal. Appl. (2013). doi:10.1186/1029-242X-2013-100
17. Araci, S., Acikgoz, M., Park, K.H., Jolany, H.: On the unification of two families of multiple twisted type polynomials by using p -adic q-integral on $\mathbb{Z}_p$ at $q = -1$, Bull. Malays. Math. Sci. Soc. (2) 37(2) (2014), 543–554
18. Araci, S.: Novel identities involving Genocchi numbers and polynomials arising from applications from umbral calculus. Appl. Math. Comput. **233**, 599–607 (2014)
19. Araci, S., Erdal, D., Seo, J.J.: A study on the fermionic p-adic q-integral representation on $\mathbb{Z}_p$ associated with weighted q-Bernstein and q-Genocchi polynomials. Abstr. Appl. Anal. **2011** (2011). Art. ID:649248
20. Araci, S., Seo, J.J., Erdal, D.: New construction weighted (h,q)-Genocchi numbers and polynomials related to Zeta type function. Discret Dyn. Nat. Soc. **2011** (2011). Art. ID:487490. doi:10.1155/2011/487490
21. Araci, S., Sen, E., Acikgoz, M.: Theorems on Genocchi polynomials of higher order arising from Genocchi basis. Taiwan. J. Math. **18**(2), 473–482 (2014)
22. Simsek, Y.: q-analogue of twisted l-series and q -twisted Euler numbers. J. Num. Theory **110**(2), 267–278 (2005)

23. Simsek, Y.: q-Dedekind type sums related to q-zeta function and basic L-series. J. Math. Anal. Appl. **318**(1), 333–351 (2006)
24. Simsek, Y.: Twisted (h, q)-Bernoulli numbers and polynomials related to twisted (h, q)-zeta function and L -function. J. Math. Anal. Appl. **324**(2), 790–804 (2006)
25. Srivastava, H.M.: Some generalizations and basic (or q-) extensions of the Bernoulli, Euler and Genocchi polynomials. Appl. Math. Inform. Sci. **5**, 390–444 (2011)
26. Srivastava, H.M., Kurt, B., Simsek, Y.: Some families of Genocchi type polynomials and their interpolation functions. Integral Transform Spec. Funct. **23**, 919–938 (2012); see also Corrigendum. Integral Transform Spec. Funct. **23**, 939–940 (2012)
27. Choi, J., Anderson, P.J., Srivastava, H.M.: Some q -extensions of the Apostol-Bernoulli and the Apostol-Euler polynomials of order n, and the multiple Hurwitz zeta function. Appl. Math. Comput. **199**, 723–737 (2008)
28. Choi, J., Anderson, P.J., Srivastava, H.M.: Carlitz's q-Bernoulli and q-Euler numbers and polynomials and a class of q-Hurwitz zeta functions. Appl. Math. Comput. **215**, 1185–1208 (2009)

7

Homotopy analysis method for fuzzy Boussinesq equation

Amir Fallahzadeh[1] · **Mohammad Ali Fariborzi Araghi**[1]

Abstract In this work, the fuzzy Boussinesq equation is considered to solve via the homotopy analysis method (HAM). For this purpose, a theorem is proved to illustrate the convergence of the proposed method. Also, two sample examples are solved by applying the HAM to verify the efficiency and importance of the method.

Keywords Homotopy analysis method · Fuzzy Boussinesq equation · Fuzzy numbers · Convergence

Introduction

In recent years, some numerical and analytical methods were proposed in order to solve fuzzy differential equations [1–8, 10, 18, 19]. One of the powerful semi-analytical methods to solve differential equations is the homotopy analysis method (HAM). In [14], the authors applied this method to solve the Boussinesq equation in crisp case. In this work, we consider the fuzzy form of Boussinesq equation as follows:

$$\widetilde{u}_{tt} + \alpha\widetilde{u}_{xx} + \beta(\widetilde{u}^2)_{xx} - \widetilde{u}_{xxxx} = \widetilde{0}, \quad 0 \le t \le T,\ x > 0. \tag{1}$$

With the following initial conditions,

$$\widetilde{u}(x,0) = \widetilde{f}(x), \tag{2}$$

$$\widetilde{u}_t(x,0) = \widetilde{g}(x), \tag{3}$$

where $\widetilde{u}$ is unknown fuzzy function, α and β are crisp constant coefficients and $\widetilde{f}$ and $\widetilde{g}$ are known fuzzy functions.

In order to solve Eq. (1), we apply the HAM in fuzzy case as an important and efficient method to find the solution of differential equations. The HAM, proposed by Liao, [16, 17], is a semi-analytical method which the solution is obtained as a series form according to a recursive relation stems from a deformation equation [13, 14].

In Sect. 2, we remind some fuzzy concepts briefly. In Sect. 3, we apply the HAM to solve the fuzzy Boussinesq equation and we prove a theorem to show the convergence of the proposed method. In Sect. 4, we solve two sample fuzzy Boussinesq equations and we obtain a series solution by this method.

Preliminaries

In this section, we recall some basic definitions of fuzzy sets theory [22].

Definition 2.1 A fuzzy parametric number u is a pair $(\underline{u}(r), \overline{u}(r))$, $0 \le r \le 1$, which satisfies the following requirements:

1. $\underline{u}(r)$ is a bounded left continuous non-decreasing function over [0, 1],
2. $\overline{u}(r)$ is a bounded left continuous non-increasing function over [0, 1],
3. $\underline{u}(r) \le \overline{u}(r)$, $0 \le r \le 1$.

✉ Amir Fallahzadeh
amir_falah6@yahoo.com

Mohammad Ali Fariborzi Araghi
fariborzi.araghi@gmail.com; m_fariborzi@iauctb.ac.ir

[1] Department of Mathematics, Islamic Azad University, Central Tehran Branch, P.O. Box 13185.768, Tehran, Iran

The set of all these fuzzy numbers is denoted by $\mathbb{E}^1$. For $u = (\underline{u}, \overline{u}), v = (\underline{v}, \overline{v}) \in \mathbb{E}^1, \quad k \in \mathbb{R}$ the addition, multiplication, and the scaler multiplication of fuzzy numbers are defined by

$$\begin{aligned}
&(\underline{u+v})(r) = \underline{u}(r) + \underline{v}(r),\\
&(\overline{u+v})(r) = \overline{u}(r) + \overline{v}(r),\\
&(\underline{u.v})(r) = \min\{\underline{u}(r).\underline{v}(r), \underline{u}(r).\overline{v}(r), \overline{u}(r).\underline{v}(r), \overline{u}(r).\overline{v}(r)\},\\
&(\overline{u.v})(r) = \max\{\underline{u}(r).\underline{v}(r), \underline{u}(r).\overline{v}(r), \overline{u}(r).\underline{v}(r), \overline{u}(r).\overline{v}(r)\},\\
&\underline{ku}(r) = k\underline{u}(r),\ \overline{ku}(r) = k\overline{u}(r), \quad k \geq 0,\\
&\underline{ku}(r) = k\overline{u}(r),\ \overline{ku}(r) = k\underline{u}(r), \quad k \leq 0.
\end{aligned}$$

Definition 2.2 A fuzzy parametric number $\widetilde{u}$ is positive (negative) if and only if $\underline{u}(r) \geq 0\, (\overline{u}(r) \leq 0) \quad \forall r \in [0,1]$.

Remark 2.3 If fuzzy parametric numbers $\widetilde{u}$ and $\widetilde{v}$ are positive, then $\widetilde{u}.\widetilde{v}(r) = (\underline{u}(r)\underline{v}(r), \overline{u}(r)\overline{v}(r))$.

Definition 2.4 A function $f : \mathbb{R}^1 \longrightarrow \mathbb{E}^1$ is called a fuzzy function. If for arbitrary fixed $t_0 \in \mathbb{E}^1$ and $\varepsilon > 0$ such that, $|t - t_0| < \delta \Longrightarrow D(f(t), f(t_0)) < \varepsilon$ exists, f is said to be continuous [20, 21].

Definition 2.5 Let $u, v \in \mathbb{E}^1$. If there exists $w \in \mathbb{E}^1$ such that $u = v + w$, then w is called the H-difference of u, v and it is denoted by $u \ominus v$ [9].

Definition 2.6 Let $a, b \in \mathbb{R}$ and $f : (a,b) \rightarrow \mathbb{E}^1$ and $t_0 \in (a,b)$. We define the n-th order differential of f as follows: We say that f is strongly generalized differentiable of n-th order at t_0, if there exists an element $f^{(s)}(t_0) \in \mathbb{E}^1 \quad \forall s = 1, \ldots, n$ such that

(i) for all $h > 0$ sufficiently close to 0, there exist $f^{(s-1)}(t_0 + h) \ominus f^{(s-1)}(t_0)$, $f^{(s-1)}(t_0) \ominus f^{(s-1)}(t_0 - h)$ and the limits

$$\lim_{h \to 0^+} \frac{f^{(s-1)}(t_0 + h) \ominus f^{(s-1)}(t_0)}{h} = \lim_{h \to 0^+} \frac{f^{(s-1)}(t_0) \ominus f^{(s-1)}(t_0 - h)}{h} = f^{(s)}(t_0),$$

or

(ii) for all $h > 0$ sufficiently close to 0, there exist $f^{(s-1)}(t_0 - h) \ominus f^{(s-1)}(t_0)$, $f^{(s-1)}(t_0) \ominus f^{(s-1)}(t_0 + h)$ and the limits

$$\lim_{h \to 0^+} \frac{f^{(s-1)}(t_0 - h) \ominus f^{(s-1)}(t_0)}{-h} = \lim_{h \to 0^+} \frac{f^{(s-1)}(t_0) \ominus f^{(s-1)}(t_0 + h)}{-h} = f^{(s)}(t_0),$$

or

(iii) for all $h > 0$ sufficiently close to 0, there exist $f^{(s-1)}(t_0 + h) \ominus f^{(s-1)}(t_0)$, $f^{(s-1)}(t_0 - h) \ominus f^{(s-1)}(t_0)$ and the limits

$$\lim_{h \to 0^+} \frac{f^{(s-1)}(t_0 + h) \ominus f^{(s-1)}(t_0)}{h} = \lim_{h \to 0^+} \frac{f^{(s-1)}(t_0 - h) \ominus f^{(s-1)}(t_0)}{-h} = f^{(s)}(t_0),$$

or

(iv) for all $h > 0$ sufficiently close to 0, there exist $f^{(s-1)}(t_0) \ominus f^{(s-1)}(t_0 + h)$, $f^{(s-1)}(t_0) \ominus f^{(s-1)}(t_0 - h)$ and the limits

$$\lim_{h \to 0^+} \frac{f^{(s-1)}(t_0) \ominus f^{(s-1)}(t_0 + h)}{-h} = \lim_{h \to 0^+} \frac{f^{(s-1)}(t_0) \ominus f^{(s-1)}(t_0 - h)}{h} = f^{(s)}(t_0),$$

(h and $(-h)$ at denominators mean $\frac{1}{h}$. and $-\frac{1}{h}$. respectively $\forall s = 1, \ldots, n$) [9, 11, 12].

Theorem 2.7 *Let $f : (a,b) \rightarrow \mathbb{E}^1$ be strongly generalized differentiable on each point $t \in (a,b)$ in the sense of Definition 2.5, (iii) or (iv). Then $f'(x) \in \mathbb{R}$ for all $t \in (a,b)$ (see [9]).*

Theorem 2.8 *Let $f : \mathbb{R}^1 \longrightarrow \mathbb{E}^1$ be a function and denote $f(t) = (\underline{f}(t,r), \overline{f}(t,r))$, for each $r \in [0,1]$. Then*

(1) *If f is differentiable in the first form (i), then $\underline{f}(t,r)$ and $\overline{f}(t,r)$ are differentiable functions and $f'(t) = (\underline{f}'(t,r), \overline{f}'(t,r))$,*

(2) *If f is differentiable in the second form (ii), then $\underline{f}(t,r)$ and $\overline{f}(t,r)$ are differentiable functions and $f'(t) = (\overline{f}'(t,r), \underline{f}'(t,r))$ (see [11]).*

Remark 2.9 Note that by the above definition, a fuzzy function is i-differentiable or ii-differentiable of order n if $f^{(s)}$ for $s = 1, \ldots, n$ is i-differentiable or ii-differentiable. It is possible that the different orders have different kind i or ii differentiability.

Main idea

In order to describe the HAM for Eq. (1), we consider the following equation:

$$\widetilde{N}[\widetilde{u}(x,t)] = \widetilde{u}_{tt} + \alpha \widetilde{u}_{xx} + \beta(\widetilde{u}^2)_{xx} - \widetilde{u}_{xxxx} = \widetilde{0}, \tag{4}$$

According to the parametric form of fuzzy numbers, we consider Eq. (4) in the following form:

$$N = \begin{pmatrix} \underline{N}[u(x,t)] \\ \overline{N}[u(x,t)] \end{pmatrix} = \begin{pmatrix} 0 \\ 0 \end{pmatrix}, \quad u(x,t) = \begin{pmatrix} \underline{u}(x,t) \\ \overline{u}(x,t) \end{pmatrix}.$$

At first, we construct the zeroth-order deformation system.

$$(I-Q)L[\phi(x,t,r;Q)-u_0(x,t,r)] = QHhN[\phi(x,t,r;Q)], \quad (5)$$

where I is the identity matrix, $L=\begin{pmatrix} L_1 & 0 \\ 0 & L_2 \end{pmatrix}$ is an auxiliary linear operator matrix, $H(x,t,r)=\begin{pmatrix} H_1(x,t,r) & 0 \\ 0 & H_2(x,t,r) \end{pmatrix}$ is an auxiliary function matrix, $h=\begin{pmatrix} h_1 & 0 \\ 0 & h_2 \end{pmatrix}$ is an auxiliary parameter matrix, $\phi(x,t;Q)=\begin{pmatrix} \underline{\phi}(x,t,r;q) \\ \overline{\phi}(x,t,r;q) \end{pmatrix}$ is an unknown function matrix, $u_0(x,t,r)$ is an initial guess of the vector $u(x,\ t,\ r)$ and $Q=\begin{pmatrix} q & 0 \\ 0 & q \end{pmatrix}$, $0\leq q\leq 1$, is a diagonal matrix which denotes the embedding parameter matrix. It is obvious, when the q, increases from 0 to 1 or in other word the embedding parameter matrix changes from $Q=\overline{0}$ to $Q=I$, the solution of system of equations (5) changes from $\phi(x,t,r;\overline{0})=u_0(x,t,r)$ to $\phi(x,t,r;I)=u(x,t,r)$. Therefore, $\phi(x,t,r)$ varies from the initial guess $u_0(x,t,r)$ to the exact solution $u(x,\ t,\ r)$ of the system.

We consider $\phi(x,t,r;Q)$ in the following matrix expansion form,

$$\phi(x,t,r;Q)=u_0(x,t,r)+\sum_{m=1}^{+\infty} Q^m u_m(x,t,r), \quad (6)$$

where

$$u_m(x,t,r)=\frac{1}{m!}\begin{pmatrix} \dfrac{\partial^m \underline{\phi}(x,t,r;q)}{\partial q^m}\Big|_{q=0} \\ \dfrac{\partial^m \overline{\phi}(x,t,r;q)}{\partial q^m}\Big|_{q=0}. \end{pmatrix}. \quad (7)$$

The convergence of the vector series (6) depends upon the auxiliary parameter matrix h, if it is convergent at $Q=I$, we have

$$u(x,t,r)=u_0(x,t,r)+\sum_{m=1}^{+\infty} u_m(x,t,r). \quad (8)$$

Now, we define the vectors,

$$\overrightarrow{u}_k(x,t,r)=\{u_0(x,t,r),\ldots,u_k(x,t,r)\}, \quad (9)$$

where

$$u_i=\begin{pmatrix} \underline{u}_i(x,t,r) \\ \overline{u}_i(x,t,r) \end{pmatrix}, \quad i=0,\ldots,k. \quad (10)$$

The m-th order deformation system can be written as

$$L[u_m(x,t,r)-\chi_m u_{(m-1)}(x,t,r)] = hHR_m(\overrightarrow{u}_{m-1}(x,t,r)), \quad (11)$$

where

$$R_m(\overrightarrow{u}_{m-1})=\frac{1}{(m-1)!}\begin{pmatrix} \dfrac{\partial^{m-1}\underline{N}[\phi(x,t;Q)]}{\partial q^{m-1}}\Big|_{Q=\overline{0}} \\ \dfrac{\partial^{m-1}\overline{N}[\phi(x,t;Q)]}{\partial q^{m-1}}\Big|_{Q=\overline{0}} \end{pmatrix}, \quad (12)$$

$$\chi_m=\begin{cases} \overline{0}, & m\leq 1, \\ I, & m>1. \end{cases}$$

If we consider $L=\begin{pmatrix} \dfrac{\partial^2}{\partial t^2} & 0 \\ 0 & \dfrac{\partial^2}{\partial t^2} \end{pmatrix}$, then we have,

$$\begin{aligned} u_m(t,r) &= \chi_m u_{m-1}(t,r)+L^{-1}[HhR_m(\overrightarrow{u}_{m-1})] \\ &= \chi_m u_{m-1}(t,r) \\ &\quad +\begin{pmatrix} h_1 & 0 \\ 0 & h_2 \end{pmatrix}\begin{pmatrix} \int_0^\tau\int_0^t d\theta d\tau & 0 \\ 0 & \int_0^\tau\int_0^t d\theta d\tau \end{pmatrix} \\ &\quad \times\left[\begin{pmatrix} H_1 & 0 \\ 0 & H_2 \end{pmatrix} R_m(\overrightarrow{u}_{m-1})\right]. \end{aligned} \quad (13)$$

Theorem 2.10 *If the series solution* (8) *of problem* (1) *obtained from the HAM and also the series* $\sum_{m=0}^{+\infty}\frac{\partial^2}{\partial t^2}\underline{u}_m$, $\sum_{m=0}^{+\infty}\frac{\partial^2}{\partial x^2}\underline{u}_m$, $\sum_{m=0}^{+\infty}\frac{\partial^2}{\partial x^2}\underline{u}_m^2$, $\sum_{m=0}^{+\infty}\frac{\partial^4}{\partial x^4}\underline{u}_m$, $\sum_{m=0}^{+\infty}\frac{\partial^2}{\partial t^2}\overline{u}_m$, $\sum_{m=0}^{+\infty}\frac{\partial^2}{\partial x^2}\overline{u}_m$, $\sum_{m=0}^{+\infty}\frac{\partial^2}{\partial x^2}\overline{u}_m^2$, *and* $\sum_{m=0}^{+\infty}\frac{\partial^4}{\partial x^4}\overline{u}_m$ *are convergent, and also* $\underline{0}_m$ *and* $\overline{0}_m$ *converge to the* $\underline{0}$ *and* $\overline{0}$ *respectively, then* (8) *converges to the exact solution of the problem* (1).

Proof Without loss of generality, we suppose u be i-differentiable with respect to the x, t and also it be a positive fuzzy number ($\forall t\in[0,T]$). Therefore, we can write Eq. (1) in the following form:

$$\widetilde{u}_{tt}+(\alpha^+-\alpha^-)\widetilde{u}_{xx}+(\beta^+-\beta^-)(\widetilde{u}^2)_{xx}-\widetilde{u}_{xxxx}=\widetilde{0}, \quad 0\leq t\leq T,\ x>0, \quad (14)$$

where $\alpha^+,\alpha^-,\beta^+,\beta^-\geq 0$.

Therefore, we have

$$\begin{aligned}\begin{pmatrix} \underline{N}[u(x,t,r)] \\ \overline{N}[u(x,t,r)] \end{pmatrix} &= \begin{pmatrix} \underline{u}_{tt}+\alpha^+\underline{u}_{xx}-\alpha^-\overline{u}_{xx}+\beta^+\underline{u}^2_{xx}-\beta^-\overline{u}^2_{xx}-\overline{u}_{xxxx}-\underline{0} \\ \overline{u}_{tt}+\alpha^+\overline{u}_{xx}-\alpha^-\underline{u}_{xx}+\beta^+\overline{u}^2_{xx}-\beta^-\underline{u}^2_{xx}-\underline{u}_{xxxx}-\overline{0} \end{pmatrix} \\ &= \begin{pmatrix} 0 \\ 0 \end{pmatrix}. \end{aligned} \quad (15)$$

If the series

$$\sum_{m=0}^{+\infty} u_m(x,t,r) = \begin{pmatrix} \sum_{m=0}^{+\infty} \underline{u}_m(x,t,r) \\ \sum_{m=0}^{+\infty} \overline{u}_m(x,t,r) \end{pmatrix}$$

converges, we assume

$$u(x,t,r) = \sum_{m=0}^{+\infty} u_m(x,t,r),$$

whereIn general, the series

$$\lim_{m\to+\infty} u_m(x,t,r) = \vec{0}. \tag{16}$$

We write

$$\sum_{m=1}^{n}[u_m(x,t,r) - \chi_m u_{m-1}(x,t,r)]$$
$$= u_1 + (u_2 - u_1) + (u_3 - u_2) + \cdots + (u_n - u_{n-1})$$
$$= u_n(x,t,r),$$

using (16), we have

$$\sum_{m=1}^{+\infty}[u_m(x,t,r) - \chi_m u_{m-1}(x,t,r)] = \lim_{n\to+\infty} u_n(x,t,r) = \vec{0}.$$

According to the definition of the operator L, we can write

$$\sum_{m=1}^{+\infty} L[u_m(x,t,r) - \chi_m u_{m-1}(x,t,r)]$$
$$= L\sum_{m=1}^{+\infty}[u_m(x,t,r) - \chi_m u_{m-1}(x,t,r)] = \vec{0}.$$

From above expression and Eq. (13), we obtain

$$\sum_{m=1}^{+\infty} L[u_m(x,t,r) - \chi_m u_{m-1}(x,t,r)] = hH(x,t)\sum_{m=1}^{+\infty}[R_m(\vec{u}_{m-1})].$$

Since $h \neq 0$ and $H(x,t) \neq 0$, we have

$$\sum_{m=1}^{+\infty}[R_m(\vec{u}_{m-1})] = \vec{0}. \tag{17}$$

From (12), it holds

$$\sum_{m=1}^{+\infty}[R_m(\vec{u}_{m-1})] =$$

$$\begin{pmatrix} \frac{\partial^2}{\partial t^2}\sum_{m=1}^{+\infty}\underline{u}_{m-1} + \alpha^+\frac{\partial^2}{\partial x^2}\sum_{m=1}^{+\infty}\underline{u}_{m-1} - \alpha^-\frac{\partial^2}{\partial x^2}\sum_{m=1}^{+\infty}\overline{u}_{m-1} + \beta^+\frac{\partial^2}{x^2}\left[\sum_{m=1}^{+\infty}\sum_{i=0}^{m-1}\underline{u}_i\underline{u}_{m-i-1}\right] \\ -\beta^-\frac{\partial^2}{x^2}\left[\sum_{m=1}^{+\infty}\sum_{i=0}^{m-1}\overline{u}_i\overline{u}_{m-i-1}\right] - \frac{\partial^4}{\partial x^4}\sum_{m=1}^{+\infty}\overline{u}_{m-1} - \sum_{m=0}^{+\infty}\underline{0}_m \\ \frac{\partial^2}{\partial t^2}\sum_{m=1}^{+\infty}\overline{u}_{m-1} + \alpha^+\frac{\partial^2}{\partial x^2}\sum_{m=1}^{+\infty}\overline{u}_{m-1} - \alpha^-\frac{\partial^2}{\partial x^2}\sum_{m=1}^{+\infty}\underline{u}_{m-1} + \beta^+\frac{\partial^2}{x^2}\left[\sum_{m=1}^{+\infty}\sum_{i=0}^{m-1}\overline{u}_i\overline{u}_{m-i-1}\right] \\ -\beta^-\frac{\partial^2}{x^2}\left[\sum_{m=1}^{+\infty}\sum_{i=0}^{m-1}\underline{u}_i\underline{u}_{m-i-1}\right] - \frac{\partial^4}{\partial x^4}\sum_{m=1}^{+\infty}\underline{u}_{m-1} - \sum_{m=0}^{+\infty}\overline{0}_m \end{pmatrix}.$$

We consider $\sum_{m=1}^{+\infty}\sum_{i=0}^{m-1}\underline{u}_i\underline{u}_{m-i-1}$, we have

$$\sum_{m=1}^{+\infty}\sum_{i=0}^{m-1}\underline{u}_i\underline{u}_{m-i-1} = \sum_{i=0}^{+\infty}\sum_{m=i+1}^{+\infty}\underline{u}_i\underline{u}_{m-i-1} = \sum_{i=0}^{+\infty}\underline{u}_i\sum_{m=i+1}^{+\infty}\underline{u}_{m-i-1}$$
$$= \sum_{i=0}^{+\infty}\underline{u}_i\sum_{m=0}^{+\infty}\underline{u}_m.$$

Similarly for next elements. Finally,

$$\sum_{m=1}^{+\infty}[R_m(\vec{u}_{m-1})]$$

$$
= \begin{pmatrix} \frac{\partial^2}{\partial t^2}\sum_{m=0}^{+\infty}\underline{u}_m + \alpha^+\frac{\partial^2}{\partial x^2}\sum_{m=0}^{+\infty}\underline{u}_m - \alpha^-\frac{\partial^2}{\partial x^2}\sum_{m=0}^{+\infty}\overline{u}_m + \beta^+\frac{\partial^2}{x^2}\left[\sum_{i=0}^{+\infty}\underline{u}_i\sum_{m=0}^{+\infty}\underline{u}_m\right] \\ -\beta^-\frac{\partial^2}{x^2}\left[\sum_{i=0}^{+\infty}\overline{u}_i\sum_{m=0}^{+\infty}\overline{u}_m\right] - \frac{\partial^4}{\partial x^4}\sum_{m=0}^{+\infty}\overline{u}_m - \sum_{m=0}^{+\infty}\underline{0}_m \\ \frac{\partial^2}{\partial t^2}\sum_{m=0}^{+\infty}\overline{u}_m + \alpha^+\frac{\partial^2}{\partial x^2}\sum_{m=0}^{+\infty}\overline{u}_m - \alpha^-\frac{\partial^2}{\partial x^2}\sum_{m=0}^{+\infty}\underline{u}_m + \beta^+\frac{\partial^2}{x^2}\left[\sum_{i=0}^{+\infty}\overline{u}_i\sum_{m=0}^{+\infty}\overline{u}_m\right] \\ -\beta^-\frac{\partial^2}{x^2}\left[\sum_{i=0}^{+\infty}\underline{u}_i\sum_{m=0}^{+\infty}\underline{u}_m\right] - \frac{\partial^4}{\partial x^4}\sum_{m=0}^{+\infty}\underline{u}_m - \sum_{m=0}^{+\infty}\overline{0}_m \end{pmatrix}
$$

$$
= \overrightarrow{0}.
$$

Therefore,

$$
\begin{pmatrix} \underline{u}_{tt} + \alpha^+\underline{u}_{xx} - \alpha^-\overline{u}_{xx} + \beta^+\underline{u}_{xx}^2 - \beta^-\overline{u}_{xx}^2 - \overline{u}_{xxxx} - \underline{0} \\ \overline{u}_{tt} + \alpha^+\overline{u}_{xx} - \alpha^-\underline{u}_{xx} + \beta^+\overline{u}_{xx}^2 - \beta^-\underline{u}_{xx}^2 - \underline{u}_{xxxx} - \overline{0} \end{pmatrix} = \begin{pmatrix} 0 \\ 0 \end{pmatrix},
$$

and it means that

$$
\widetilde{u}_{tt} + \alpha\widetilde{u}_{xx} + \beta(\widetilde{u}^2)_{xx} - \widetilde{u}_{xxxx} = \widetilde{0}.
$$

□

Test examples

In this section, we solve two sample examples to illustrate the applicability of the proposed method. The results are provided by Maple.

Example 3.1 We consider the following fuzzy Boussinesq equation

$$
\widetilde{u}_{tt} - \widetilde{u}_{xx} + (\widetilde{u}^2)_{xx} - \widetilde{u}_{xxxx} = \widetilde{0},
$$

with the initial conditions:

$$
\widetilde{u}(x,0) = (r, 3-2r)\frac{6}{x^2}, \ \widetilde{u}_t(x,0) = (r, 3-2r)\frac{-12}{x^3},
$$

where $\widetilde{0} = (3r-3, 3-3r)\widehat{u}_{xx} + (r^2-(3-2r), (3-2r)^2-(r))\widehat{u}_{xxxx}$ and $\widehat{u}$ is the solution of crisp case of the equation.

We suppose, $\widetilde{u}_t$ be i-differentiable with respect to the t and $\widetilde{u}_x, \widetilde{u}_x^2$ and $\widetilde{u}_{xxx}$ are i-differentiable with respect to the x. Also $\widetilde{u}$ be a positive fuzzy number ($\forall t \in [0,T], x > 0$), therefore we have

$$
\begin{pmatrix} \underline{u}_{tt} - \overline{u}_{xx} + (\underline{u}^2)_{xx} - \overline{u}_{xxxx} - (3r-3)\widehat{u}_{xx} - (r^2-(3-2r))\widehat{u}_{xxxx} \\ \overline{u}_{tt} - \underline{u}_{xx} + (\overline{u}^2)_{xx} - \underline{u}_{xxxx} - (3-3r)\widehat{u}_{xx} - ((3-2r)^2-r)\widehat{u}_{xxxx} \end{pmatrix} = \begin{pmatrix} 0 \\ 0 \end{pmatrix}.
$$

We consider $H = I$ and $h = -I$, and also, we choose the initial approximate as $\begin{pmatrix} r\frac{6x-12t}{x^3} \\ (3-2r)\frac{6x-12t}{x^3} \end{pmatrix}$. According to Eq. (13), we have

$$
R_m(\overrightarrow{u}_{m-1})] = \begin{pmatrix} \frac{\partial^2}{\partial t^2}\underline{u}_{m-1} - \frac{\partial^2}{\partial x^2}\sum_{m=1}^{+\infty}\overline{u}_{m-1} + \frac{\partial^2}{x^2}\left[\sum_{i=0}^{m-1}\underline{u}_i\underline{u}_{m-i-1}\right] - \frac{\partial^4}{\partial x^4}\overline{u}_{m-1} \\ -(3r-3)\frac{\partial^2}{x^2}\widehat{u} - (r^2-(3-2r))\frac{\partial^4}{\partial x^4}\widehat{u} \\ \frac{\partial^2}{\partial t^2}\overline{u}_{m-1} - \frac{\partial^2}{\partial x^2}\sum_{m=1}^{+\infty}\underline{u}_{m-1} + \frac{\partial^2}{x^2}\left[\sum_{i=0}^{m-1}\overline{u}_i\overline{u}_{m-i-1}\right] - \frac{\partial^4}{\partial x^4}\underline{u}_{m-1} \\ -(3-3r)\frac{\partial^2}{x^2}\widehat{u} - ((3-2r)^2-r)\frac{\partial^4}{\partial x^4}\widehat{u} \end{pmatrix}.
$$

Therefore,

$$
u_0 = \begin{pmatrix} r\left(\frac{6}{x^2} - \frac{12t}{x^3}\right) \\ (3-2r)\left(\frac{6}{x^2} - \frac{12t}{x^3}\right) \end{pmatrix}
$$

$$
u_1 = \begin{pmatrix} -\frac{18rt^2}{x^4} - \frac{24rt^3}{x^5} - \frac{504r^2t^4}{x^8} \\ -\frac{(72r-108)t^2}{2x^4} - \frac{(-144r+216)t^3}{3x^5} - \frac{3(1512+672r^2-2016r)t^4}{x^8} \end{pmatrix}
$$

$$
u_2 = \begin{pmatrix} -\frac{30rt^4}{x^6} - \frac{24rt^5}{x^7} + \frac{504r^2t^4}{x^8} + \cdots \\ -\frac{(240r-360)t^4}{4x^6} - \frac{(540-360r)t^5}{5x^7} + \frac{3(1512+672r^2-2016r)t^4}{x^8} + \cdots \end{pmatrix}
$$

$$
\vdots
$$

In general, the series solution is given by

$$u(x,t)=\sum_{n=0}^{\infty}u_n(x,t)=\begin{pmatrix} r\left(\frac{6}{x^2}-\frac{12}{x^3}t+\frac{18}{x^4}t^2-\frac{24}{x^5}t^3+\frac{30}{x^6}t^4+\cdots\right)\\ (3-2r)\left(\frac{6}{x^2}-\frac{12}{x^3}t+\frac{18}{x^4}t^2-\frac{24}{x^5}t^3+\frac{30}{x^6}t^4+\cdots\right)\end{pmatrix}.$$

That gives the exact solution

$$\begin{pmatrix}\dfrac{6r}{(x+t)^2}\\ \dfrac{6(3-2r)}{(x+t)^2}\end{pmatrix}.$$

Therefore, $\widetilde{u}=(r,3-2r)\frac{6}{(x+t)^2}$ is the exact solution of the fuzzy differential equation.

Example 3.2 We consider the following fuzzy Boussinq equation

$$\widetilde{u}_{tt}-\widetilde{u}_{xx}-\left(\widetilde{u}^2\right)_{xx}-\widetilde{u}_{xxxx}=\widetilde{0},$$

with the initial conditions:

$$\widetilde{u}(x,0)=\left(\frac{r}{2}+\frac{1}{2},2-r\right)\frac{-6}{x^2},$$

$$\widetilde{u}_t(x,0)=\left(\frac{r}{2}+\frac{1}{2},2-r\right)\frac{12}{x^3},$$

where $\widetilde{0}=(\frac{3r}{2}-\frac{3}{2},-\frac{3r}{2}+\frac{3}{2})(-\widehat{u}_{xx})+((\frac{r}{2}+\frac{1}{2})-(2-r)^2,(2-r)-(\frac{r}{2}+\frac{1}{2})^2)(-\widehat{u}_{xxxx})$ and $\widehat{u}$ is the solution of crisp case of the equation.

We suppose $\widetilde{u}_t$ be ii-differentiable with respect to the t, $\widetilde{u}_x,\widetilde{u}_{xxx}$ are ii-differentiable with respect to the x and $\widetilde{u}_x^2$ is i-differentiable with respect to the x. Also $\widetilde{u}$ be a negative fuzzy number $(\forall t\in[0,T],x>0)$, therefore we have

$$\begin{pmatrix}\overline{u}_{tt}-\underline{u}_{xx}-\left(\overline{u}^2\right)_{xx}-\underline{u}_{xxxx}-\left(\frac{3r}{2}-\frac{3}{2}\right)(-\widehat{u}_{xx})-\left(\left(\frac{r}{2}+\frac{1}{2}\right)-(2-r)^2\right)(-\widehat{u}_{xxxx})\\ \underline{u}_{tt}-\overline{u}_{xx}-\left(\underline{u}^2\right)_{xx}-\overline{u}_{xxxx}-\left(\frac{-3r}{2}+\frac{3}{2}\right)(-\widehat{u}_{xx})-\left((2-r)-\left(\frac{r}{2}+\frac{1}{2}\right)^2\right)(-\widehat{u}_{xxxx})\end{pmatrix}=\begin{pmatrix}0\\0\end{pmatrix}.$$

We consider $H=I$ and $h=-I$, and also, we choose the initial approximate as $\begin{pmatrix}\left(\frac{r}{2}+\frac{1}{2}\right)\frac{-6x+12t}{x^3}\\ (2-r)\frac{-6x+12t}{x^3}\end{pmatrix}$. According to Eq. (13), we have

$$R_m(\vec{u}_{m-1})]=\begin{pmatrix}\frac{\partial^2}{\partial t^2}\underline{u}_{m-1}-\frac{\partial^2}{\partial x^2}\sum_{m=1}^{+\infty}\overline{u}_{m-1}-\frac{\partial^2}{x^2}\left[\sum_{i=0}^{m-1}\overline{v}_i\overline{v}_{m-i-1}\right]-\frac{\partial^4}{\partial x^4}\underline{u}_{m-1}\\ -\left(\frac{3}{2}r-\frac{3}{2}\right)\frac{\partial^2}{x^2}(-\widehat{u})-\left(\left(\frac{r}{2}+\frac{1}{2}\right)-(2-r)^2\right)\frac{\partial^4}{\partial x^4}(-\widehat{u})\\ \frac{\partial^2}{\partial t^2}\underline{u}_{m-1}-\frac{\partial^2}{\partial x^2}\sum_{m=1}^{+\infty}\overline{u}_{m-1}-\frac{\partial^2}{x^2}\left[\sum_{i=0}^{m-1}\underline{u}_i\underline{u}_{m-i-1}\right]+\frac{\partial^4}{\partial x^4}\overline{u}_{m-1}\\ -\left(\frac{3}{2}-\frac{3}{2}r\right)\frac{\partial^2}{x^2}(-\widehat{u})-\left((2-r)-\left(\frac{r}{2}+\frac{1}{2}\right)^2\right)\frac{\partial^4}{\partial x^4}(-\widehat{u})\end{pmatrix}.$$

Therefore,

$$u_0=\begin{pmatrix}\left(\frac{r}{2}+\frac{1}{2}\right)\left(\frac{6}{x^2}-\frac{12t}{x^3}\right)\\ (2-r)\left(\frac{6}{x^2}-\frac{12t}{x^3}\right)\end{pmatrix}$$

$$u_1 = \begin{pmatrix} -\dfrac{(18r+18)t^2}{2x^4} - \dfrac{(36-36r)t^3}{3x^5} + \dfrac{3(42+48r+42r^2)t^4}{x^8} \\ -\dfrac{(-36r+72)t^2}{2x^4} - \dfrac{(144r+72)t^3}{3x^5} + \dfrac{3(1344-1344r+336r^2)t^4}{2x^8} \end{pmatrix}$$

$$u_2 = \begin{pmatrix} -\dfrac{(60r+60)t^4}{4x^6} - \dfrac{(90r-90)t^5}{5x^7} - \dfrac{3(42+48r+42r^2)t^4}{x^8} + \cdots \\ -\dfrac{(240-120r)t^4}{4x^6} - \dfrac{(180r-360)t^5}{5x^7} - \dfrac{3(1344-1344r+336r^2)t^4}{2x^8} + \cdots \end{pmatrix}$$

$$\vdots$$

In general, the series solution is given by

$$u(x,t) = \sum_{n=0}^{\infty} u_n(x,t) = \begin{pmatrix} \left(\frac{r}{2}+\frac{1}{2}\right)\left(-\frac{6}{x^2}+\frac{12}{x^3}t-\frac{18}{x^4}t^2+\frac{24}{x^5}t^3-\frac{30}{x^6}t^4+\cdots\right) \\ (2-r)\left(-\frac{6}{x^2}+\frac{12}{x^3}t-\frac{18}{x^4}t^2+\frac{24}{x^5}t^3-\frac{30}{x^6}t^4+\cdots\right) \end{pmatrix}.$$

That gives the exact solution

$$\begin{pmatrix} \left(\frac{r}{2}+\frac{1}{2}\right)\dfrac{-6}{(x+t)^2} \\ (2-r)\dfrac{-6}{(x+t)^2} \end{pmatrix}.$$

Therefore, $\widetilde{u} = \left(\frac{r}{2}+\frac{1}{2}, 2-r\right)\left(\frac{-6}{(x+t)^2}\right)$ is the exact solution of the fuzzy differential equation.

Conclusion

In this work, we applied the fuzzy HAM in order to solve the fuzzy Boussinesq equation. For this aim, we considered the parametric form of a fuzzy number and established the deformation equations for two crisp Bousinnesq equations obtained from the proposed method. Also, we presented a theorem to warrant the convergence of the proposed method too. Similar to the discussion in this work, the HAM can be used in order to solve other kinds of fuzzy differential equations as an efficient and proper method.

Acknowledgments The authors would like to thank the anonymous reviewers for their careful reading and suggestions to improve the quality of this work and also are thankful to the Islamic Azad University, Central Tehran branch for their support during this research.

References

1. Abbasbandy, S., Allahviranloo, T.: Numerical solution of fuzzy differential equation by Taylor method. J. Comput. Methods Appl. Math. **2**, 113–124 (2002)
2. Abbasbandy, S., Allahviranloo, T., Loez-Pouso, O., Nieto, J.J.: Numerical methods for fuzzy differential inclusions. J. Comput. Math. Appl. **48**, 1633–1641 (2004)
3. Allahviranloo, T., Ahmadi, N., Ahmadi, E.: Numerical solution of fuzzy differential equations by predictor-corrector method. Inf. Sci. **177**, 1633–1647 (2007)
4. Allahviranloo, T., Ahmadi, E., Ahmadi, A.: *N*-th fuzzy differential equations. Inf. Sci. **178**, 1309–1324 (2008)
5. Allahviranloo, T., Kiani, N.A., Motamedi, N.: Solving fuzzy differential equations by differential transformation method. Inf. Sci. **179**(7), 956–966 (2009)
6. Allahviranloo, T., Chehlabi, M.: Solving fuzzy differential equations based on the length function properties. Soft Comput. **19**, 307–320 (2015)
7. Allahviranloo, T., Salahshour, S.: Euler method for solving hybrid fuzzy differential equation. Soft Comput. **15**, 1247–1253 (2011)
8. Buckley, J.J., Feuring, T.: Fuzzy differential equations. Fuzzy Sets Syst. **110**, 43–54 (2000)
9. Bede, B., Gal, S.G.: Generalizations of the differentiability of fuzzy number value functions with applications to fuzzy differential equations. Fuzzy Sets Syst. **151**, 581–599 (2005)
10. Chalco-Cano, Y., Roman-Flores, H.: Comparision between some approaches to solve the fuzzy differential equations. Fuzzy Sets Syst. **160**, 1517–1562 (2009)
11. Chalco-Cano, Y., Roman-Flores, H.: On new solutions of fuzzy differential equations. Chaos Solitons Fractals **38**, 112–119 (2006)
12. Cong-xin, W., Ming, M.: Embedding problem of fuzzy number space: I. Fuzzy Sets Syst. **44**, 33–38 (1991)
13. Araghi, M.A.F., Fallahzadeh, A.: On the convergence of the homotopy analysis method for solving the Schrodinger equation. J. Basic Appl. Sci. Res. **2**(6), 6076–6083 (2012)
14. Araghi, M.A.F., Fallahzadeh, A.: Explicit series solution of Boussinesq equation by homotopy analysis method. J. Am. Sci. **8**(11), 448–452 (2012)

15. Goetschel, R., Voxman, W.: Elementary fuzzy calculus. Fuzzy Sets Syst. **18**, 31–43 (1986)
16. Liao, S.J.: Beyond Pertubation: Introduction to the Homotopy Analysis Method. Chapman and Hall/CRC Press, Boca Raton (2003)
17. Liao, S.J.: Notes on the homotopy analysis method: some definitions and theorems. Commun. Nonlinear Sci. Numer. Simul. **14**, 983–997 (2009)
18. Lupulescu, V.: On a class of fuzzy functional differential equations. Fuzzy Sets Syst. **160**, 1547–1562 (2009)
19. Palligkinis, SCh., Papageorgiou, G., Famelis, ITh: Runge–Kutta methods for fuzzy differential equations. Appl. Math. Comput. **209**, 97–105 (2009)
20. Puri, M.L., Ralescu, D.: Fuzzy random variables. J. Math. Anal. Appl. **114**, 409–422 (1986)
21. Seikkala, S.: On the fuzzy initial value problem. Fuzzy Set Syst. **24**, 319–330 (1987)
22. Zimmerman, H.J.: Fuzzy Set Theory and Its Applications. Kluwer Academic, New York (1996)

8

Contra g-α- and g-β-preirresolute functions on GTS's

A. Acikgoz[1] · N. A. Tas[1] · M. S. Sarsak[2]

Abstract In this present paper, we define g-α-preirresolute, g-β-preirresolute, contra g-α-preirresolute and contra g-β-preirresolute functions on generalized topological spaces. We give some examples of this definitions. We investigate some properties and characterizations of this functions.

Keywords g-α-preirresolute · g-β-preirresolute · Contra g-α-preirresolute · Contra g-β-preirresolute

Mathematics Subject Classification 54A05 · 54C08

Introduction

Császár [2] introduced generalized open sets in 1997. Subsequently, he [3] defined generalized topology and generalized continuity in 2002. Also, (g_X, g_Y)-open functions [4] were introduced in 2003 and strong generalized topology [5] was presented in 2004. g-semi-open sets, g-preopen sets, g-α-open sets and g-β-open sets [6] were introduced by Császár in 2005. Also he [7] showed how the definition of the product of generalized topologies in 2009. In 2012, Jayanthi [8] introduced contra continuity on generalized topological space. Furthermore, Min [9] defined (α, g_Y)-continuous functions, (σ, g_Y)—continuous functions, (π, g_Y)-continuous functions and (β, g_Y)-continuous functions on generalized topological spaces in 2009. Additionally, Bai and Zuo [1] introduced g-α-irresolute functions in 2011. In 2009, Shen [10] studied the relationship between the product and some operations (σ, π, α and β) of generalized topologies. Our aim in this paper, is to introduce g-α-preirresolute, g-β-preirresolute, contra g-α-preirresolute, contra g-β-preirresolute on generalized topological spaces. Also we obtain some properties and characterizations of this functions.

Preliminaries

Definition 2.1 [3] Let $X \neq \emptyset$ and $g \subseteq X$. Then g is called a generalized topology (briefly; GT) on X iff $\emptyset \in g$ and $G_i \in g$ for $i \in I \neq \emptyset$ implies $G = \bigcup_{i \in I} G_i \in g$. The pair (X, g) is called a generalized topological space (briefly; GTS) on X. The elements of g are called g-open sets and their complements are called g-closed sets.

Definition 2.2 [3] Let (X, g) be a generalized topological space and $A \subseteq X$.

(1) The closure of A is defined as follows:

$$c_g(A) = \bigcap\{F : F \, is \, g\text{-}closed, \, A \subseteq F\}.$$

(2) The interior of A is defined as follows:

$$i_g(A) = \bigcup\{G : G \, is \, g\text{-}open, G \subseteq A\}.$$

Theorem 2.3 [3] *Let (X, g) be a generalized topological space. Then the following hold:*

✉ N. A. Tas
nihalarabacioglu@hotmail.com

A. Acikgoz
ahuacikgoz@gmail.com

M. S. Sarsak
sarsak@hu.edu.jo

1 Department of Mathematics, Balikesir University, 10145 Balikesir, Turkey

2 Department of Mathematics, The Hashemite University, P.O. Box 150459, Zarqa 13115, Jordan

(1) $c_g(A) = X - i_g(X - A)$.
(2) $i_g(A) = X - c_g(X - A)$.

Definition 2.4 [6] Let (X, g) be a generalized topological space and $A \subseteq X$. A is said to be

(1) g-semi-open if $A \subseteq c_g(i_g(A))$;
(2) g-preopen if $A \subseteq i_g(c_g(A))$;
(3) g-α-open if $A \subseteq i_g(c_g(i_g(A)))$;
(4) g-β-open if $A \subseteq c_g(i_g(c_g(A)))$.

The complement of g-semi-open (resp. g-preopen, g-α-open, g-β-open) is said to be g-semi-closed (resp. g -preclosed, g-α-closed, g-β-closed). The set of all g-semi-open sets (resp. g-preopen sets, g-α-open sets, g-β-open sets) is denoted by $\sigma(g)$ (resp. $(\pi(g), \alpha(g), \beta(g))$.

The closure of g-semi-closed (resp. g-preclosed, g-α-closed, g-β-closed) sets is denoted by $c_\sigma(X)$ (resp. $c_\pi(X)$, $c_\alpha(X)$, $c_\beta(X)$). Also the interior of g-semi-open (resp. g-preopen, g-α-open, g-β-open) sets is denoted by $i_\sigma(X)$ (resp. $i_\pi(X)$, $i_\alpha(X)$, $i_\beta(X)$).

Definition 2.5 [4] Let (X, g_X) and (Y, g_Y) be GTS's. Then a function $f : X \to Y$ is said to be (g_X, g_Y)-open if $f(U) \in g_Y$ for each $U \in g_X$.

Definition 2.6 [3] Let (X, g_X) and (Y, g_Y) be GTS's. Then a function $f : X \to Y$ is said to be (g_X, g_Y)-continuous if $f^{-1}(V) \in g_X$ for each $V \in g_Y$.

Definition 2.7 [9] Let (X, g_X) and (Y, g_Y) be GTS's. Then a function $f : X \to Y$ is said to be

(1) (α, g_Y)-continuous if $f^{-1}(V)$ is g-α-open in X for each g-open set V in Y;
(2) (σ, g_Y)-continuous if $f^{-1}(V)$ is g-semi-open in X for each g-open set V in Y.
(3) (π, g_Y)-continuous if $f^{-1}(V)$ is g-preopen in X for each g-open set V in Y.
(4) (β, g_Y)-continuous if $f^{-1}(V)$ is g-β-open in X for each g-open set V in Y.

Definition 2.8 [8] Let (X, g_X) and (Y, g_Y) be GTS's. Then a function $f : X \to Y$ is said to be

(1) contra (g_X, g_Y)-continuous if $f^{-1}(V)$ is g-closed in X for each $V \in g_Y$.
(2) contra (α, g_Y)-continuous if $f^{-1}(V)$ is g-α-closed in X for each g-open set V in Y.
(3) contra (σ, g_Y)-continuous if $f^{-1}(V)$ is g-semi-closed in X for each g-open set V in Y.
(4) contra (π, g_Y)-continuous if $f^{-1}(V)$ is g-preclosed in X for each g-open set V in Y.
(5) contra (β, g_Y)-continuous if $f^{-1}(V)$ is g-β-closed in X for each g-open set V in Y.

Definition 2.9 [5] Let g be a GT on a set $X \neq \emptyset$. Then g is said to be strong if $X \in g$.

Definition 2.10 [7] Let $K \neq \emptyset$ be an index set, $X_k \neq \emptyset$ for $k \in K$ and $X = \prod_{k \in K} X_k$ the cartesian product of the sets X_k. Also $p_k : X \to X_k$ is the projection.

Let g_k be a given GT on X_k for $k \in K$. Then g is called the product of the GT's g_k.

Proposition 2.11 [10] *If every g_{X_k} is strong then each p_k is (g_X, g_{X_k})-continuous (resp. $(\alpha(g_X), \alpha(g_{X_k}))$-continuous, $(\sigma(g_X), \sigma(g_{X_k}))$-continuous, $(\pi(g_X), \pi(g_{X_k}))$-continuous, $(\beta(g_X), \beta(g_{X_k}))$-continuous) for $k \in K$.*

Theorem 2.12 [7] *Let $G = \prod_{k \in K} G_k$. Then*

(1) *If K is finite and every G_k is g-semi-open, then G is g-semi-open set.*
(2) *If K is finite and every G_k is g-preopen, then G is g-preopen set.*
(3) *If K is finite and every G_k is g-α-open, then G is g-α-open set.*
(4) *If K is finite and every G_k is g-β-open, then G is g-β-open set.*

Definition 2.13 [1] A function $f : X \to Y$ is said to be g-α-irresolute if $f^{-1}(V)$ is g-α-open in X for every g-α-open set V of Y.

g-α-Preirresolute and g-β-preirresolute functions

Definition 3.1 Let (X, g_X) and (Y, g_Y) be GTS's. Then a function $f : X \to Y$ is said to be g-α-preirresolute if $f^{-1}(V)$ is g-α-open in X for every g-preopen set V of Y.

Example 3.2 Let $X = \{x, y\}$, $Y = \{a, b\}$, $g_X = P(X)$ and $g_Y = \{\emptyset, \{a\}\}$. Then we obtain $\pi(g_Y) = \{\emptyset, \{a\}\}$.

$f : (X, g_X) \to (Y, g_Y)$ such that $f(x) = a, f(y) = b$.

Since $f^{-1}(\emptyset) = \emptyset$ and $f^{-1}(\{a\}) = \{x\}$ are g-α-open subsets of X, then f is g-α-preirresolute.

Definition 3.3 Let (X, g_X) and (Y, g_Y) be GTS's. Then a function $f : X \to Y$ is said to be g-β-preirresolute if $f^{-1}(V)$ is g-β-open in X for every g-preopen set V of Y.

Example 3.4 Let $X = \{x, y\}$, $Y = \{a, b, c\}$, $g_X = \{\emptyset, \{x\}\}$ and $g_Y = \{\emptyset, \{a\}, \{a, b\}\}$. Then we obtain $\pi(g_Y) = \{\emptyset, \{a\}, \{a, b\}\}$.

$f : (X, g_X) \to (Y, g_Y)$ such that $f(x) = f(y) = a$.

Since $f^{-1}(\emptyset) = \emptyset$, $f^{-1}(\{a\}) = X$ and $f^{-1}(\{a, b\}) = X$ are g-β-open subsets of X, then f is g-β-preirresolute.

Definition 3.5 Let (X, g_X) and (Y, g_Y) be GTS's. Then a function $f : X \to Y$ is said to be g-α-preirresolute at $x \in X$

if there exists a g-α-open set U of X containing x such that $f(U) \subseteq V$ for each g-preopen set V of Y containing $f(x)$.

Definition 3.6 Let (X, g_X) and (Y, g_Y) be GTS's. Then a function $f : X \to Y$ is said to be g-β-preirresolute at $x \in X$ if there exists a g-β-open set U of X containing x such that $f(U) \subseteq V$ for each g-preopen set V of Y containing $f(x)$.

Theorem 3.7 *Let (X, g_X), (Y, g_Y) be GTS's and $f : X \to Y$ be a function. The following conditions are equivalent:*

(1) *f is g-α-preirresolute;*
(2) *For each $x \in X$ and each g-preopen set V of Y containing $f(x)$, there exists a g-α-open set U of X containing x such that $f(U) \subseteq V$;*
(3) *$f^{-1}(V) \subseteq i_g(c_g(i_g(f^{-1}(V))))$ for every g-preopen set V of Y;*
(4) *$f^{-1}(V)$ is g-α-closed in X for every g-preclosed set V of Y;*
(5) *$c_g(i_g(c_g(f^{-1}(V)))) \subseteq f^{-1}(c_\pi(V))$ for every subset V of Y;*
(6) *$f(c_g(i_g(c_g(U)))) \subseteq c_\pi(f(U))$ for every subset U of X.*

Proof (1) $\Rightarrow$ (2). Let $x \in X$ and V be any g-preopen set of Y containing $f(x)$. By hypothesis, $f^{-1}(V)$ is g-α-open in X and contains x. Suppose $U = f^{-1}(V)$, then U is g-α-open set in X containing x and $f(U) \subseteq V$.

(2) $\Rightarrow$ (3). Let V be any g-preopen set of Y and $x \in f^{-1}(V)$. By hypothesis, there exists a g-α-open set U of X such that $f(U) \subseteq V$. Hence we obtain

$$x \in U \subseteq i_g(c_g(i_g(U))) \subseteq i_g(c_g(i_g(f^{-1}(V)))).$$

As a consequence, $f^{-1}(V) \subseteq i_g(c_g(i_g(f^{-1}(V))))$.

(3) $\Rightarrow$ (4). Let V be any g-preclosed of Y. Then $U = Y - V$ is g-preopen in Y. By (3), we have $f^{-1}(U) \subseteq i_g(c_g(i_g(f^{-1}(U))))$. Therefore

$$\begin{aligned} f^{-1}(U) &= f^{-1}(Y - V) = X - f^{-1}(V) \subseteq i_g(c_g(i_g(f^{-1}(U)))) \\ &= X - c_g(i_g(c_g(f^{-1}(V)))). \end{aligned}$$

As a consequence, we obtain $f^{-1}(V)$ is g-α-closed set in X.

(4) $\Rightarrow$ (5). Let V be any subset of Y. Since $c_\pi(V)$ is g-preclosed subset of Y, then $f^{-1}(c_\pi(V))$ is g-α-closed in X by (4). Hence

$$c_g(i_g(c_g(f^{-1}(c_\pi(V))))) \subseteq f^{-1}(c_\pi(V)).$$

Therefore we obtain $c_g(i_g(c_g(f^{-1}(V)))) \subseteq f^{-1}(c_\pi(V))$.

(5) $\Rightarrow$ (6). Let U be any subset of X. By hypothesis, we have

$$c_g(i_g(c_g(U))) \subseteq c_g(i_g(c_g(f^{-1}(f(U))))) \subseteq f^{-1}(c_\pi(f(U))).$$

As a consequence, $f(c_g(i_g(c_g(U)))) \subseteq c_\pi(f(U))$.

(6) $\Rightarrow$ (1). Let V be any g-preopen subset of Y. $f^{-1}(Y - V) = X - f^{-1}(V)$ is a subset of X and by hypothesis, we obtain

$$\begin{aligned} f(c_g(i_g(c_g(f^{-1}(Y - V))))) &\subseteq c_\pi(f(f^{-1}(Y - V))) \\ &\subseteq c_\pi(Y - V) = Y - i_\pi(V) \\ &= Y - V \end{aligned}$$

and so

$$\begin{aligned} &X - i_g(c_g(i_g(f^{-1}(V)))) = c_g(i_g(c_g(X - f^{-1}(V)))) = \\ &c_g(i_g(c_g(f^{-1}(Y - V)))) \subseteq f^{-1}(f(c_g(i_g(c_g(f^{-1}(Y - V)))))) \\ &\subseteq f^{-1}(Y - V) = X - f^{-1}(V). \end{aligned}$$

Thus $f^{-1}(V) \subseteq i_g(c_g(i_g(f^{-1}(V))))$ and $f^{-1}(V)$ is g-α-open set in X. As a consequence, f is g-α-preirresolute. □

Theorem 3.8 *Let (X, g_X), (Y, g_Y) be GTS's and $f : X \to Y$ be a function. The following conditions are equivalent:*

(1) *f is g-β-preirresolute;*
(2) *For each $x \in X$ and each g-preopen set V of Y containing $f(x)$, there exists a g-β-open set U of X containing x such that $f(U) \subseteq V$;*
(3) *$f^{-1}(V) \subseteq c_g(i_g(c_g(f^{-1}(V))))$ for every g-preopen set V of Y;*
(4) *$f^{-1}(V)$ is g-β-closed in X for every g-preclosed set V of Y;*
(5) *$i_g(c_g(i_g(f^{-1}(V)))) \subseteq f^{-1}(c_\pi(V))$ for every subset V of Y;*
(6) *$f(i_g(c_g(i_g(U)))) \subseteq c_\pi(f(U))$ for every subset U of X.*

Proof It is proved similar to the proof of Theorem 3.7. □

Theorem 3.9 *Let (X, g_X), (Y, g_Y) be GTS's and $f : X \to Y$ be a function. The following conditions are equivalent:*

(1) *f is g-α-preirresolute;*
(2) *$f^{-1}(F)$ is g-α-closed in X for every g-preclosed set F of Y;*
(3) *$f(c_\alpha(A)) \subseteq c_\pi(f(A))$ for every subset A of X;*
(4) *$c_\alpha(f^{-1}(B)) \subseteq f^{-1}(c_\pi(B))$ for every subset B of Y;*
(5) *$f^{-1}(i_\pi(B)) \subseteq i_\alpha(f^{-1}(B))$ for every subset B of Y;*
(6) *f is g-α-preirresolute at every $x \in X$.*

Proof (1) $\Rightarrow$ (2). It is obvious from Theorem 3.7.

(2) $\Rightarrow$ (3). Let $A \subseteq X$. Then $c_\pi(f(A))$ is a g-preclosed set of Y. By hypothesis, $f^{-1}(c_\pi(f(A)))$ is a g-α-closed set. Now $c_\alpha(A) \subseteq c_\alpha(f^{-1}(f(A))) \subseteq c_\alpha(f^{-1}(c_\pi(f(A)))) = f^{-1}(c_\pi(f(A)))$. Hence $f(c_\alpha(A)) \subseteq c_\pi(f(A))$.

(3) $\Rightarrow$ (4). Let $B \subseteq Y$. Then $f^{-1}(B) \subseteq X$. By hypothesis, $f(c_\alpha(f^{-1}(B))) \subseteq c_\pi(f(f^{-1}(B))) \subseteq c_\pi(B)$. Hence $c_\alpha(f^{-1}(B))$

$\subseteq f^{-1}(f(c_\alpha(f^{-1}(B)))) \subseteq f^{-1}(c_\pi(B))$. So we obtain $c_\alpha(f^{-1}(B)) \subseteq f^{-1}(c_\pi(B))$.

(4) $\Rightarrow$ (5). It is obvious from the complement of (4).

(5) $\Rightarrow$ (1). Let V be any g-preopen set of Y, then $V = i_\pi(V)$. By hypothesis, $f^{-1}(V) = f^{-1}(i_\pi(V)) \subseteq i_\alpha(f^{-1}(V)) \subseteq f^{-1}(V)$. Hence $f^{-1}(V) = i_\alpha(f^{-1}(V))$. Thus $f^{-1}(V)$ is a g-α-open set of X. As a consequence, f is g-α-preirresolute.

(1) $\Rightarrow$ (6). Let f is g-α-preirresolute, $x \in X$ and any g-preopen set V of Y containing $f(x)$. Then $x \in f^{-1}(V)$ and $f^{-1}(V)$ is g-α-open set in X. Suppose $U = f^{-1}(V)$, then U is a g-α-open set of X and $f(U) \subseteq V$. Therefore f is g-α-preirresolute for each $x \in X$. □

Theorem 3.10 *Let (X, g_X), (Y, g_Y) be GTS's and $f : X \to Y$ be a function. The following conditions are equivalent:*

(1) *f is g-β-preirresolute;*
(2) *$f^{-1}(F)$ is g-β-closed in X for every g-preclosed set F of Y;*
(3) *$f(c_\beta(A)) \subseteq c_\pi(f(A))$ for every subset A of X;*
(4) *$c_\beta(f^{-1}(B)) \subseteq f^{-1}(c_\pi(B))$ for every subset B of Y;*
(5) *$f^{-1}(i_\pi(B)) \subseteq i_\beta(f^{-1}(B))$ for every subset B of Y;*
(6) *f is g-β-preirresolute at every $x \in X$.*

Proof It is proved by a similar way in Theorem 3.9. □

Theorem 3.11 *Let $f : X \to Y$ be a function from two GTS's. Then f is g-α-preirresolute if $f^{-1}(V) \subseteq i_g(c_g(i_g(f^{-1}(i_\pi(V)))))$ for every g-preopen subset V of Y.*

Proof Let V be g-preclosed set of Y. Then $Y - V$ is g-preopen set in Y. By hypothesis, $f^{-1}(Y - V) = X - f^{-1}(V) \subseteq i_g(c_g(i_g(f^{-1}(i_\pi(Y - V))))) = i_g(c_g(i_g(f^{-1}(Y - V)))) = X - c_g(i_g(c_g(f^{-1}(V))))$. Hence we obtain $c_g(i_g(c_g(f^{-1}(V)))) \subseteq f^{-1}(V)$. Therefore $f^{-1}(V)$ is g-α-closed set in X. As a consequence, f is g-α-preirresolute from Theorem 3.7(4). □

Theorem 3.12 *Let $f : X \to Y$ be a function from two GTS's. Then f is g-β-preirresolute if $f^{-1}(V) \subseteq c_g(i_g(c_g(f^{-1}(i_\pi(V)))))$ for every g-preopen subset V of Y.*

Proof It is similar to Theorem 3.11. □

Theorem 3.13 *Let $f : X \to Y$ be a function from two GTS's and g_X be a strong. f is g-α-preirresolute if the graph function $g : X \to X \times Y$ defined by $g(x) = (x, f(x))$ for each $x \in X$, is g-α-preirresolute.*

Proof Let $x \in X$ and V be any g-preopen set of Y containing $f(x)$. Then $X \times V$ is a g-preopen set of $X \times Y$ by Theorem 2.12 and contains $g(x)$. Since g is g-α-preirresolute, there exists a g-α-open U of X containing x such that $g(U) \subseteq X \times V$ and so $f(U) \subseteq V$. Thus f is g-α-preirresolute. □

Theorem 3.14 *Let $f : X \to Y$ be a function from two GTS's and g_X be a strong. f is g-β-preirresolute if the graph function $g : X \to X \times Y$ defined by $g(x) = (x, f(x))$ for each $x \in X$, is g-β-preirresolute.*

Proof The proof is similar to that of Theorem 3.13 □

Theorem 3.15 *Let g_{Y_k} be a given GT on Y_k for $k \in K$ and g_{Y_k} be a strong. If a function $f : X \to \prod Y_k$ is g-α-preirresolute, then $p_k \circ f : X \to Y_k$ is g-α-preirresolute for each $k \in K$, where p_k is the projection of $\prod Y_k$ onto Y_k.*

Proof Let V_k be any g-preopen set of Y_k. p_k is $(\pi(g_Y), \pi(g_{Y_k}))$-continuous from Proposition 2.11 since g_{Y_k} is strong and so $p_k^{-1}(V_k)$ is g-preopen set. Since f is g-α-preirresolute, $f^{-1}(p_k^{-1}(V_k)) = (p_k \circ f)^{-1}(V_k)$ is a g-α-open. As a consequence, we have $p_k \circ f$ is g-α-preirresolute for each $k \in K$.

□

Theorem 3.16 *Let g_{Y_k} be a given GT on Y_k for $k \in K$ and g_{Y_k} be a strong. If a function $f : X \to \prod Y_k$ is g-β-preirresolute, then $p_k \circ f : X \to Y_k$ is g-β-preirresolute for each $k \in K$, where p_k is the projection of $\prod Y_k$ onto Y_k.*

Proof It is proved similar to that of Theorem 3.15.

□

Theorem 3.17 *If the function $f : \prod X_k \to \prod Y_k$ defined by $f(\{x_k\}) = \{f_k(x_k)\}$ for each $\{x_k\} \in \prod X_k$, is g-α-preirresolute, then $f_k : X_k \to Y_k$ is g-α-preirresolute for each $k \in K$.*

Proof Let $k_0 \in K$ be an arbitrary fixed index and V_{k_0} be any g-preopen set of Y_{k_0}. Then $\prod Y_m \times V_{k_0}$ is g-preopen in $\prod Y_k$ by Theorem 2.12, where $k_0 \neq m \in K$. Since f is g-α-preirresolute, $f^{-1}(\prod Y_m \times V_{k_0}) = \prod X_m \times f_{k_0}^{-1}(V_{k_0})$ is g-α-open in $\prod X_k$ and $f_{k_0}^{-1}(V_{k_0})$ is g-α-open in X_{k_0}. As a consequence, f_{k_0} is g-α-preirresolute. □

Theorem 3.18 *If the function $f : \prod X_k \to \prod Y_k$ defined by $f(\{x_k\}) = \{f_k(x_k)\}$ for each $\{x_k\} \in \prod X_k$, is g-β-preirresolute, then $f_k : X_k \to Y_k$ is g-β-preirresolute for each $k \in K$.*

Proof It is proved by a similar way in Theorem 3.17. □

Theorem 3.19 *If $f : X \to Y$ is g-α-preirresolute and A is a g-α-open in X , then the restriction $f|A : A \to Y$ is g-α-preirresolute.*

Proof Let V be any g-preopen set in Y. Then we have $f^{-1}(V)$ is a g-α-open set in Y. Since the set A is a g-α-open set, we have $(f|A)^{-1}(V) = A \cap f^{-1}(V)$ is g-α-open. Therefore $f|A$ is g-α-preirresolute. □

Theorem 3.20 *If $f : X \to Y$ is g-β-preirresolute and A is g-open in X, then the restriction $f|A : A \to Y$ is g-β-preirresolute.*

Proof It is proved by a similar way of that of Theorem 3.19. □

Definition 3.21 Let (X, g_X) and (Y, g_Y) be GTS's. Then a function $f : X \to Y$ is said to be g-preirresolute if $f^{-1}(V)$ is g-preopen in X for every g-preopen set V of Y.

Theorem 3.22 *Let (X, g_X), (Y, g_Y) and (Z, g_Z) be GTS's. If $f : X \to Y$ is g-α-preirresolute and $g : Y \to Z$ is g-preirresolute, then the composition $g \circ f : X \to Z$ is g-α-preirresolute.*

Proof Let V be any g-preopen subset of Z. Since g is g-preirresolute, $g^{-1}(V)$ is g-preopen in Y. Since f is g-α-preirresolute, then $f^{-1}(g^{-1}(V)) = (g \circ f)^{-1}(V)$ is g-α-open in X. As a consequence, $g \circ f$ is g-α-preirresolute. □

Theorem 3.23 *Let (X, g_X), (Y, g_Y) and (Z, g_Z) be GTS's. If $f : X \to Y$ is g-β-preirresolute and $g : Y \to Z$ is g-preirresolute, then the composition $g \circ f : X \to Z$ is g-β-preirresolute.*

Proof It is similar to that of Theorem 3.22 □

Definition 3.24 Let (X, g_X) and (Y, g_Y) be GTS's. Then a function $f : X \to Y$ is said to be g-α-pre-continuous if $f^{-1}(V)$ is g-preopen in X for every g-α-open set V of Y.

Definition 3.25 Let (X, g_X) and (Y, g_Y) be GTS's. Then a function $f : X \to Y$ is said to be almost g-α-irresolute if $f^{-1}(V)$ is g-β-open in X for every g-α-open set V of Y.

From the definitions stated above, we obtain the following diagram:

g-α-preirresolute ⟶ g-preirresolute ⟶ g-β-preirresolute
↘ ↓ ↙
g-α-irresolute ⟶ g-α-pre-continuity ⟶ almost g-α-irresolute

Remark 3.26 The following examples enables us to realize that none of these implications is reversible.

Example 3.27 Let $X = Y = \{a, b, c, d\}$, $g_X = \{\emptyset, \{a\}, \{d\}, \{b, c\}, \{a, b, c\}, \{b, c, d\}, \{a, d\}\}$ and $g_Y = \{\emptyset, Y, \{b\}\}$. The identity function $f : X \to Y$ is g-α-pre-continuous function, but it is not g-α-irresolute. Also, f is g-preirresolute, but it is not g-α-preirresolute.

Example 3.28 Let $X = Y = \{a, b, c, d\}$, $g_X = \{\emptyset, X, \{a\}, \{c\}, \{a, c\}\}$ and $g_Y = \{\emptyset, Y, \{c, d\}\}$. The identity function $f : X \to Y$ is almost g-α-irresolute function, but it is neither g-α-pre-continuous nor g-β-preirresolute.

Example 3.29 Let $X = Y = \{a, b\}$ and $g_X = g_Y = \{\emptyset, \{a\}\}$. We define the function $f : X \to Y$ such that $f(a) = f(b) = a$. Then f is g-β-preirresolute function, but it is not g-preirresolute.

Example 3.30 Let $X = Y = \{a, b, c\}$ and $g_X = g_Y = \{\emptyset, X, \{a\}, \{b, c\}\}$. The identity function $f : X \to Y$ is g-α-irresolute function, but it is not g-α-preirresolute.

Contra g-α-preirresolute and contra g-β-preirresolute functions

Definition 4.1 Let (X, g_X) and (Y, g_Y) be GTS's. Then a function $f : X \to Y$ is said to be contra g-α-preirresolute if $f^{-1}(V)$ is g-α-closed in X for every g-preopen V of Y.

Example 4.2 Let $X = \{x, y\}$, $Y = \{a, b, c\}$, $g_X = \{\emptyset, \{y\}, X\}$ and $g_Y = \{\emptyset, \{a\}, \{c\}, \{a, c\}\}$. Then we obtain $\pi(g_Y) = \{\emptyset, \{a\}, \{c\}, \{a, c\}\}$.

$f : (X, g_X) \to (Y, g_Y)$ such that

$f(x) = a, f(y) = b.$

Since $f^{-1}(\emptyset) = \emptyset$, $f^{-1}(\{a\}) = \{x\}$, $f^{-1}(\{c\}) = \emptyset$ and $f^{-1}(\{a, c\}) = \{x\}$ are g-α-closed subsets of X, then f is contra g-α-preirresolute.

Definition 4.3 Let (X, g_X) and (Y, g_Y) be GTS's. Then a function $f : X \to Y$ is said to be contra g-β-preirresolute if $f^{-1}(V)$ is g-β-closed in X for every g-preopen V of Y.

Example 4.4 Let $X = \{x, y, z\}$, $Y = \{a, b\}$, $g_X = \{\emptyset, \{x\}\}$ and $g_Y = \{\emptyset, \{a\}\}$. Then we obtain $\pi(g_Y) = \{\emptyset, \{a\}\}$.

$f : (X, g_X) \to (Y, g_Y)$ such that

$f(x) = a, f(y) = f(z) = b.$

Since $f^{-1}(\emptyset) = \emptyset$ and $f^{-1}(\{a\}) = \{x\}$ are g-β-closed subsets of X, then f is contra g-β-preirresolute.

Definition 4.5 Let (X, g_X) and (Y, g_Y) be GTS's. Then a function $f : X \to Y$ is said to be contra g-α-preirresolute at $x \in X$ if there exists a g-α-open set U containing x such that $f(U) \subseteq V$ for each g-preclosed V of Y containing $f(x)$.

Definition 4.6 Let (X, g_X) and (Y, g_Y) be GTS's. Then a function $f : X \to Y$ is said to be contra g-β-preirresolute at $x \in X$ if there exists a g-β-open set U containing x such that $f(U) \subseteq V$ for each g-preclosed V of Y containing $f(x)$.

Theorem 4.7 *Let $f : X \to Y$ be a function from two GTS's. Then the following are equivalent*:

(1) *f is contra g-α-preirresolute*;
(2) *$f^{-1}(F)$ is g-α-open set in X for each g-preclosed set F of Y*;
(3) For each $x \in X$ *and each g-preopen set V of Y with $f(x) \notin V$, there exists g-α-closed set U in X such that $x \notin U$ and $f^{-1}(V) \subseteq U$*;
(4) *f is contra g-α-preirresolute at any $x \in X$*;
(5) *$f^{-1}(V) \subseteq i_\alpha(f^{-1}(V))$ for any g-preclosed set V of Y*;
(6) *$c_\alpha(f^{-1}(U)) \subseteq f^{-1}(U)$ for any g-preopen set U of Y*;
(7) *$c_\alpha(f^{-1}(i_\pi(A))) \subseteq f^{-1}(i_\pi(A))$ for any $A \subseteq Y$*;
(8) *$f^{-1}(c_\pi(A)) \subseteq i_\alpha(f^{-1}(c_\pi(A)))$ for any $A \subseteq Y$.*

Proof (1) $\Rightarrow$ (2). Let F be a g-preclosed set in Y. Then $Y - F$ is a g-preopen set in Y. By (1), $f^{-1}(Y - F) = X - f^{-1}(F)$ is a g-α-closed set in X. Hence $f^{-1}(F)$ is a g-α-open set in X.

(1) $\Rightarrow$ (3). Let $x \in X$ and V be a g-preopen set of Y with $f(x) \notin V$. Then $x \notin f^{-1}(V)$. By (1), $f^{-1}(V)$ is a g-α-closed set in X. Suppose $U = f^{-1}(V)$. Then $f^{-1}(V) \subseteq U$ and $x \notin U$.

(3) $\Rightarrow$ (1). Let V be a g-preopen set of Y. For each $x \in f^{-1}(Y - V), f(x) \notin V$. By (3), there exists a g-α-closed set U_x in X such that $x \notin U_x$ and $f^{-1}(V) \subseteq U_x$. Then $X - U_x \subseteq X - f^{-1}(V) = f^{-1}(Y - V)$. Hence we have

$$\bigcup_{x \in f^{-1}(Y-V)} \{x\} \subseteq \bigcup_{x \in f^{-1}(Y-V)} (X - U_x) \subseteq f^{-1}(Y - V).$$

Thus $f^{-1}(Y - V) = \bigcup_{x \in f^{-1}(Y-V)} (X - U_x)$ is a g-α-open set in X. As a consequence, $f^{-1}(V)$ is a g-α-closed set in X and so f is g-α-preirresolute.

(2) $\Rightarrow$ (4). Let $x \in X$ and V be a g-preclosed set of Y containing $f(x)$. By (2), $f^{-1}(V)$ is a g-α-open set in X containing x. Put $U = f^{-1}(V)$. Thus we obtain U is a g-α-open set in X containing x and $f(U) \subseteq V$.

(4) $\Rightarrow$ (5). Let V be a g-preclosed set of Y. For each $x \in f^{-1}(V), f(x) \in V$. By (4), there exists a g-α-open set U in X containing x such that $f(U) \subseteq V$. Since $x \in U \subseteq f^{-1}(V)$, we obtain $x \in i_\alpha(f^{-1}(V))$. Thus $f^{-1}(V) \subseteq i_\alpha(f^{-1}(V))$.

(5) $\Rightarrow$ (6). Let U be a g-preopen set of Y. Then $Y - U$ is a g-preclosed set of Y. By (5), $X - f^{-1}(U) = f^{-1}(Y - U) \subseteq i_\alpha(f^{-1}(Y - U)) = i_\alpha(X - f^{-1}(U)) = X - c_\alpha(f^{-1}(U))$. Thus $c_\alpha(f^{-1}(U))) \subseteq f^{-1}(U)$.

(6) $\Rightarrow$ (7). Let $A \subseteq Y$. Since $i_\pi(A)$ is a g-preopen set of Y, by (6), we obtain $c_\alpha(f^{-1}(i_\pi(A))) \subseteq f^{-1}(i_\pi(A))$.

(7) $\Rightarrow$ (8). Let $A \subseteq Y$. Then $Y - A \subseteq Y$. By (7), $c_\alpha(f^{-1}(i_\pi(Y - A))) = c_\alpha(f^{-1}(Y - c_\pi(A))) = c_\alpha(X - f^{-1}(c_\pi(A))) = X - i_\alpha(f^{-1}(c_\pi(A))) \subseteq f^{-1}(i_\pi(Y - A)) = f^{-1}(Y - c_\pi(A)) = X - f^{-1}(c_\pi(A))$. Thus $f^{-1}(c_\pi(A)) \subseteq i_\alpha(f^{-1}(c_\pi(A)))$.

(8) $\Rightarrow$ (1). Let V be a g-preopen set of Y. Then $Y - V$ is g-preclosed set of Y. By (8), $f^{-1}(c_\pi(Y - V)) = f^{-1}(Y - V) = X - f^{-1}(V) \subseteq i_\alpha(f^{-1}(c_\pi(Y - V))) = i_\alpha(f^{-1}(Y - V)) = i_\alpha(X - f^{-1}(V)) = X - c_\alpha(f^{-1}(V))$. Thus we obtain $c_\alpha(f^{-1}(V)) \subseteq f^{-1}(V)$. As a consequence, $f^{-1}(V)$ is a g-α-closed set in X and f is contra g-α-preirresolute. □

Theorem 4.8 *Let $f : X \to Y$ be a function from two GTS's. Then the following are equivalent:*

(1) *f is contra g-β-preirresolute*;
(2) *$f^{-1}(F)$ is g-β-open set in X for each g-preclosed set F of Y*;
(3) For each $x \in X$ *and each g-preopen set V of Y with $f(x) \notin V$, there exists g-β-closed set U in X such that $x \notin U$ and $f^{-1}(V) \subseteq U$*;
(4) *f is contra g-β-preirresolute at any $x \in X$*;
(5) *$f^{-1}(V) \subseteq i_\beta(f^{-1}(V))$ for any g-preclosed set V of Y*;
(6) *$c_\beta(f^{-1}(U)) \subseteq f^{-1}(U)$ for any g-preopen set U of Y*;
(7) *$c_\beta(f^{-1}(i_\pi(A))) \subseteq f^{-1}(i_\pi(A))$ for any $A \subseteq Y$*;
(8) *$f^{-1}(c_\pi(A)) \subseteq i_\beta(f^{-1}(c_\pi(A)))$ for any $A \subseteq Y$.*

Proof It is similar to that of Theorem 4.7 □

Theorem 4.9 *Let $f : X \to Y$ be a function from two GTS's. Then the following are equivalent:*

(1) *f is contra g-α-preirresolute*;
(2) For each *g-preclosed set F of $Y, f^{-1}(F)$ is g-α-open in X*;
(3) *$f^{-1}(B) \subseteq i_g(c_g(i_g(f^{-1}(c_\pi(B)))))$ for every subset B of Y.*

Proof (1) $\Leftrightarrow$ (2) : It is obvious from Definition 4.1 and Theorem 4.7.

(2) $\Rightarrow$ (3) : Let $B \subseteq Y$. Since the set $c_\pi(B)$ is g-preclosed in $Y, f^{-1}(c_\pi(B))$ is g-α-open and so

$$f^{-1}(c_\pi(B)) \subseteq i_g(c_g(i_g(f^{-1}(c_\pi(B))))).$$

As a consequence, we obtain

$$f^{-1}(B) \subseteq i_g(c_g(i_g(f^{-1}(c_\pi(B))))).$$

(3) $\Rightarrow$ (1) : Let V be a g-preopen in Y. Then $Y - V$ is a subset of Y. By (3),

$$f^{-1}(Y - V) \subseteq i_g(c_g(i_g(f^{-1}(c_\pi(Y - V))))).$$

Hence we obtain

$$c_g(i_g(c_g(f^{-1}(V)))) = c_g(i_g(c_g(f^{-1}(i_\pi(V))))) \subseteq f^{-1}(V).$$

As a consequence, $f^{-1}(V)$ is g-α-closed. □

Theorem 4.10 *Let $f: X \to Y$ be a function from two GTS's. Then the following are equivalent:*

(1) *f is contra g-β-preirresolute;*
(2) *For each g-preclosed set F of Y, $f^{-1}(F)$ is g-β-open in X;*
(3) *$f^{-1}(B) \subseteq c_g(i_g(c_g(f^{-1}(c_\pi(B)))))$ for every subset B of Y.*

Proof It is proved by a similar way of that of Theorem 4.9. □

Theorem 4.11 *Let $f: X \to Y$ be a function from two GTS's. Suppose one of the following conditions hold:*

(1) *$f(c_\alpha(A)) \subseteq i_\pi(f(A))$ for each subset A in X.*
(2) *$c_\alpha(f^{-1}(B)) \subseteq f^{-1}(i_\pi(B))$ for each subset B in Y.*
(3) *$f^{-1}(c_\pi(B)) \subseteq i_\alpha(f^{-1}(B))$ for each subset B in Y.*

Then f is contra g-α-preirresolute.

Proof (1) ⇒ (2) : Let $B \subseteq Y$. Then $f^{-1}(B) \subseteq X$. By hypothesis, $f(c_\alpha(f^{-1}(B))) \subseteq i_\pi(f(f^{-1}(B))) \subseteq i_\pi(B)$. Then $f^{-1}(f(c_\alpha(f^{-1}(B)))) \subseteq f^{-1}(i_\pi(B))$. Hence $c_\alpha(f^{-1}(B)) \subseteq f^{-1}(f(c_\alpha(f^{-1}(B)))) \subseteq f^{-1}(i_\pi(B))$. As a consequence, (2) is obtained.

(2) ⇒ (3): It is obvious from the complement of (2).

Suppose (3) holds: Let $B \subseteq Y$ be g-preclosed. Then by (3), $f^{-1}(c_\pi(B)) \subseteq i_\alpha(f^{-1}(B))$. Thus $f^{-1}(B) = f^{-1}(c_\pi(B)) \subseteq i_\alpha(f^{-1}(B))$. Hence $f^{-1}(B)$ is a g-α-open in X. As a consequence, we obtain f is contra g-α-preirresolute. □

Theorem 4.12 *Let $f: X \to Y$ be a function from two GTS's. Suppose one of the following conditions hold:*

(1) *$f(c_\beta(A)) \subseteq i_\pi(f(A))$ for each subset A in X.*
(2) *$c_\beta(f^{-1}(B)) \subseteq f^{-1}(i_\pi(B))$ for each subset B in Y.*
(3) *$f^{-1}(c_\pi(B)) \subseteq i_\beta(f^{-1}(B))$ for each subset B in Y.*

Then f is contra g-β-preirresolute.

Proof It is similar to proof of Theorem 4.11 □

Theorem 4.13 *Let $f: X \to Y$ be a function from two GTS's and g_X is a strong. f is contra g-α-preirresolute if the graph function $g: X \to X \times Y$ defined by $g(x) = (x, f(x))$ for each $x \in X$, is contra g-α-preirresolute.*

Proof Let $x \in X$ and V be g-preopen containing $f(x)$ in Y. Then $X \times V$ is a g-preopen set of $X \times Y$ by Theorem 2.12 and contains $g(x)$. Then $g^{-1}(X \times V)$ is a g-α-closed set in X. Since $g^{-1}(X \times V) = f^{-1}(V)$, $f^{-1}(V)$ is a g-α-closed set in X. As a consequence, f is contra g-α-preirresolute. □

Theorem 4.14 *Let $f: X \to Y$ be a function from two GTS's and g_X is a strong. f is contra g-β-preirresolute if the graph function $g: X \to X \times Y$ defined by $g(x) = (x, f(x))$ for each $x \in X$, is contra g-β-preirresolute.*

Proof It is proved similar to that of Theorem 4.13. □

Theorem 4.15 *Let g_{Y_k} be a given GT on Y_k for $k \in K$ and g_{Y_k} be a strong. If a function $f: X \to \prod Y_k$ is contra g-α-preirresolute, then $p_k \circ f: X \to Y_k$ is contra g-α-preirresolute for each $k \in K$, where p_k is the projection of $\prod Y_k$ onto Y_k.*

Proof Let V_k be any g-preopen set of Y_k. p_k is $(\pi(g_Y), \pi(g_{Y_k}))$-continuous from Proposition 2.11 since g_{Y_k} is strong and so $p_k^{-1}(V_k)$ is g-preopen set. Since f is contra g-α-preirresolute, $f^{-1}(p_k^{-1}(V_k)) = (p_k \circ f)^{-1}(V_k)$ is a g-α-closed. As a consequence, we have $p_k \circ f$ is contra g-α-preirresolute for each $k \in K$. □

Theorem 4.16 *Let g_{Y_k} be a given GT on Y_k for $k \in K$ and g_{Y_k} be a strong. If a function $f: X \to \prod Y_k$ is contra g-β-preirresolute, then $p_k \circ f: X \to Y_k$ is contra g-β-preirresolute for each $k \in K$, where p_k is the projection of $\prod Y_k$ onto Y_k.*

Proof It is similar to that of Theorem 4.15 □

Theorem 4.17 *If the function $f: \prod X_k \to \prod Y_k$ defined by $f(\{x_k\}) = \{f_k(x_k)\}$ for each $\{x_k\} \in \prod X_k$, is contra g-α-preirresolute, then $f_k: X_k \to Y_k$ is contra g-α-preirresolute for each $k \in K$.*

Proof Let $k_0 \in K$ be an arbitrary fixed index and V_{k_0} be any g-preopen set of Y_{k_0}. Then $\prod Y_m \times V_{k_0}$ is g-preopen in $\prod Y_k$ by Theorem 2.12, where $k_0 \neq m \in K$. Since f is contra g-α-preirresolute, $f^{-1}(\prod Y_m \times V_{k_0}) = \prod X_m \times f_{k_0}^{-1}(V_{k_0})$ is g-α-closed in $\prod X_k$ and $f_{k_0}^{-1}(V_{k_0})$ is g-α-closed in X_{k_0}. As a consequence, f_{k_0} is contra g-α-preirresolute. □

Theorem 4.18 *If the function $f: \prod X_k \to \prod Y_k$ defined by $f(\{x_k\}) = \{f_k(x_k)\}$ for each $\{x_k\} \in \prod X_k$, is contra g-β-preirresolute, then $f_k: X_k \to Y_k$ is contra g-β-preirresolute for each $k \in K$.*

Proof It is proved similar to Theorem 4.17. □

Theorem 4.19 *If $f: X \to Y$ is contra g-α-preirresolute and A is a g-αclosed in-X, then the restriction $f|A: A \to Y$ is contra g-α-preirresolute.*

Proof Let V be any g-preopen set in Y. Then we have $f^{-1}(V)$ is a g-α-closed set in Y. Since the set A is g-α-closed set, we have $(f|A)^{-1}(V) = A \cap f^{-1}(V)$ is g-α-closed. Therefore $f|A$ is contra g-α-preirresolute. □

Theorem 4.20 *If $f: X \to Y$ is contra g-β-preirresolute and A is a g-β-closed in X, then the restriction $f|A: A \to Y$ is contra g-β-preirresolute.*

Proof It is proved by a similar way as that of Theorem 4.19. □

Theorem 4.21 *Let (X, g_X), (Y, g_Y) and (Z, g_Z) be GTS's. If $f : X \to Y$ is contra g-α-preirresolute and $g : Y \to Z$ is g-preirresolute, then the composition $g \circ f : X \to Z$ is contra g-α-preirresolute.*

Proof Let V be any g-preopen subset of Z. Since g function is g-preirresolute, $g^{-1}(V)$ is g-preopen in Y. Since f is contra g-α-preirresolute, then $f^{-1}(g^{-1}(V)) = (g \circ f)^{-1}(V)$ is g-α-closed in X. As a consequence, $g \circ f$ is contra g-α-preirresolute.

□

Theorem 4.22 *Let (X, g_X), (Y, g_Y) and (Z, g_Z) be GTS's. If $f : X \to Y$ is contra g-β-preirresolute and $g : Y \to Z$ is g-preirresolute, then the composition $g \circ f : X \to Z$ is contra g-β-preirresolute.*

Proof It is proved similar to that of Theorem 4.21. □

Conclusion

The concepts of g-α-preirresolute, g-β-preirresolute, contra g-α-preirresolute, contra g-β-preirresolute have been introduced on generalized topological spaces and some properties of this continuity have been investigated. These concepts may be used in other topological spaces and can be defined in different forms.

References

1. Bai, S.-Z., Zuo, S.-Z.: On g-α—irresolute functions. Acta Math. Hung. **130**(4), 382–389 (2011)
2. Császár, Á.: Generalized open sets. Acta Math. Hung. **72**(1–2), 65–87 (1997)
3. Császár, Á..: Generalized topology, generalized continuity. Acta Math. Hung. **96**, 351–357 (2002)
4. Császár, Á..: γ-connected sets. Acta Math. Hung. **101**, 273–279 (2003)
5. Császár, Á..: Extremally disconnected generalized topologies. Ann. Univ. Sci. Budapest **47**, 91–96 (2004)
6. Császár, Á..: Generalized open sets in generalized topologies. Acta Math. Hung. **106**, 53–66 (2005)
7. Császár, Á..: Product of generalized topologies. Acta Math. Hung. **123**(1–2), 127–132 (2009)
8. Jayanthi, D.: Contra continuity on generalized topological spaces. Acta Math. Hung. **134**(4), 263–271 (2012)
9. Min, W. K.: Generalized continuous functions defined by generalized open sets on generalized topological spaces. Acta Math. Hung. **128** (2009). doi:10.1007/s10474-009-9037-6
10. Shen, R.: Remarks on products of generalized topologies. Acta Math. Hung. **124**(4), 363–369 (2009)

9

How and how much to invest for fighting cheaters: from an ODE to a Cellular Automata model

Luca Meacci[1]

Abstract In this paper, we present a study which is the completion of the problem left open in the appendix of the paper Nuño et al. (Eur J Appl Math 21:459–478, 2010). The goal of the paper is to offer a deeper analysis of the type of the ODE system featured in the paper above concerning the evolution of the total wealth and cheater population and, mainly, the purpose is to generalize this one to a Cellular Automata model introducing spatial and local effects. The work presented shows the original interaction between an Ordinary Differential Equations model and a Cellular Automata model.

Keywords Mathematical modeling · ODE equations · Cellular Automata model · Tax evasion

Mathematics Subject Classification 34A34 · 37B15 · 68Q80 · 68U20 · 91D10 · 91D25

Introduction

An ideal society is composed by citizens whom all contribute to the common wealth. The common wealth can be used for education, public buildings and for all the needs of the population. But unfortunately in each real society there are always people that do not contribute to the common wealth but benefit from it, the cheaters. This is a form of criminal behavior that can create serious problems in each society (see [1, 2]).

How can a society organize itself to fight this social threat? In particular, how and how much can the common wealth be spent to fight cheaters?

In this paper, we try to answer these questions using simple models. Recent publications show the growing interest in the applications of mathematics in preventing tax evasion and crime [3, 4]. The present work is organized essentially in two parts. In the first part, mathematical modeling provides a system of Ordinary Differential Equations, as in the appendix of work [5]. We studied the system by tuning the parameters and solved it. In the second part, we generalized the model in a 2D system using the method of cellular automaton, a sort of agent-based model. We confronted the results with the previous model, and then we studied the effects that arose from assuming different polities in different zones of the world.

Our decisions depend on the global factors that are related to the whole society. Examples are the social forces, defined as external factors which have bias opinions, that include political advertisements, news reports, laws and sanctions. These elements are called "macro-sociological" components [6], because their influence is a characteristic of the entire system. These factors are those which are involved in a homogeneous system, of which we take into account in an ODE system. On the other hand, each member of a population is influenced by the other people whom each one interacts with. For example, our behavior is influenced by the opinions and the actions of our neighbors and friends. These influences no longer belong to the whole system, but they are local elements involved in the so-called "micro-sociological" level [7]. For these reasons, we have to take in account the effects of the so-called "micro-sociology" and "macro-sociology", to completely describe the evolution of the society. What happens when we introduce these local influences? Our

✉ Luca Meacci
luca.meacci@gmail.com

[1] Dipartimento di Matematica, "Ulisse Dini", Università degli Studi di Firenze, Viale Morgagni, 67/A, 50134 Florence, Italy

aim is to build a Cellular Automata model, consistent with the ODE model, which allows it to describe the spatial effects due to local influences at the "micro-sociological" level.

Homogeneous case

Let us assume that a society is composed by the following two classes:

1. $X :=$ people that contribute to common wealth (taxpayers),
2. $Y :=$ people that do not contribute (cheaters).

We suppose that the total population is constant, i.e.,

$$N = X + Y. \tag{1}$$

Let us define the common wealth, W, like wealth sustained by the contributions of law-abiding citizens and from which all people benefit. This common wealth is also used to convince cheaters to contribute to W with policies of repression and education. The time evolution of W is given by

$$\dot{W} = a(N - Y) - \theta W - \phi W. \tag{2}$$

The common wealth increases in proportion to the number of taxpayers (a is the constant of proportion) and decreases in proportion with the expenditures. A part of the expenditures is due to fight cheaters. The parameter ϕ takes into account these kinds of measures while all other expenditures of the society are expressed by parameter θ.

The time evolution of cheater population, according to the classical population dynamics [8], is

$$\dot{Y} = \tau(N - Y) - \alpha Y. \tag{3}$$

The rates τ and α are, respectively, influenced by social promotion and police repression (arrests). This can be expressed postulating

$$\tau = \frac{\tau_0}{1 + \tau_1\, s\, \phi\, W} \tag{4}$$

and

$$\alpha = \alpha_0 + \alpha_1 (1 - s)\, \phi\, W \tag{5}$$

where $\phi\, W$ is the fraction of the total wealth devoted to fight cheaters.

With these definitions, the dynamical system that describes completely the system behavior is:

$$\begin{aligned} \dot{Y} &= \frac{\tau_0}{1 + \tau_1\, s\, \phi\, W} (N - Y) - (\alpha_0 + \alpha_1\, (1 - s)\, \phi\, W)\, Y \\ \dot{W} &= a(N - Y) - \theta\, W - \phi\, W \end{aligned} \tag{6}$$

After the normalization

$$y = \frac{Y}{N} \quad w = \frac{W}{aN} \tag{7}$$

and the re-definitions

$$\tau_1 = \tau_1 aN \quad \alpha_1 = \alpha_1 Na \tag{8}$$

the dynamical system that describes completely the system behavior is

$$\begin{aligned} \dot{y} &= \frac{\tau_0}{1 + \tau_1\, s\, \phi\, w} (1 - y) - (\alpha_0 + \alpha_1\, (1 - s)\, \phi\, w)\, y \\ \dot{w} &= (1 - y) - (\theta + \phi)\, w \end{aligned} \tag{9}$$

The question we are facing is the following: given a social state, that is characterized by a particular setup of parameters, which is the optimal social expenditure that assures the greatest total wealth with the lowest cheaters' population?

This question can be solved by studying the steady state of the dynamical system as a function of ϕ and s. In particular, the steady cheaters' number is implicitly given by the isocline equation

$$\tau_0\, (1 - y) - (1 + \tau_1\, s\, \phi\, w)\, (\alpha_0 + \alpha_1\, (1 - s)\, \phi\, w)\, y = 0 \tag{10}$$

and steady total wealth is

$$\bar{w} = \frac{1 - y}{\theta + \phi}. \tag{11}$$

Let's take a look at the following remarks:

- ϕ is the parameter that describes *how much* of the common wealth the state decides to invest.
- s is the parameter that is related to *how* the society decides to invest its money to defeat criminality. In fact, when $s = 0$ we consider only police repression while when $s = 1$ society's efforts are totally directed to social promotion and education.

Figures 1 and 2 depict the tridimensional plots of both the stationary cheater population and the corresponding total wealth as a function of ϕ (how much) and s (how). In our case, the setup of parameters is

$$\alpha_0 = 0.01 \quad \alpha_1 = 1 \quad \tau_0 = 0.01 \quad \tau_1 = 100 \quad \theta = 1 \tag{12}$$

As it can be seen, the cheater population decreases monotonically as ϕ increases for all $s \in [0; 1]$, the total wealth presents a peak for a given value of ϕ that is a function of s.

Consequently, the simultaneous optimization of the cheater population and the total wealth is not possible. To overcome this drawback, we can define auxiliary objective functions and we can look for the couple $(\phi; s)$ that optimizes the strategy to fight cheaters. For our current

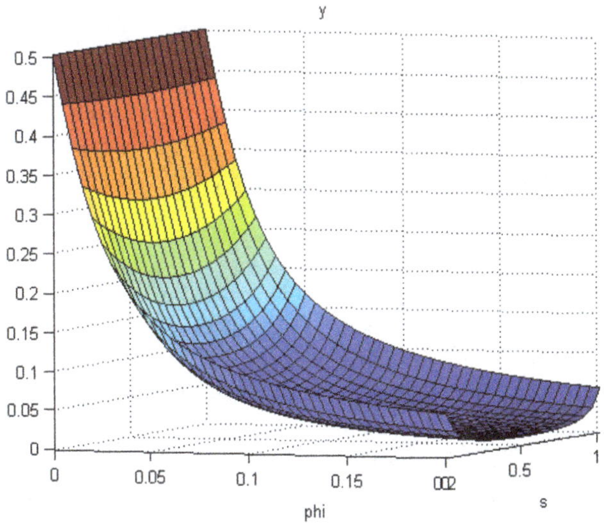

Fig. 1 3D graph of y as a function of ϕ and s

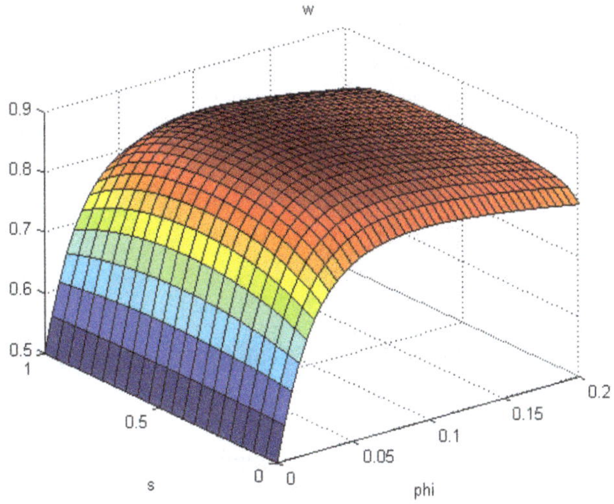

Fig. 2 3D graph of w as a function of ϕ and s

purpose, a simple, but still useful choice is to study the function:

$$L = \log\left(\frac{1}{\sqrt{(w-1)^2 + y^2}}\right) \tag{13}$$

Figure 3 shows the 3D graph of L as a function of ϕ and s.

2D Cellular Automata

We generalize the previous model to a spatial system. The spatial and temporal evolution of collective behavior in a population can be simulated with a cellular automaton model (see [9–11]).

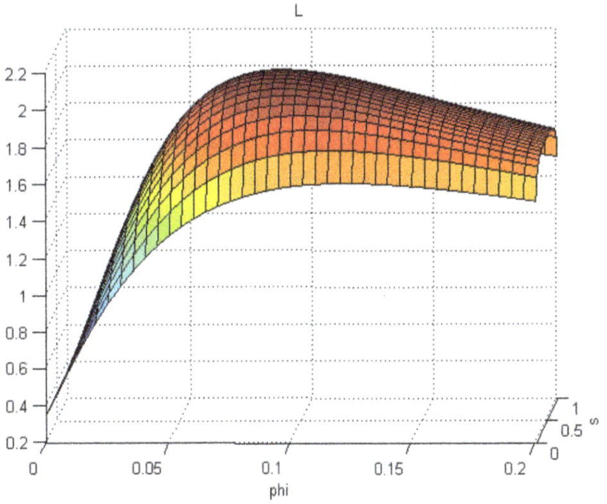

Fig. 3 3D graph of L as a function of ϕ and s

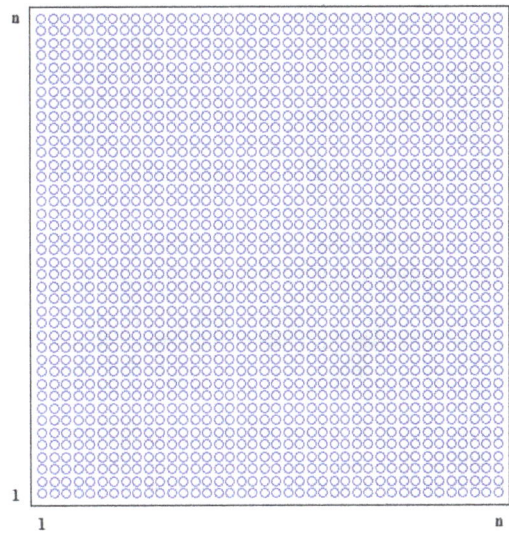

Fig. 4 The grid is a scheme of the population distributed in a *square plane*

Let us suppose that the population is spread in a square grid $n \times n$ as we schematized in the Fig. 4. The total number N of individuals in the population is

$$N = n \times n. \tag{14}$$

Each member of the society can be a taxpayer or a cheater, that means, as we have just defined, a person who does not contribute to the common wealth but benefits from it. In the grid, if an individual is a taxpayer the point is "white" while if he is a cheater the point is "blue".

Each member of a population is influenced by the other people whom each one interacts with. Our behavior is influenced by the opinions and the actions of our neighbors and friends. On the other hand, our decisions depend on the global factors that are related to the whole society. We are referring to the social forces, defined as external factors which have bias opinions, that include political advertisements, news reports, laws and sanctions. Because of this, we have to take in account the effects of the so-called "micro-sociology" and "macro-sociology", to completely describe the evolution of the society. Let us first consider the "macro-sociology" influences.

The "macro-sociology" problem is related to social forces that influence the whole world. It is a sort of a global field. In fact, in its definition, we can introduce the possible actions of a state to control this form of criminality. Essentially, the possibility of a nation to fight cheaters can be divided into two strategies:

A To invest money from the common wealth to improve education and social awareness.

B To make stronger police repression (arrests). Naturally, to beef up police forces has a cost.

As we can see from the dynamic system from the previous section, the strategy A is related to the parameter τ to prevent through social promotion the flux from taxpayers to cheaters while strategy B is linked to the factor α, forcing the flux to taxpayers trough police repression. Meacci et al. [12] introduced a probability of choosing an opinion that incorporates the global factors of the dynamic system.[1] We referred to this approach to link our macro-sociology parameters with the probabilities of changing behavior.

We have to define two probabilities that take into account macro-sociology parameters:

1. $P_M^{X\to Y}$ The probability that a taxpayer becomes a cheater;
2. $P_M^{Y\to X}$ The probability that a cheater becomes a taxpayer;

In the case of a taxpayer, the probability to become a cheater is defined:

$$P_M^{X\to Y} = \tau \quad (\tau \leq 1) \tag{15}$$

where τ is the same parameter of Eq. (4), according to the strategy A. Naturally, it is clear that the probability P_t that a taxpayer remains a taxpayer is

$$P_M^{X\to X} = 1 - P_M^{X\to Y}. \tag{16}$$

In the other case, if we considered a cheater, we can define the probability to become a taxpayer as

$$P_M^{Y\to X} = \alpha \quad (\alpha \leq 1) \tag{17}$$

where α is the same parameter of Eq. (5). In this case, the social force which pushes a cheater towards a honest behavior is related to the B strategy of police repression. The probability $P_M^{Y\to Y}$ that a cheater remains a cheater is

$$P_M^{Y\to Y} = 1 - P_M^{Y\to X}. \tag{18}$$

The results under the assumptions are the same as of the ODE problem. For instance, Fig. 5 shows the goal function (13) calculated with the CA model.

Let us focus on the "micro-sociology" problem. In a case of a certain probability P, with the notation $P(i, j)$, we indicate the probability related to the individual that lives in the position of row i and column j in the grid. The local probability to change opinion (or behavior) is proportional to the number of neighbors that already have different opinions (see [7]). We consider that each individual (i, j) is influenced on his own and by the other 8 people close to him, as we see in Fig. 6.

These effects are called the effects of "micro-sociology" which lead to the definition of the local probability. For instance, let us suppose to define the local probability $P_l^{*\to Y}$ to become a cheater

$$P_l^{*\to Y}(i,j) = \frac{n_l^c(i,j)}{N_l} \tag{19}$$

where

- $n_l^c(i,j)$ is the number of people (including himself) which are the neighbors of individual (i, j) who are already cheaters.
- N_l is the total number of neighbors plus himself. It follows that $N_l = 9$.[2]

Obviously, the local probability $P_l^{*\to X}$ to become a taxpayer is

$$P_l^{*\to X}(i,j) = 1 - P_l^{*\to Y}(i,j). \tag{20}$$

For instance, in the case of Fig. 7 the local probabilities are $P_l^{*\to Y}(i,j) = 4/9$ and $P_l^{*\to X}(i,j) = 5/9$.

Let us therefore define the probability of changing the current state as the convex combination of these two contributions. In particular, the total probability[3] that a taxpayer became a cheater is

$$P_{\text{TOT}}^{X\to Y} = \gamma \, l \, P_l^{*\to Y} + P_M^{X\to Y} \tag{21}$$

[1] Differently from the present work, which begins its study from a particular ODE model and generalizes this one to a CA model introducing spatial and local effects, the paper cited, starts by defining an original CA model, compares it with the corresponding ODE model with effects of contagion and shows the advantages of the cooperation to fight tax evasion.

[2] We remark that people on the boundaries feel the influence of the people of the other side of the grid. Under this assumption the world is a sort of toros.

[3] We remark that the parameters will be chosen in such a way that it is a well-defined probability.

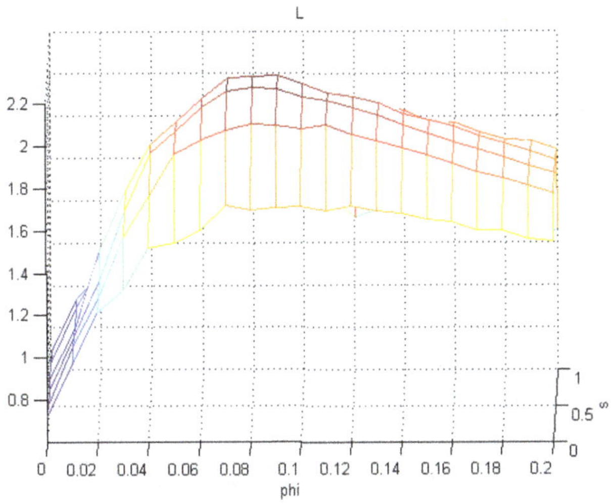

Fig. 5 3D graph of L as a function of ϕ and s calculated with CA model

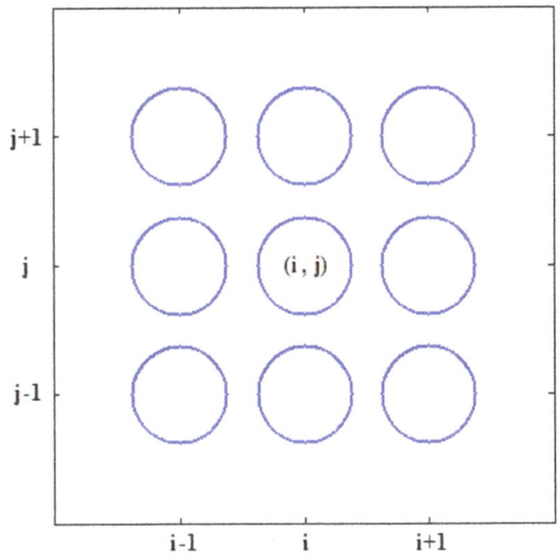

Fig. 6 An individual in position (i, j) is influenced by himself and his eight neighbors

and the total probability that a cheater becomes a taxpayer is

$$P_{\text{TOT}}^{Y\to X} = \gamma\, k\, P_l^{*\to X} + P_M^{Y\to X} \tag{22}$$

where $\gamma \in [0, 1]$ is a parameter that allows to weigh the importance of local contributions. Essentially, tuning γ we are exploring different types of societies according with the effectiveness of "micro-sociology" interactions. If nothing else is specified it is assumed that $\gamma = 1$. Besides, $l \in [0, 1]$ and $k \in [0, 1]$ are 2 factors that allow to determine whether increasing the influence of contagion positive or negative. Obviously, the probability of remaining in the same state is given by $1 - P_{\text{TOT}}$ in both cases.

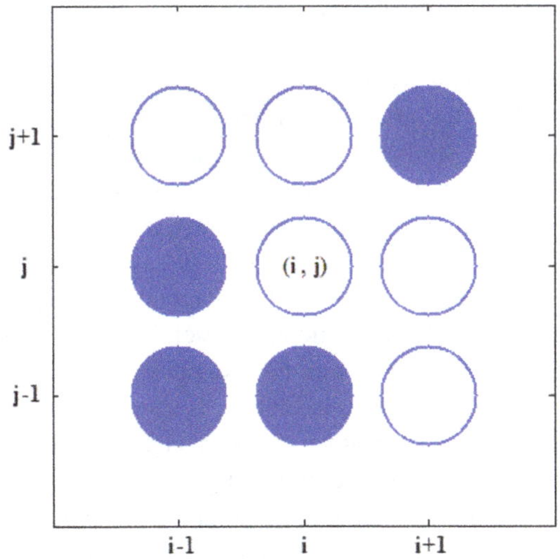

Fig. 7 The local probabilities are $P_l^{*\to Y}(i,j) = 4/9$ and $P_l^{*\to X}(i,j) = 5/9$

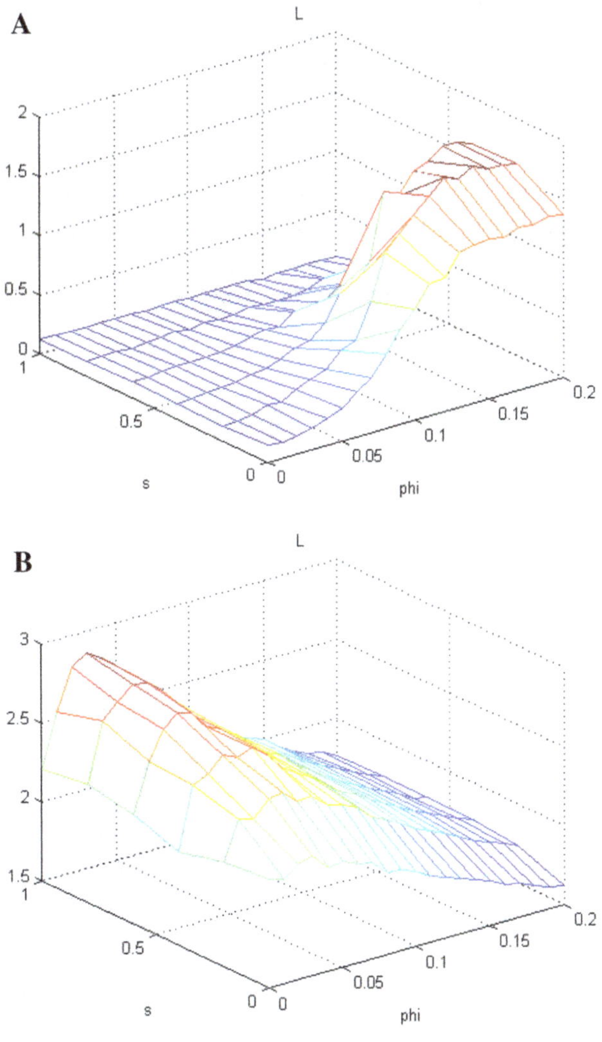

Fig. 8 L goal function in case **a** with $l = 0.40$ and $k = 0.30$ and in case **b** with $l = 0.30$ and $k = 0.40$

Results and discussion

In this section, we show the results of the CA model. These results show that the local effects have an impact on the forms of the solutions, and then the policies of optimal choice. In particular, the system is very sensitive if we decide to promote the positive or negative influence. We show in Fig. 8 the graphs of L in both cases, respectively, with $l > k$ (negative case) and $l<k$ (positive case).

The CA model is particularly useful for studying the spatial effects of the system. For example, Fig. 9 shows full screenshot of the system evolution. Cheaters are identified in blue. It is evident that we can see a cluster of cheaters in the area where we have defined a lighter contrast policy (ϕ lower than in other regions). You can also appreciate the contagion effect in the border area of low contrast.

The effect on the system due to the influence (positive or negative) is clearly evident from Fig. 10 that shows the number of cheaters for position in a world where on the left is applied a policy of strong contrast and on the right of low contrast. The effect of infection is evident from the blunt shape of the curve, especially in the vicinity of the areas of policy change.

In fact, even if we apply different parameters in different areas of the system, by reason of the phenomenon of contagion, the effects are propagated throughout the system. Let us consider, for instance, the test shown in Fig. 11. The graph shows the density of cheaters in the world 2 where it is accompanied by the world 1. Notice how the increase of ϕ in the first world, it produces a positive effect in the second world.

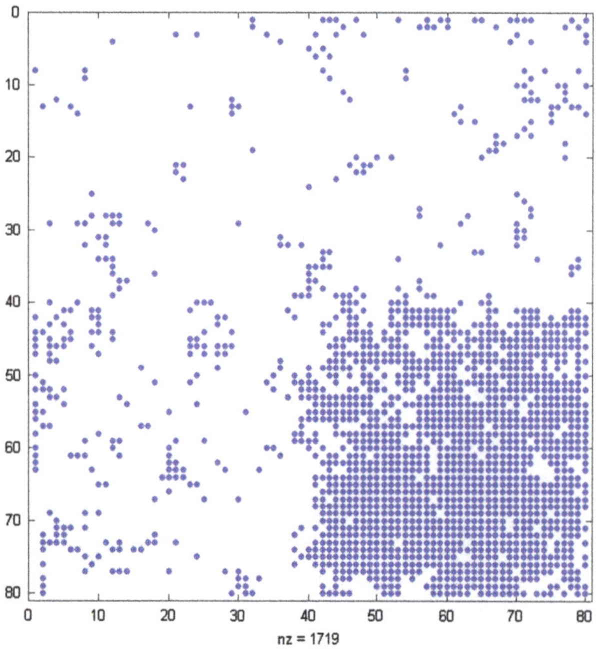

Fig. 9 A screenshot of the evolution of the system in a world where in the *lower right corner* we have applied a lower value of ϕ

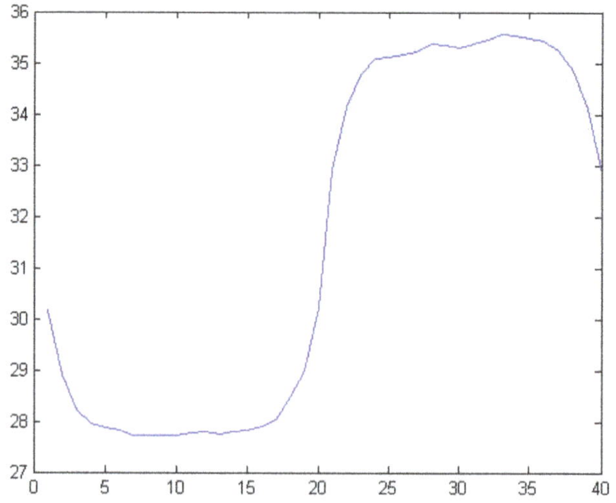

Fig. 10 Graph of the number of cheaters for position in a world in which we applied a low ϕ in the *right area*. We can appreciate the curvature of the curve due to the effect of contagion

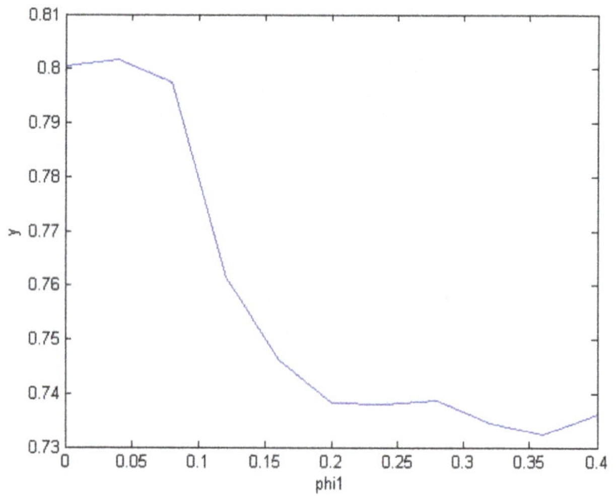

Fig. 11 Graph of the density of cheaters in the world with increasing ϕ of the attached world. When the ϕ of the attached world grows it produces a positive effect also on the world reported in the graph as a result of contagion

Conclusion

In this paper, we built a Cellular Automata model, which is consistent with the homogeneous case that corresponds to the ODE model. In particular, the CA model takes into account the "micro-sociological" effects or local effects due to the influence of contagion between individuals. These influences are particularly relevant and deeply affect the behavior of the system. Thanks to these influences we can study the spatial effects as shown in the "Results" section.

Acknowledgments The author wishes to thank Prof. Mario Primicerio and Prof. Juan Carlos Nuño for their contribution to bring out the ideas of the present work. The author is also grateful to Rita Adornato for checking the English of this manuscript. The paper is dedicated to Madrid, for what happened and for what will never happen.

References

1. Bloomquist, K.: Tax compliance as an evolutionary coordination game: an agent-based approach. Financ. Rev. **39**(1), 25–49 (2011)
2. Zaklan, G., Westerhoff, F., Stauffer, D.: Analysing tax evasion dynamics via the Ising model. J. Econ. Interact. Coord. **4**, 1–14 (2009)
3. Muñoz, F., Nuño, J.C., Primicerio, M.: Effects of inspections in small world social networks with different contagion rules. Physica A: Stat. Mech. Appl. **432**, 76–86 (2015)
4. D'Orsogna, M.R., Perc, M.: Statistical physics of crime: a review. Phys. Life Rev. **12**, 1–21 (2015)
5. Nuño, J.C., Herrero, M.A., Primicerio, M.: Fighting cheaters: how and how much to invest. Eur. J. Appl. Math. **21**(4–5), 459–478 (2010)
6. Bahr, D.B.: Statistical mechanics of opinion formation and collective behavior: micro-sociology. J. Math. Sociol. **23**(1), 1–27 (1973a)
7. Bahr, B.D.: Statistical mechanics of collective behavior: macro-sociology. J. Math. Sociol. **23**(1), 29–49 (1973b)
8. Bacaër, N.: A Short History of Mathematical Population Dynamics. Springer-Verlag, London ld (2011)
9. Wolfram, S. (ed.): Theory and Applications of Cellular Automata. World Scientific Press, Singapore (1986)
10. Kari, J.: Theory of cellular automata: a survey. Theor. Comput. Sci. **334**, 3–33 (2005)
11. Chopart, B., Droz, M.: Cellular Automata Modeling of Physical Systems. Cambridge University Press, Cambridge (1998)
12. Meacci, L., Nuño, J.C., Primicerio, M.: Fighting tax evasion: a cellular automata approach. Adv. Math. Sci. Appl. **22**(2), 597–610 (2012)

10

Hyers–Ulam–Rassias stability of Cauchy–Jensen functional equation in RN-spaces

H. Azadi Kenary[1] · H. Keshavarz[1]

Abstract Using the fixed point and direct methods, we prove the Hyers–Ulam stability of the following Cauchy–Jensen functional equation.

$$2f\left(\frac{\sum_{i=1}^{p} x_i + \sum_{j=1}^{q} y_j}{2} + \sum_{k=1}^{d} z_k\right) = \sum_{i=1}^{p} f(x_i) + \sum_{j=1}^{q} f(y_j) + 2\sum_{k=1}^{d} f(z_k)$$

where p, q, d are positive integers, in random normed spaces.

Keywords Hyers–Ulam–Rassias stability · Cauchy–Jensen functional equation · Fixed point method · Random normed spaces

Mathematics Subject Classification Primary 39B55

Introduction and preliminaries

A classical question in the theory of functional equations is the following: "When is it true that a function which approximately satisfies a functional equation must be close to an exact solution of the equation?". If the problem accepts a solution, we say that the equation is *stable*. The first stability problem concerning group homomorphisms was raised by Ulam [32] in 1940. In the next year, Hyers [14] gave a positive answer to the above question for additive groups under the assumption that the groups are Banach spaces. In 1978, Rassias [25] proved a generalization of Hyers's theorem for additive mappings. This new concept is known as generalized Hyers–Ulam stability or Hyers–Ulam–Rassias stability of functional equations. Furthermore, in 1994, a generalization of Rassias's theorem was obtained by Găvruta [13] by replacing the bound $\epsilon(\|x\|^p + \|y\|^p)$ by a general control function $\phi(x, y)$.

In 1983, a generalized Hyers–Ulam stability problem for the quadratic functional equation was proved by Skof [31] for mappings $f : X \to Y$, where X is a normed space and Y is a Banach space. In 1984, Cholewa [7] noticed that the theorem of Skof is still true if the relevant domain X is replaced by an Abelian group and, in 2002, Czerwik [9] proved the generalized Hyers–Ulam stability of the quadratic functional equation. The reader is referred to [1–29] and references therein for detailed information on stability of functional equations. In the sequel, we adopt the usual terminology, notions and conventions of the theory of random normed spaces as in [30].

Throughout this paper, let Γ^+ denote the set of all probability distribution functions $F : R \cup [-\infty, +\infty] \to [0, 1]$ such that F is left-continuous and nondecreasing on $\mathbb{R}$ and $F(0) = 0, F(+\infty) = 1$. It is clear that the set $D^+ = \{F \in \Gamma^+ : l^- F(+\infty) = 1\}$, where $l^- f(x) = \lim_{t \to x^-} f(t)$, is a subset of Γ^+. The set Γ^+ is partially ordered by the usual point-wise ordering of functions, that is, $F \leq G$ if and only if $F(t) \leq G(t)$ for all $t \in \mathbb{R}$. For any $a \geq 0$, the element $H_a(t)$ of D^+ is defined by

$$H_a(t) = \begin{cases} 0 & \text{if} \quad t \leq a, \\ 1 & \text{if} \quad t > a. \end{cases}$$

We can easily show that the maximal element in Γ^+ is the distribution function $H_0(t)$.

✉ H. Azadi Kenary
azadi@mail.yu.ac.ir

1 Department of Mathematics, College of Sciences, Yasouj University, Yasouj 75914-353, Iran

Definition 1.1 A function $T : [0,1]^2 \to [0,1]$ is a *continuous triangular norm* (briefly, a t-norm) if T satisfies the following conditions:

(a) T is commutative and associative;
(b) T is continuous;
(c) $T(x,1) = x$ for all $x \in [0,1]$;
(d) $T(x,y) \le T(z,w)$ whenever $x \le z$ and $y \le w$ for all $x,y,z,w \in [0,1]$.

Three typical examples of continuous t-norms are as follows: $T(x,y) = xy, T(x,y) = \max\{a+b-1,0\}$, $T(x,y) = \min(a,b)$.

Recall that, if T is a t-norm and $\{x_n\}$ is a sequence in $[0,1]$, then $T^n_{i=1}x_i$ is defined recursively by $T^1_{i=1}x_1 = x_1$ and $T^n_{i=1}x_i = T(T^{n-1}_{i=1}x_i, x_n)$ for all $n \ge 2$. $T^\infty_{i=n}x_i$ is defined by $T^\infty_{i=1}x_{n+i}$.

Definition 1.2 A *random normed space* (briefly, RN-space) is a triple (X,μ,T), where X is a vector space, T is a continuous t-norm and $\mu : X \to D^+$ is a mapping such that the following conditions hold:

(a) $\mu_x(t) = H_0(t)$ for all $x \in X$ and $t > 0$ if and only if $x = 0$;
(b) $\mu_{\alpha x}(t) = \mu_x\left(\frac{t}{|\alpha|}\right)$ for all $\alpha \in \mathbb{R}$ with $\alpha \ne 0$, $x \in X$ and $t \ge 0$;
(c) $\mu_{x+y}(t+s) \ge T(\mu_x(t), \mu_y(s))$ for all $x,y \in X$ and $t,s \ge 0$.

Definition 1.3 Let (X,μ,T) be an RN-space.

(a) A sequence $\{x_n\}$ in X is said to be *convergent* to a point $x \in X$ (write $x_n \to x$ as $n \to \infty$) if $\lim_{n\to\infty} \mu_{x_n - x}(t) = 1$ for all $t > 0$.
(b) A sequence $\{x_n\}$ in X is called a *Cauchy sequence* in X if $\lim_{n\to\infty} \mu_{x_n - x_m}(t) = 1$ for all $t > 0$.
(c) The RN-space (X,μ,T) is said to be *complete* if every Cauchy sequence in X is convergent.

Theorem 1.4 [30] *If (X,μ,T) is an RN-space and $\{x_n\}$ is a sequence such that $x_n \to x$, then $\lim_{n\to\infty} \mu_{x_n}(t) = \mu_x(t)$.*

Definition 1.5 Let X be a set. A function $d : X \times X \to [0,\infty]$ is called a generalized metric on X if d satisfies the following conditions:

(a) $d(x,y) = 0$ if and only if $x = y$ for all $x,y \in X$;
(b) $d(x,y) = d(y,x)$ for all $x,y \in X$;
(c) $d(x,z) \le d(x,y) + d(y,z)$ for all $x,y,z \in X$.

Theorem 1.6 *Let (X,d) be a complete generalized metric space and $J : X \to X$ be a strictly contractive mapping with Lipschitz constant $L < 1$. Then, for all $x \in X$, either*

$$d(J^n x, J^{n+1}x) = \infty \tag{1.1}$$

for all nonnegative integers n or there exists a positive integer n_0 such that

(a) $d(J^n x, J^{n+1}x) < \infty$ *for all* $n_0 \ge n_0$;
(b) *the sequence* $\{J^n x\}$ *converges to a fixed point* y^* *of* J;
(c) y^* is the unique fixed point of J in the set $Y = \{y \in X : d(J^{n_0}x, y) < \infty\}$;
(d) $d(y,y^*) \le \frac{1}{1-L} d(y, Jy)$ for all $y \in Y$.

In 1996, Hyers et al. [15] were the first to provide applications of stability theory of functional equations for the proof of new fixed point theorems with applications. By using fixed point methods, the stability problems of several functional equations have been extensively investigated by a number of authors (see [10, 22–24]).

This paper is organized as follows: In "Stability of the Cauchy–Jensen functional equation : a directmethod", using direct method, we prove the Hyers–Ulam–Rassias stability of the following functional equation that we call Cauchy–Jensen mapping

$$2f\left(\frac{\sum_{i=1}^p x_i + \sum_{j=1}^q y_j}{2} + \sum_{k=1}^d z_k\right) = \sum_{i=1}^p f(x_i) + \sum_{j=1}^q f(y_j) + 2\sum_{k=1}^d f(z_k) \tag{1.2}$$

where $x_i, y_j, z_k \in X$, in random normed space. In "Stability of the Cauchy–Jensen functional equation : a fixed pointapproach", using the fixed point method, we prove the Hyers–Ulam–Rassias stability of the Cauchy–Jensen functional equation (1.2) in random normed spaces.

Stability of the Cauchy–Jensen functional equation: a direct method

For a given mapping $f : X \to Y$, we define

$$AQ_f(x,y,z) = 2f\left(\frac{\sum_{i=1}^p x_i + \sum_{j=1}^q y_j}{2} + \sum_{k=1}^d z_k\right) - \sum_{i=1}^p f(x_i) - \sum_{j=1}^q f(y_j) - 2\sum_{k=1}^d f(z_k)$$

for all $x_i, y_j, z_k \in X$.

In this section, using direct method, we prove the generalized Hyers–Ulam–Rassias stability of the Cauchy–Jensen additive functional equation (1.2) in random space .

Theorem 2.1 Let X be a real linear space $(Z,\mu',\min)$ be an RN-space and $\phi : X^{p+q+d} \to Z$ be a function such that there exists $0 < \alpha < \frac{2}{p+q+2d}$ satisfying

$$\mu'_{\phi\left(\frac{2x_i}{p+q+2d},\frac{2y_j}{p+q+2d},\frac{2z_k}{p+q+2d}\right)}(t) \geq \mu'_{\alpha\phi(x_i,y_j,z_k)}(t) \tag{2.1}$$

for all $x_i, y_j, z_k \in X$ and $t > 0$ and

$$\lim_{n\to\infty} \mu'_{\phi\left(\frac{2^n x_i}{(p+q+2d)^n},\frac{2^n y_j}{(p+q+2d)^n},\frac{2^n z_k}{(p+q+2d)^n}\right)}\left(\frac{2^n t}{(p+q+2d)^n}\right) = 1$$

for all $x_i, y_j, z_k \in X$ and $t > 0$. Let $(Y, \mu, \min)$ be a complete RN-space. If $f : X \to Y$ be a mapping with $f(0) = 0$ and satisfying

$$\mu_{2f\left(\frac{\sum_{i=1}^{p} x_i + \sum_{j=1}^{q} y_j}{2} + \sum_{k=1}^{d} z_k\right) - \sum_{i=1}^{p} f(x_i) - \sum_{j=1}^{q} f(y_j) - 2\sum_{k=1}^{d} f(z_k)}(t) \geq \mu'_{\phi(x_i,y_j,z_k)}(t) \tag{2.2}$$

for all $x_i, y_j, z_k \in X$ and $t > 0$, then the limit

$$L(x) = \lim_{n\to\infty} \frac{(p+q+2d)^n}{2^n} f\left(\frac{2^n x}{(p+q+2d)^n}\right)$$

exists for all $x \in X$ and defines a unique Cauchy–Jensen mapping $L : X \to Y$ such that

$$\mu_{f(x)-L(x)}(t) \geq \mu'_{\phi(x,x,\ldots,x)}\left(\frac{(2-(p+q+2d)\alpha)t}{\alpha}\right) \tag{2.3}$$

for all $x \in X$ and $t > 0$.

Proof Putting $x_i = y_j = z_k = x$ in (2.2), we see that

$$\mu_{2f\left(\frac{p+q+2d}{2}x\right)-(p+q+2d)f(x)}(t) \geq \mu'_{\phi_{(x,x,\ldots,x)}}(t) \tag{2.4}$$

for all $x \in X$ and all $t > 0$. Replacing x by $\frac{2x}{p+q+2d}$ in (2.4), we obtain

$$\mu_{f(x)-\frac{p+q+2d}{2}f\left(\frac{2x}{p+q+2d}\right)}(t) \geq \mu'_{\phi\left(\frac{2x}{p+q+2d},\frac{2x}{p+q+2d},\ldots,\frac{2x}{p+q+2d}\right)}(2t) \tag{2.5}$$

for all $x \in X$ and all $t > 0$. Replacing x by $\frac{2^n x}{(p+q+2d)^n}$ in (2.5) and using (2.1), we obtain

$$\begin{aligned}
&\mu_{\frac{(p+q+2d)^{n+1}}{2^{n+1}}f\left(\frac{2^{n+1}x}{(p+q+2d)^{n+1}}\right)-\frac{(p+q+2d)^n}{2^n}f\left(\frac{2^n x}{(p+q+2d)^n}\right)}(t)\\
&\geq \mu'_{\phi\left(\frac{2^{n+1}x}{(p+q+2d)^{n+1}},\frac{2^{n+1}x}{(p+q+2d)^{n+1}},\ldots,\frac{2^{n+1}x}{(p+q+2d)^{n+1}}\right)}\left(\frac{2^{n+1}t}{(p+q+2d)^n}\right)\\
&\geq \mu'_{\phi(x,x,\ldots,x)}\left(\frac{2^{n+1}t}{(p+q+2d)^n\alpha^{n+1}}\right)
\end{aligned}$$

and so

$$\begin{aligned}
&\mu_{\frac{(p+q+2d)^n}{2^n}f\left(\frac{2^n x}{(p+q+2d)^n}\right)-f(x)}\left(\sum_{k=0}^{n-1}\frac{t(p+q+2d)^k\alpha^{k+1}}{2^{k+1}}\right)\\
&=\mu_{\sum_{k=0}^{n-1}2^{k+1}\frac{(p+q+2d)^{k+1}}{2^{k+1}}f\left(\frac{2^{k+1}x}{(p+q+2d)^{k+1}}\right)-\frac{(p+q+2d)^k}{2^k}f\left(\frac{2^k x}{(p+q+2d)^k}\right)}\\
&\quad\times\left(\sum_{k=0}^{n-1}\frac{t(p+q+2d)^k\alpha^{k+1}}{2^{k+1}}\right)\\
&\geq T_{k=0}^{n-1}\left(\mu'_{\frac{(p+q+2d)^{k+1}}{2^{k+1}}f\left(\frac{2^{k+1}x}{(p+q+2d)^{k+1}}\right)-\frac{(p+q+2d)^k}{2^k}f\left(\frac{2^k x}{(p+q+2d)^k}\right)}\left(\frac{t(p+q+2d)^k\alpha^{k+1}}{2^{k+1}}\right)\right)\\
&\geq T_{k=0}^{n-1}\left(\mu'_{\phi(x,x,\ldots,x)}(2t)\right)=\mu'_{\phi(x,x,\ldots,x)}(2t).
\end{aligned} \tag{2.6}$$

This implies that

$$\mu_{\frac{(p+q+2d)^n}{2^n}f\left(\frac{2^n x}{(p+q+2d)^n}\right)-f(x)}(t) \geq \mu'_{\phi(x,x,\ldots,x)}\left(\frac{2t}{\sum_{k=0}^{n-1}\frac{(p+q+2d)^k\alpha^{k+1}}{2^k}}\right). \tag{2.7}$$

Replacing x by $\frac{2^l x}{(p+q+2d)^l}$ in (2.7), we obtain

$$\begin{aligned}
&\mu_{\frac{(p+q+2d)^{n+l}}{2^{n+l}}f\left(\frac{2^{n+l}x}{(p+q+2d)^{n+l}}\right)-\frac{(p+q+2d)^l}{2^l}f\left(\frac{2^l x}{(p+q+2d)^l}\right)}(t)\\
&\geq \mu'_{\phi(x,x,\ldots,x)}\left(\frac{2t}{\sum_{k=l}^{n+l-1}\frac{(p+q+2d)^k\alpha^{k+1}}{2^k}}\right).
\end{aligned} \tag{2.8}$$

Since

$$\lim_{l,n\to\infty}\mu'_{\phi(x,x,\ldots,x)}\left(\frac{2t}{\sum_{k=l}^{n+l-1}\frac{(p+q+2d)^k\alpha^{k+1}}{2^k}}\right) = 1$$

it follows that $\left\{\frac{(p+q+2d)^n}{2^n}f\left(\frac{2^n x}{(p+q+2d)^n}\right)\right\}_{n\geq 1}$ is a Cauchy sequence in a complete RN-space $(Y, \mu, \min)$ and so there exists a point $L(x) \in Y$ such that

$$\lim_{n\to\infty}\frac{(p+q+2d)^n}{2^n}f\left(\frac{2^n x}{(p+q+2d)^n}\right) = L(x).$$

Fix $x \in X$ and put $p = 0$ in (2.8). Then we obtain

$$\mu_{\frac{(p+q+2d)^n}{2^n}f\left(\frac{2^n x}{(p+q+2d)^n}\right)-f(x)}(t) \geq \mu'_{\phi(x,x,\ldots,x)}\left(\frac{2t}{\sum_{k=0}^{n-1}\frac{(p+q+2d)^k\alpha^{k+1}}{2^k}}\right)$$

and so, for any $\epsilon > 0$,

$$\mu_{L(x)-f(x)}(t+\epsilon)\geq T\left(\mu_{L(x)-\frac{(p+q+2d)^n}{2^n}f\left(\frac{2^n x}{(p+q+2d)^n}\right)}(\epsilon),\mu_{\frac{(p+q+2d)^n}{2^n}f\left(\frac{2^n x}{(p+q+2d)^n}\right)-f(x)}(t)\right)$$
$$\geq T\left(\mu_{L(x)-\frac{(p+q+2d)^n}{2^n}f\left(\frac{2^n x}{(p+q+2d)^n}\right)}(\epsilon),\mu'_{\phi(x,x,\ldots,x)}\left(\frac{2t}{\sum_{k=0}^{n-1}\frac{(p+q+2d)^k\alpha^{k+1}}{2^k}}\right)\right). \tag{2.9}$$

Taking $n\to\infty$ in (2.9), we get

$$\mu_{L(x)-f(x)}(t+\epsilon)\geq \mu'_{\phi(x,x,\ldots,x)}\left(\frac{2t}{\sum_{k=0}^{\infty}\frac{(p+q+2d)^k\alpha^{k+1}}{2^k}}\right). \tag{2.10}$$

Since ϵ is arbitrary, by taking $\epsilon\to 0$ in (2.10), we get

$$\mu_{L(x)-f(x)}(t)\geq \mu'_{\phi(x,x,\ldots,x)}\left(\frac{(2-(p+q+2d)\alpha)t}{\alpha}\right).$$

Replacing x_i, y_j and z_k by $\frac{2^n x_i}{(p+q+2d)^n}$, $\frac{2^n y_j}{(p+q+2d)^n}$ and $\frac{2^n z_k}{(p+q+2d)^n}$ in (2.2), respectively, we get

$$\mu_{\frac{(p+q+2d)^n}{2^n}AQ_f\left(\frac{2^n x_i}{(p+q+2d)^n},\frac{2^n y_j}{(p+q+2d)^n},\frac{2^n z_k}{(p+q+2d)^n}\right)}(t)$$
$$\geq \mu'_{\phi\left(\frac{2^n x_i}{(p+q+2d)^n},\frac{2^n y_j}{(p+q+2d)^n},\frac{2^n z_k}{(p+q+2d)^n}\right)}\left(\frac{2^n t}{(p+q+2d)^n}\right)$$

for all $x_i, y_j, z_k\in X$ and $t>0$. Since

$$\lim_{n\to\infty}\mu'_{\phi\left(\frac{2^n x_i}{(p+q+2d)^n},\frac{2^n y_j}{(p+q+2d)^n},\frac{2^n z_k}{(p+q+2d)^n}\right)}\left(\frac{2^n t}{(p+q+2d)^n}\right)=1$$

we conclude that L satisfies (1.2).

To prove the uniqueness of the additive mapping L, assume that there exists another mapping $M:X\to Y$ which satisfies (2.3). Then we have

$$\mu_{L(x)-M(x)}(t)$$
$$=\lim_{n\to\infty}\mu_{\frac{(p+q+2d)^n}{2^n}L\left(\frac{2^n x}{(p+q+2d)^n}\right)-\frac{(p+q+2d)^n}{2^n}M\left(\frac{2^n x}{(p+q+2d)^n}\right)}(t)$$
$$\geq \lim_{n\to\infty}\min\left\{\mu_{\frac{(p+q+2d)^n}{2^n}L\left(\frac{2^n x}{(p+q+2d)^n}\right)-\frac{(p+q+2d)^n}{2^n}f\left(\frac{2^n x}{(p+q+2d)^n}\right)}\left(\frac{t}{2}\right),\right.$$
$$\left.\mu_{\frac{(p+q+2d)^n}{2^n}f\left(\frac{2^n x}{(p+q+2d)^n}\right)-\frac{(p+q+2d)^n}{2^n}L\left(\frac{2^n x}{(p+q+2d)^n}\right)}\left(\frac{t}{2}\right)\right\}$$
$$\geq \lim_{n\to\infty}\mu'_{\phi\left(\frac{2^n x}{(p+q+2d)^n},\frac{2^n x}{(p+q+2d)^n},\ldots,\frac{2^n x}{(p+q+2d)^n}\right)}$$
$$\times\left(\frac{2^n(2-(p+q+2d)\alpha)t}{2(p+q+2d)^n\alpha}\right)$$
$$\geq \lim_{n\to\infty}\mu'_{\phi(x,x,\ldots,x)}\left(\frac{2^n(2-(p+q+2d)\alpha)t}{2(p+q+2d)^n\alpha^{n+1}}\right).$$

Since

$$\lim_{n\to\infty}\frac{2^n(2-(p+q+2d)\alpha)t}{2(p+q+2d)^n\alpha^{n+1}}=\infty$$

we get

$$\lim_{n\to\infty}\mu'_{\phi(x,x,\ldots,x)}\left(\frac{2^n(2-(p+q+2d)\alpha)t}{2(p+q+2d)^n\alpha^{n+1}}\right)=1.$$

Therefore, it follows that $\mu_{L(x)-M(x)}(t)=1$ for all $t>0$ and so $L(x)=M(x)$. This completes the proof. □

Corollary 2.2 Let X be a real normed linear space $(Z,\mu',\min)$ be an RN-space and $(Y,\mu,\min)$ be a complete RN-space. Let r is a positive real number with $0<r<1$, $z_0\in Z$ and $f:X\to Y$ be a mapping with $f(0)=0$ and satisfying

$$\mu_{2f\left(\frac{\sum_{i=1}^p x_i+\sum_{j=1}^q y_j}{2}+\sum_{k=1}^d z_k\right)-\sum_{i=1}^p f(x_i)-\sum_{j=1}^q f(y_j)-2\sum_{k=1}^d f(z_k)}(t)$$
$$\geq\mu'_{\left(\sum_{i=1}^p\|x_i\|^r+\sum_{j=1}^q\|y_j\|^r+\sum_{k=1}^d\|z_k\|^r\right)z_0}(t) \tag{2.11}$$

for all $x_i,y_j,z_k\in X$ and $t>0$. Then the limit $L(x)=\lim_{n\to\infty}\frac{(p+q+2d)^n}{2^n}f\left(\frac{2^n x}{(p+q+2d)^n}\right)$ exists for all $x\in X$ and defines a unique Cauchy–Jensen mapping $L:X\to Y$ such that

$$\mu_{f(x)-L(x)}(t)\geq\mu'_{\|x\|^r z_0}\left(\frac{(2(p+q+2d)^r-2^r(p+q+2d))t}{2^r(p+q+d)}\right)$$

for all $x\in X$ and $t>0$.

Proof Let $\alpha=\left(\frac{2}{p+q+2d}\right)^r$ and $\phi:X^{p+q+d}\to Z$ be a mapping defined by

$$\phi(x_i,y_j,z_k)=\left(\sum_{i=1}^p\|x_i\|^r+\sum_{j=1}^q\|y_j\|^r+\sum_{k=1}^d\|z_k\|^r\right)z_0.$$

Then, from Theorem 2.1, the conclusion follows. □

Theorem 2.3 Let X be a real linear space $(Z,\mu',\min)$ be an RN-space and $\phi:X^{p+q+d}\to Z$ be a function such that there exists $0<\alpha<\frac{p+q+2d}{2}$ satisfying

$$\mu'_{\phi\left(\frac{(p+q+2d)x_i}{2},\frac{(p+q+2d)y_j}{2},\frac{(p+q+2d)z_k}{2}\right)}(t)\geq\mu'_{\alpha\phi(x_i,y_j,z_k)}(t) \tag{2.12}$$

for all $x_i,y_j,z_k\in X$ and $t>0$ and

$$\lim_{n\to\infty}\mu'_{\phi\left(\frac{(p+q+2d)^n x_i}{2^n},\frac{(p+q+2d)^n y_j}{2^n},\frac{(p+q+2d)^n z_k}{2^n}\right)}\left(\frac{(p+q+2d)^n t}{2^n}\right)=1$$

for all $x_i,y_j,z_k\in X$ and $t>0$. Let $(Y,\mu,\min)$ be a complete RN-space. If $f:X\to Y$ be a mapping with $f(0)=0$ and satisfying (2.2). Then the limit

$$L(x) = \lim_{n\to\infty} \frac{2^n}{(p+q+2d)^n} f\left(\frac{(p+q+2d)^n x}{2^n}\right)$$

exists for all $x \in X$ and defines a unique Cauchy–Jensen mapping $L: X \to Y$ such that

$$\mu_{f(x)-L(x)}(t) \geq \mu'_{\phi(x,x,\ldots,x)}((p+q+2d)-2\alpha)t). \tag{2.13}$$

for all $x \in X$ and $t > 0$.

Proof By (2.4), we have

$$\mu_{\frac{2}{p+q+2d}f\left(\frac{p+q+2d}{2}x\right)-f(x)}(t) \geq \mu'_{\phi(x,x,\ldots,x)}((p+q+2d)t) \tag{2.14}$$

for all $x \in X$ and all $t > 0$. Replacing x by $\frac{(p+q+2d)^n x}{2^n}$ in (2.14) and using (2.12), we obtain

$$\mu_{\frac{2^{n+1}}{(p+q+2d)^{n+1}}f\left(\frac{(p+q+2d)^{n+1}x}{2^{n+1}}\right)-\frac{2^n}{(p+q+2d)^n}f\left(\frac{(p+q+2d)^n x}{2^n}\right)}(t)$$

$$\geq \mu'_{\phi\left(\frac{(p+q+2d)^n x}{2^n},\frac{(p+q+2d)^n x}{2^n},\ldots,\frac{(p+q+2d)^n x}{2^n}\right)}\left(\frac{(p+q+2d)^{n+1}t}{2^n}\right)$$

$$\geq \mu'_{\phi(x,x,\ldots,x)}\left(\frac{(p+q+2d)^{n+1}t}{2^n\alpha^n}\right).$$

The rest of the proof is similar to the proof of Theorem 2.1. □

Corollary 2.4 Let X be a real normed linear space $(Z, \mu', \min)$ be an RN-space and $(Y, \mu, \min)$ be a complete RN-space. Let r is a positive real number with $r > 1$, $z_0 \in Z$ and $f : X \to Y$ be a mapping with $f(0) = 0$ and satisfying (2.11). Then the limit $L(x) = \lim_{n\to\infty} \frac{2^n}{(p+q+2d)^n} f\left(\frac{(p+q+2d)^n x}{2^n}\right)$ exists for all $x \in X$ and defines a unique Cauchy–Jensen mapping $L : X \to Y$ such that

$$\mu_{f(x)-L(x)}(t) \geq \mu'_{\|x\|^r z_0}\left(\frac{((p+q+2d)-2^r(p+q+2d)^{1-r})t}{(p+q+d)}\right)$$

for all $x \in X$ and $t > 0$.

Proof Let $\alpha = \left(\frac{p+q+2d}{2}\right)^{1-r}$ and $\phi : X^{p+q+d} \to Z$ be a mapping defined by

$$\phi(x_i, y_j, z_k) = \left(\sum_{i=1}^{p} \|x_i\|^r + \sum_{j=1}^{q} \|y_j\|^r + \sum_{k=1}^{d} \|z_k\|^r\right) z_0.$$

Then, from Theorem 2.3, the conclusion follows. □

Stability of the Cauchy–Jensen functional equation: a fixed point approach

In the rest of the paper, by $\Phi(x_i, y_j, z_k)$, we mean that $\Phi(x_1, \ldots, x_p, y_1, \ldots, y_q, z_1, \ldots, z_d)$. Using the fixed point method, we prove the generalized Hyers–Ulam–Rassias stability of the functional equation $AQ_f(x, y, z) = 0$ in random normed spaces.

Theorem 3.1 Let X be a linear space (Y, μ, T_M) be a complete RN-space and Φ be a mapping from X^{p+q+d} to $D^+(\Phi(x_i, y_j, z_k)$ is denoted by $\Phi_{x_i,y_j,z_k})$ such that there exists $0<\alpha<\frac{2}{p+q+2d}$ satisfying

$$\Phi_{\frac{2x_i}{p+q+2d},\frac{2y_j}{p+q+2d},\frac{2z_k}{p+q+2d}}(\alpha t) \geq \Phi_{x_i,y_j,z_k}(t) \tag{3.1}$$

for all $x_i, y_j, z_k \in X$ and $t > 0$. Let $f : X \to Y$ be a mapping with $f(0) = 0$ and satisfying

$$\mu_{2f\left(\frac{\sum_{i=1}^{p} x_i+\sum_{j=1}^{q} y_j}{2}+\sum_{k=1}^{d} z_k\right)-\sum_{i=1}^{p} f(x_i)-\sum_{j=1}^{q} f(y_j)-2\times\sum_{k=1}^{d} f(z_k)}(t) \geq \Phi_{x_i,y_j,z_k}(t) \tag{3.2}$$

for all $x_i, y_j, z_k \in X$ and $t > 0$. Then the limit

$$\lim_{n\to\infty} \frac{(p+q+2d)^n}{2^n} f\left(\frac{2^n x}{(p+q+2d)^n}\right) = L(x)$$

exists for all $x \in X$ and $L : X \to Y$ is a unique mapping such that

$$\mu_{f(x)-L(x)}(t) \geq \Phi_{\underbrace{x, x, \ldots, x}_{(p+q+d)-\text{times}}}\left(\frac{2(1-\alpha)t}{\alpha}\right) \tag{3.3}$$

for all $x \in X$ and $t > 0$.

Proof Putting $x_i = y_j = z_k = x$ in (2.2), we get

$$\mu_{2f\left(\frac{p+q+2d}{2}x\right)-(p+q+2d)f(x)}(t) \geq \Phi_{\underbrace{x, x, \ldots, x}_{(p+q+d)-\text{times}}}(t) \tag{3.4}$$

for all $x \in X$. Replacing x by $\frac{2x}{p+q+2d}$ in (3.4), we obtain

$$\mu_{f(x)-\frac{p+q+2d}{2}f\left(\frac{2x}{p+q+2d}\right)}(t) \geq \Phi_{\underbrace{\frac{2x}{p+q+2d},\frac{2x}{p+q+2d},\ldots,\frac{2x}{p+q+2d}}_{(p+q+d)-\text{times}}}(2t)$$

$$\geq \Phi_{\underbrace{x, x, \ldots, x}_{(p+q+d)-\text{times}}}\left(\frac{2t}{\alpha}\right) \tag{3.5}$$

for all $x \in X$ and all $t > 0$.

Consider the set

$$S := \{h : X \to Y;\ h(0) = 0\}$$

and introduce the generalized metric on S:

$$d(g,h) = \inf_{u\in(0,+\infty)} \left\{\mu_{g(x)-h(x)}(ut) \geq \Phi_{\underbrace{x,x,\ldots,x}_{(p+q+d)-\text{times}}}(t),\ \forall x \in X\right\},$$

where, as usual, $\inf\phi = +\infty$. It is easy to show that (S,d) is complete (see [18]). Now we consider the linear mapping $J : (S,d) \to (S,d)$ such that

$$Jg(x) := \frac{p+q+2d}{2} g\left(\frac{2x}{p+q+2d}\right)$$

for all $x \in X$. First we prove that J is a strictly contractive mapping with the Lipschitz constant $\frac{p+q+2d}{2}\alpha$. In fact, Let $g, h \in S$ be given such that $d(g,h) < \varepsilon$. Then

$$\mu_{g(x)-h(x)}(\varepsilon t) \geq \Phi_{x,x,\ldots,x}(t)$$

for all $x \in X$ and all $t > 0$. Hence

$$\begin{aligned}
&\mu_{Jg(x)-Jh(x)}\left(\frac{(p+q+2d)\alpha\varepsilon t}{2}\right) \\
&= \mu_{\frac{p+q+2d}{2}g\left(\frac{2x}{p+q+2d}\right)-\frac{p+q+2d}{2}h\left(\frac{2x}{p+q+2d}\right)}\left(\frac{(p+q+2d)\alpha\varepsilon t}{2}\right) \\
&= \mu_{g\left(\frac{2x}{p+q+2d}\right)-h\left(\frac{2x}{p+q+2d}\right)}(\alpha\varepsilon t) \\
&\geq \Phi_{\underbrace{\frac{2x}{p+q+2d},\ldots,\frac{2x}{p+q+2d}}_{(p+q+d)-\text{times}}}(\alpha t) \\
&\geq \Phi_{\underbrace{x,x,\ldots,x}_{(p+q+d)-\text{times}}}(t)
\end{aligned}$$

for all $x \in X$. So $d(g,h) < \varepsilon$ implies that $d(Jg, Jh) \leq \frac{(p+q+2d)\alpha\varepsilon}{2}$. This means that

$$d(Jg, Jh) \leq \frac{(p+q+2d)\alpha}{2} d(g,h)$$

for all $g, h \in S$.

It follows from (3.5) that $d(f, Jf) \leq \frac{\alpha}{2}$.

By Theorem 1.6, there exists a mapping $L : X \to Y$ satisfying the following:

(1) L is a fixed point of J, i.e.,

$$\frac{2L}{p+q+2d} = L\left(\frac{2x}{p+q+2d}\right) \tag{3.6}$$

for all $x \in X$. The mapping L is a unique fixed point of J in the set

$$M = \{g \in S : d(h,g) < \infty\}.$$

This implies that L is a unique mapping satisfying (3.6) such that there exists a $u \in (0, \infty)$ satisfying

$$\mu_{g(x)-h(x)}(ut) \geq \Phi_{\underbrace{x,x,\ldots,x}_{(p+q+d)-\text{times}}}(t)$$

for all $x \in X$ and all $t > 0$;

(2) $d(J^n f, L) \to 0$ as $n \to \infty$. This implies the equality

$$\lim_{n\to\infty} \frac{(p+q+2d)^n}{2^n} f\left(\frac{2^n x}{(p+q+2d)^n}\right) = L(x)$$

for all $x \in X$;

(3) $d(f, L) \leq \frac{1}{1-\alpha} d(f, Jf)$, which implies the inequality

$$d(f, L) \leq \frac{\alpha}{2(1-\alpha)}$$

and so

$$\mu_{f(x)-L(x)}\left(\frac{\alpha}{2(1-\alpha)}\right) \geq \Phi_{x,x,\ldots,x}(t)$$

for all $x \in X$ and all $t > 0$. This implies that the inequalities (3.3) hold. It follows from (3.1) and (3.2) that

$$\begin{aligned}
&\mu_{\frac{(p+q+2d)^n}{2^n}AQ_f\left(\frac{2^n x}{(p+q+2d)^n},\frac{2^n y}{(p+q+2d)^n},\frac{2^n z}{(p+q+2d)^n}\right)}(t) \\
&\geq \Phi_{\frac{2^n x_i}{(p+q+2d)^n},\frac{2^n y_j}{(p+q+2d)^n},\frac{2^n z_k}{(p+q+2d)^n}}\left(\frac{2^n t}{(p+q+2d)^n}\right) \\
&\geq \Phi_{x_i,y_j,z_k}\left(\frac{2^n t}{(p+q+2d)^n \alpha^n}\right)
\end{aligned}$$

for all $x_i, y_j, z_k \in X$ and all $t > 0$. Since

$$\lim_{n\to\infty} \Phi_{x_i,y_j,z_k}\left(\frac{2^n t}{(p+q+2d)^n \alpha^n}\right) = 1$$

for all $x_i, y_j, z_k \in X$ and all $t > 0$. So

$$\begin{aligned}
&\mu_{2L\left(\frac{\sum_{i=1}^{p} x_i + \sum_{j=1}^{q} y_j + 2\sum_{k=1}^{d} z_k}{2}\right) - \sum_{i=1}^{p} L(x_i) - \sum_{j=1}^{q} L(y_j) - 2\sum_{k=1}^{d} L(z_k)}(t) \\
&= 1
\end{aligned}$$

for all $x_i, y_j, z_k \in X$ and all $t > 0$. Hence $L : X \to Y$ is an Cauchy–Jensen mapping and we get desired results. □

Corollary 3.2 Let X be a real normed space, θ be a positive real number and r is a real number with $r > 1$. Let $f : X \to Y$ be a mapping with $f(0) = 0$ and satisfying

$$\begin{aligned}
&\mu_{2f\left(\frac{\sum_{i=1}^{p} x_i + \sum_{j=1}^{q} y_j}{2} + \sum_{k=1}^{d} z_k\right) - \sum_{i=1}^{p} f(x_i) - \sum_{j=1}^{q} f(y_j) - 2\sum_{k=1}^{d} f(z_k)}(t) \\
&\geq \frac{t}{t + \theta\left(\sum_{i=1}^{p} \|x_i\|^r + \sum_{j=1}^{q} \|y_j\|^r + \sum_{k=1}^{d} \|z_k\|^r\right)}
\end{aligned} \tag{3.7}$$

for all $x_i, y_j, z_k \in X$ and all $t > 0$. Then the limit $\lim_{n\to\infty} \frac{(p+q+2d)^n}{2^n} f\left(\frac{2^n x}{(p+q+2d)^n}\right) = L(x)$ exists for all $x \in X$ and defines a unique Cauchy–Jensen mapping $L : X \to Y$ such that

$$\mu_{f(x)-L(x)}(t) \geq \frac{((p+q+2d)^r - 2^r)t}{((p+q+2d)^r - 2^r)t + 2^{r-1}(p+q+d)\theta\|x\|^r}$$

for all $x \in X$ and all $t > 0$.

Proof The proof follows from Theorem 2.1 by taking

$$\Phi_{x_i,y_j,z_k}(t) = \frac{t}{t+\theta\left(\sum_{i=1}^{p}\|x_i\|^r+\sum_{j=1}^{q}\|y_j\|^r+\sum_{k=1}^{d}\|z_k\|^r\right)}$$

for all $x_i, y_j, z_k \in X$ and all $t > 0$. Then we can choose $\alpha = \left(\frac{2}{p+q+2d}\right)^r$ and we get the desired result. □

Theorem 3.3 Let X be a linear space (Y, μ, T_M) be a complete RN-space and Φ be a mapping from X^{p+q+d} to D^+ ($\Phi(x_i, y_j, z_k)$ is denoted by Φ_{x_i,y_j,z_k}) such that there exists $0<\alpha<\frac{2}{p+q+2d}$ satisfying

$$\Phi_{\frac{2x_i}{p+q+2d},\frac{2y_j}{p+q+2d},\frac{2z_k}{p+q+2d}}(\alpha t) \geq \Phi_{x_i,y_j,z_k}(t)$$

for all $x_i, y_j, z_k \in X$ and all $t > 0$. Let $f : X \to Y$ be a mapping with $f(0) = 0$ and satisfying (3.2). Then the limit

$$\lim_{n\to\infty}\frac{2^n}{(p+q+2d)^n}f\left(\frac{(p+q+2d)^n x}{2^n}\right) = L(x)$$

exists for all $x \in X$ and $L : X \to Y$ is a unique Cauchy–Jensen additive mapping such that

$$\mu_{f(x)-L(x)}(t) \geq \Phi_{\underbrace{x, x, \ldots, x}_{(p+q+d)-\text{times}}}((p+q+2d)(1-\alpha)t) \tag{3.8}$$

for all $x \in X$.

Proof Let (S, d) be the generalized metric space defined in the proof of Theorem 2.1.

Now we consider the linear mapping $J : (S, d) \to (S, d)$ such that

$$Jg(x) := \frac{2}{p+q+2d}g\left(\frac{p+q+2d}{2}x\right)$$

for all $x \in X$.

It follows from (3.4) that $d(f, Jf) \leq \frac{1}{p+q+2d}$. By Theorem 1.6, there exists a mapping $L : X \to Y$ satisfying the following:

(1) L is a fixed point of J, i.e.,

$$\frac{(p+q+2d)L}{2} = L\left(\frac{(p+q+2d)x}{2}\right) \tag{3.9}$$

for all $x \in X$. The mapping L is a unique fixed point of J in the set

$$M = \{g \in S : d(h, g) < \infty\}.$$

This implies that L is a unique mapping satisfying (3.9) such that there exists a $u \in (0, \infty)$ satisfying

$$\mu_{g(x)-h(x)}(ut) \geq \Phi_{x,x,\ldots,x}(t)$$

for all $x \in X$ and all $t > 0$;

(2) $d(J^n f, L) \to 0$ as $n \to \infty$. This implies the equality

$$\lim_{n\to\infty}\frac{2^n}{(p+q+2d)^n}f\left(\frac{(p+q+2d)^n x}{2^n}\right) = L(x)$$

for all $x \in X$;

(3) $d(f, L) \leq \frac{1}{1-\alpha}d(f, Jf)$, which implies the inequality

$$d(f, L) \leq \frac{1}{(p+q+2d)(1-\alpha)}$$

and so

$$\mu_{f(x)-L(x)}\left(\frac{t}{(p+q+2d)(1-\alpha)}\right) \geq \Phi_{\underbrace{x, x, \ldots, x}_{(p+q+d)-\text{times}}}(t)$$

for all $x \in X$ and all $t > 0$. This implies that the inequalities (3.8) hold. The rest of the proof is similar to the proof of Theorem 2.1. □

Corollary 3.4 Let X be a real normed space, θ be a positive real number and r be a real number with $0<r<1$. Let $f : X \to Y$ be a mapping with $f(0) = 0$ and satisfying (3.7). Then there exists a unique Cauchy–Jensen mapping $L : X \to Y$ such that

$$\mu_{f(x)-L(x)}(t) \geq \frac{(p+q+2d)(2^r-(p+q+2d)^r)t}{(p+q+2d)(2^r-(p+q+2d)^r)t+2^r(p+q+d)\theta\|x\|^r}$$

for all $x \in X$ and all $t > 0$.

Proof The proof follows from Theorem 3.3 by taking

$$\Phi_{x_i,y_j,z_k}(t) = \frac{t}{t+\theta\left(\sum_{i=1}^{p}\|x_i\|^r+\sum_{j=1}^{q}\|y_j\|^r+\sum_{k=1}^{d}\|z_k\|^r\right)}$$

for all $x_i, y_j, z_k \in X$ and all $t > 0$. Then we can choose $\alpha = \left(\frac{p+q+2d}{2}\right)^r$ and we get the desired result. □

References

1. Aoki, T.: On the stability of the linear transformation in Banach spaces. J. Math. Soc. Jap. **2**, 64–66 (1950)
2. Arriola, L.M., Beyer, W.A.: Stability of the Cauchy functional equation over *p*-adic fields. Real Anal. Exch. **31**, 125–132 (2005)
3. Azadi Kenary, H.: On the stability of a cubic functional equation in random normed spaces. J. Math. Ext. **4**, 1–11 (2009)
4. Kenary, H.A.: On the Hyres–Rassias stability of the Pexiderial functional equation. Ital. J. Pure Appl. Math. (In press)
5. Kenary, H.A.: The probabilistic stability of a Pexiderial functional equation in random normed spaces. Rendiconti Del Circolo Mathematico Di Palermo (In press)
6. Cădariu, L., Radu, V.: Fixed points and the stability of Jensen's functional equation. J. Inequal. Pure Appl. Math. **4**(1), Art. ID 4 (2003)
7. Cholewa, P.W.: Remarks on the stability of functional equations. Aequationes Math. **27**, 76–86 (1984)
8. Chung, J., Sahoo, P.K.: On the general solution of a quartic functional equation. Bull. Korean Math. Soc. **40**, 565–576 (2003)

9. Czerwik, S.: Functional Equations and Inequalities in Several Variables. World Scientific, River Edge (2002)
10. Cădariu, L., Radu, V.: On the stability of the Cauchy functional equation: a fixed point approach. Grazer Math. Ber. **346**, 43–52 (2004)
11. Eshaghi-Gordji, M., Abbaszadeh, S., Park, C.: On the stability of a generalized quadratic and quartic type functional equation in quasi-Banach spaces. J. Inequal. Appl. 2009, Article ID 153084, 26 pages (2009)
12. Eshaghi-Gordji, M., Kaboli-Gharetapeh, S., Park, C., Zolfaghri, S.: Stability of an additive-cubic-quartic functional equation. Adv. Differ. Euqations 2009, Article ID 395693, 20 pages (2009)
13. Găvruta, P.: A generalization of the Hyers–Ulam–Rassias stability of approximately additive mappings. J. Math. Anal. Appl. **184**, 431–436 (1994)
14. Hyers, D.H.: On the stability of the linear functional equation. Proc. Nat. Acad. Sci. USA **27**, 222–224 (1941)
15. Hyers, D.H., Isac, G., Rassias, ThM: Stability of Functional Equations in Several Variables. Birkhäuser, Basel (1998)
16. Jun, K., Kim, H., Rassias, J.M.: Extended Hyers–Ulam stability for Cauchy–Jensen mappings. J. Differ. Equ. Appl. **13**, 1139–1153 (2007)
17. Lee, S., Im, S., Hwang, I.: Quartic functional equations. J. Math. Anal. Appl. **307**, 387–394 (2005)
18. Mihet, D., Radu, V.: On the stability of the additive Cauchy functional equation in random normed spaces. J. Math. Anal. Appl. **343**, 567–572 (2008)
19. Mohammadi, M., Cho, Y.J., Park, C., Vetro, P., Saadati, R.: Random stability of an additive-quadratic-quartic functional equation. J. Inequal. Appl. 2010, Article ID 754210, 18 pages (2010)
20. Najati, A., Park, C.: The Pexiderized Apollonius–Jensen type additive mapping and isomorphisms between C^*-algebras. J. Differ. Equ. Appl. **14**, 459–479 (2008)
21. Park, C.: Generalized Hyers–Ulam–Rassias stability of n-sesquilinear-quadratic mappings on Banach modules over C^*-algebras. J. Comput. Appl. Math. **180**, 279–291 (2005)
22. Park, C.: Fixed points and Hyers–Ulam–Rassias stability of Cauchy–Jensen functional equations in Banach algebras. Fixed Point Theory Appl. 2007, Art. ID 50175 (2007)
23. Park, C.: Generalized Hyers–Ulam–Rassias stability of quadratic functional equations: a fixed point approach. Fixed Point Theory Appl. 2008, Art. ID 493751 (2008)
24. Radu, V.: The fixed point alternative and the stability of functional equations. Fixed Point Theory **4**, 91–96 (2003)
25. Rassias, ThM: On the stability of the linear mapping in Banach spaces. Proc. Am. Math. Soc. **72**, 297–300 (1978)
26. Rätz, J.: On inequalities associated with the Jordan-von Neumann functional equation. Aequationes Math. **66**, 191–200 (2003)
27. Saadati, R., Park, C.: Non-Archimedean $\mathcal{L}$-fuzzy normed spaces and stability of functional equations (In press)
28. Saadati, R., Vaezpour, M., Cho, Y.J.: A note to paper "On the stability of cubic mappings and quartic mappings in random normed spaces". J. Ineqal. Appl. **2009**, Article ID 214530. doi: 10.1155/2009/214530
29. Saadati, R., Zohdi, M.M., Vaezpour, S. M.: Nonlinear L-Random stability of an ACQ functional equation. J. Inequal. Appl. **2011**, Article ID 194394, 23 pages. doi:10.1155/2011/194394
30. Schewizer, B., Sklar, A.: Probabilistic Metric Spaces. In: North-Holland Series in Probability and Applied Mathematics. North-Holland, New York (1983)
31. Skof, F.: Local properties and approximation of operators. Rend. Sem. Mat. Fis. Milano **53**, 113–129 (1983)
32. Ulam, S.M.: Problems in Modern Mathematics. In: Science Editions. Wiley, New York (1964)

11

Dynamics in a periodic two-species predator–prey system with pure delays

Rouzimaimaiti Mahemuti · Ahmadjan Muhammadhaji · Zhidong Teng

Abstract A class of non-autonomous two-species Lotka–Volterra predator–prey system with pure discrete time delays is discussed. Some sufficient conditions on the boundedness, permanence, extinction, positive periodic solution and global attractivity of the system are established by means of the comparison method, coincidence degree theory and Liapunov functional.

Keywords Lotka–Volterra predator–prey system · Discrete time delay · Liapunov functional · Global attractivity · Positive periodic solution.

Introduction

In the real world, there are many types of interactions between two species. Predator–prey relations are among the most common ecological interactions. Remarkably, the whole field of mathematical ecology began with the studies of population dynamics subject to the predator–prey interaction, that is, with the classical works by Lotka [1] and Volterra [2]. Traditional two-species non-autonomous Lotka–Volterra predator–prey systems take the form

$$\begin{aligned} \frac{dx_1(t)}{dt} &= x_1(t)\big[r_1(t) - a_{11}(t)x_1(t) - a_{12}(t)x_2(t)\big], \\ \frac{dx_2(t)}{dt} &= x_2(t)\big[-r_2(t) + a_{21}(t)x_1(t) - a_{22}(t)x_2(t)\big]. \end{aligned} \tag{1}$$

Recently, the properties of system (1) are discussed by many scholars [3–7]. Some sufficient conditions are obtained for the persistence, permanence and extinction of the species, the existence and uniqueness of periodic solutions or almost periodic solutions, and the global stability of solutions for system (1).

However, in the real world, the growth rate of a natural species will not often respond immediately to changes in its own population or that of an interacting species, but will rather do so after a time lag [8]. Time delays have a great destabilizing influence on the species population; this result was put forwarded by May [9].

Therefore, we should introduce time delay into model foundation, which will have more resemblance to the real ecosystem.

In this paper, we investigate the following two-species Lotka–Volterra type predator–prey systems with pure discrete time delays

$$\begin{aligned} \dot{x}_1(t) &= x_1(t)\big[r_1(t) - a_{11}(t)x_1(t-\tau_1) - a_{12}(t)x_2(t-\tau_1)\big], \\ \dot{x}_2(t) &= x_2(t)\big[-r_2(t) + a_{21}(t)x_1(t-\tau_2) - a_{22}(t)x_2(t-\tau_2)\big]. \end{aligned} \tag{2}$$

Our main purpose is to establish some sufficient conditions on the boundedness, permanence, extinction, positive periodic solution and global attractivity of the system (2).

The organization of this paper is as follows. In the next section, we will present some basic assumptions and main lemmas. In "Main results", we will consider conditions for the boundedness, permanence, extinction, positive periodic solution and global attractivity of the system. In the final section, as an application, one special case of the system is considered.

Preliminaries

In system (2), we have that $x_1(t)$ is the prey population density and $x_2(t)$ is the predator population density, $r_1(t)$ and

R. Mahemuti · A. Muhammadhaji (✉) · Z. Teng
College of Mathematics and System Sciences, Xinjiang University, Urumqi 830046, People's Republic of China
e-mail: ahmadjanm@gmail.com

$a_{11}(t)$ are the intrinsic growth rate and density-dependent coefficient of the prey, respectively, $r_2(t)$ is the intrinsic growth rate of the predator, $a_{12}(t)$ is the capturing rate of the predator and $a_{21}(t)$ is the rate of conversion of nutrients into the reproduction of the predator. Throughout this paper, for system (2) we introduce the following hypotheses.

(H_1) $\tau_i > 0 (i = 1,2)$ are positive constants, $r_i(t)(i = 1,2)$ are continuous ω-periodic functions with $\int_0^{\omega} r_i(t)dt > 0$. $a_{ij}(t)(i,j = 1,2)$ are continuous positive ω-periodic functions.

From the viewpoint of mathematical biology, in this paper for system (2), we only consider the solution with the following initial condition:

$$x_i(t) = \phi_i(t) \quad \text{for all} \quad t \in [-\tau, 0), \quad i = 1,2, \qquad (3)$$

where $\phi_i(t)$ $(i = 1,2)$ are nonnegative continuous functions defined on $[-\tau, 0)$ satisfying $\phi_i(0) > 0$ $(i = 1,2)$ and $\tau = \max\{\tau_1, \tau_2\}$.

In this paper, for any ω-periodic continuous function $f(t)$, we denote

$$f^L = \min_{t\in[0,\omega]} f(t), \quad f^M = \max_{t\in[0,\omega]} f(t), \quad \bar{f} = \frac{1}{\omega}\int_0^{\omega} f(t)dt.$$

To obtain the existence of positive ω-periodic solutions of system (2), we will use the continuation theorem. For the reader's convenience, we will introduce the continuation theorem in the following. First we define some definitions.

Let X and Z be two normed vector spaces. Let L : Dom $L \subset X \to Z$ be a linear operator and $N : X \to Z$ be a continuous operator. The operator L is called a Fredholm operator of index zero, if $\dim \operatorname{Ker} L = \operatorname{codim} \operatorname{Im} L < \infty$ and $\operatorname{Im} L$ is a closed set in Z. If L is a Fredholm operator of index zero, then there exist continuous projectors $P : X \to X$ and $Q : Z \to Z$ such that $\operatorname{Im} P = \operatorname{Ker} L$ and $\operatorname{Im} L = \operatorname{Ker} Q = \operatorname{Im}(I - Q)$. It follows that $L|\operatorname{Dom} L \cap \operatorname{Ker} P : \operatorname{Dom} L \cap \operatorname{Ker} P \to \operatorname{Im} L$ is invertible and its inverse is denoted by K_P and denoted by $J : ImQ \to KerL$ an isomorphism of ImQ onto $KerL$. Let Ω be a bounded open subset of X, we say that the operator N is L-compact on $\overline{\Omega}$, where $\overline{\Omega}$ denotes the closure of Ω in X, if $QN(\overline{\Omega})$ is bounded and $K_P(I - Q)N : \overline{\Omega} \to X$ is compact. Such definitions can be found in [10–12].

Now, we present some useful lemmas.

Lemma 1 Set $R^2_+ = \{(x_1, x_2) : x_i > 0, i = 1,2\}$ is positively invariant for system (2).

The proof of Lemma 1 is simple, and here we omit it.

Lemma 2 [(see,[13])] Consider the following equation: $\dot{u}(t) = u(t)(d_1 - d_2u(t))$, where $d_2 > 0$, we have If $d_1 > 0$, *then* $\lim_{t\to+\infty} u(t) = d_1/d_2$. If $d_1 < 0$, *then* $\lim_{t\to+\infty} u(t) = 0$.

Lemma 3 (see [4]) Let L be a Fredholm operator of index zero and let N be L-compact on $\overline{\Omega}$. If

(a) for each $\lambda \in (0,1)$ and $x \in \partial\Omega \cap \operatorname{Dom} L$, $Lx \neq \lambda Nx$;
(b) for each $x \in \partial\Omega \cap \operatorname{Ker} L$, $QNx \neq 0$;
(c) $\deg\{JQN, \Omega \cap \operatorname{Ker} L, 0\} \neq 0$,

then the operator equation $Lx = Nx$ has at least one solution lying in $\operatorname{Dom} L \cap \overline{\Omega}$.

Main results

In this section, we will obtain some sufficient conditions for the boundedness, existence of periodic solution, global attractivity, permanence, and extinction of system (2).

Theorem 1 Suppose that assumption **(H1)** holds, then there exist positive constants $M_i (i = 1,2)$ such that

$$x_i(t) \leq M_i,$$

for any positive solution $x_i(t)$ of system (2).

Proof Let $(x_1(t), x_2(t))$ be a solution of system (2). Firstly, it follows from the first equation of system (2) that for $t > \tau$, we have

$$\begin{aligned}\frac{dx_1(t)}{dt} &= x_1(t)[r_1(t) - a_{11}(t)x_1(t-\tau_1) - a_{12}(t)x_2(t-\tau_1)] \\ &\leq x_1(t)[r_1^M - a_{11}^L e^{-r_1^M\tau_1} x_1(t)] \quad \text{for} \quad t > \tau.\end{aligned}$$

We consider the following auxiliary equation

$$\frac{du(t)}{dt} = u(t)[r_1^M - a_{11}^L e^{-r_1^M\tau_1} u(t)].$$

By Lemma 2, we derive

$$\lim_{t\to+\infty} u(t) = \frac{r_1^M e^{r_1^M\tau_1}}{a_{11}^L} =: M_1.$$

By comparison, there exists a $T_1 > \tau$ such that $x_1(t) \leq M_1$ for $t \geq T_1$.

Next, from the second equation of system (2) for $t > T_1$, we have

$$\begin{aligned}\frac{dx_2(t)}{dt} &\leq x_2(t)[a_{21}^M M_1 - a_{22}^L x_2(t-\tau_2)] \\ &\leq x_2(t)[a_{21}^M M_1 - a_{22}^L e^{-a_{21}^M M_1 \tau_2} x_2(t)] \quad \text{for} \quad t > T_1.\end{aligned}$$

We consider the following auxiliary equation

$$\frac{du(t)}{dt} = u(t)[a_{21}^M M_1 - a_{22}^L e^{-a_{21}^M M_1\tau_2} u(t)].$$

By Lemma 2, we derive

$$\lim_{t\to+\infty} u(t) = \frac{a_{21}^M M_1 e^{a_{21}^M M_1\tau_2}}{a_{22}^L} =: M_2.$$

By comparison, there exists a $T_2 > \tau$ such that $x_2(t) \le M_2$ for $t \ge T_2$. This completes the proof. □

Theorem 2 Suppose that assumption **(H1)** holds and $\bar{r}_1 - (\frac{a_{11}}{a_{21}})^M \bar{r}_2 > 0$. Then system (2) has at least one positive $\omega-$ periodic solution.

Proof Let

$$x_1(t) = \exp\{u_1(t)\} \quad \text{and} \quad x_2(t) = \exp\{u_2(t)\}.$$

Then system (2) is rewritten in the following system

$$\begin{aligned} \dot{u}_1(t) &= r_1(t) - a_{11}(t)\exp\{u_1(t-\tau_1)\} - a_{12}(t)\exp\{u_2(t-\tau_1)\}, \\ \dot{u}_2(t) &= -r_2(t) + a_{21}(t)\exp\{u_1(t-\tau_2)\} - a_{22}(t)\exp\{u_2(t-\tau_2)\}. \end{aligned} \tag{4}$$

To apply Lemma 3 to system (4), we introduce the normed vector spaces X and Z as follows. Let $C(R, R^2)$ denote the space of all continuous functions $u(t) = (u_1(t), u_2(t)) : R \to R^2$. We take

$$X = Z = \{u(t) \in C(R, R^2) : u(t) \text{ an } \omega\text{-periodic function}\}$$

with norm

$$\| u \| = \max_{t\in[0,\omega]} |u_1(t)| + \max_{t\in[0,\omega]} |u_2(t)|.$$

It is obvious that X and Z are the Banach spaces. We define a linear operator $L : \operatorname{Dom} L \subset X \to Z$ and a continuous operator $N : X \to Z$ as follows.

$$Lu(t) = \dot{u}(t)$$

and

$$Nu(t) = (Nu_1(t), Nu_2(t)). \tag{5}$$

where

$$\begin{aligned} Nu_1(t) &= r_1(t) - a_{11}(t)\exp\{u_1(t-\tau_1)\} - a_{12}(t)\exp\{u_2(t-\tau_1)\}, \\ Nu_2(t) &= -r_2(t) + a_{21}(t)\exp\{u_1(t-\tau_2)\} - a_{22}(t)\exp\{u_2(t-\tau_2)\}. \end{aligned}$$

Further, we define continuous projectors $P : X \to X$ and $Q : Z \to Z$ as follows.

$$Pu(t) = \frac{1}{\omega}\int_0^\omega u(t)dt, \quad Qv(t) = \frac{1}{\omega}\int_0^\omega v(t)dt.$$

We easily see $\operatorname{Im} L = \{v \in Z : \int_0^\omega v(t)dt = 0\}$ and $\operatorname{Ker} L = R^2$. It is obvious that $\operatorname{Im} L$ is closed in Z and $\dim\operatorname{Ker} L = 2$. Since for any $v \in Z$ there are unique $v_1 \in R^n$ and $v_2 \in \operatorname{Im} L$ with

$$v_1 = \frac{1}{\omega}\int_0^\omega v(t)dt, \quad v_2(t) = v(t) - v_1,$$

such that $v(t) = v_1 + v_2(t)$, we have $\operatorname{codim}\operatorname{Im} L = 2$. Therefore, L is a Fredholm mapping of index zero. Furthermore, the generalized inverse (to L) $K_p : \operatorname{Im} L \to \operatorname{Ker} P \cap \operatorname{Dom} L$ is given in the following form:

$$K_p v(t) = \int_0^t v(s)ds - \frac{1}{\omega}\int_0^\omega \int_0^t v(s)dsdt.$$

For convenience, we denote $F(t) = (F_1(t), F_2(t))$ as follows

$$\begin{aligned} F_1(t) &= r_1(t) - a_{11}(t)\exp\{u_1(t-\tau_1)\} - a_{12}(t)\exp\{u_2(t-\tau_1)\}, \\ F_2(t) &= -r_2(t) + a_{21}(t)\exp\{u_1(t-\tau_2)\} - a_{22}(t)\exp\{u_2(t-\tau_2)\}. \end{aligned} \tag{6}$$

Thus, we have

$$QNu(t) = \frac{1}{\omega}\int_0^\omega F(t)dt \tag{7}$$

and

$$\begin{aligned} K_p(I-Q)Nu(t) &= K_p INu(t) - K_p QNu(t) \\ &= \int_0^t F(s)ds - \frac{1}{\omega}\int_0^\omega \int_0^t F(s)dsdt \\ &\quad + \left(\frac{1}{2} - \frac{t}{\omega}\right)\int_0^\omega F(s)ds. \end{aligned} \tag{8}$$

From formulas (7) and (8), we easily see that QN and $K_p(I-Q)N$ are continuous operators. Furthermore, it can be verified that $\overline{K_p(I-Q)N(\overline{\Omega})}$ is compact for any open-bounded set $\Omega \subset X$ by using Arzela–Ascoli theorem and $QN(\overline{\Omega})$ is bounded. Therefore, N is L-compact on $\overline{\Omega}$ for any open-bounded subset $\Omega \subset X$.

Now, we reach the position to search for an appropriate open-bounded subset Ω for the application of the continuation theorem (Lemma 3) to system (2).

Corresponding to the operator equation $Lu(t) = \lambda Nu(t)$ with parameter $\lambda \in (0,1)$, we have

$$\dot{u}_i(t) = \lambda F_i(t), \quad i = 1, 2, \tag{9}$$

where $F_i(t)$ $(i = 1, 2)$ are given in Eq. 6.

Assume that $u(t) = (u_1(t), u_2(t)) \in X$ is a solution of system (9) for some parameter $\lambda \in (0,1)$. By integrating system (9) over the interval $[0, \omega]$, we obtain

$$\begin{aligned} &\int_0^\omega \big[r_1(t) - a_{11}(t)\exp\{u_1(t-\tau_1)\} - a_{12}(t)\exp\{u_2(t-\tau_1)\}\big]dt = 0, \\ &\int_0^\omega \big[-r_2(t) + a_{21}(t)\exp\{u_1(t-\tau_2)\} - a_{22}(t)\exp\{u_2(t-\tau_2)\}\big]dt = 0. \end{aligned} \tag{10}$$

By (10) we get ,

$$\begin{aligned} &\int_0^\omega \big[a_{11}(t)\exp\{u_1(t-\tau_1)\} + a_{12}(t)\exp\{u_2(t-\tau_1)\}\big]dt = \bar{r}_1\omega, \\ &\int_0^\omega \big[a_{21}(t)\exp\{u_1(t-\tau_2)\} - a_{22}(t)\exp\{u_2(t-\tau_2)\}\big]dt = \bar{r}_2\omega. \end{aligned} \tag{11}$$

For each $i,j = 1,2$, we have

$$\begin{aligned}
&\int_0^{\omega} a_{ij}(t)\exp\{u_i(t-\tau_i)\}dsdt \\
&= \int_{-\tau_i}^{\omega-\tau_i} a_{ij}(s+\tau_i)\exp\{u_i(s)\}ds \\
&= \int_0^{\omega} a_{ij}(s+\tau_i)\exp\{u_i(s)\}ds \\
&= \int_0^{\omega} a_{ij}(t+\tau_i)\exp\{u_i(t)\}dt.
\end{aligned} \tag{12}$$

From the continuity of $u(t) = (u_1(t), u_2(t))$, there exist constants $\xi_i, \eta_i \in [0, \omega]$ $(i = 1, 2)$ such that

$$u_i(\xi_i) = \max_{t\in[0,\omega]} u_i(t), \quad u_i(\eta_i) = \min_{t\in[0,\omega]} u_i(t), \quad i = 1,2. \tag{13}$$

From (11)–(13) and the condition of Theorem 2, we further obtain

$$\begin{aligned}
&u_i(\eta_i) \le \ln\left(\frac{\bar{r}_1}{\bar{A}_i}\right), \quad u_1(\xi_1) \ge \ln\left(\frac{\bar{r}_2}{\bar{A}_3}\right) =: B_1, \\
&\quad u_2(\xi_2) \ge \ln A_4 =: B_2,
\end{aligned} \tag{14}$$

where

$$A_i(t) = a_{1i}(t+\tau_1), \ A_3(t) = a_{22}(t+\tau_2),$$
$$A_4 = \frac{\bar{r}_1 - (\frac{a_{11}}{a_{21}})^M \bar{r}_2}{\bar{A}_2 + (\frac{a_{11}}{a_{21}})^M \bar{A}_3}, \quad i = 1,2.$$

From (4) and (11) we have

$$\begin{aligned}
\int_0^{\omega} |\dot{u}_1(t)|dt &\le \int_0^{\omega} \big[r_1(t) + a_{11}(t)\exp\{u_1(t-\tau_1)\} \\
&\quad + a_{12}(t)\exp\{u_2(t-\tau_1)\}\big]dt \\
&\le 2\bar{r}_1\omega =: C_1,
\end{aligned} \tag{15}$$

$$\begin{aligned}
\int_0^{\omega} |\dot{u}_2(t)|dt &\le \int_0^{\omega} \big[r_2(t) + a_{21}(t)\exp\{u_1(t-\tau_2)\} \\
&\quad + a_{22}(t)\exp\{u_2(t-\tau_2)\}\big]dt \\
&\le \frac{2a_{21}^M \bar{r}_1\omega}{a_{11}^L} =: C_2.
\end{aligned} \tag{16}$$

By (14)–(16), we have

$$\begin{aligned}
u_i(t) &\le u_i(\eta_i) + \int_0^{\omega} |\dot{u}_i(t)|dt \le \ln\left(\frac{\bar{r}_1}{A_i}\right) \\
&\quad + C_i =: M_i \quad i = 1,2,
\end{aligned} \tag{17}$$

and

$$u_i(t) \ge u_i(\xi_i) - \int_0^{\omega} |\dot{u}_i(t)|dt \ge B_i - C_i =: N_i \quad i = 1,2. \tag{18}$$

Therefore, from (17), (18) we have

$$\max_{t\in[0,\,\omega]} |u_i(t)| \le \max\left\{|M_i|, |N_i|\right\} =: H_i, \qquad i = 1,2.$$

It can be seen that the constants $H_i\,(i = 1,2)$ are independent of parameter $\lambda \in (0,1)$.

For any $u = (u_1, u_2) \in R^2$, from (5) we can obtain

$$QNu = (QNu_1, QNu_2).$$

where

$$QNu_1 = \bar{r}_1 - \bar{a}_{11}\exp\{u_1\} - \bar{a}_{12}\exp\{u_2\},$$
$$QNu_2 = -\bar{r}_2 + \bar{a}_{21}\exp\{u_1\} - \bar{a}_{22}\exp\{u_2\}.$$

We consider the following system of algebraic equations

$$\begin{aligned}
&\bar{r}_1 - \bar{a}_{11}v_1 - \bar{a}_{12}v_2 = 0, \\
&-\bar{r}_2 + \bar{a}_{21}v_1 - \bar{a}_{22}v_2 = 0.
\end{aligned}$$

By direct calculation we can get

$$v_1^* = \frac{\bar{r}_1\bar{a}_{22} + \bar{r}_2\bar{a}_{12}}{\bar{a}_{11}\bar{a}_{22} + \bar{a}_{12}\bar{a}_{21}},$$
$$v_2^* = \frac{\bar{r}_1\bar{a}_{21} - \bar{r}_2\bar{a}_{11}}{\bar{a}_{11}\bar{a}_{22} + \bar{a}_{12}\bar{a}_{21}} > \frac{\bar{a}_{21}[\bar{r}_1 - (\frac{a_{11}}{a_{21}})^M \bar{r}_2]}{\bar{a}_{11}\bar{a}_{22} + \bar{a}_{12}\bar{a}_{21}}.$$

From the assumption of Theorem 2, the system of algebraic equations has a unique positive solution $v^* = (v_1^*, v_2^*)$. Hence, the equation $QNu = 0$ has a unique solution $u^* = (u_1^*, u_2^*) = (\ln v_1^*, \ln v_2^*, \ln) \in R^2$.

Choosing constant $H > 0$ large enough such that $|u_1^*| + |u_2^*| < H$ and $H > H_1 + H_2$, we define a bounded open set $\Omega \subset X$ as follows

$$\Omega = \{u \in X : \|u\| < H\}.$$

It is clear that Ω satisfies conditions (a) and (b) of Lemma 3. On the other hand, by directly calculating we can obtain

$$\begin{aligned}
&\deg\{JQN, \Omega \cap \text{ Ker } L, (0,0)\} \\
&\quad = \text{sgn}\begin{vmatrix} -\bar{a}_{11}K_1 - \bar{a}_{12}K_2 \\ \bar{a}_{21}K_1 - \bar{a}_{22}K_2 \end{vmatrix},
\end{aligned}$$

where $K_i = \exp\{u_i\} (i = 1,2)$.

From the assumption of Theorem 2, we have

$$\begin{vmatrix} -\bar{a}_{11}K_1 - \bar{a}_{12}K_2 \\ \bar{a}_{21}K_1 - \bar{a}_{22}K_2 \end{vmatrix} \neq 0.$$

From this, we finally have

$$\deg\{JQN, \Omega \cap \text{ Ker } L, (0,0)\} \neq 0.$$

This shows that Ω satisfies condition (c) of Lemma 3. Therefore, system (4) has a ω-periodic solution $u^*(t) = (u_1^*(t), u_2^*(t)) \in \bar{\Omega}$. Hence, system (2) has a positive ω-periodic solution $x^*(t) = (x_1^*(t), x_2^*(t))$. □

Theorem 3 Suppose that assumptions of Theorem 2 hold. Further suppose that the following (H_2) holds.

(H_2) There exists a constant $\mu_i > 0$ $(i = 1, 2)$ such that

$$\liminf_{t\to\infty} A_i(t) > 0, \quad i = 1, 2,$$

where

$$\begin{aligned} A_1(t) &= \mu_1 a_{11}(t) - \mu_1 \int_{t-\tau_1}^{t} a_{11}(u+\tau_1)du[r_1(t) \\ &\quad + (a_{11}(t) + a_{12}(t))M] \\ &\quad - M\sum_{j=1}^{2} \mu_j \tau_j a_{jj}^M a_{j1}(t+\tau_j) - \mu_2 a_{21}(t+\tau_2), \\ A_2(t) &= \mu_2 a_{22}(t) - \mu_2 \int_{t-\tau_2}^{t} a_{22}(u+\tau_2)du[r_2(t) \\ &\quad + (a_{21}(t) + a_{22}(t))M] \\ &\quad - M\sum_{j=1}^{2} \mu_j \tau_j a_{jj}^M a_{j2}(t+\tau_j) - \mu_1 a_{12}(t+\tau_1), \end{aligned}$$

where $M = \max\{M_1, M_2\}$ and M_1, M_2 are defined in Theorem 1. Then system (2) has a positive periodic solution which is globally attractive.

Proof From Theorem 2, we can obtain that system (2) has a positive periodic solution.

Let $x(t) = (x_1(t), x_2(t))$ be a positive periodic solution of system (2) and $(y_1(t), y_2(t))$ be a any positive solution of system (2). From Theorem 1, choose positive constants $m_i > 0$, $M_i > 0$ such that

$$m \le x_i(t) \le M, \quad i = 1, 2, \tag{19}$$

where $M = \max\{M_1, M_2\}$ and $m = \min\{m_1, m_2\}$ for all $t \ge T$. Let

$$W_1(t) = \sum_{i=1}^{2} \mu_i |\ln x_i(t) - \ln y_i(t)|.$$

Calculating the upper right derivation of $W_1(t)$ along system (2) for all $t \ge T$, we have

$$\begin{aligned} D^+W_1(t) &= \mu_1 \text{ sign } (x_1(t) - y_1(t))[-a_{11}(t)(x_1(t-\tau_1) \\ &\quad - y_1(t-\tau_1)) - a_{12}(t)(x_2(t-\tau_1) - y_2(t-\tau_1))] \\ &\quad + \mu_2 \text{ sign } (x_2(t) - y_2(t)) \\ &\quad \times [a_{21}(t)(x_1(t-\tau_2) - y_1(t-\tau_2)) \\ &\quad - a_{22}(t)(x_2(t-\tau_2) - y_2(t-\tau_2))] \\ &= \mu_1 \text{ sign } (x_1(t) - y_1(t))[-a_{11}(t)(x_1(t) - y_1(t)) \\ &\quad - a_{12}(t)(x_2(t-\tau_1) - y_2(t-\tau_1)) \\ &\quad + a_{11}(t)\int_{t-\tau_1}^{t} (\dot{x}_1(u) - \dot{y}_1(u))du] \\ &\quad + \mu_2 \text{ sign } (x_2(t) - y_2(t))[-a_{22}(t)(x_2(t) - y_2(t)) \\ &\quad + a_{21}(t)(x_1(t-\tau_2) - y_1(t-\tau_2)) \\ &\quad + a_{22}(t)\int_{t-\tau_2}^{t} (\dot{x}_2(u) - \dot{y}_2(u))du] \\ &= \mu_1 \text{ sign } (x_1(t) - y_1(t))[-a_{11}(t)(x_1(t) - y_1(t)) \\ &\quad - a_{12}(t)(x_2(t-\tau_1) - y_2(t-\tau_1)) \\ &\quad + a_{11}(t)\int_{t-\tau_1}^{t} (x_1(u)[r_1(u) - a_{11}(u)x_1(u-\tau_1) \\ &\quad - a_{12}(u)x_2(u-\tau_1)] - y_1(u)[r_1(u) \\ &\quad - a_{11}(u)y_1(u-\tau_1) - a_{12}(u)y_2(u-\tau_1)])du] \\ &\quad + \mu_2 \text{ sign } (x_2(t) - y_2(t))[-a_{22}(t)(x_2(t) - y_2(t)) \\ &\quad + a_{21}(t)(x_1(t-\tau_2) - y_1(t-\tau_2)) \\ &\quad + a_{22}(t)\int_{t-\tau_2}^{t} (x_2(u)[r_2(u) + a_{21}(u)x_1(u-\tau_2) \\ &\quad - a_{22}(u)x_2(u-\tau_2)] - y_2(u)[r_2(u) \\ &\quad + a_{21}(u)y_1(u-\tau_2) - a_{22}(u)y_2(u-\tau_2)])du] \\ &= \mu_1 \text{ sign } (x_1(t) - y_1(t))[-a_{11}(t)(x_1(t) - y_1(t)) \\ &\quad - a_{12}(t)(x_2(t-\tau_1) - y_2(t-\tau_1)) \\ &\quad + a_{11}(t)\int_{t-\tau_1}^{t} ((x_1(u) - y_1(u))[r_1(u) - a_{11}(u)y_1(u-\tau_1) \\ &\quad - a_{12}(u)y_2(u-\tau_1)] \\ &\quad + x_1(u)[-a_{11}(u)(x_1(u-\tau_1) - y_1(u-\tau_1)) \\ &\quad - a_{12}(u)(x_2(u-\tau_1) - y_2(u-\tau_1))])du] \\ &\quad + \mu_2 \text{ sign } (x_2(t) - y_2(t))[-a_{22}(t)(x_2(t) - y_2(t)) \\ &\quad + a_{21}(t)(x_1(t-\tau_2) - y_1(t-\tau_2)) \\ &\quad + a_{22}(t)\int_{t-\tau_2}^{t} ((x_2(u) - y_2(u))[r_2(u) + a_{21}(u)y_1(u-\tau_2) \\ &\quad - a_{22}(u)y_2(u-\tau_2)] + x_2(u)[a_{21}(u)(x_1(u-\tau_2) \\ &\quad - y_1(u-\tau_2)) - a_{22}(u)(x_2(u-\tau_2) - y_2(u-\tau_2))])du] \\ &\le -\sum_{i=1}^{2} \mu_i a_{ii}(t)|x_i(t) - y_i(t)| + \mu_1 a_{12}(t)|x_2(t-\tau_1) - y_2(t-\tau_1)| \\ &\quad + \mu_2 a_{21}(t)|x_1(t-\tau_2) - y_1(t-\tau_2)| \\ &\quad + \mu_1 a_{11}(t)\int_{t-\tau_1}^{t} (|x_1(u) - y_1(u)|[r_1(u) \\ &\quad + a_{11}(u)y_1(u-\tau_1) + a_{12}(u)y_2(u-\tau_1)] + x_1(u) \\ &\quad [a_{11}(u)|x_1(u-\tau_1) - y_1(u-\tau_1)| \\ &\quad + a_{12}(u)|x_2(u-\tau_1) - y_2(u-\tau_1)|])du \\ &\quad + \mu_2 a_{22}(t)\int_{t-\tau_2}^{t} (|x_2(u) - y_2(u)|[r_2(u) + a_{21}(u)y_1(u-\tau_2) \\ &\quad + a_{22}(u)y_2(u-\tau_2)] + x_2(u)[a_{21}(u)|x_1(u-\tau_2) \\ &\quad - y_1(u-\tau_2)| + a_{22}(u)|x_2(u-\tau_2) - y_2(u-\tau_2)|])du. \end{aligned} \tag{20}$$

Define

$$W_2(t) = \mu_1 V_1(t) + \mu_2 V_2(t),$$

where

$$V_1(t) = \int_{t-\tau_1}^{t}\int_{u}^{t} a_{11}(u+\tau_1)\big(\big[r_1(s)+a_{11}(s)y_1(s-\tau_1) + a_{12}(s)y_2(s-\tau_1)\big]|x_1(s)-y_1(s)| + x_1(s)\big[a_{11}(s)|x_1(s-\tau_1)-y_1(s-\tau_1)| + a_{12}(s)|x_2(s-\tau_1)-y_2(s-\tau_1)|\big)dsdu,$$

$$V_2(t) = \int_{t-\tau_2}^{t}\int_{u}^{t} a_{22}(u+\tau_2)\big(\big[r_2(s)+a_{21}(s)y_1(s-\tau_2) + a_{22}(s)y_2(s-\tau_2)\big]|x_2(s)-y_2(s)| + x_2(s)\big[a_{21}(s)|x_1(s-\tau_2)-y_1(s-\tau_2)| + a_{22}(s)|x_2(s-\tau_2)-y_2(s-\tau_2)|\big)dsdu,$$

Calculating the upper right derivative, from (20) we have

$$\begin{aligned}\sum_{i=1}^{2} D^+W_i(t) \le & -\sum_{i=1}^{2}\mu_i a_{ii}(t)|x_i(t)-y_i(t)| \\ & + \mu_1 a_{12}(t)|x_2(t-\tau_1)-y_2(t-\tau_1)| \\ & + \mu_2 a_{21}(t)|x_1(t-\tau_2)-y_1(t-\tau_2)| \\ & + \mu_1\int_{t-\tau_1}^{t} a_{11}(u+\tau_1)du\big[r_1(t) \\ & + (a_{11}(t)+a_{12}(t))M\big]|x_1(t)-y_1(t)| \\ & + \mu_1\tau_1 a_{11}^M M\sum_{i=1}^{2} a_{1i}(t) \\ & \times |x_i(t-\tau_1)-y_i(t-\tau_1)| \\ & + \mu_2\int_{t-\tau_2}^{t} a_{22}(u+\tau_2)du\big[r_2(t) \\ & + (a_{21}(t)+a_{22}(t))M\big]|x_2(t)-y_2(t)| \\ & + \mu_2\tau_2 a_{22}^M M\sum_{i=1}^{2} a_{2i}(t) \\ & \times \sum_{i=1}^{2} a_{2i}(t)|x_i(t-\tau_2)-y_i(t-\tau_2)|. \end{aligned} \tag{21}$$

Define

$$W_3(t) = \mu_1 V_3(t) + \mu_2 V_4(t),$$

where

$$V_3(t) = \mu_1\tau_1 a_{11}^M M\sum_{i=1}^{2}\int_{t-\tau_1}^{t} a_{1i}(u+\tau_1)|x_i(u) - y_i(u)|du + \int_{t-\tau_1}^{t} a_{12}(u+\tau_1)|x_2(u)-y_2(u)|du,$$

$$V_4(t) = \mu_2\tau_2 a_{22}^M M\sum_{i=1}^{2}\int_{t-\tau_1}^{t} a_{2i}(u+\tau_2)|x_i(u)-y_i(u)|du + \int_{t-\tau_2}^{t} a_{21}(u+\tau_2)|x_1(u)-y_1(u)|du.$$

Further, we define a Liapunov function as follows

$$V(t) = \sum_{i=1}^{3} W_i(t).$$

Calculating the upper right derivation of $V(t)$, from (20) and (21) we finally can obtain for all $t \ge T$

$$D^+V(t) \le -\sum_{i=1}^{2} A_i(t)|x_i(t)-y_i(t)|. \tag{22}$$

From assumption (H_2), there exists a constant $\alpha > 0$ and $T^* \ge T$ such that for all $t \ge T^*$ we have

$$A_i(t) \ge \alpha > 0, \quad i = 1,2. \tag{23}$$

Integrating from T^* to t on both sides of (22) and by (23) produces

$$V(t) + \alpha\int_{T^*}^{t}\left(\sum_{i=1}^{2}|x_i(s)-y_i(s)|\right)ds \le V(T^*), \tag{24}$$

then

$$\int_{0}^{t}\left(\sum_{i=1}^{2}|x_i(s)-y_i(s)|\right)ds \le \frac{V(T^*)}{\alpha}, \quad t \ge T^*. \tag{25}$$

By the definition of $V(t)$ and (24) we have

$$\sum_{i=1}^{2}\mu_i|\ln x_i(t) - \ln y_i(t)| \le V(t) \le V(T^*), \quad t \ge 0. \tag{26}$$

Therefore, for $i = 1,2$ we have

$$\mu_i|\ln x_i(t) - \ln y_i(t)| \le V(T^*), \quad t \ge 0. \tag{27}$$

which, together with (19), lead to

$$m_i\exp\{-V(T^*)/\mu_i\} \le y_i(t) \le M_i\exp\{V(T^*)/\mu_i\}, \quad i = 1,2. \tag{28}$$

From the boundedness of $x_i(t)$ and (25), it follows that $y_i(t)(i=1,2)$ are bounded for $t \ge 0$. From the boundedness of $x_i(t)$ and $y_i(t)$ we know that the derivatives $\dot{x}_i(t)$ and $\dot{y}_i(t)$ are bounded. Furthermore, we can obtain that $x_i(t) - y_i(t)$ and their derivatives remain bounded on $[0,+\infty)$. Therefore $\sum_{i=1}^{2}|x_i(t)-y_i(t)|$ is uniformly continuous on $[0,+\infty)$. By Barbalat's theorem it follows that:

$$\lim_{t\to+\infty}\sum_{i=1}^{2}|x_i(t)-y_i(t)| = 0.$$

Therefore,

$$\lim_{t\to+\infty}(x_i(t)-y_i(t)) = 0, \quad i = 1,2.$$

This completes the proof of Theorem 3. □

From the global attractivity of bounded positive solutions, we have the following result.

Corollary 1 Suppose that the conditions of Theorem 3 hold, then system (2) is permanent.

As a direct corollary of Lemma 2, we have

Corollary 2 Suppose that $a_{21}^M M_1 < r_2^L$, then the predator species of system (2) goes to extinction.

Application

In this section, we will apply the results in Sect. 3 to the following predator–prey system with pure delays

$$\dot{x}_1(t) = x_1(t)\left[r_1(t) - a_{11}(t)x_1(t-\tau_1) - a_{12}(t)x_2(t-\tau_1)\right],$$
$$\dot{x}_2(t) = x_2(t)\left[r_2(t) + a_{21}(t)x_1(t-\tau_2) - a_{22}(t)x_2(t-\tau_2)\right]. \quad (29)$$

Corollary 3 Suppose that assumption (**H1**) holds, then there exist positive constants $N_i (i = 1, 2)$ such that

$$x_i(t) \leq N_i,$$

for any positive solution $x_i(t)$ of system (29).

Corollary 4 Suppose that assumption (**H1**) holds and $\bar{r}_1 - (\frac{a_{12}}{a_{22}})^M \bar{r}_2 > 0$. Then system (29) has at least one positive $\omega-$ periodic solution.

Corollary 5 Suppose that assumptions of Corollary 4 hold. Further suppose that the following (H_3) holds.

(H_3) There exists a constant $v_i > 0$ $(i = 1, 2)$ such that

$$\liminf_{t\to\infty} B_i(t) > 0, \quad i = 1, 2,$$

where

$$B_1(t) = v_1 a_{11}(t) - v_1 \int_{t-\tau_1}^{t} a_{11}(u+\tau_1)du[r_1(t) + (a_{11}(t) + a_{12}(t))N] - N\sum_{j=1}^{2} v_j \tau_j a_{jj}^M a_{j1}(t+\tau_j) - v_2 a_{21}(t+\tau_2),$$

$$B_2(t) = v_2 a_{22}(t) - v_2 \int_{t-\tau_2}^{t} a_{22}(u+\tau_2)du[r_2(t) + (a_{21}(t) + a_{22}(t))N] - N\sum_{j=1}^{2} v_j \tau_j a_{jj}^M a_{j2}(t+\tau_j) - v_1 a_{12}(t+\tau_1),$$

where $N = \max\{N_1, N_2\}$. Then system (29) has a positive periodic solution which is globally attractive.

Corollary 6 Suppose that the conditions of Theorem 3 hold, then system (29) is permanent.

Acknowledgments This work was supported by the National Natural Science Foundation of China (Grant Nos. 11401509,11271312, 11261056, 11261058).

References

1. Lotka, J.A.: Elements of Physical Biology. Williams and Wilkins, Baltimore (1925)
2. Volterra, V.: Fluctuations in the abundance of a species considered mathematically. Nature **118**, 558–560 (1926)
3. H.I. Freedman, Deterministic mathematical models in population ecology, in: Monogr. Text-books Pure Appl. Math., vol. 57, Marcel Dekker, New York, 1980.
4. Teng, Z.: Uniform persistence of the periodic predator-prey Lotka-Volterra systems. Appl. Anal. **72**, 339–352 (1999)
5. Teng, Z., Li, Z., Jiang, H.: Pemanence criteria in non-autonomous predator-prey Kolmogorov systems and its applications. Dynam. Syst. **19**, 171–194 (2004)
6. Teng, Z., Yu, Y.: The extinction in nonautonomous prey-predator Lotka-Volterra systems. Acta. Math. Appl. Sinica. **15**, 401–408 (1999)
7. Zhao, J., Jiang, J.: Permanence in nonautonomous Lotka-Volterra system with predator-prey. Appl. Math. Comput. **152**, 99–109 (2004)
8. Muhammadhaji, A.; Teng, Z.; Zhang, L.: Permanence in general nonautonomous predator-prey Lotka-Volterra systems with distributed delays and impulses, J. Biol. Syst. 21(2), DOI: 10.1142/S0218339013500125. (2013)
9. May, R.M.: Time-delay versus stability in population models with two and three trophic levels. Ecology **54**, 315–325 (1973)
10. Muhammadhaji, A., Teng, Z.: Positive periodic solutions of n-species Lotka-Volterra cooperative systems with delays. Vietnam J. Math. **40**(4), 453–467 (2012)
11. A. Muhammadhaji, Z. Teng, Global attractivity of a periodic delayed *N*-Species model of facultative mutualism, Discrete Dynamics in Nature and Society, Vol. 2013, Article ID 580185, p 11
12. Niyaz T., Muhammadhaji A.: Positive periodic solutions of cooperative systems with delays and feedback controls, Int. J. Differ. Equ., **2013**, 9, Article ID 502963 (2013)
13. Ma, Z., Li, Z., Wang, S., Li, T., Zhang, F.: Permanence of a predator-prey system with stage structure and time delay. Appl. Math. Comput. **201**, 65–71 (2008)
14. Gaines, R.E., Mawhin, J.L.: Lecture notes in mathematics, vol. 568. Springer, Berlin (1977)

12

Classical string field mechanics with non-standard Lagrangians

Rami Ahmad El-Nabulsi[1]

Abstract We reformulate the string classical mechanics by substituting the standard Lagrangian by a non-standard exponential Lagrangian where higher-order derivative terms occur naturally in the equations of motion. Our motivation is based on the accumulating evidence that higher-order derivatives play a leading role in string field theories. Since non-standard Lagrangians generate higher-order derivatives in a usual way, it will be of interest to explore their roles in classical string field mechanics. It was observed that replacing standard by non-standard Lagrangians gives another possibility to obtain new aspects which may have interesting physical effects.

Keywords String theory · Non-standard Lagrangians · Modified Euler–Lagrange equations

Mathematics Subject Classification Primary 70S05 · Secondary 83E30

Introduction

String theory is a quantum theory of one-dimensional objects called strings which come in two different types: open and closed. Geometrically, open strings are characterized by free endpoints, whereas closed strings are characterized by connected endpoints [19]. String theory arise in different forms depending on is associated quantum field theory. There exist many physical results which make string theory an appealing approach to describe a number of fundamental aspects in theoretical physics, mainly the particle theory and the unification problem as the main aim of string theory is to unify the standard electroweak model with a quantum theory of gravity. Some of these results include: the occurrence of graviton in closed string theories, the emergence of non-abelian gauge fields and chiral fermions in open string theories, the appearance of microscopic black holes at high energy limit, the Einstein–Hilbert action contained in the perturbative string theory and so on. From mathematical points of view, string theory joins algebraic geometry and differential geometry with theoretical physics and this is quite amazing. However, there is no clear physical reason why higher-order derivative curvature terms could not be present in string actions, yet there are some recent arguments which prove that higher-derivative terms are fundamentally important [12, 16, 17], i.e. higher-order derivative quantum corrections to supergravity. In fact, higher-derivative corrections in string theories are significantly investigated in a number of ways, e.g. nonlinear sigma model, duality, scattering amplitude and so on [1, 13, 14 and references therein]. These higher-order derivative corrections may be also added to the Einstein–Hilbert action and hence represent an additional way to describe gravitational string theories. This will help us to construct a perturbative low-energy effective action [20]. In fact, the simplest theory which describes the emergence of gravity in the string theory model is known as the bosonic string theory, which is formulated by the Polyakov action that is nothing but the action of the non-linear sigma model in two-dimensional conformal field theory [15]. The aim of this paper is to modify string actions, mainly the Nambu–Goto and the Polyakov actions by replacing the string standard Lagrangian by a string non-standard Lagrangian and to derive the corresponding

✉ Rami Ahmad El-Nabulsi
nabulsiahmadrami@yahoo.fr

1 College of Mathematics and Information Science, Neijiang Normal University, Neijiang 641112, Sichuan, China

equations of motion and study their main consequences. Our motivation to non-standardize the string Lagrangian arises from the observation that higher-order derivatives arise naturally in non-standard Lagrangian (NSL) theories and not by implementing these later by hand. In fact, NSL plays an important role in different branches of theoretical physics and applied mathematics and is in general characterized by a deformed kinetic energy term and a deformed potential function [2–11, 21, 22, 24, 25]. The follow-on modified Euler–Lagrange equation that results from the standard calculus of variations leads to equations of motion that correspond to physically interesting non-linear dynamical systems. NSL comes in different forms, yet in this paper we choose the exponential NSL (ENSL) and prove that many interesting consequences will arise in bosonic string theory. We will deal with a bosonic string embedded in a Minkowskian flat spacetime of signature $(1, D-1)$.

The paper is organized as follows: in "The modified Nambu–Goto string action and equations of motion", we introduce basic concepts of the modified Nambu–Goto string actions and derive the corresponding equations of motion; in "The modified Polyakov string action and equations of motion", we discuss the modified Polyakov string action and the main consequences of NSL formulation; finally conclusions and perspectives are given in "Conclusions and perspectives".

The modified Nambu–Goto string action and equations of motion

In general, the action for a relativistic string must be a functional of the string trajectory. When a particle moves through a spacetime, it traces out a world line, whereas a string would trace out a surface, the worldsheet. There exist two different types of strings: the open string which traces out a flat sheet and the closed string which traces out a closed tube-like surface. When dealing with action functional, it is notable that the string action is proportional to the worldsheet proper area, in contrast to the relativistic particle where the action is proportional to the proper distance of the world line. The string action is recognized as the Nambu–Goto action. Usually, the Nambu–Goto action (NGA) which is proportional to the proper area is defined as follows [18]: we consider a scalar parameter τ and the following Lagrangian coordinates $\{X^\mu(\sigma), h^{ab}(\sigma)\}$ assumed to be classical fields in the curved $1+1$ worldsheet geometry $\mathcal{V}_2$ spanned by (σ, τ) and characterized by the worldsheet pure gauge metric $h_{ab}(\sigma, \tau)$. The NGA is then defined by $S = -\frac{T}{c}\int\sqrt{(\dot{X}\cdot X')^2 - (\dot{X})^2(X')^2}\mathrm{d}\tau\mathrm{d}\sigma$, where the dot refers to the time derivative and the prime refers to the space derivative. Here, T and c are, respectively, the tension of the string and the celerity of light. In our approach, the following definition holds:

Definition 2.1 Let $X^\mu(\sigma)$ be string coordinates and (σ, τ) be coordinates on the worldsheet augmented by the constraint $\dot{X}_0(\text{endpoints}) \neq 0$. We define the exponentially non-standard Nambu–Goto action by

$$\mathbb{S} = -\frac{\xi T}{c}\int e^{\zeta\sqrt{(\dot{X}\cdot X')^2 - (\dot{X})^2(X')^2}}\mathrm{d}\tau\mathrm{d}\sigma, \tag{1}$$

where ξ is a parameter introduced to take into account the dimensional problem, c is the celerity of light and ζ is a constant. The NSL-density along the string is $L = -\frac{\xi T}{c}e^{\zeta\sqrt{(\dot{X}\cdot X')^2 - (\dot{X})^2(X')^2}} \equiv -\frac{\xi T}{c}e^{\mathbb{L}}$.

Theorem 2.1 *The modified Euler–Lagrange equations which correspond to the action function* (1) *are:*

$$\frac{\mathrm{d}}{\mathrm{d}\tau}\left(\frac{\partial\mathbb{L}}{\partial\dot{X}^\mu}\right) + \frac{\mathrm{d}}{\mathrm{d}\sigma}\left(\frac{\partial\mathbb{L}}{\partial X'^\mu}\right) = -\zeta\left(\ddot{X}^\mu\left(\frac{\partial\mathbb{L}}{\partial\dot{X}^\mu}\right)^2 + X''^\mu\left(\frac{\partial\mathbb{L}}{\partial X'^\mu}\right)^2\right). \tag{2}$$

Proof The variation of the action (1) gives the full set:

$$\begin{aligned}
\delta\mathbb{S} = &-\frac{\xi T}{c}\int_{\tau_{\text{initial}}}^{\tau_{\text{final}}}\int_0^{\sigma_1}\mathrm{d}\sigma\mathrm{d}\tau e^{-\zeta\mathbb{L}}\left(\frac{\mathrm{d}}{\mathrm{d}\tau}\left(\delta X^\mu\frac{\partial e^{\zeta\mathbb{L}}}{\partial\dot{X}^\mu}\right)\right.\\
&\left.+\frac{\mathrm{d}}{\mathrm{d}\sigma}\left(\delta X^\mu\frac{\partial e^{\zeta\mathbb{L}}}{\partial X'^\mu}\right) - \left(\frac{\mathrm{d}}{\mathrm{d}\tau}\frac{\partial e^{\zeta\mathbb{L}}}{\partial\dot{X}^\mu} + \frac{\mathrm{d}}{\mathrm{d}\sigma}\frac{\partial e^{\zeta\mathbb{L}}}{\partial X'^\mu}\right)\delta X^\mu\right),\\
= &-\frac{\xi T}{c}\int_0^{\sigma_1}\mathrm{d}\sigma\delta X^\mu\frac{\partial e^{\zeta\mathbb{L}}}{\partial\dot{X}^\mu}\bigg|_{\tau_{\text{initial}}}^{\tau_{\text{final}}} - \xi T\int_{\tau_{\text{initial}}}^{\tau_{\text{final}}}\mathrm{d}\sigma\delta X^\mu\frac{\partial e^{\zeta\mathbb{L}}}{\partial X'^\mu}\bigg|_0^{\sigma_1}\\
&+\frac{\xi T}{c}\int_{\tau_{\text{initial}}}^{\tau_{\text{final}}}\int_0^{\sigma_1}\mathrm{d}\sigma\mathrm{d}\tau e^{-\xi\mathbb{L}}\left(\frac{\mathrm{d}}{\mathrm{d}\tau}\frac{\partial e^{\zeta\mathbb{L}}}{\partial\dot{X}^\mu} + \frac{\mathrm{d}}{\mathrm{d}\sigma}\frac{\partial e^{\zeta\mathbb{L}}}{\partial X'^\mu}\right)\delta X^\mu.
\end{aligned}$$

Using the boundary conditions $\delta X^\mu(\tau_{initial}, \sigma) = \delta X^\mu(\tau_{final}, \sigma) = 0$, the equation of motion for an arbitrarily parameterized string reads:

$$\frac{\mathrm{d}}{\mathrm{d}\tau}\frac{\partial e^{\zeta\mathbb{L}}}{\partial\dot{X}^\mu} + \frac{\mathrm{d}}{\mathrm{d}\sigma}\frac{\partial e^{\zeta\mathbb{L}}}{\partial X'^\mu} = 0.$$

This equation may be written explicitly as:

$$\frac{\mathrm{d}}{\mathrm{d}\tau}\left(\frac{\partial\mathbb{L}}{\partial\dot{X}^\mu}\right) + \frac{\mathrm{d}}{\mathrm{d}\sigma}\left(\frac{\partial\mathbb{L}}{\partial X'^\mu}\right) = -\zeta\left(\frac{\mathrm{d}\mathbb{L}}{\mathrm{d}\tau}\frac{\partial\mathbb{L}}{\partial\dot{X}^\mu} + \frac{\mathrm{d}\mathbb{L}}{\mathrm{d}\sigma}\frac{\partial\mathbb{L}}{\partial X'^\mu}\right).$$

Using the chain rules

$$\begin{aligned}
\frac{\mathrm{d}\mathbb{L}}{\mathrm{d}\tau} &= \frac{\partial\mathbb{L}}{\partial\tau} + \dot{X}^\mu\frac{\partial\mathbb{L}}{\partial X^\mu} + \ddot{X}^\mu\frac{\partial\mathbb{L}}{\partial\dot{X}^\mu} = \ddot{X}^\mu\frac{\partial\mathbb{L}}{\partial\dot{X}^\mu},\\
\frac{\mathrm{d}\mathbb{L}}{\mathrm{d}\sigma} &= \frac{\partial\mathbb{L}}{\partial\sigma} + X'^\mu\frac{\partial\mathbb{L}}{\partial X^\mu} + X''^\mu\frac{\partial\mathbb{L}}{\partial X'^\mu} = X''^\mu\frac{\partial\mathbb{L}}{\partial X'^\mu},
\end{aligned}$$

$$\frac{d\mathbb{L}}{d\sigma}=\frac{\partial\mathbb{L}}{\partial\sigma}+X'^{\mu}\frac{\partial\mathbb{L}}{\partial X^{\mu}}+X''^{\mu}\frac{\partial\mathbb{L}}{\partial X'^{\mu}}=X''^{\mu}\frac{\partial\mathbb{L}}{\partial X'^{\mu}},$$

the equation of motion reads as:

$$\frac{\mathrm{d}}{\mathrm{d}\tau}\left(\frac{\partial\mathbb{L}}{\partial\dot{X}^{\mu}}\right)+\frac{\mathrm{d}}{\mathrm{d}\sigma}\left(\frac{\partial\mathbb{L}}{\partial X'^{\mu}}\right)=-\zeta\left(\ddot{X}^{\mu}\left(\frac{\partial\mathbb{L}}{\partial\dot{X}^{\mu}}\right)^{2}+X''^{\mu}\left(\frac{\partial\mathbb{L}}{\partial X'^{\mu}}\right)^{2}\right).\quad\square$$

Remark 2.1 When $\zeta=0$, Eq. (2) is reduced to the standard Nambu–Goto equations of motion. Obviously, in Eq. (2) higher-order derivative terms appear and these new terms will modify string dynamics accordingly. These higher-order derivative terms are coupled to the parameter ζ.

To simplify the equations of motion, we follow the standard arguments by using temporal and σ-parameterizations in Minkowski spacetime (cdt, dx, dy, dz) [1]. For the case of temporal parameterization, we set $\tau=t$ which is the coordinate time, as it was for the point particle.

In the temporal gauge, we can choose $X'^{\mu}=\begin{pmatrix}0\\X'\end{pmatrix}$ and $\dot{X}^{\mu}=\begin{pmatrix}c\\\dot{X}\end{pmatrix}$. We can write Eq. (1) as:

$$\mathbb{S}=-\frac{\xi T}{c}\int e^{\zeta\sqrt{(\dot{X}\cdot X')^{2}-(-c^{2}+\dot{X}\cdot\dot{X})^{2}(X'\cdot\dot{X})}}\mathrm{d}t\mathrm{d}\sigma.$$

In the σ-parameterization gauge, we can choose $s=\sigma$, i.e. arc-length parameterization which gives $|X'|=1$. We can use this to define the transverse component of the string velocity $\mathbf{v}_{\perp}=\dot{X}-(\dot{X}\cdot X')X'$, where $\dot{X}=\mathrm{d}X/\mathrm{d}t$ and $X'=\mathrm{d}X/\mathrm{d}s$. Then we can write $\mathbf{v}_{\perp}^{2}=\dot{X}\cdot\dot{X}-(\dot{X}\cdot X')^{2}$, i.e.

$$\mathbb{S}=-\frac{\xi T}{c}\int e^{\zeta\sqrt{1-v_{\perp}^{2}}}\mathrm{d}t\mathrm{d}s.$$

Corollary 2.1 *Consider the boundary condition in this parameterization, together with our pre-gauge fixed expression for* $\partial\mathbb{L}/\partial X^{\sigma}$, *then the equations of motion are given by:*

$$\frac{1}{c^{2}}\left(1+\zeta\frac{v^{2}}{c^{2}}\right)\ddot{X}-\left(1-\zeta\left(1-\frac{v^{2}}{c^{2}}\right)\right)X''=0.\tag{3}$$

Proof In fact, we have:

$$\frac{\partial\mathbb{L}}{\partial\dot{X}^{\mu}}=-\frac{\xi T}{c}\frac{(\dot{X}\cdot X')X'_{\mu}-(X'\cdot X')\dot{X}_{\mu}}{\sqrt{(\dot{X}_{\mu}X'_{\mu})^{2}-(\dot{X}_{\mu}\dot{X}^{\mu})^{2}(X'\cdot X')}},$$

$$\frac{\partial\mathbb{L}}{\partial X'^{\mu}}=-\frac{\xi T}{c}\frac{(\dot{X}\cdot X')\dot{X}_{\mu}-(\dot{X}\cdot\dot{X})X'_{\mu}}{\sqrt{(\dot{X}_{\mu}X'_{\mu})^{2}-(\dot{X}_{\mu}\dot{X}^{\mu})^{2}(X'\cdot X')}}.$$

Assuming $\tau=t$, i.e. static gauge and choosing a σ parameter in a way that $\dot{X}\cdot X'=0$, an identification made on a two-dimensional grid [26], then we can simplify these later equations respectively to:

$$\frac{\partial\mathbb{L}}{\partial\dot{X}^{\mu}}=\frac{\xi T}{c}\frac{(X'\cdot X')\dot{X}_{\mu}}{\sqrt{(c^{2}-v^{2})(X'\cdot X')}},$$

$$\frac{\partial\mathbb{L}}{\partial X'^{\mu}}=-\frac{\xi T}{c}\frac{(c^{2}-v^{2})X'_{\mu}}{\sqrt{(c^{2}-v^{2})(X'\cdot X')}}.$$

As we have $X'_{0}=0$, $\dot{X}_{0}=-1$ and $\ddot{X}_{0}=0$, the equation of motion is $\frac{\mathrm{d}}{\mathrm{d}\tau}(\xi T(X'\cdot X')\big/\sqrt{(c^{2}-v^{2})(X'\cdot X')}=0$, and for the case of an arc length, we found $\sqrt{X'\cdot X'}=f(\sigma)\sqrt{c^{2}-v^{2}}\big/\xi T$, where $f(\sigma)$ is an arbitrary function of σ.

With the choice $f(\sigma)=f_{0}=\frac{\xi T}{c}$(a constant), $X'\cdot X'=1-\frac{1}{c^{2}}\dot{X}\cdot\dot{X}=1-\frac{v^{2}}{c^{2}}$, $\frac{\partial\mathbb{L}}{\partial\dot{X}^{\mu}}=\frac{\xi T}{c^{2}}\dot{X}$ and $\frac{\partial\mathbb{L}}{\partial X'^{\mu}}=-\xi TX'$, we find:

$$\frac{1}{c^{2}}\left(1+\zeta\frac{v^{2}}{c^{2}}\right)\ddot{X}-\left(1-\zeta\left(1-\frac{v^{2}}{c^{2}}\right)\right)X''=0.\quad\square$$

Remark 2.2 This is a modified wave equation which is reduced to its standard form when $\zeta=0$.

Corollary 2.2 *In the exponentially non-standard Nambu–Goto framework, the effective celerity of light* c *is:*

$$\frac{1}{\mathsf{c}^{2}}=\frac{1}{c^{2}}\frac{1+\zeta\frac{v^{2}}{c^{2}}}{1-\zeta\left(1-\frac{v^{2}}{c^{2}}\right)}.\tag{4}$$

For $\zeta=1$, we find $\mathsf{c}^{2}=\frac{c^{2}v^{2}}{c^{2}+v^{2}}$, which means that for $v=\alpha c$ where α is a positive constant, $\mathsf{c}=\frac{\alpha c}{\sqrt{1+\alpha^{2}}}<c\,\forall\alpha$.

However, for $\zeta=-1$, we find $\mathsf{c}^{2}=c^{2}\frac{2c^{2}-v^{2}}{c^{2}-v^{2}}$, and for $v=\alpha c$, we find $\mathsf{c}=c\sqrt{\frac{2-\alpha^{2}}{1-\alpha^{2}}}$, and hence for $0<\alpha<1$, we find $\mathsf{c}>c$, i.e. the effective celerity of light is greater than the celerity of light. For $\alpha\geq\sqrt{2}$, $\mathsf{c}<c$, but $v>c$.

Corollary 2.3 *For the case of a rotating string, the solution of Eq.* (3) *is given by:*

$$\begin{aligned}\mathrm{X}(\tau,\sigma)&=\frac{\sigma_{\text{final}}}{\pi}\cos\left(\frac{\pi\sigma}{\sigma_{\text{final}}}\right)\left(\cos\left(\frac{\pi\mathsf{c}t}{\sigma_{\text{final}}}\right)\hat{x}+\sin\left(\frac{\pi\mathsf{c}t}{\sigma_{\text{final}}}\right)\hat{y}\right),\\&\equiv\frac{\sigma_{\text{final}}}{\pi}\cos\left(\frac{\pi\sigma}{\sigma_{\text{final}}}\right)\left(\cos\left(\frac{\pi ct}{\sigma_{\text{final}}}\sqrt{\frac{1-\zeta\left(1-\frac{v^{2}}{c^{2}}\right)}{1+\zeta\frac{v^{2}}{c^{2}}}}\right)\hat{x}\right.\\&\quad\left.+\sin\left(\frac{\pi ct}{\sigma_{\text{final}}}\sqrt{\frac{1-\zeta\left(1-\frac{v^{2}}{c^{2}}\right)}{1+\zeta\frac{v^{2}}{c^{2}}}}\right)\hat{y}\right),\\&\equiv\mathrm{X}_{x}(\tau,\sigma)\hat{x}+\mathrm{X}_{y}(\tau,\sigma)\hat{y},\end{aligned}\tag{5}$$

and the perpendicular velocity which corresponds to the transverse motion is

$$\mathrm{v}_{\perp} = \sqrt{\dot{X}\cdot\dot{X}} = \mathrm{c}\cos\left(\frac{\pi\sigma}{\sigma_{\text{final}}}\right)$$
$$\equiv c\sqrt{\frac{1-\zeta\left(1-\frac{\mathrm{v}^2}{c^2}\right)}{1+\zeta\frac{\mathrm{v}^2}{c^2}}}\cos\left(\frac{\pi\sigma}{\sigma_{\text{final}}}\right). \tag{6}$$

Proof The proof follows directly from Eq. (3) after considering a rotating string with constant angular velocity and assuming a general solution of the motion describing left and right motion of the form [26]: $X(\tau,\sigma) = \beta(A\cos\chi(\gamma-ct) + B\cos\chi(\gamma+ct))\hat{x} + \beta(C\sin\chi(\gamma-ct) + D\sin\chi(\gamma+ct))\hat{y}$, $\beta,\gamma,\chi\in\mathbb{R}$, where (A,B,C,D) are constants to be determined from boundary conditions and $(\hat{x},\hat{y})$ are unit vectors. Using the boundary condition $X'(\tau,\sigma=0) = X'(\tau,\sigma=\sigma_{\text{final}}) = 0$ gives, respectively, $A=B, D=-C$ and $\chi = n\pi/\sigma_{\text{final}}$. For $n=1$ and using the fact that $\dot{X}\cdot X' = 0$ gives $C=-A$, finally the condition $X'\cdot X' = 1 - \dot{X}\cdot\dot{X}/c^2$ gives $A = \pm\sigma_{\text{final}}/2\beta\pi$. Consequently, Eq. (5) is obtained after using Eq. (4). Equation (6) is a straightforward derivation. □

For $\zeta=-1$ and $\mathrm{v}=\alpha c$, we have $\mathsf{c} = c\sqrt{\frac{2-\alpha^2}{1-\alpha^2}}$ and then Eq. (6) takes the form:

$$\mathrm{v}_{\perp} = c\sqrt{\frac{2-\alpha^2}{1-\alpha^2}}\cos\left(\frac{\pi\sigma}{\sigma_{\text{final}}}\right).$$

Hence for $0<\alpha<1$, the end of the string travels at a speed larger than the velocity of light in contrast to the case $\zeta=1$ where the end of the string moves at a speed lower than the velocity of light. These results show the main differences between the standard Lagrangian and the NSL approach in string classical mechanics. We plot in Figs. 1 and 2, respectively, the variations of $X_x(\tau,\sigma)$ and $X_y(\tau,\sigma)$ for $\zeta=1, \sigma_{\text{final}}=\pi$ and $\mathsf{v}=c=1$(for illustration purpose) and in Figs. 3 and 4, respectively, the variations of $X_x(\tau,\sigma)$ and $X_y(\tau,\sigma)$ for $\zeta=-1, \sigma_{\text{final}}=\pi$ and $\mathsf{v}=\alpha c$ for $\alpha=\frac{1}{2}$.

It is obvious that the dynamics between cases $\zeta=1$ and $\zeta=-1$ differs. The rotating string oscillates more rapidly for the case $\zeta=-1$ than $\zeta=1$, which is due to the fact that the end of the string moves at a velocity larger than the velocity of light. For $\zeta\ll1$, we find the standard result and Figs. 5 and 6 illustrate the variations of $X_x(\tau,\sigma)$ and $X_y(\tau,\sigma)$.

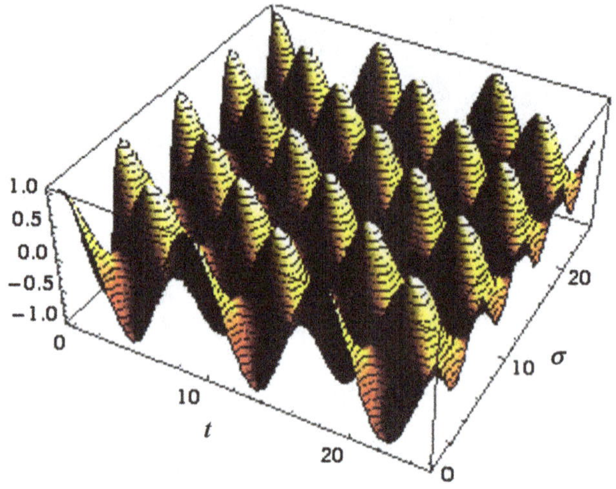

Fig. 1 Plot of $X_x(\tau,\sigma)$ for $\zeta=1$

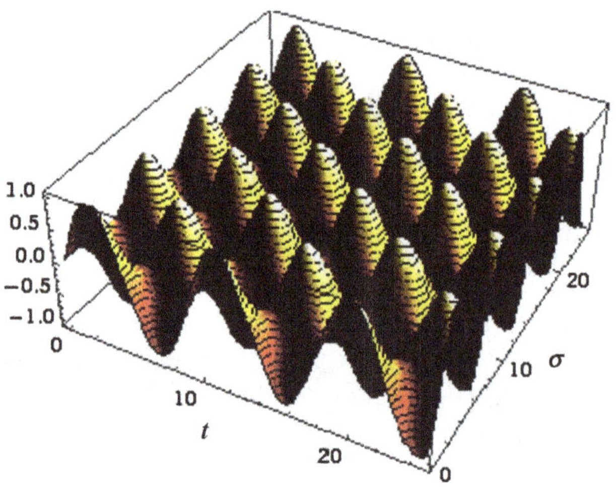

Fig. 2 Plot of $X_y(\tau,\sigma)$ for $\zeta=1$

The modified Polyakov string action and equations of motion

In the standard approach, the Nambu–Goto action functional is somewhat difficult due to the occurrence of the square root in the Lagrangian density. However, one way to remove the square root is to introduce an auxiliary field in the curved 1 + 1 worldsheet geometry $\mathcal{V}_2$ known as the pure gauge metric $h_{ab}(\sigma,\tau)$ [18]. The spacetime is assumed to be flat with metric $\eta=(+,-,-,-)$ and its relation with the worldsheet geometry is through the constraints: $\eta_{\mu\nu}\partial_\tau X^\mu\partial_\tau X^\nu<0$ and $\eta_{\mu\nu}\partial_\sigma X^\mu\partial_\sigma X^\nu>0$. These conditions which correspond, respectively, for space-like and time-like tangent vectors are accompanied by the initial conditions $X(\sigma,\tau_0)=X_0(\sigma)$ and $\partial_\tau X(\sigma,\tau_0)=Y_0(\sigma)$. This will lead to the standard-Polyakov action being defined by $S = -\frac{T}{2c}\eta_{\mu\nu}\int d\sigma\sqrt{h}h^{ab}\partial_a X^\mu\partial_b X^\nu$, $(a,b=0,1)$, $(\mu,\nu=0,1,\ldots,D)$, which is the starting point for the path integral quantization. Here, $h=\det h_{ab}$ and $(h^{-1})^{ab}=h_{ab}$. In our approach, the following definition holds.

Definition 3.1 Let $X^\mu(\sigma)$ be string coordinates, (σ,τ) be coordinates on the worldsheet geometry $\mathcal{V}_2$ with pure

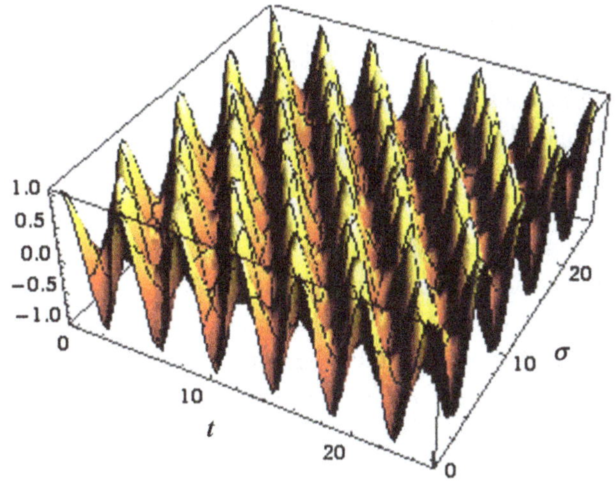

Fig. 3 Plot of $X_x(\tau,\sigma)$ for $\zeta=-1$

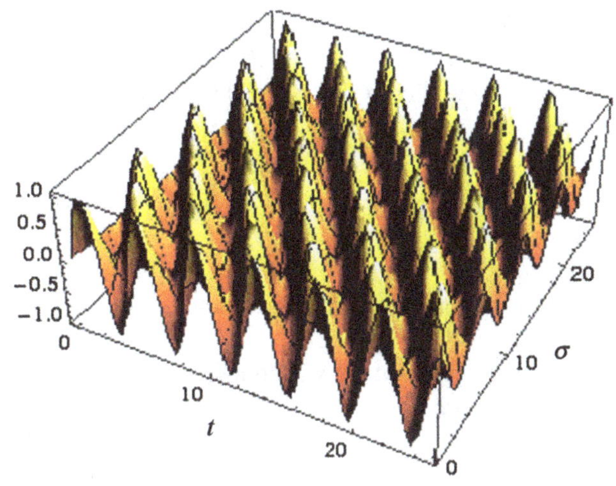

Fig. 4 Plot of $X_y(\tau,\sigma)$ for $\zeta=-1$

gauge metric $h_{ab}(\sigma,\tau)$ augmented by the constraints $\eta_{\mu\nu}\partial_\tau X^\mu \partial_\tau X^\nu < 0$ and $\eta_{\mu\nu}\partial_\sigma X^\mu \partial_\sigma X^\nu > 0$. We define the exponentially non-standard-Polyakov action by

$$S=-\frac{\xi T}{2c}\eta_{\mu\nu}\int e^{\zeta\sqrt{h}h^{ab}\partial_a X^\mu \partial_b X^\nu}\mathrm{d}\sigma, \tag{7}$$

with the Lagrangian density along the string given by $L=-\frac{\xi T}{2c}\eta_{\mu\nu}e^{\zeta\sqrt{h}h^{ab}\partial_a X^\mu \partial_b X^\nu}\equiv -\frac{\xi T}{2c}\eta_{\mu\nu}e^{\mathbb{L}}$.

Corollary 3.1 *The Lagrangian density for the exponentially non-standard-Polyakov action is written as*:

$$\begin{aligned} L &= -\frac{\xi T}{2c}\eta_{\mu\nu}e^{\zeta\sqrt{h}h^{ab}\partial_a X^\mu \partial_b X^\nu} \\ &= -\frac{\xi T}{2c}\eta_{\mu\nu}e^{\zeta\sqrt{h}(h^{\tau\tau}\dot X^\mu X^\nu + 2h^{\tau\sigma}\dot X^\mu X'^\nu + h^{\sigma\sigma}X'^\mu X'^\nu)}, \end{aligned} \tag{8}$$

where the dot refers to the time derivative and the prime refers to the space derivative.

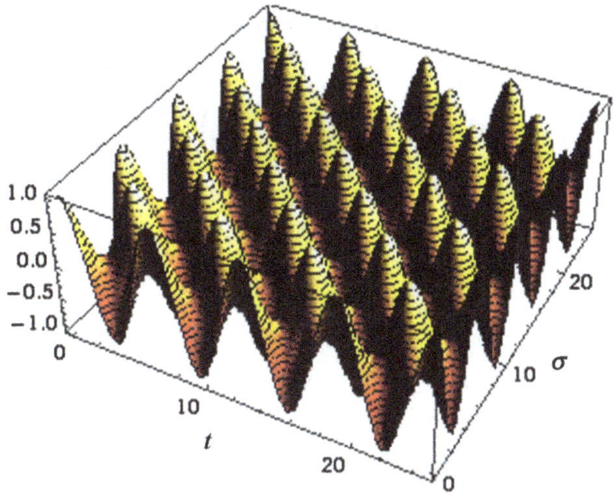

Fig. 5 Plot of $X_x(\tau,\sigma)$ for $\zeta << 1$

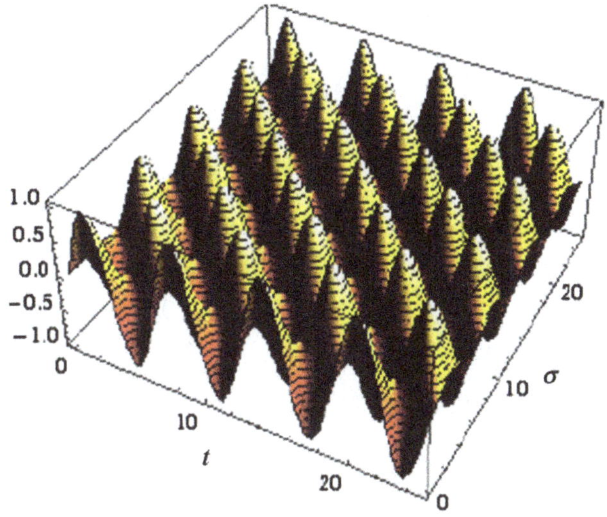

Fig. 6 Plot of $X_y(\tau,\sigma)$ for $\zeta << 1$

Usually, for the case of ENSL $S=\int e^{\zeta\mathbb{L}(\dot x,x,t)}\mathrm{d}t$, the Euler–Lagrange equation is modified and $\frac{\partial\mathbb{L}}{\partial x}-\frac{\mathrm{d}}{\mathrm{d}_t}\left(\frac{\partial\mathbb{L}}{\partial\dot x}\right)=\zeta\frac{\partial\mathbb{L}}{\partial\dot x}\frac{\mathrm{d}\mathbb{L}}{\mathrm{d}_t}$, where $\frac{\mathrm{d}\mathbb{L}}{\mathrm{d}_t}=\frac{\partial\mathbb{L}}{\partial t}+\dot x\frac{\partial\mathbb{L}}{\partial x}+\ddot x\frac{\partial\mathbb{L}}{\partial\dot x}$ is the total derivative operator [5].

Theorem 3.1 *For the case of an explicitly time-independent Lagrangian density, the string coordinates $X^\mu(\sigma)$ and the gauge metric $h_{ab}(\sigma,\tau)$, respectively, obey the following modified Euler–Lagrange equations*:

$$\frac{\partial\mathbb{L}}{\partial X^\mu}-\partial_a\frac{\partial\mathbb{L}}{\partial(\partial_a X^\mu)}=\zeta\frac{\partial\mathbb{L}}{\partial(\partial_a X^\mu)}\left(\partial_a X^\mu\frac{\partial\mathbb{L}}{\partial X^\mu}+\partial_a\partial_a X^\mu\frac{\partial\mathbb{L}}{\partial(\partial_a X^\mu)}\right), \tag{9}$$

$$\frac{\partial\mathbb{L}}{\partial h^{bc}}-\partial_a\frac{\partial\mathbb{L}}{\partial(\partial_a h^{bc})}=\zeta\frac{\partial\mathbb{L}}{\partial(\partial_a h^{bc})}\left(\partial_a h^{bc}\frac{\partial\mathbb{L}}{\partial h^{bc}}+\partial_a\partial_a h^{bc}\frac{\partial\mathbb{L}}{\partial(\partial_a h^{bc})}\right). \tag{10}$$

Corollary 3.2 *The equations of motion for the action functional* (7) *are*:

$$\partial_a\partial^a X_\mu - \frac{1}{2}h_{cd}\partial_a h^{cd}\partial_a X_\mu = \frac{\zeta\xi T}{c}\sqrt{h}(\partial^a X_\mu)^2\partial_a\partial_a X^\mu, \quad (11)$$

$$\partial_b X^\mu \partial_c X_\mu - \frac{h_{bc}}{2}h^{ef}\partial_e X^\mu \partial_f X_\mu = 0. \quad (12)$$

Proof From Eq. (8), we find that $\frac{\partial \mathbb{L}}{\partial X^\mu} = 0$ and $\frac{\partial \mathbb{L}}{\partial(\partial_a X^\mu)} = -\frac{\xi T}{c}\sqrt{h}\partial^a X_\mu$. Then from Eq. (9), we find:

$$\partial_a\left(\sqrt{h}\partial^a X_\mu\right) = \frac{\zeta\xi T}{c}\sqrt{h}\partial^a X_\mu\left(\sqrt{h}\partial^a X_\mu \partial_a\partial_a X^\mu\right).$$

Using the fact that $\partial_a(\sqrt{h}\partial^a X_\mu) = \sqrt{h}\partial_a\partial^a X_\mu + \partial_a\sqrt{h}\partial^a X_\mu = \sqrt{h}\partial_a\partial^a X_\mu - \frac{1}{2}\sqrt{h}h_{cd}\partial_a h^{cd}\partial_a X_\mu$, we find:

$$\partial_a\partial^a X_\mu - \frac{1}{2}h_{cd}\partial_a h^{cd}\partial_a X_\mu = \frac{\zeta\xi T}{c}\sqrt{h}(\partial^a X_\mu)^2\partial_a\partial_a X^\mu.$$

Again from Eq. (8), $\frac{\partial \mathbb{L}}{\partial \mathrm{h}^{bc}} = -\frac{\xi T}{2c}(\partial_b X^\mu \partial_c X_\mu - \frac{\mathrm{h}_{bc}}{2}\mathrm{h}^{ef}\partial_e X^\mu \partial_f X_\mu)$ and $\frac{\partial \mathbb{L}}{\partial(\partial_a \mathrm{h}^{bc})} = 0$. Then Eq. (10) gives:

$$\partial_b X^\mu \partial_c X_\mu - \frac{h_{bc}}{2}h^{ef}\partial_e X^\mu \partial_f X_\mu = 0. \qquad \square$$

Remark 3.1 In the NSL approach, only Eq. (11) which corresponds to string coordinates is modified. This equation holds higher-order derivative terms. Equation (12) is similar to the one derived in the standard approach.

Lemma 3.1 *Let* $G_{bc} = \partial_b X^\mu \partial_c X_\mu$, *then* $S = -\frac{\xi T}{2c}\int d\tau \int d\sigma e^{\zeta\sqrt{G}}$.

Equation (12) is equivalent to $G_{bc} = \frac{h_{bc}}{2}tr\|G\|$ for $G_{bc} = \partial_b X^\mu \partial_c X_\mu$ which is written after some algebra as $\sqrt{h}h^{ab}\partial_a X^\mu \partial_b X_\mu = 2G$ [18], and hence we find the require results. $\square$

In fact, Lemma 3.1 states that "the exponentially non-standard Nambu–Goto action is obtained from the exponentially non-standard-Polyakov action using Euler–Lagrange equations for the worldsheet metric". However, in our approach, the equation of motion (11) for the string coordinates $X^\mu(\sigma)$ is different.

Remark 3.2 Using diffeomorphisms and Weyl transformation [23], one can write in the conformal gauge Eq. (7) as

$$S = -\frac{\xi T}{2c}\int d^2\sigma e^{\zeta\left(X'^2 - \frac{1}{c^2}\dot{X}^2\right)},$$

and in particular in the Minkowski spacetime after making the choice $h_{ab} = \eta_{ab}$, where η_{ab} is the Minkowski signature, i.e. conformal gauge and c is the effective celerity of light defined by Eq. (4).

The corresponding equation of motion for X is then given by Eq. (2) which is rewritten in that case as:

$$\frac{1}{\mathrm{c}^2}\ddot{X}(1 - 2\zeta^2\dot{X}^2) - X''(1 + 2\zeta^2 X'^2) = 0.$$

This equation has to be consistent with the equation of motion (12) for h^{ab}, which is $\mathrm{T}_{ab} = \partial_a X \partial_b X - \frac{1}{2}\eta_{ab}\eta^{ef}\partial_e X \partial_f X = 0$ and gives $\mathrm{T}_{01} = X' \cdot \dot{X} = 0$ and $\mathrm{T}_{00} = \mathrm{T}_{11} = \frac{1}{2}(\dot{X}^2 + X'^2) = 0$, known as the Virasoro constraints. Accordingly, we can write the equation of motion as $(1 - 2\zeta^2\dot{X}^2)(\ddot{X} - c^2 X'') = 0$ and then $\ddot{X} - \mathrm{c}^2 X'' = 0$ or $1 - 2\zeta^2\dot{X}^2 = 0$. So the equation of motion implies a harmonic wave equation $\ddot{X} - \mathrm{c}^2 X'' = 0$ or a solution of the form $\dot{X}^2 = -X'^2 = 1/2\zeta^2$. This last solution gives $X(\sigma) = \pm\sqrt{-\frac{1}{2\zeta}\sigma + X_{\sigma 0}}$ and $X(\tau) = \pm\sqrt{\frac{1}{2\zeta}\tau + X_{\tau 0}}$, where $X_{\sigma 0}$ and $X_{\tau 0}$ are initial solutions. For $\zeta < 0(>0)$, we find a complexified (real) evolution of $X(\tau)$ and a real (complexified) evolution of $X(\sigma)$. In other words, these solutions show that we have a complexified linear evolution in time, which is not realistic unless we perform a Wick rotation $\tau \to i\tau, i = \sqrt{-1} \in \mathbb{C}$ for $\zeta < 0$, i.e. complexified action. For $X_{\sigma 0} = X_{\tau 0} = 0$, we have $X^2(\sigma) - X^2(\tau) = -\frac{1}{2\zeta}(\tau^2 + \sigma^2)$ which means that for $\zeta < 0(>0)$, we have $X^2(\sigma) - X^2(\tau) > (<)0$ that corresponds, respectively, to stable solutions and unstable solutions.

Conclusions and perspectives

The objective of this work was to discuss the impacts of exponentially non-standard Lagrangians in string classical mechanics. In reality, NSL naturally generates higher-order derivatives in the equations of motion; therefore, our basic motivation was to explore the main consequences of these higher-order terms in string theory as there exist many arguments which proved that these terms played an important role in mainly all string theories. To do this, we gave in this work the basic setups where two exponentially non-standard-string Lagrangian densities were discussed: the Nambu–Goto and the Polyakov actions. It was observed that the non-standard formulation of both fundamental actions gave some new insights into classical string theory.

For the case of an exponentially non-standard Nambu–Goto action, the wave equation is characterized by an effective celerity of light which depends on the sign of the parameter ζ. For positive value of ζ, the effective celerity of light is lower than c, whereas for the case of a negative value of ζ, the effective celerity of light is greater than c which means that the end of the string moves at a velocity larger than the celerity of light, in particular when the velocity of motion is less than the celerity of light, $\mathrm{v} = \alpha c$ with $0 < \alpha < 1$. Moreover, for the case of a rotating string, it was observed that the dynamics depends also on the sign of the parameter ζ. Oscillations are faster for the negative case than for the positive case.

For the case of an exponentially non-standard-Polyakov action, the equation of motion for the string coordinates differs from the standard case where higher-order derivative terms occur, whereas it is identical to the standard case for the gauge metric. However, the exponentially non-standard Numbu–Goto action may be derived from the exponentially non-standard-Polyakov action using Euler–Lagrange equations for the worldsheet metric. Using diffeomorphisms and Weyl transformation in the Minkowski spacetime, we have introduced a new exponentially non-standard-Polyakov action where the celerity of light is replaced by an effective one. In that way, the equation of motion yields the harmonic wave equation characterized as in the exponentially non-standard Nambu–Goto action by an effective celerity of light, which depends on the sign of the parameter ζ.

These results must be used to explore their main consequences in different string theoretical aspects, mainly their quantization aspects. Work in this direction is under progress.

References

1. Becker, K., Becker, M., Schwarz, J.: String Theory and M-Theory: A Modern Introduction. Cambridge University Press, Cambridge (2007)
2. El-Nabulsi, R.A.: Non-standard fractional Lagrangians. Nonlinear Dyn. **74**, 381–394 (2013)
3. El-Nabulsi, R.A.: Fractional oscillators from non-standard Lagrangians and time-dependent fractional exponent. Comp. Appl. Math. **33**, 163–179 (2014)
4. El-Nabulsi, R.A.: Non-standard non-local-in-time Lagrangian in classical mechanics. Qual. Theory Dyn. Syst. **13**, 149–160 (2014)
5. El-Nabulsi, R.A.: Nonlinear dynamics with non-standard Lagrangians. Qual. Theory Dyn. Syst. **13**, 273–291 (2013)
6. El-Nabulsi, R.A.: Modified Proca equation and modified dispersion relation from a power-law Lagrangian functional. Indian J. Phys. **87**, 465–470 (2013); Erratum Indian J. Phys. **87**, 1059 (2013)
7. El-Nabulsi, R.A.: Quantum field theory from an exponential action functional. Indian J. Phys. **87**, 379–383 (2013)
8. El-Nabulsi, R.A.: Generalizations of the Klein–Gordon and the Dirac equations from non-standard Lagrangians. Proc. Natl. Acad. Sci. India Sect. A Phys. Sci. **83**, 383–387 (2013)
9. El-Nabulsi, R.A.: A generalized nonlinear oscillator from non-standard degenerate Lagrangians and its consequent Hamiltonian formalism. Proc. Natl. Acad. Sci. India Sect. A Phys. Sci **84**, 563–569 (2014)
10. El-Nabulsi, R.A.: Electrodynamics of relativistic particles through non-standard Lagrangian. J. At. Mol. Sci. **5**(3), 268–279 (2014)
11. El-Nabulsi, R.A.: Non-standard power-law Lagrangians in classical and quantum dynamics. Appl. Math. Lett. doi:10.1016/j.aml.2014.12.002
12. Forger, K., Ovrut, B.A., Theisen, S.J., Waldram, D.: Higher-derivative gravity in string theory. Phys. Lett. B **388**, 512–520 (1996)
13. Green, M.B., Schwarz, J.H., Witten, E.: Superstring Theory, Vol. 1: Introduction, Cambridge Monographs on Mathematical Physics. Cambridge University Press, Cambridge (1987)
14. Green, M.B., Schwarz, J.H., Witten, E.: Superstring Theory, Vol. 1: Loop Amplitudes, Anomalies and Phenomenology, Cambridge Monographs on Mathematical Physics. Cambridge University Press, Cambridge (1987)
15. Guns, S.: Low energy effective actions from string theory. Proseminar CFT & Strings 2013, ETH Zurich, 29 April (2013)
16. Hindawi, A., Ovrut, B.A., Waldram, D.: Two-dimensional higher-derivative supergravity and a new mechanism for supersymmetry breaking. Nucl. Phys. B **471**, 409–429 (1996)
17. Hindawi, A., Ovrut, B.A., Waldram, D.: Four-dimensional higher-derivative supergravity and spontaneous symmetry breaking. Nucl. Phys. B **476**, 175–199 (1996)
18. Materassi, M.: Polyakov's string classical mechanics, hep-th/9905028
19. McAllister, L.: Lectures on String Theory, notes on String Theory from Liam McAllister's Physics 7683: String Theory course at Cornell University in Spring (2010)
20. Moura, F., Schiappa, R.: Higher-derivative corrected black holes: perturbative stability and absorption cross-section in heterotic string theory. Class. Quantum Grav. **24**, 361–386 (2007)
21. Musielak, Z.E.: Standard and non-standard Lagrangians for dissipative dynamical systems with variable coefficients. J. Phys. A Math. Theor. **41**, 055205–055222 (2008)
22. Musielak, Z.E.: General conditions for the existence of non-standard Lagrangians for dissipative dynamical systems. Chaos Solitons Fractals **42**, 2645–2652 (2009)
23. Polchinski, J.: What is String Theory? hep-th/9411028; see also String Theory (Cambridge Monographs on Mathematical Physics) (Volume 1), Cambridge University Press, Cambridge (1998), ISBN-13: 978-0521633031
24. Saha, A., Talukdar, B.: On the non-standard Lagrangian equations. arXiv: 1301.2667
25. Saha, A., Talukdar, B.: Inverse variational problem for non-standard Lagrangians. arXiv: 1305.6386
26. Thomas, B.: Relativistic string solution, Lecture 37, lectures given at Reed College, December 5th (2007)

13

Differentiation matrices in polynomial bases

A. Amiraslani[1]

Abstract Explicit differentiation matrices in various polynomial bases are presented in this work. The idea is to avoid any change of basis in the process of polynomial differentiation. This article concerns both degree-graded polynomial bases such as orthogonal bases, and non-degree-graded polynomial bases including the Lagrange and Bernstein bases.

Keywords Polynomial interpolation · Polynomial bases · Differentiation

Mathematics Subject Classification 41A05 · 26A24

Introduction

Consider a scalar (complex) polynomial $P(x)$ of degree n and a basis given by $\{B_0(x), B_1(x), \ldots, B_{n-1}(x), B_n(x)\}$. This basis determines the representation $P(x) = \sum_{j=0}^{n} a_j B_j(x)$, where $a_j \in C$ and, regardless of the basis, the coefficient of x^n in the expression is nonzero. Alternatively, this polynomial can be written as

$$P(x) = \begin{bmatrix} a_0 & a_1 & \ldots & a_{n-1} & a_n \end{bmatrix} \mathbf{I} \begin{bmatrix} B_0(x) \\ B_1(x) \\ \vdots \\ B_{n-1}(x) \\ B_n(x) \end{bmatrix}, \quad (1.1)$$

where $\mathbf{I}$ is the unit matrix of size $n+1$.

We want to find a matrix $\mathbf{D}$, the differentiation matrix, of size $n+1$ such that the kth order derivative of $P(x)$, shown by $\frac{\mathrm{d}^k P(x)}{\mathrm{d}x^k}$ or $P^{(k)}(x)$, can be written as

$$P^{(k)}(x) = \begin{bmatrix} a_0 & a_1 & \ldots & a_{n-1} & a_n \end{bmatrix} \mathbf{D}^k \begin{bmatrix} B_0(x) \\ B_1(x) \\ \vdots \\ B_{n-1}(x) \\ B_n(x) \end{bmatrix}. \quad (1.2)$$

For a polynomial of degree n, $\mathbf{D}$ has to be a nilpotent matrix of degree $n+1$.

$\mathbf{D}$ has a well-known structure and can be easily found for the monomial basis. For convenience, let us assume $n = 5$ and the generalizations for all positive n will be clear. If

$$P(x) = \begin{bmatrix} a_0 & a_1 & a_2 & a_3 & a_4 & a_5 \end{bmatrix} \mathbf{I} \begin{bmatrix} 1 \\ x \\ x^2 \\ x^3 \\ x^4 \\ x^5 \end{bmatrix}, \quad (1.3)$$

then

✉ A. Amiraslani
aamirasl@hawaii.edu

[1] STEM Department, University of Hawaii-Maui College, 310 W Kaahumanu Ave, Kahului, HI 96732, USA

$$\mathbf{D} = \begin{bmatrix} 0 & 0 & 0 & 0 & 0 & 0 \\ 1 & 0 & 0 & 0 & 0 & 0 \\ 0 & 2 & 0 & 0 & 0 & 0 \\ 0 & 0 & 3 & 0 & 0 & 0 \\ 0 & 0 & 0 & 4 & 0 & 0 \\ 0 & 0 & 0 & 0 & 5 & 0 \end{bmatrix}. \qquad (1.4)$$

Differentiation in bases other than the monomial basis has been occasionally studied. One of the most important applications of polynomial differentiation in other bases is in spectral methods like collocation method [14]. Differentiation matrices for Chebyshev and Jacobi polynomials were computed [13]. The differentiation of Jacobi polynomials through Bernstein basis was studied [12]. Chirikalov [2] computed the differentiation matrix for the Hermite basis.

In this paper, we present explicit formulas for **D** in different polynomial bases. Constructing **D** is fairly straightforward and having **D**, we can easily find derivatives of higher order by raising **D** to higher powers accordingly. Another important advantage of having a formula for **D** in a basis is that we do not need to change the basis—often to the monomial basis— to differentiate $P(x)$. Conversion between bases has been exhaustively studied in [6], but it can be unstable [8].

Section "Degree-graded bases" of this paper considers degree-graded bases and finds **D** in general for them. Orthogonal bases are all among degree-graded bases. Section "Degree-graded bases" then discusses other important special cases such as the monomial and Newton bases as well as the Hermite basis. Section "Bernstein basis" and "Lagrange basis" concern the Bernstein and Lagrange bases, respectively, and find **D** for them.

The Bernstein (Bézier) basis and the Lagrange basis are most useful in computer-aided geometric design (see [5], for example). For some problems in partial differential equations with symmetries in the boundary conditions Legendre polynomials can be successfully used the most natural. Finally, in approximation theory, Chebyshev polynomials have a special place due to their minimum-norm property (see e.g., [11]).

Degree-graded bases

Real polynomials $\{\phi_n(x)\}_{n=0}^{\infty}$ with $\phi_n(x)$ of degree n which are orthonormal on an interval of the real line (with respect to some nonnegative weight function) necessarily satisfy a three-term recurrence relation (see Chapter 10 of [3], for example). These relations can be written in the form

$$x\phi_j(x) = \alpha_j\phi_{j+1}(x) + \beta_j\phi_j(x) + \gamma_j\phi_{j-1}(x), \qquad j = 0, 1, 2, \ldots, \qquad (2.1)$$

where the α_j, β_j, γ_j are real, $\alpha_j \neq 0$, $\phi_{-1}(x) \equiv 0$, $\phi_0(x) \equiv 1$.

The choices of coefficients α_j, β_j, γ_j defining three well-known sets of orthogonal polynomials (associated with the names of Chebyshev and Legendre) are summarized in Table 1.

Orthogonal polynomials have well-established significance in mathematical physics and numerical analysis (see e.g., [7]). More generally, any sequence of polynomials $\{\phi_j(x)\}_{j=0}^{\infty}$ with $\phi_j(x)$ of degree j is said to be *degree-graded* and obviously forms a linearly independent set; but is not necessarily orthogonal.

A scalar polynomial of degree n can now be written in terms of a set of degree-graded polynomials $P(x) = \sum_{j=0}^{n} a_j\phi_j(x)$, where $a_j \in \mathcal{C}$ and $a_n \neq 0$. We can then write

$$P(x) = [a_0 \quad a_1 \quad \ldots \quad a_{n-1} \quad a_n]\mathbf{I}\begin{bmatrix} \phi_0(x) \\ \phi_1(x) \\ \vdots \\ \phi_{n-1}(x) \\ \phi_n(x) \end{bmatrix}. \qquad (2.2)$$

Table 1 Three well-known orthogonal polynomials

Polynomial	$T_n(x)$	$P_n(x)$	$U_n(x)$
Name of polynomial	Chebyshev (1st kind)	Legendre (Spherical)	Chebyshev (2nd kind)
Weight function	$(1-x^2)^{-\frac{1}{2}}$	1	$(1-x^2)^{\frac{1}{2}}$
Orthogonality interval	$[-1, 1]$	$[-1, 1]$	$[-1, 1]$
Leading coefficient k_n	2^{n-1}	$\frac{(2n)!}{2^n(n!)^2}$	2^n
α_n	1 for $n = 0$; $\frac{1}{2}$ otherwise	$\frac{n+1}{2n+1}$	$\frac{1}{2}$
β_n	0	0	0
γ_n	$\frac{1}{2}$	$\frac{n}{2n+1}$	$\frac{1}{2}$

Lemma 1 *If P(x) is given by (2.2), then*

$$P^{(k)}(x) = [a_0 \quad a_1 \quad \dots \quad a_{n-1} \quad a_n]\mathbf{D}^k \begin{bmatrix} \phi_0(x) \\ \phi_1(x) \\ \vdots \\ \phi_{n-1}(x) \\ \phi_n(x) \end{bmatrix}, \tag{2.3}$$

where

$$\mathbf{D} = \begin{bmatrix} 0 & 0 & \dots & 0 \\ & & & \vdots \\ & \mathbf{Q} & & \\ & & & 0 \end{bmatrix}. \tag{2.4}$$

Q *is a size n lower triangular matrix that has the following structure for* $i = 1, \dots, n$.

$$q_{i,j} = \begin{cases} \dfrac{i}{\alpha_{i-1}}, & i = j \\ \dfrac{1}{\alpha_{i-1}}((\beta_{j-1} - \beta_{i-1})q_{i-1,j} + \alpha_{j-2}q_{i-1,j-1} + \gamma_j q_{i-1,j+1} - \gamma_{i-1}q_{i-2,j}). & i > j \end{cases} \tag{2.5}$$

Any entry, q, with a negative or zero index is set to 0 in the above formula.

Proof We start by differentiating (2.1) to get

$$\phi_j(x) = \alpha_j\phi'_{j+1}(x) + (\beta_j - x)\phi'_j(x) + \gamma_j\phi'_{j-1}(x), \qquad j = 0, 1, \dots. \tag{2.6}$$

We can write this equation in a matrix-vector form. Without loss of generality, we assume $n = 4$.

$$\begin{bmatrix} 1 & 0 & 0 & 0 & 0 \\ 0 & 1 & 0 & 0 & 0 \\ 0 & 0 & 1 & 0 & 0 \\ 0 & 0 & 0 & 1 & 0 \\ 0 & 0 & 0 & 0 & 0 \end{bmatrix} \begin{bmatrix} \phi_0(x) \\ \phi_1(x) \\ \phi_2(x) \\ \phi_3(x) \\ \phi_4(x) \end{bmatrix} = \begin{bmatrix} 0 & \alpha_0 & 0 & 0 & 0 \\ \gamma_1 & \beta_1 - x & \alpha_1 & 0 & 0 \\ 0 & \gamma_2 & \beta_2 - x & \alpha_2 & 0 \\ 0 & 0 & \gamma_3 & \beta_3 - x & \alpha_3 \\ 0 & 0 & 0 & 0 & 0 \end{bmatrix} \begin{bmatrix} \phi'_0(x) \\ \phi'_1(x) \\ \phi'_2(x) \\ \phi'_3(x) \\ \phi'_4(x) \end{bmatrix}. \tag{2.7}$$

Given that $\phi'_0(x) = 0$ and $\phi_4(x)$ does not appear in the system, we can eliminate them and rewrite the above system as

$$\begin{bmatrix} \phi_0(x) \\ \phi_1(x) \\ \phi_2(x) \\ \phi_3(x) \end{bmatrix} = \mathbf{H} \begin{bmatrix} \phi'_1(x) \\ \phi'_2(x) \\ \phi'_3(x) \\ \phi'_4(x) \end{bmatrix}, \tag{2.8}$$

where

$$\mathbf{H} = \begin{bmatrix} \alpha_0 & 0 & 0 & 0 \\ \beta_1 - x & \alpha_1 & 0 & 0 \\ \gamma_2 & \beta_2 - x & \alpha_2 & 0 \\ 0 & \gamma_3 & \beta_3 - x & \alpha_3 \end{bmatrix}. \tag{2.9}$$

Since $\alpha_i \neq 0$, $\mathbf{H}^{-1}$ exists and is a lower triangular matrix that has the following row structure in general for $i = 1, \dots, n$.

$$h^{(*)}_{i,j} = \begin{cases} \dfrac{1}{\alpha_{i-1}}, & j = i \\ \dfrac{x - \beta_j}{\alpha_{j-1}}(h^{(*)}_{i,j+1}) - \dfrac{\gamma_{j+1}}{\alpha_{j-1}}(h^{(*)}_{i,j+2}), & j = (i-1), \dots, 1 \end{cases} \tag{2.10}$$

Now

$$\begin{bmatrix} \phi'_1(x) \\ \phi'_2(x) \\ \phi'_3(x) \\ \phi'_4(x) \end{bmatrix} = \mathbf{H}^{-1} \begin{bmatrix} \phi_0(x) \\ \phi_1(x) \\ \phi_2(x) \\ \phi_3(x) \end{bmatrix}. \tag{2.11}$$

It is obvious from (2.10) that the variable, x, appears in entries of $\mathbf{H}^{-1}$. Through a fairly straightforward, but rather tedious process and using (2.1) to eliminate x from $\mathbf{H}^{-1}$, we get **Q** as given by (2.5), and we have

$$\begin{bmatrix} \phi'_1(x) \\ \phi'_2(x) \\ \phi'_3(x) \\ \phi'_4(x) \end{bmatrix} = \mathbf{Q} \begin{bmatrix} \phi_0(x) \\ \phi_1(x) \\ \phi_2(x) \\ \phi_3(x) \end{bmatrix}. \tag{2.12}$$

From there, adding a **0** row and a **0** column, we find **D** and have

$$\begin{bmatrix} \phi_0'(x) \\ \phi_1'(x) \\ \phi_2'(x) \\ \phi_3'(x) \\ \phi_4'(x) \end{bmatrix} = \begin{bmatrix} 0 & 0 & \dots & 0 \\ & & & \vdots \\ & \mathbf{Q} & & \\ & & & 0 \end{bmatrix} \begin{bmatrix} \phi_0(x) \\ \phi_1(x) \\ \phi_2(x) \\ \phi_3(x) \\ \phi_4(x) \end{bmatrix}. \tag{2.13}$$

Using these results, we can write the first derivative of a generic first kind Chebyshev polynomial of degree 4 (see Table 1) as

$$P'(x) = [a_0 \quad a_1 \quad a_2 \quad a_3 \quad a_4] \begin{bmatrix} 0 & 0 & 0 & 0 & 0 \\ 1 & 0 & 0 & 0 & 0 \\ 0 & 4 & 0 & 0 & 0 \\ 3 & 0 & 6 & 0 & 0 \\ 0 & 8 & 0 & 8 & 0 \end{bmatrix} \begin{bmatrix} T_0(x) \\ T_1(x) \\ T_2(x) \\ T_3(x) \\ T_4(x) \end{bmatrix}. \tag{2.14}$$

Similarly, for a second kind Chebyshev polynomial of degree 4, we can write the first derivative as

$$P'(x) = [a_0 \quad a_1 \quad a_2 \quad a_3 \quad a_4] \begin{bmatrix} 0 & 0 & 0 & 0 & 0 \\ 2 & 0 & 0 & 0 & 0 \\ 0 & 4 & 0 & 0 & 0 \\ 2 & 0 & 6 & 0 & 0 \\ 0 & 4 & 0 & 8 & 0 \end{bmatrix} \begin{bmatrix} U_0(x) \\ U_1(x) \\ U_2(x) \\ U_3(x) \\ U_4(x) \end{bmatrix}. \tag{2.15}$$

$$P'(x) = [a_0 \quad a_1 \quad a_2 \quad a_3 \quad a_4] \begin{bmatrix} 0 & 0 & 0 & 0 & 0 \\ 1 & 0 & 0 & 0 & 0 \\ 0 & 3 & 0 & 0 & 0 \\ 1 & 0 & 5 & 0 & 0 \\ 0 & 3 & 0 & 7 & 0 \end{bmatrix} \begin{bmatrix} P_0(x) \\ P_1(x) \\ P_2(x) \\ P_3(x) \\ P_4(x) \end{bmatrix}. \tag{2.16}$$

Special degree-graded bases

As mentioned above, the family of degree-graded polynomials with recurrence relations of the form (2.1) include all the orthogonal bases, but are not limited to them. Here, we discuss some of the famous non-orthogonal bases of this kind and, consequently, for which we find the differentiation matrix, **D**, formulas. In particular, if in (2.1), we let $\alpha_j = 1$ and $\beta_j = \gamma_j = 0$, it will become the monomial basis. Using (2.4) and (2.5), we can easily verify that in this case, **D** has a form like (1.4).

Another important basis of this kind is the Newton basis. Let a polynomial $P(x)$ be specified by the data $\{(z_j, P_j)\}_{j=0}^{n}$ where the z_js are distinct. If the "Newton polynomials" are defined by setting $N_0(x) = 1$ and, for $k = 1, \dots, n$,

$$N_k(x) = \prod_{j=0}^{k-1}(x - z_j), \tag{2.17}$$

then

$$P(x) = [a_0 \quad a_1 \quad \dots \quad a_{n-1} \quad a_n]\mathbf{I} \begin{bmatrix} N_0(x) \\ N_1(x) \\ \vdots \\ N_{n-1}(x) \\ N_n(x) \end{bmatrix} \tag{2.18}$$

For $j = 0, \dots, n$, the a_js can be found by divided differences as follows.

$$a_j = [P_0, P_1, \dots, P_{j-1}], \tag{2.19}$$

where we have $[P_j] = P_j$, and

$$[P_i, \dots, P_{i+j}] = \frac{[P_{i+1}, \dots, P_{i+j}] - [P_i, \dots, P_{i+j-1}]}{z_{i+j} - z_i}. \tag{2.20}$$

If in (2.1), we let $\alpha_j = 1$, $\beta_j = z_j$ and $\gamma_j = 0$, it will become the Newton basis. For $n = 4$, **D**, as given by (2.4), has the following form.

$$\mathbf{D} = \begin{bmatrix} 0 & 0 & 0 & 0 & 0 \\ 1 & 0 & 0 & 0 & 0 \\ z_0 - z_1 & 2 & 0 & 0 & 0 \\ (z_0 - z_2)(z_0 - z_1) & -2z_2 + z_1 + z_0 & 3 & 0 & 0 \\ (z_0 - z_3)(z_0 - z_2)(z_0 - z_1) & (z_1 - z_3)(z_1 - 2z_2 + z_0) + (z_0 - z_2)(z_0 - z_1) & -3z_3 + z_2 + z_1 + z_0 & 4 & 0 \end{bmatrix}. \tag{2.21}$$

The confluent case

Suppose that a polynomial $P(x)$ of degree n as well as its derivatives are sampled at k nodes, i.e., distinct (finite) points $z_0, z_1, \ldots, z_{k-1}$. We write $P_j := P(z_j), P_j' := P'(z_j), \ldots, P_j^{(s_j)} := P^{(s_j)}(z_j)$; $j = 0, \ldots, k-1$. Here $s = (s_0, s_1, \ldots, s_{k-1})$ shows the confluencies (i.e., the orders of the derivatives) associated with the nodes and we have $\sum_{i=0}^{k-1} s_i = n + 1 - k$. If for $j = 0, \ldots, k-1$, all $s_j = 0$, then $k = n+1$ and we have the Lagrange interpolation.

This is an interesting polynomial interpolation that deserves a better consideration: the "Hermite interpolation" (See e.g., [2, 10]). It is basically similar to the Lagrange interpolation, but at each node, we have the value of $P(x)$ as well as its derivatives up to a certain order.

Now, we assume that at each node, z_j, we have the value and the derivatives of $P(x)$ up to the s_jth order. The nodes at which the derivatives are given are treated as extra nodes. In fact we pretend that we have $s_j + 1$ nodes, z_j, at which the value is P_j and remember that $\sum_{i=0}^{k-1} s_i = n + 1 - k$. In fact, the first $s_0 + 1$ nodes are z_0, the next $s_1 + 1$ nodes are z_1 and so on.

Using the divided differences technique, as given by (2.20), to find a_js, whenever we get $[P_j, P_j, \ldots, P_j]$ where P_j is repeated m times, we have

$$[P_j, P_j, \ldots, P_j] = \frac{P_j^{(m-1)}}{(m-1)!}, \tag{2.22}$$

and all the values P_j' to $P_j^{(s_j)}$ for $j = 0, \ldots, k-1$ are given. For more details see e.g., [10].

The bottom line is that the Hermite basis can be seen as a special case of the Newton basis, thus a degree-graded basis. For the Hermite basis, like the Newton basis, $\alpha_j = 1$, $\beta_j = z_i$, and $\gamma_j = 0$, but some of the β_js are repeated. Other than that, the differentiation matrix, $\mathbf{D}$, can be similarly found for the Hermite basis.

For a data set like $\{(z_0, P(z_0)), (z_0, P'(z_0)), (z_1, P(z_1)), (z_2, P(z_2)), (z_3, P(z_3)\}$, $\beta_0 = z_0$, we have $\beta_1 = z_0$, $\beta_2 = z_1$, $\beta_3 = z_2$, and $\beta_4 = z_3$. In this case, (2.21) becomes

$$\mathbf{D} = \begin{bmatrix} 0 & 0 & 0 & 0 & 0 \\ 1 & 0 & 0 & 0 & 0 \\ 0 & 2 & 0 & 0 & 0 \\ 0 & -2z_1 + 2z_0 & 3 & 0 & 0 \\ 0 & (z_0 - z_2)(2z_0 - 2z_1) & -3z_3 + z_1 + 2z_0 & 4 & 0 \end{bmatrix}. \tag{2.23}$$

Bernstein basis

A Bernstein polynomial (also called Bézier polynomial) defined over the interval $[a, b]$ has the form

$$b_j(x) = \binom{n}{j} \frac{(x-a)^j (b-x)^{n-j}}{(b-a)^n}, \quad j = 0, \ldots, n. \tag{3.1}$$

This is not a typical scaling of the Bernstein polynomials; however, this scaling makes matrix notations related to this basis slightly easier to write.

The Bernstein polynomials are nonnegative in $[a, b]$, i.e., $b_j(x) \geq 0$ for all $x \in [a, b]$ $(j = 0, \ldots, n)$. Bernstein polynomials are widely used in computer-aided geometric design (e.g., see [4]).

A polynomial $P(x)$ written in the Bernstein basis is of the form

$$P(x) = [a_0 \quad a_1 \quad \ldots \quad a_{n-1} \quad a_n]\mathbf{I} \begin{bmatrix} b_0(x) \\ b_1(x) \\ \vdots \\ b_{n-1}(x) \\ b_n(x) \end{bmatrix}, \tag{3.2}$$

where the a_js are sometimes called the Bézier coefficients $(j = 0, \ldots, n)$.

Lemma 2 *If $P(x)$ is given by (3.2), then*

$$P^{(k)}(x) = [a_0 \quad a_1 \quad \ldots \quad a_{n-1} \quad a_n]\mathbf{D}^k \begin{bmatrix} b_0(x) \\ b_1(x) \\ \vdots \\ b_{n-1}(x) \\ b_n(x) \end{bmatrix}, \tag{3.3}$$

where $\mathbf{D}$ is a size $n+1$ tridiagonal matrix that has the following structure for $i = 1, \ldots, n+1$.

$$d_{i,j} = \begin{cases} \dfrac{n - 2(i-1)}{a-b}, & i = j \\ \dfrac{i}{a-b}, & i = j-1 \\ -\dfrac{(n-i+1)}{a-b} & i = j+1 \end{cases} \tag{3.4}$$

Proof A little computation using (3.1) shows that

$$b_k'(x) = \frac{k-1}{a-b} b_{k-1}(x) + \frac{n-2k}{a-b} b_k(x) + \frac{n-k}{a-b} b_{k+1}(x), \tag{3.5}$$

for $k = 0, \ldots, n$. Any $b_i(x)$ with either a negative or larger than n index is set to 0 in (3.5). This is why $\mathbf{D}$ is tridiagonal and form here, it is easy to derive (3.4) for $\mathbf{D}$.

For $n = 4$, the differentiation matrix is as follows.

$$\mathbf{D} = \begin{bmatrix} \frac{4}{a-b} & \frac{1}{a-b} & 0 & 0 & 0 \\ -\frac{4}{a-b} & \frac{2}{a-b} & \frac{2}{a-b} & 0 & 0 \\ 0 & -\frac{3}{a-b} & 0 & \frac{3}{a-b} & 0 \\ 0 & 0 & -\frac{2}{a-b} & -\frac{2}{a-b} & \frac{4}{a-b} \\ 0 & 0 & 0 & -\frac{1}{a-b} & -\frac{4}{a-b} \end{bmatrix}. \tag{3.6}$$

□

Lagrange basis

Lagrange polynomial interpolation is traditionally viewed as a tool for theoretical analysis; however, recent work reveals several advantages to computation using new polynomial interpolation techniques in the Lagrange basis (see e.g., [1, 9]). Suppose that a polynomial $P(x)$ of degree n is sampled at $n+1$ distinct points $z_0, z_1, \ldots, z_n$, and write $p_j := P(z_j)$. Lagrange polynomials are defined by

$$L_j(x) = \frac{\ell(x) w_j}{x - z_j}, \qquad j = 0, 1, \ldots, n \tag{4.1}$$

where the "weights" w_j are

$$w_j = \prod_{m=0, m \neq j}^{n} \frac{1}{z_j - z_m}, \tag{4.2}$$

and

$$\ell(x) = \prod_{m=0}^{n} (x - z_m). \tag{4.3}$$

Then $P(x)$ can be expressed in terms of its samples in the following form.

$$P(x) = [p_0 \quad p_1 \quad \cdots \quad p_{n-1} \quad p_n]\mathbf{I} \begin{bmatrix} L_0(x) \\ L_1(x) \\ \vdots \\ L_{n-1}(x) \\ L_n(x) \end{bmatrix} \tag{4.4}$$

Lemma 3 *If $P(x)$ is given by (4.4), then*

$$P^{(k)}(x) = [p_0 \quad p_1 \quad \cdots \quad p_{n-1} \quad p_n]\mathbf{D}^k \begin{bmatrix} L_0(x) \\ L_1(x) \\ \vdots \\ L_{n-1}(x) \\ L_n(x) \end{bmatrix}, \tag{4.5}$$

where $\mathbf{D}$ is a size $n+1$ matrix that has the following structure.

$$d_{i,j} = \begin{cases} \sum_{k=0, k \neq i-1}^{n} \frac{1}{z_{i-1} - z_k}, & i = j \\ \frac{w_{i-1}}{w_{j-1}(z_{j-1} - z_{i-1})}, & i \neq j \end{cases} \tag{4.6}$$

Proof A little computation using (4.1) shows that

$$L_i'(x) = \sum_{j=0, j \neq i}^{n} \frac{w_i}{w_j(z_j - z_i)} L_j(x) + \sum_{k=0, k \neq i}^{n} \frac{1}{z_i - z_k} L_i(x), \tag{4.7}$$

for $i = 0, \ldots, n$. Form here, it is easy to derive (4.6) for $\mathbf{D}$. For $n = 3$, the differentiation matrix is as follows.

$$\mathbf{D} = \begin{bmatrix} \frac{1}{z_0 - z_1} + \frac{1}{z_0 - z_2} + \frac{1}{z_0 - z_3} & \frac{w_0}{w_1(z_1 - z_0)} & \frac{w_0}{w_2(z_2 - z_0)} & \frac{w_0}{w_3(z_3 - z_0)} \\ \frac{w_1}{w_0(z_0 - z_1)} & \frac{1}{z_1 - z_0} + \frac{1}{z_1 - z_2} + \frac{1}{z_1 - z_3} & \frac{w_1}{w_2(z_2 - z_1)} & \frac{w_1}{w_3(z_3 - z_1)} \\ \frac{w_2}{w_0(z_0 - z_2)} & \frac{w_2}{w_1(z_1 - z_2)} & \frac{1}{z_2 - z_0} + \frac{1}{z_2 - z_1} + \frac{1}{z_2 - z_3} & \frac{w_2}{w_3(z_3 - z_2)} \\ \frac{w_3}{w_0(z_0 - z_3)} & \frac{w_3}{w_1(z_1 - z_3)} & \frac{w_3}{w_2(z_2 - z_3)} & \frac{1}{z_3 - z_0} + \frac{1}{z_3 - z_1} + \frac{1}{z_3 - z_2} \end{bmatrix}. \tag{4.8}$$

□

Concluding remarks

A differentiation matrix, $\mathbf{D}$, of size $n+1$ in a certain basis is a nilpotent matrix of degree $n+1$. As such, all of its eigenvalues are zero. Assuming $n=5$—and the generalizations for all positive n will be clear—we can fairly easily verify that the Jordan form for $\mathbf{D}$ in any basis is

$$\mathbf{J}=\begin{bmatrix} 0 & 1 & 0 & 0 & 0 & 0\\ 0 & 0 & 1 & 0 & 0 & 0\\ 0 & 0 & 0 & 1 & 0 & 0\\ 0 & 0 & 0 & 0 & 1 & 0\\ 0 & 0 & 0 & 0 & 0 & 1\\ 0 & 0 & 0 & 0 & 0 & 0 \end{bmatrix}. \tag{5.1}$$

We can write $\mathbf{J}=\mathbf{T}^{-1}\mathbf{D}\mathbf{T}$. The jth column of the transformation matrix, $\mathbf{T}$, is the first column of $\mathbf{D}^{n+1-j}$ for $j=1,\ldots,n+1$. For example, for the $\mathbf{D}$ given by (1.4), we have

$$\mathbf{T}=\begin{bmatrix} 0 & 0 & 0 & 0 & 0 & 1\\ 0 & 0 & 0 & 0 & 1 & 0\\ 0 & 0 & 0 & 2 & 0 & 0\\ 0 & 0 & 6 & 0 & 0 & 0\\ 0 & 24 & 0 & 0 & 0 & 0\\ 120 & 0 & 0 & 0 & 0 & 0 \end{bmatrix}. \tag{5.2}$$

Now, if we consider two different polynomial bases for which the differentiation matrices are $\mathbf{D}_1$ and $\mathbf{D}_2$, then we can write $\mathbf{J}=\mathbf{T}_1^{-1}\mathbf{D}_1\mathbf{T}_1$ and $\mathbf{J}=\mathbf{T}_2^{-1}\mathbf{D}_2\mathbf{T}_2$. From here it is easy to check that

$$\mathbf{D}_2=(\mathbf{T}_1\mathbf{T}_2)^{-1}\mathbf{D}_1(\mathbf{T}_1\mathbf{T}_2), \tag{5.3}$$

which shows, as expected, the differentiation matrices in different bases are similar.

In this paper, we have found explicit formulas for the differentiation matrix, $\mathbf{D}$, in various polynomial bases. The most important advantage of having $\mathbf{D}$ explicitly is that there is no need to go from one basis to another (normally monomial) to differentiate a polynomial in a given basis. Moreover, having $\mathbf{D}$, we can easily find higher order derivatives of any polynomial in its original basis. One may hope that new and more efficient polynomial-related algorithms, such as root-finding methods, can be developed using $\mathbf{D}$.

These results can be easily extended to matrix polynomials in different bases.

References

1. Berrut, J., Trefethen, L.: Barycentric Lagrange interpolation. SIAM Rev. **46**(3), 501–517 (2004)
2. Chirikalov, V.I.: Computation of the differentiation matrix for the Hermite interpolating polynomials. J. Math. Sci. **68**(6), 766–770 (1994)
3. Davis, P.J.: Interpolation and Approximation. Blaisdell, New York (1963)
4. Farin, G.: Curves and Surfaces for Computer-Aided Geometric Design. Acadmeic Press, San Diego (1997)
5. Farouki, R.T., Goodman, T.N.T., Sauer, T.: Construction of orthogonal bases for polynomials in Bernstein form on triangular simplex domains. Comput. Aided Geom. Design **20**, 209–230 (2003)
6. Gander, W.: Change of basis in polynomial interpolation. Numer. Linear Algebra Appl. **12**(8), 769–778 (2002)
7. Gautschi, W.: Orthogonal Polynomials: Computation and Approximation. Clarendon, Oxford (2004)
8. Hermann, T.: On the stability of polynomial transformations between Taylor, Bézier Hermite forms. Numer. Algo. **13**, 307–320 (1996)
9. Higham, N.: The numerical stability of barycentric Lagrange interpolation. IMA J. Numer. Anal. **24**, 547–556 (2004)
10. Lorentz, G., Jetter, K., Riemenschneider, S.D.: Birkhoff Interpolation. Addison Wesley Publishing Company, Boston (1983)
11. Ravlin, T.: Chebyshev Polynomials. Wiley, New York (1990)
12. Rabbah, A., Al-Refai, M., Al-Jarrah, R.: Computing derivatives of Jacobi polynomials using Bernstein transformation and differentiation matirx. Numer. Funct. Anal. Optim. **29**(5–6), 660–673 (2008)
13. Shen, J., Tang, T., Wang, L.: Spectral Methods: Algorithms, Analysis and Applications. Springer Series in Computational Mathematics, vol. 41. Springer, New York (2011)
14. Wang, L., Samson, D., Zhao, X.: A well-conditioned collocation method using a pseudospectral integration matrix. SIAM J. Sci. Comput. **36**(3), 907–929 (2014)

14

Asymptotic efficiency and small sample power of a locally most powerful linear rank test for the log-logistic distribution

Hidetoshi Murakami

Abstract Assume that two independent random samples are distributed according to a log-logistic distribution (LLD). In this study, the score functions for the locally most powerful rank test were derived for the location and scale parameters. The Wilcoxon rank-sum test was shown to be locally most powerful rank test for the LLD. The asymptotic efficiency of the Wilcoxon rank-sum test was derived and compared with that of the modified Wilcoxon rank-sum test for the LLD.

Keywords Asymptotic efficiency · Locally most powerful rank test · Log-logistic distribution · Modified Wilcoxon rank sum test · Wilcoxon rank sum test

Introduction

Testing hypotheses is one of the most important problems in performing nonparametric statistics. Various nonparametric statistics have been shown to be a locally most powerful rank test (LMPRT) for a specific distribution over the course of many years. For example, [9] obtained the LMPRTs for comparing two possibly censored samples for a given alternative, by deriving scores for censored and uncensored observations. In finance literature, the Lévy distribution arises as a special case for describing security price returns by mixtures of distributions. Runde [13] derived the score functions of LMPRTs for the location and scale parameters with the Lévy distribution. Hájek et al. [5] discussed the LMPRTs for various distributions. The most powerful test for correlation is well-known, and the LMPRT uses Fisher-Yates expected normal scores. However, the bivariate normal distribution does not fit some types of data. Conover [2] considered the LMPRTs for correlation with four examples. Pandit [11] investigated the LMPRT for testing independence against a weighted contamination alternative. Few studies have discussed the LMPRT when only a fraction of treated subjects respond to treatment; however, [12] examined the LMPRT for several of these case types.

Let $\{X_{ij}; i = 1, 2, j = 1, \ldots, n_i\}$ be two random samples of size n_1 and n_2 independent observations, each of which have a continuous distribution described as F_1 and F_2, respectively. Assume that X_{ij} is distributed according to a log-logistic distribution (LLD); see e.g., [14], with the probability density function (pdf) $f(x)$ and the cumulative distribution function (cdf) $F(x)$ as follows:

$$f_i(x) = \frac{\beta_i\{(x-\mu_i)/\alpha_i\}^{\beta_i-1}}{\alpha_i[1+\{(x-\mu_i)/\alpha_i\}^{\beta_i}]^2},$$

$$F_i(x) = \frac{(x-\mu_i)^{\beta}}{\alpha^{\beta}+(x-\mu_i)^{\beta}}, \quad \alpha_i,\ \beta_i > 0, \quad x > \mu_i,$$

respectively. The parameters μ, α and β denote the location, scale and shape parameters, respectively. Note that the median of the LLD is $\alpha + \mu$. This distribution is a special case of Burr's type-XII distribution; see e.g., [1]. In addition, the LLD is a special case of the kappa distributions introduced by [6]. The LLD is often applied to economics as a simple model for wealth or income distribution [3]. Note that the LLD is known as the Fisk distribution in economics. Additionally, the LLD is used in survival analysis as a parametric model for events whose hazard rate increases initially and decreases later.

On the basis of these data types, we proposed the following hypothesis:

H. Murakami (✉)
Department of Mathematics, National Defense Academy,
1-10-20 Hashirimizu, Yokosuka, Kanagawa 239-8686, Japan
e-mail: murakami@gug.math.chuo-u.ac.jp

$$H_0: F_1(x) = F_2(x) \quad \text{against} \quad H_L: F_1(x) = F_2(x-\mu), \quad \mu \neq 0$$

or

$$H_0: F_1(x) = F_2(x) \quad \text{against} \quad H_S: F_1(x) = F_2(x/\alpha), \quad \alpha > 0,\ \alpha \neq 1.$$

For a recent comparison study of many nonparametric tests for scale, see [8]. To test these hypotheses, we developed a linear rank statistic. Let $V_j = 0$ if the jth smallest of the $N = n_1 + n_2$ observations is from X_{1j}; otherwise, $V_j = 1$. This provides a general two-sample linear rank statistic, as follows:

$$T = \sum_{j=1}^{N} a(j)V_j = \sum_{j=1}^{n_2} a(R_j).$$

Herein, R_j denotes the rank of sample X_{2j}. In Sect. 2, we derive a LMPRT for H_L and H_S. Since finite sample sizes are used in practice, we investigate a small sample power of the linear rank tests for the LLD in Sect. 3. Finally, conclusions are presented in Sect. 4.

Locally most powerful linear rank test

In this section, we derive the LMPRT for H_L and H_S. Assume that f is absolutely continuous and that

$$\int |f'(x)|dx < \infty.$$

The LMPRT for H_L is given by [5], as follows:

$$T_L = \sum_{j=1}^{N} a_N(j,f)V_j = \sum_{j=1}^{N} \mathrm{E}\{\psi_L(U_{(j:N)}, f)\}V_j = \sum_{j=1}^{N} \mathrm{E}\left\{-\frac{f'(F^{-1}(U_{(j:N)}))}{f(F^{-1}(U_{(j:N)}))}\right\}V_j,$$

where $U_{(1:N)} < \cdots < U_{(N:N)}$ are ordered statistics from the uniform distribution in the interval $[0, 1]$. In addition, the LMPRT for H_S is given by [5] as

$$T_S = \sum_{j=1}^{N} a_N(j,f)V_j = \sum_{j=1}^{N} \mathrm{E}\{\psi_S(U_{(j:N)}, f)\}V_j = \sum_{j=1}^{N} \mathrm{E}\left\{-1 - F^{-1}(U_{(j:N)})\frac{f'(F^{-1}(U_{(j:N)}))}{f(F^{-1}(U_{(j:N)}))}\right\}V_j.$$

By Theorem 3.1.2.4 in [5], the score function $a_N(j,f)$ can be rewritten as

$$a_N(j,f) = N\binom{N-1}{j-1}\int_0^1 \psi(u,f)u^{j-1}(1-u)^{N-j}du. \tag{1}$$

LMPRT for the location parameter

In this section, the score function of the LMPRT is derived for the LLD for the location parameter. The standard LLD is given as follows:

$$f(x) = \frac{1}{(1+x)^2} \quad \text{and} \quad F(x) = \frac{x}{1+x}. \tag{2}$$

From (2), the first derivative of pdf and the inverse of cdf for the standard LLD are, respectively, given by

$$f'(x) = -\frac{2}{(1+x)^3} \quad \text{and} \quad F^{-1}(x) = \frac{x}{1-x}.$$

The score function $\psi_L(U_{(j:N)}, f)$ for H_L is obtained from the following:

$$\psi_L(U_{(j:N)}, f) = 2(1 - U_{(j:N)}), \tag{3}$$

and

$$\sum_{j=1}^{N} a_N(U_{(j:N)}, f)V_j = \frac{2}{N+1}\sum_{j=1}^{n_2}(N+1-R_j) \tag{4}$$

using (1). (4) denotes the inverse function of R_j. Thus, the Wilcoxon rank-sum test is the LMPRT for H_L with LLDs.

Note that the LMPRT is valid in the neighborhood close to the null hypothesis. In many cases, the asymptotic efficiency of the LMPRT is the highest for the adjusted distribution. However, moderate to large sample sizes are required to assume a specific distribution for deriving the LMPRT. Herein, we consider the asymptotic efficiency of both the original and modified Wilcoxon rank-sum tests for the LLD. Tamura [15] proposed the modified Wilcoxon rank-sum test to raise the asymptotic relative efficiency as follows:

$$T(p) = \sum_{i=1}^{N} i^p V_i = \sum_{i=1}^{n_2} R_i^p, \quad p \in \mathbb{R}^+. \tag{5}$$

Note that the test statistic $T(p)$ is the original Wilcoxon rank-sum test when $p = 1$. By applying the idea of [15] to the score function (3), we can obtain the score function of the inverse rank test:

$$\psi_L^*(U_{(j:N)}, f) = (p+1)(1 - U_{(j:N)})^p. \tag{6}$$

The asymptotic efficiency of the score function (6) for the location parameter is given as follows:

Table 1 Asymptotic efficiency for the location parameter with the LLD

p	0.1	0.5	1	1.5	2	2.5	3
$AEL(p,f)$	0.250	0.720	1.333	2.041	2.813	3.630	4.480
$ARE(p,1,f)$	0.188	0.540	1.000	1.531	2.110	2.723	3.361

$$\text{AEL}(p,f) = \lambda(1-\lambda)(2p+1)(p+1)^3 \left(\int_{-\infty}^{\infty} f(x)^2\{1-F(x)\}^{p-1}dx\right)^2,$$

where $\lambda = \lim_{N\to\infty}\lambda_N = \lim_{N\to\infty} n_1/N$. The asymptotic efficiency $AEL(p,f)$ and asymptotic relative efficiency $ARE(p, 1, f)$ to the original Wilcoxon rank-sum test for the LLD are listed in Table 1.

The results shown in Table 1 indicate that the asymptotic efficiency of the $T(p)$ test is higher than that of the original Wilcoxon rank-sum test for LLDs when $p > 1$.

LMPRT for the scale parameter

Consider the score function of the LMPRT for H_S. Using a similar procedure as that applied to the location case, we obtain the score function, given below:

$$\psi_S(U_{(i:N)}, f) = 2U_{(i:N)} - 1.$$

Therefore,

$$\sum_{j=1}^{N} a_N(U_{(j:N)}, f)V_j = \frac{2}{N+1}\left(\sum_{j=1}^{n_2} R_i - \frac{n_2(N+1)}{2}\right) \tag{7}$$

using (1). Thus, (7) reveals that the Wilcoxon rank-sum test is the LMPRT for H_S with a LLD. Note that the score function of the modified Wilcoxon rank-sum test is given as [4]:

$$\psi_S^*(U_{(i:N)}, f) = (p+1)U_{(i:N)}^p - 1.$$

From this, the asymptotic efficiency of the modified Wilcoxon rank-sum test for the scale parameter is given by

$$AES(p,f) = \lambda(1-\lambda)(2p+1)(p+1)^2 \times \left(\int_{-\infty}^{\infty} xf(x)^2F(x)^{p-1}dx\right)^2.$$

The asymptotic efficiency $AES(p,f)$ and the asymptotic relative efficiency ARE(p, 1, f) to the original Wilcoxon rank-sum test for the LLD are listed in Table 2.

Therefore, the results shown in Table 2 reveal that the original Wilcoxon rank-sum test is suitable for testing LLDs.

Herein, we consider another distribution, as follows:

$$f^*(x) = \frac{2^q - 1}{x^{q+1}\{1 + (2^q - 1)x^{-q}\}^{(1+q)/q}},$$

$$F^*(x) = \frac{1}{\{1 + (2^q - 1)x^{-q}\}^{1/q}}, \quad (x > 0).$$

For the case $q = 1$, this distribution is equivalent to that of the standard LLD. Thus, f^* and F^* is another extension of the standard LLD. Using a procedure similar to the one presented earlier in this section, the score function $\psi_S(U_{(i:N)}, f)$ is given by

$$\psi_S(U_{(i:N)}, f^*) = (q+1)U_{(i:N)}^q - 1.$$

Therefore, the modified Wilcoxon rank-sum test is the LMPRT for H_S with f^*. Additionally, we compared the asymptotic efficiency of the modified Wilcoxon rank-sum test with that of the original Wilcoxon rank-sum test for f^* in Table 3.

Table 3 shows that the modified Wilcoxon rank-sum test is more efficient than is the original Wilcoxon rank-sum test for f^*.

Powers of small sample sizes

In this section, we present our investigation of the behavior of the original Wilcoxon rank-sum test and the modified Wilcoxon rank-sum test for the LLD $f(x)$. In a previous section, we obtained asymptotic results. However, because finite sample sizes are used in practice, we investigated a small sample power of the original and modified Wilcoxon rank-sum tests. Generally, the location, scale, location–scale, and shape parameters of the X_{1j} and X_{2j} samples are unequal. Here, we considered a two-sample problem in which the hypothesis, $H_0 : F_1(x) = F_2(x)$, was tested against $H_1 : F_1(x) \neq F_2(x)$. The following assumptions were made for LLDs, as follows:

Table 2 Asymptotic efficiency for the scale parameter with the LLD

p	0.1	0.5	1	1.5	2	2.5	3
AES(p, f)	0.272	0.320	0.333	0.327	0.313	0.296	0.280
ARE(p, 1, f)	0.817	0.961	1.000	0.982	0.940	0.889	0.841

Table 3 Asymptotic efficiency for the scale parameter with the f^*

q	0.1	0.5	1	1.5	2	2.5	3
AES(q, f^*)	0.008	0.125	0.333	0.562	0.800	1.042	1.286
AES(1, f^*)	0.007	0.120	0.333	0.551	0.750	0.926	1.080
ARE(q, 1, f)	1.225	1.042	1.000	1.021	1.067	1.125	1.190

Table 4 Simulated power: $n_1 = n_2 = 10$ for $\alpha = 0.05$

μ_2	α_2			
	1.0	2.0	3.0	4.0
LLD(0, 1, 1) and LLD(μ_2, α_2, 1)				
0.0				
T(0.5)	0.044	0.312	0.748	0.959
T(0.75)	0.044	0.312	0.757	0.964
T(1)	0.044	0.318	0.761	0.965
T(2)	0.044	0.305	0.736	0.954
0.5				
T(0.5)	0.115	0.438	0.819	0.973
T(0.75)	0.106	0.426	0.816	0.974
T(1)	0.097	0.413	0.809	0.973
T(2)	0.074	0.360	0.764	0.958
1.0				
T(0.5)	0.209	0.545	0.864	0.981
T(0.75)	0.187	0.522	0.855	0.981
T(1)	0.167	0.499	0.845	0.979
T(2)	0.113	0.414	0.788	0.962
1.5				
T(0.5)	0.310	0.628	0.895	0.986
T(0.75)	0.276	0.599	0.885	0.985
T(1)	0.244	0.570	0.872	0.983
T(2)	0.158	0.463	0.810	0.966
LLD(0, 1, 1) and LLD(μ_2, α_2, 0.5)				
0.0				
T(0.5)	0.070	0.522	0.933	0.997
T(0.75)	0.059	0.489	0.918	0.996
T(1)	0.050	0.450	0.900	0.994
T(2)	0.027	0.322	0.802	0.976
0.5				
T(0.5)	0.142	0.599	0.944	0.998
T(0.75)	0.118	0.557	0.929	0.996
T(1)	0.098	0.515	0.911	0.994
T(2)	0.050	0.367	0.813	0.974
1.0				
T(0.5)	0.227	0.662	0.952	0.998
T(0.75)	0.189	0.616	0.938	0.996
T(1)	0.158	0.571	0.921	0.995
T(2)	0.080	0.408	0.825	0.975
1.5				
T(0.5)	0.313	0.711	0.958	0.998
T(0.75)	0.264	0.664	0.944	0.997
T(1)	0.223	0.618	0.928	0.995
T(2)	0.116	0.445	0.834	0.975

Table 4 continued

μ_2	α_2			
	1.0	2.0	3.0	4.0
LLD(0, 1, 1) and LLD(μ, α_2, 1.5)				
0.0				
T(0.5)	0.036	0.195	0.517	0.813
T(0.75)	0.041	0.217	0.554	0.844
T(1)	0.046	0.236	0.584	0.865
T(2)	0.066	0.289	0.645	0.897
0.5				
T(0.5)	0.105	0.332	0.652	0.884
T(0.75)	0.103	0.337	0.666	0.896
T(1)	0.102	0.341	0.676	0.904
T(2)	0.102	0.348	0.689	0.913
1.0				
T(0.5)	0.206	0.460	0.747	0.923
T(0.75)	0.192	0.451	0.747	0.927
T(1)	0.180	0.441	0.747	0.929
T(2)	0.147	0.409	0.730	0.925
1.5				
T(0.5)	0.315	0.567	0.811	0.946
T(0.75)	0.289	0.548	0.804	0.947
T(1)	0.266	0.530	0.798	0.946
T(2)	0.200	0.467	0.762	0.936

Table 5 Simulated power: $n_1 = 5$, $n_2 = 10$ for $\alpha = 0.05$

μ_2	α_2			
	1.0	2.0	3.0	4.0
LLD(0, 1, 1) and LLD(μ_2, α_2, 1)				
0.0				
T(0.5)	0.050	0.253	0.610	0.878
T(0.75)	0.050	0.255	0.614	0.881
T(1)	0.050	0.255	0.612	0.876
T(2)	0.049	0.241	0.573	0.833
0.5				
T(0.5)	0.121	0.351	0.676	0.900
T(0.75)	0.113	0.343	0.671	0.899
T(1)	0.104	0.329	0.659	0.892
T(2)	0.080	0.285	0.599	0.840
1.0				
T(0.5)	0.201	0.434	0.724	0.915
T(0.75)	0.184	0.417	0.714	0.912
T(1)	0.166	0.395	0.696	0.903
T(2)	0.115	0.324	0.620	0.846
1.5				
T(0.5)	0.276	0.502	0.760	0.926
T(0.75)	0.253	0.480	0.747	0.921
T(1)	0.228	0.452	0.726	0.912
T(2)	0.149	0.358	0.639	0.852

Table 5 continued

μ_2	α_2			
	1.0	2.0	3.0	4.0
LLD(0, 1, 1) and LLD(μ_2, α_2, 0.5)				
0.0				
T(0.5)	0.092	0.428	0.813	0.965
T(0.75)	0.084	0.407	0.795	0.959
T(1)	0.073	0.376	0.766	0.948
T(2)	0.048	0.273	0.620	0.847
0.5				
T(0.5)	0.160	0.487	0.829	0.967
T(0.75)	0.145	0.462	0.811	0.961
T(1)	0.126	0.427	0.782	0.951
T(2)	0.073	0.300	0.630	0.849
1.0				
T(0.5)	0.228	0.537	0.842	0.969
T(0.75)	0.208	0.510	0.724	0.963
T(1)	0.183	0.472	0.796	0.953
T(2)	0.100	0.325	0.638	0.850
1.5				
T(0.5)	0.292	0.579	0.853	0.970
T(0.75)	0.269	0.551	0.835	0.964
T(1)	0.241	0.512	0.807	0.954
T(2)	0.125	0.348	0.645	0.851
LLD(0, 1, 1) and LLD(μ, α_2, 1.5)				
0.0				
T(0.5)	0.033	0.154	0.406	0.693
T(0.75)	0.036	0.166	0.429	0.718
T(1)	0.040	0.178	0.451	0.738
T(2)	0.052	0.207	0.485	0.757
0.5				
T(0.5)	0.103	0.265	0.518	0.766
T(0.75)	0.099	0.266	0.527	0.779
T(1)	0.095	0.264	0.532	0.785
T(2)	0.086	0.258	0.528	0.778
1.0				
T(0.5)	0.190	0.365	0.602	0.813
T(0.75)	0.177	0.356	0.601	0.817
T(1)	0.162	0.344	0.596	0.818
T(2)	0.126	0.306	0.565	0.795
1.5				
T(0.5)	0.274	0.450	0.665	0.845
T(0.75)	0.253	0.433	0.658	0.846
T(1)	0.230	0.413	0.646	0.842
T(2)	0.166	0.349	0.596	0.809

LLD(μ, α, β) with location parameter μ, scale parameter α and shape parameter β.

Table 4 and Table 5 show the analysis results of the power of the $T(p)$ for $(n_1, n_2) = (10, 10)$ and (5,10). The significance level was 5 %, and the simulations were repeated 1,000,000 times. In this study, we assumed $p =$ 0.5, 0.75, 1 and 2. The exact probability of the original Wilcoxon rank-sum test $T(1)$ is 0.044 for $n_1 = n_2 = 10$ under the null hypothesis. Thus, we adopted an exact probability of 0.044 for the $T(p)$ test.

The results indicated that a smaller p in the $T(p)$ test was more powerful than a larger p in the $T(p)$ test for equal and unequal sample sizes, with $\beta \leq 1$ for the shifted location, scale, and location–scale parameters. For the case of $\beta > 1$, a larger p in the $T(p)$ test was more efficient than a small p in the $T(p)$ test for the changed scale parameter, but not the location and location–scale parameters. The results indicated $T(0.5)$ and $T(0.75)$ tests were suitable for a wide range of LLD. Therefore, a smaller p in the $T(p)$ test is suitable for parameters associated with LLDs.

Conclusion and discussions

In this paper, we considered the LMPRT for the location and scale parameters with the LLD. We showed that the Wilcoxon rank-sum test is suitable for hypothesis testing with the LLD. In addition, the asymptotic efficiency of the modified Wilcoxon rank-sum test was higher than that of the original Wilcoxon rank-sum test for the location parameter with a LLD. For the scale parameter, the asymptotic efficiency of the original Wilcoxon rank-sum test was highest for the LLD. Additionally, the score function of another extension of the LLD was derived as the modified Wilcoxon rank-sum test. However, because finite sample sizes are used in practice, we investigated a small sample power of the original and modified Wilcoxon rank-sum tests. The results indicated that a smaller p in the $T(p)$ test was suitable for parameters associated with LLDs. Future research directions should consider the two-sample location–scale problem—i.e., where two null hypotheses H_0 are simultaneously tested, see; e.g., [7, 10]—for the LLD.

Acknowledgments The author would like to thank the editor and the referee for their valuable comments and suggestions. The author appreciates that this research is supported by the Grant-in-Aid for Young Scientists B of JSPS, KAKENHI Number 26730025.

References

1. Burr, I.W.: Cumulative frequency functions. Annal Math. Stat. **13**, 215–232 (1942)
2. Conover, W.J.: Some locally most powerful rank tests for correlation. J Modern Appl Stat Methods **1**, 19–23 (2002)
3. Fisk, P.R.: The graduation of income distributions. Econometrika **29**, 171–185 (1961)
4. Goria, M.N.: Some locally most powerful generalized rank tests. Biometrika **67**, 497–500 (1980)
5. Hájek, J., Šidák, Z., Sen, P.K.: Theory of rank tests, 2nd edn. Academic Press, San Diego (1999)
6. Mielke, P.W., Johnson, E.S.: Three-parameter kappa distribution maximum likelihood estimates and likelihood ratio tests. Mon Weather Rev **101**, 701–709 (1973)
7. Marozzi, M.: Some notes on the location-scale Cucconi test. J Nonparametric Stat. **21**, 629–647 (2009)
8. Marozzi, M.: Levene type tests for the ratio of two scales. J Stat. Comput. Simul **81**, 815–826 (2011)
9. Moreau, T., Maccario, J., Lellouch, J., Huber, C.: Weighted log rank statistics for comparing two distributions. Biometrika **79**, 195–198 (1992)
10. Murakami, H.: Lepage type statistic based on the modified Baumgartner statistic. Comput. Stat. Data Anal **51**, 5061–5067 (2007)
11. Pandit, P.V.: Locally most powerful and other rank tests for independence - with a contaminated weighted alternative. Metrika **64**, 379–387 (2006)
12. Rosenbaum, P.R.: Confidence intervals for uncommon but dramatic responses to treatment. Biometrics **63**, 1164–1171 (2007)
13. Runde, R.: Locally most powerful two-sample rank tests for Lévy distributions. Stat. Papers **39**, 179–188 (1998)
14. Shoukri, M.M., Mian, I.U.M., Tracy, D.S.: Sampling Properties of Estimators of the Log-Logistic Distribution with Application to Canadian Precipitation Data. Can. J. Stat. **16**, 223–236 (1988)
15. Tamura, R.: On a modification of certain rank tests. Ann. Math. Stat. **34**, 1101–1103 (1963)

15

Numerical calculation of the Riemann zeta function at odd-integer arguments: a direct formula method

Qiang Luo · Zhidan Wang

Abstract In this article, we introduce a recurrence formula which only involves two adjacent values of the Riemann zeta function at integer arguments. Based on the formula, an algorithm to evaluate ζ-values (i.e., the values of Riemann zeta function) at odd integers from the two nearest ζ-values at even integers is posed and proved. The behavior of the error bound is $O(10^{-n})$ approximately where n is the argument. Our method is especially powerful for the calculation of Riemann zeta function at large argument, while for smaller ones, it can also reach spectacular accuracies such as more than ten decimal places.

Keywords Riemann zeta function · Numerical analysis · Algorithm

Mathematics Subject Classification Primary 11Y16 · Secondary 11M06

Introduction

Zeta functions of various kinds, such as Hurwitz zeta function, Epstein zeta function and Dirichlet L-function, are all-pervasive objects in modern mathematics, especially in analytical number theory, and among which the prototype zeta function is the famous Riemann zeta function. It is classically defined as the sum of the infinite series [1–3]

$$\zeta(s) = \sum_{n=1}^{\infty} \frac{1}{n^s} \tag{1}$$

with the complex variable $s = \sigma + it$. Specially, the series converges if $\sigma = \mathrm{Re}\, s > 1$. We can extend $\zeta(s)$ from s with $\mathrm{Re}\, s > 1$ to s with $\mathrm{Re}\,(s) > 0, s \neq 1$ by the following formula

$$\eta(s) = \sum_{n=1}^{\infty} \frac{(-1)^{n+1}}{n^s} = \left(1 - 2^{1-s}\right)\zeta(s) \tag{2}$$

where $\eta(s)$ is the Dirichlet eta function or alternating eta function.

Historically, people prefer to study the closed form of the Riemann zeta function at positive integer arguments in that those special values seem to dictate the properties of the objects they associated. In condensed matter physics for instance, the famous Sommerfeld expansion, which is useful for the calculation of particle number and internal energy of electrons, involves Riemann zeta function at even integers [4], while the spin–spin correlation functions of isotropic spin-1/2 Heisenberg model are expressed by $\ln 2$ and Riemann zeta functions with odd-integer arguments [5]. It was, without doubt, a profound discovery of Euler in 1736 to work out the prolonged Basel problem [6]

$$\zeta(2) = \frac{\pi^2}{6} \tag{3}$$

superbly. It is well known that for positive even-integer arguments, the Riemann zeta function can be expressed explicitly as [7]

Q. Luo and Z. Wang have contributed equally to this work.

Q. Luo (✉)
Department of Physics, Renmin University of China,
Zhongguancun Street, Beijing 100872, China
e-mail: qiangluo@ruc.edu.cn

Z. Wang
School of Mathematical Science, Yangzhou University,
Siwangting Road, Yangzhou 225002, China
e-mail: zhitanwang@gmail.com

$$\zeta(2n) = \frac{(-1)^{n+1}(2\pi)^{2n}}{2(2n)!}B_{2n} \tag{4}$$

in terms of the Bernoulli numbers B_n. On the contrary, however, the explicit formula for Riemann zeta function at odd values is difficult if not fundamentally impossible to obtain. Euler himself once conjectured that $\zeta(2n+1) = c(n)\pi^{2n+1}$ and c involve the irrational constant $\eta(1) = \ln 2$ [8]. This suggests that Riemann zeta function at odd integers produces a recurrence relation that is self-recursive. Even up to now, for positive odd-integer arguments, the Riemann zeta function can only be expressed by series and integral [see (35) and (36) for detail]. One possible integral expression is [9]

$$\zeta(2n+1) = \frac{(-1)^{n+1}(2\pi)^{2n+1}}{2(2n+1)!}\int_0^1 B_{2n+1}(x)\cot(\pi x)dx \tag{5}$$

where $B_{2n+1}(x)$ are Bernoulli polynomials. A relevant aspect is that, for Riemann zeta function, the celebrated Goldbach–Euler theorem[10] assumes the elegant form

$$\sum_{n=2}^{\infty} \mathbf{frac}(\zeta(n)) = 1, \tag{6}$$

where $\mathbf{frac}(x) = x - [x]$ denotes the fractional part of the real number x. It turns out that

$$\sum_{n=1}^{\infty} \mathbf{frac}(\zeta(2n)) = \frac{3}{4}, \sum_{n=1}^{\infty} \mathbf{frac}(\zeta(2n+1)) = \frac{1}{4}. \tag{7}$$

Indeed, the formulas (4) and (5), along with (7) do reveal somewhat similarity for the values of Riemann zeta function at even and odd arguments. Meanwhile, the calculation of Riemann zeta function and related series is a hot topic in computational mathematics. The traditional methods are Euler–Maclaurin formula and Riemann–Siegel formula, and algorithms are still being developed in earnest ever since [11–14]. Typically, a particular numerical method is limited to a special domain. Therefore, when concentrating on Riemann zeta function at odd integers, a special method should be constructed in view of the connection of Riemann zeta function values between odd and even integers.

In this paper, we mainly obtain a recurrence formula (22) relating to the Riemann zeta function and based on which we construct an algorithm for the calculation of the Riemann zeta function at odd integers. In addition, numerical calculation implies that the algorithm can reach considerable accuracies with small odd-integer arguments, not to speak of larger ones. Quantificationally, the behavior of the error bound is $O(10^{-n})$ where n is the argument.

Notations and preliminaries

We begin by recalling the definition of the Bernoulli polynomials $B_n(x)$ and their basic properties in a nutshell to render the paper essentially self-contained. The generating function of the Bernoulli polynomials $B_n(x)$ is [1–3]

$$\frac{te^{tx}}{e^t - 1} = \sum_{n=0}^{\infty} \frac{B_n(x)}{n!}t^n. \tag{8}$$

Taking a derivative with respect to x on both sides of (8), we find that

$$B_n'(x) = nB_{n-1}(x). \tag{9}$$

Bernoulli polynomials can also be expressed explicitly from Bernoulli numbers

$$B_n(x) = \sum_{k=0}^{n} \binom{n}{k} B_k x^{n-k}. \tag{10}$$

For convenience, we introduce two kinds of reduced Bernoulli numbers (RBNs), one relates to the even-labeled Bernoulli numbers (denoted by $+$)

$$B_n^+ = (-1)^{n+1}B_{2n}, \tag{11}$$

and another relates to the odd-labeled Bernoulli polynomials (denoted by $-$)

$$B_n^- = (-1)^{n+1}\int_0^1 B_{2n+1}(x)\cot(\pi x)dx. \tag{12}$$

In this section, we will demonstrate the asymptotic representation of the two kinds of RBNs in a uniform framework and establish their integral representation subsequently.

Asymptotic representations of RBNs

The asymptotic expressions of Bernoulli polynomials at even and odd subscript are, respectively [17]

$$(-1)^{n+1}\frac{(2\pi)^{2n}}{2(2n)!}B_{2n}(x) \sim \cos(2\pi x), \tag{13a}$$

$$(-1)^{n+1}\frac{(2\pi)^{2n+1}}{2(2n+1)!}B_{2n+1}(x) \sim \sin(2\pi x). \tag{13b}$$

Moreover, the Bernoulli polynomials can also be expressed in a stronger form based on Fourier sine and cosine series expansion [18]

$$B_{2n}(x) = (-1)^{n+1}\frac{2(2n)!}{(2\pi)^{2n}}\sum_{k=1}^{\infty}\frac{\cos(2\pi kx)}{k^{2n}}, \tag{14a}$$

$$B_{2n+1}(x) = (-1)^{n+1}\frac{2(2n+1)!}{(2\pi)^{2n+1}}\sum_{k=1}^{\infty}\frac{\sin(2\pi kx)}{k^{2n+1}}. \tag{14b}$$

Obviously, (13a, b) is just the corollary of (14a, b). Joining together we, therefore, obtain the asymptotic behavior of RBNs

$$B_n^+ = (-1)^{n+1} B_{2n}(0) \sim \frac{2(2n)!}{(2\pi)^{2n}} \tag{15a}$$

$$B_n^- = (-1)^{n+1} \int_0^1 B_{2n+1}(x)\cot(\pi x)\mathrm{d}x \sim \frac{2(2n+1)!}{(2\pi)^{2n+1}}. \tag{15b}$$

where we use the fact that $\int_0^1 \sin(2\pi x)\cot(\pi x)\mathrm{d}x = 1$.

Integral representations of RBNs

Let us consider two auxiliary integrals [7]

$$I_c(n,m) = \int_0^1 B_{2n}(t)\cos(m\pi t)\mathrm{d}t \tag{16a}$$

$$I_s(n,m) = \int_0^1 B_{2n+1}(t)\sin(m\pi t)\mathrm{d}t \tag{16b}$$

with m and n are integers and $n \geq 1$. Specially, when $n = 1$, direct computation shows that

$$I_c(1,m) = \int_0^1 \left(t^2 - t + \frac{1}{6}\right)\cos(m\pi t)\mathrm{d}t = \begin{cases} 0, & m = 1,3,5,\ldots \\ \frac{2!}{(m\pi)^2}, & m = 2,4,6,\ldots \end{cases} \tag{17a}$$

$$I_s(1,m) = \int_0^1 \left(t^3 - \frac{3}{2}t^2 + \frac{1}{2}t\right)\sin(m\pi t)\mathrm{d}t = \begin{cases} 0, & m = 1,3,5,\ldots \\ \frac{3!}{(m\pi)^3}, & m = 2,4,6,\ldots \end{cases} \tag{17b}$$

By virtue of (9) and integrating by parts twice, readily yields

$$I_c(n,m) = -\frac{(2n)(2n-1)}{(m\pi)^2} I_c(n-1,m) \tag{18a}$$

$$I_s(n,m) = -\frac{(2n+1)(2n)}{(m\pi)^2} I_s(n-1,m). \tag{18b}$$

Combining (17a, b) and (18a, b), we find that

$$I_c(n,m) = \frac{(-1)^{n+1}(2n)!}{(m\pi)^{2n}} \tag{19a}$$

$$I_s(n,m) = \frac{(-1)^{n+1}(2n+1)!}{(m\pi)^{2n+1}} \tag{19b}$$

hold if m is even. Immediately, the integral representations of the two RBNs B_n^+ and B_n^- are

$$B_n^+ \sim 2(-1)^{n+1} I_c(n,2) \tag{20a}$$

$$B_n^- \sim 2(-1)^{n+1} I_s(n,2). \tag{20b}$$

Algorithm to calculate the Riemann zeta function

Mathematically, Riemann zeta function is said to be monotonically decreasing since its values are only falling and never rising with increasing values of s with $s \geq 2$. Besides, $\zeta(2) = \frac{\pi^2}{6}$, $\zeta(+\infty) > 1$, thus $0 < \zeta(s) - 1 < 1$. Analogously, one can show that $0 < \frac{1}{\eta(s)} - 1 < 1$. For brevity, we denote the reciprocal function as below

$$\rho(s) \equiv \mathbf{frac}\left(\frac{1}{\eta(s)}\right) = \frac{1}{\eta(s)} - 1. \tag{21}$$

Now that what we concerned most is the values of the Riemann zeta function at integers for the moment, the asymptotic behavior of the ratio of the reciprocal function (21) at odd integers and even integers interests us. Motivated by (6) and (7), we manage to demonstrate a formula, the so-called recurrence formula (not in a strict sense, though), on condition that the argument is a positive integer. Motivated by the recurrence, we manage to construct an algorithm to compute the Riemann zeta function.

Demonstration of the recurrence formula

Theorem 1 *If n is a positive integer such that $n \geq 1$, the recurrence relation holds*

$$\lim_{n\to\infty} \frac{\rho(2n+1)}{\rho(2n)} = \frac{1}{2}. \tag{22}$$

Proof Using (4) and (5) and the definition of reciprocal function (21), we have

$$\frac{\rho(2n+1)}{\rho(2n)} = \frac{(2n+1)! - (2^{2n}-1)\pi^{2n+1}B_n^-}{(2n)! - (2^{2n-1}-1)\pi^{2n}B_n^+} \frac{B_n^+}{\pi B_n^-} \frac{2^{2n-1}-1}{2^{2n}-1}. \tag{23}$$

Since the limitation of the rightmost term is exactly equal to 1/2 if n is large enough, what we want to prove is

$$\frac{(2n+1)! - (2^{2n}-1)\pi^{2n+1}B_n^-}{(2n)! - (2^{2n-1}-1)\pi^{2n}B_n^+} \frac{B_n^+}{\pi B_n^-} \sim 1 \tag{24}$$

or equivalently

$$\frac{(2n+1)!}{\pi^{2n+1}} \frac{1}{B_n^-} - \frac{(2n)!}{\pi^{2n}} \frac{1}{B_n^+} \sim 2^{2n-1}. \tag{25}$$

From the asymptotic formulae (15a) and (b) of the two kinds of RBNs, we find that, without any difficulty, we have finished the demonstration of the recurrence formula of the Riemann zeta function. □

As a matter of fact, we can extend the validity of (22) from positive integers to positive real numbers straightforward. We, therefore, can obtain the asymptotic behavior of Riemann zeta function as

$$\frac{1}{\zeta(s)} \sim \frac{2^{s-1}-1}{2^{2s-3}}\left(\frac{2}{\zeta(2)}-1\right)+\frac{2^{s-1}-1}{2^{s-1}}. \tag{26}$$

using (22). The application of (26) can be diverse, here we just pick a example relating to prime number theorem. The positive integer x is s-free if and only if in the prime factorization of x, no prime number occurs more than $s-1$. Indeed, if $Q(x,s)$ denotes the number of s-free integers (e.g., 2-free integers being square-free integers) between 1 and x, one can show that[19]

$$Q(x,s)=\frac{x}{\zeta(s)}+O(\sqrt[s]{x}), \tag{27}$$

therefore, we find that the asymptotic density of s-free integers $Q(x,s)/x \sim \frac{1}{\zeta(s)}$ is nothing but (26).

Another intriguing issue is to what degree can (26) reveals its ability to obtain the ζ-values. Figure 1 is thus plotted as follow.

The fact that all the stars "*" lie on the solid curve indicates that (26) may be a suitable candidate for the calculation of Riemann zeta function. The emergence of the abnormal slope between $s=3$ and $s=4$ in the inserted figure, however, implies that any ζ-value obtained from its nearest neighbors should be much more accuracy. We, therefore, come up with a satisfactory proposal which is postponed until next subsection.

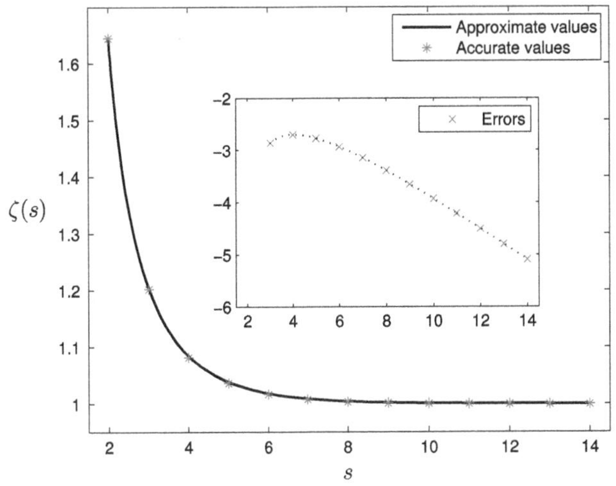

Fig. 1 Asymptotic behavior of Riemann zeta function. The *solid line* represents the approximate values ($\zeta^{\mathrm{ap}}(s)$) obtained by (26), while the *stars* represent the accurate values ($\zeta^{\mathrm{ac}}(s)$) when s is an integer. The *crosses* in the inserted figure indicate the base 10 logarithm of the absolute errors($\epsilon(s)=\lg\left(|\zeta^{\mathrm{ap}}(s)-\zeta^{\mathrm{ac}}(s)|\right)$) at integers

Basic ideas for the algorithm

Abundant methods to evaluate the $\zeta(2n)$ have appeared in the mathematical literatures from now and then ever since Euler's seminal work. In contract, the explicit formula for odd-argument ζ-values remains to be an open problem though some results shed light on it [15, 16]. By analogy to $\zeta(2n)$, several authors have established the series and integral representations of $\zeta(2n+1)$, which, to some degree, provides some perspectives on the difficulty of evaluating $\zeta(2n+1)$ as opposed to $\zeta(2n)$. From the viewpoint of numerical method, one natural way to construct the corresponding algorithm to evaluate the odd-argument Riemann zeta function is by virtue of the even-argument ζ-values near to them. In the current paper, only the two nearest ζ-values are taken into consideration currently for simplicity. When n is large enough, (22) can be rewritten as

$$\rho^l(2n+1) \sim \frac{1}{2}\rho(2n) \tag{28a}$$

$$\rho^r(2n+1) \sim 2\rho(2n+2) \tag{28b}$$

where $\rho^l(2n+1)$ and $\rho^r(2n+1)$ represent two different representations of the asymptotic behavior of $\rho(2n+1)$. Judging by appearance, One can use any of the formula above to calculate the Riemann zeta function at odd integers. When considering that those two formulae give the upper and lower bound of the zeta-values at odd integers (see Theorem 2), we come up with the idea that we can combine them together by a special method. It happens to us that there may exist a somewhat mysterious map from $\zeta(2n)$ and $\zeta(2n+2)$ to $\zeta(2n+1)$, which will ensure us to obtain the approximation values of $\zeta(2n+1)$ with higher precision. Let us give a proposition relating to Dirichlet eta function first before we move forward to give another theorem.

Lemma 1 *If n is a positive integer such that $n \geq 1$, the two inequalities hold*

$$\frac{4}{\eta(2n+2)}-\frac{1}{\eta(2n)}>3 \tag{29}$$

$$\eta(2n)>\frac{2^{2n-1}-2}{2^{2n-1}-1} \tag{30}$$

Those two inequalities are quite new to the authors because we have not seen them in any literature or monograph before. However, we are not intended to give the details here since the demonstration is rather elementary. The theorem below holds once we take advantage of Lemma 1.

Theorem 2 *If n is a positive integer such that $n \geq 1$, the inequality holds*

$$\zeta^l(2n+1) > \zeta^r(2n+1) > 1 \tag{31}$$

where $\zeta^l(2n+1)$ and $\zeta^r(2n+1)$ correspond to $\rho^l(2n+1)$ and $\rho^r(2n+1)$, respectively.

Since Riemann zeta function is a monotonic decreasing function, the exact value $\zeta(2n+1)$ is just between $\zeta^l(2n+1)$ and $\zeta^r(2n+1)$ for any given positive integer n. For the benefit of accuracy, we regard the geometric mean values of (28a) and (b) as the approximate values of the reciprocal function $\rho(2n+1)$, namely

$$\rho(2n+1) \approx \sqrt{\rho(2n)\rho(2n+2)} \tag{32}$$

which is the most valuable ingredient of our algorithm.

The basic steps for the calculation of $\zeta(2n+1)$ are presented as follow. First, $\rho(2n)$ and $\rho(2n+2)$ should be calculated from (4), (2) and (21) in sequence. Second, the value of $\rho(2n+1)$ is ready to be obtained in light of (32). Lastly, the ultimate aim, i.e., $\zeta(2n+1)$ is just at hand from (21) and (2), reversely. Our algorithm does not bother circulation of any kind, it just looks like a formula, therefore, we refer it as the direct formula method.

To start our method, we need to know some ζ-values at even integers. For example, $\zeta(2)$ and $\zeta(4)$ should be available to get $\zeta(3)$. We can obtain $\zeta(2n)$ through (4) systematically for small argument. However, it is almost impossible to obtain Bernoulli numbers by the ordinary recursive methods thus we hardly know the values of $\zeta(2n)$, if the argument is large enough. Many methods for computing Bernoulli numbers have been invented. David Harvey introduced an efficient multimodular algorithm [20] which ensures us to obtain the Bernoulli numbers B_n at $n = 10^8$. However, one can also use the intrinsic function **Zeta**[*s*] in Mathematica since it is also based on an efficient algorithm. Therefore, for convenience, our computation platform is mainly on Mathematica and we regard those values as benchmarks.

Calculation of Riemann zeta function at odd integers

The calculation of Riemann zeta function plays an essential role in the study of number theory and associated subjects such as statistical physics and condensed matter physics. Various approaches to accomplish this task have been proposed [11–13, 21], especially for the evaluation of zeta function at integer arguments or in the critical strip (for the computation of Riemann's zeros). Most of the methods available consist of using integral forms of some particular functions or recursive series forms. Quite recently, Babolian et al. transform $\zeta(s)$ to some appropriate integral forms and introduce a method to compute the Riemann zeta function based on Gauss–Hermite and Gauss–Laguerre quadratures [11]. Numerical result shows that 20 points are capable of producing an accuracy of seven-decimal place for small arguments. Besides, many rapidly converging series for $\zeta(2n+1)$ have been introduced by Srivastava in a review article [13] and by other authors [12, 14]. In this section, we first give some numerical examples according to our method to illustrate its accuracy, then we compare our method to two selected ones to show that our method is especially powerful to calculate the ζ-values at large odd-integer arguments.

Numerical test and error bound of the algorithm

We regard the ζ-values obtained by Mathematica as benchmarks. The result of the Riemann zeta function at odd integers with $n = 1, 2, \ldots, 10$ obtained by our method (approximate value) is presented in Table 1. The accuracy values and the absolute errors are also presented at the same time.

Table 1 tells us that, the idea that making the geometric mean instead of any of the upper or lower bound (see Theorem 2) be the best estimate of the Riemann zeta function dramatically reduces errors and satisfactory accuracy such as twelve decimal places in the tenth odd argument of the Riemann zeta function can be achieved. It is interesting to find that only the Apéry's constant $\zeta(3)$ sightly larger than the approximate value obtained by our method. It is also funny to see the errors present an upside-down stair configuration, which implies that the error declines about ten times as long as the argument n increase 1.

In Table 2, we present the absolute errors $\epsilon(n)$ versus n, for the purpose of exploring the error bound when the argument n is large enough.

Table 1 Comparison between accurate values $\zeta^{\mathbf{ac}}(2n+1)$ and approximate values $\zeta^{\mathbf{ac}}(2n+1)$

n	$\zeta^{\mathbf{ap}}(2n+1)$	$\zeta^{\mathbf{ac}}(2n+1)$	Errors
1	1.201335874256	1.202056903160	−0.007210289040
2	1.036972837734	1.036927755143	0.000045082590
3	1.008365209797	1.008349277382	0.000015932415
4	1.002011075857	1.002008392826	0.000002683031
5	1.000494555053	1.000494188604	0.000000364486
6	1.000122758824	1.000122713348	0.000000045476
7	1.000030593607	1.000030588236	0.000000005371
8	1.000007637815	1.000007637198	0.000000000617
9	1.000001908283	1.000001908213	0.000000000070
10	1.000000476941	1.000000476933	0.000000000008

Table 2 The errors of $\zeta(2n+1)$ based on our method

n	Errors	n	Errors
1×10^2	1.05×10^{-97}	1×10^4	1.04×10^{-9544}
2×10^2	3.94×10^{-193}	2×10^4	3.92×10^{-19087}
5×10^2	2.10×10^{-479}	5×10^4	2.08×10^{-47714}
1×10^3	1.59×10^{-956}	1×10^5	1.56×10^{-95426}
2×10^3	9.09×10^{-1911}	2×10^5	8.75×10^{-190851}
5×10^3	1.70×10^{-4773}	5×10^5	1.55×10^{-477123}

It is clear that, from Table 2, the error is of the order $O(10^{-n})$ approximately. Using least square method, we notice that

$$\lg(\epsilon(n)) = -0.9542n - 1.6884. \tag{33}$$

This formula suggests that when the argument of Riemann zeta function is large enough, our algorithm should be powerful enough to obtain the ζ-values at odd integers.

Comparison with the existed methods

In this subsection, we aim to compare our algorithm with the already existed ones, namely the Gauss–Hermite quadrature (Integral method, see [11], Corollary 3.1) and rapid converging series (Series method, see [13, eq.(3.30)]). The Gauss–Hermite quadrature formula has the form [11]

$$\int_{-\infty}^{\infty} f(x)e^{-x^2}\mathrm{d}x = \sum_{k=1}^{N} w_k f(x_k) + R_N \tag{34}$$

where x_k is one of the zeros of $H_N(x)$, the Hermite polynomial of degree N, and $w_k = -\frac{2^{N+1}N!\sqrt{\pi}}{H_N'(x_k)H_{N+1}(x_k)}$ is the corresponding weight. $R_N = \frac{N!\sqrt{\pi}}{2^N(2N)!}f^{2N}(\eta), \eta\in(-\infty,\infty)$ is, obviously, the error bound of the above integral. Riemann zeta function is such an amazing function that it can be transformed into [11]

$$\zeta(s) = \frac{\int_{-\infty}^{\infty}\left(|x|^{2s-1}\mathrm{e}^{-x^2}/(1-\mathrm{e}^{-x^2})\right)\mathrm{d}x}{\int_{-\infty}^{\infty}|x|^{2s-1}\mathrm{e}^{-x^2}\mathrm{d}x} \tag{35}$$

whose numerator and denominator are of the form presented in (34). Among all the series representations of Riemann zeta function, the series below

$$\zeta(2n+1) = \frac{(-1)^{n-1}(2\pi)^{2n}}{(2n)![2^{2n}(2n-3)-2n+1]}\cdot \times\left[\sum_{m=1}^{n-1}(-1)^m\binom{2n-1}{2m-2}\frac{(2m)!(2^{2m}-1)}{(2\pi)^{2m}}\zeta(2m+1) + 2\sum_{k=0}^{\infty}\frac{\zeta(2k)}{(2k+2n-1)(k+n)(2k+2n+1)2^{2k}}\right] \tag{36}$$

converges most rapidly as pointed out by Srivastava [13]. When $n=1$ for instance, the error bound $R_N^{(\mathrm{s})}$ of the N-th partial sum of the infinite series in (36) satisfies

$$\begin{aligned}|R_N^{(\mathrm{s})}| &= \frac{4\pi^2}{15}\sum_{k=N+1}^{\infty}\frac{\zeta(2k)}{(2k+1)(k+1)(2k+3)4^k}\\ &< \frac{4\pi^2}{15}\frac{\zeta(2N+2)}{(2N+3)(N+2)(2N+5)}\sum_{k=N+1}^{\infty}\frac{1}{4^k}\\ &= \frac{4\pi^2}{45}\frac{1}{(2N+3)(N+2)(2N+5)(4^N-\frac{1}{2})}\end{aligned} \tag{37}$$

where we have used the fact that $\zeta(s) < \frac{1}{1-2^{1-s}}$ since $\eta(s) < 1$. If $N=25$, the error bound is $|R_{25}^{(\mathrm{s})}| < 1.0\times10^{-20}$, which is superior to other rapid series $|R_{25}| < 0.9\times10^{-18}$ as noted in [8, 12]. Specially, when N is larger than some typical numbers, the asymptotic behavior of (37) reads

$$\lg(|R_N^{(\mathrm{s})}|) \sim -2\lg 2(N+1) - 3\lg N. \tag{38}$$

The accuracy of the latter two methods relies on the number of zeros (denoted as N_1) of the associated polynomial (in this occasion, it is Hermite polynomial) and the terms (denoted as N_2) of partial sum of the infinite series, respectively. We set two integers to be the same value, i.e., $N_1 = N_2 = 25$ since the corresponding methods are both efficient as have been declared by many Mathematicians.

The behaviors of the error bound of integral method and series method, as can be seen from Table 3, are totally different. When the argument increases, the errors of the former decrease exponential from a high level, while the latter maintain at a nearly constant low level despite of the variation of n. Our method exhibits the worst results for small arguments, but the errors decrease dramatically with argument increasing. It outstrips integration method and series method before $n=12$ and $n=21$, respectively. Our method

Table 3 Errors of three different methods for $\zeta(2n+1)$

n	Integral method	Series method ($\times10^{-20}$)	Our method
3	2.42×10^{-7}	3.17434484	1.50×10^{-5}
6	1.21×10^{-10}	3.14630746	4.52×10^{-8}
9	4.31×10^{-11}	3.14592124	4.99×10^{-11}
12	1.36×10^{-12}	3.14591532	9.79×10^{-14}
15	7.40×10^{-13}	3.14591522	1.35×10^{-16}
18	6.20×10^{-13}	3.14591522	1.85×10^{-19}
21	8.46×10^{-14}	3.14591522	2.54×10^{-22}
24	7.99×10^{-15}	3.14591522	3.48×10^{-25}
27	7.77×10^{-16}	3.14591522	4.78×10^{-28}
30	1.11×10^{-16}	3.14591522	6.55×10^{-31}

superior to them absolutely afterwards. To reach the accuracy obtained by our method, the number of nodes and terms in the above two methods should be augmented largely. In the series method for instance, the terms of the order n in the infinity series should be included according to (33) and (38). Obviously, it is almost impossible to carry on within the limited CPU time when n is an astronomical number.

Conclusion

In summary, we first introduce two kinds of reduced Bernoulli numbers (RBNs) and prove their asymptotic behaviors in an uniform framework, and their series and integral representations are available at the same time. What is more, we discover and prove a recurrence formula (22) of the Riemann zeta function original and construct an algorithm to evaluate the Riemann zeta function at odd integers based on it. The idea of our method is quiet simple, but it turns out to be a competent algorithm. The behavior of the error bound $\epsilon(n)$ is governed by $\lg(\epsilon(n)) = -0.9542n - 1.6884$ or $\epsilon(n) = O(10^{-n})$ approximately, which, of course, suggests that our method is especially suit for the calculation of ζ-values at large odd-integer arguments. Therefore, our results can also work as benchmarks to test the accuracy of other related algorithms. However, more works should be carried on to improve the accuracy at small arguments in future. Remarkably, the recurrence formula (22) is likely to act as a touchstone to explore the closed form of the Riemann zeta function at positive integers since it witnesses the connection between ζ-values at odd integers and even integers.

Acknowledgments The authors would like to show their appreciation to Junesang Choi, Yong Lin and Changle Liu for some useful discussions, and express their thanks to Jinlin Liu and Jiurong Han for their suggestions. Especially, they wishes to thank the anonymous referees of this paper for valuable suggestions which have improved the presentation of the paper.

References

1. Titchmarsh, E.C.: The theory of the Riemann Zeta function. Claredon Press, Oxford (1986)
2. Beals, R., Wong, R.: Special functions. Cambridge University Press, Cambridge (2010)
3. Edwards, H.M.: Riemann zeta function. Dover Publications Inc., New York (2001)
4. Ashcroft, N.W., Mermin, N.D.: Solid state physics. Saunders College, Philadelphia (1976)
5. Shiroishi, M., Takahashi, M.: Exact calculation of correlation functions for Spin-1/2 Heisenberg chain. J. Phys. Soc. Jpn. **74**, 47–52 (2005)
6. Benko, D.: The Basel problem as a telescoping series. College Math. J. **43**(3), 244–250 (2012)
7. Ciaurri, O., Navas, L.M., Ruiz, F.J., Varona, J.L.: A simple computation of $\zeta(2k)$ by using Bernoulli polynomials and a telescoping series. arXiv: 1209.5030v1
8. Scheufens, E.: From Fourier series to rapidly convergent series for zeta(3). Math. Mag. **84**(1), 26–32 (2011)
9. Cvijovic, D., Klinowski, J.: Integral representations of the Riemann zeta function for odd-integer arguments. J. Comput. Appl. Math. **142**(2), 435–439 (2002)
10. Choi, J., Srivastava, H.M.: Series involving the zeta functions and a family of generalized Goldbach-Euler deries. Am. Math. Mon. **121**(3), 229–236 (2014)
11. Babolian, E., Hajikandi, A.A.: Numerical computation of the Riemann zeta function and prime counting function by using Gauss-Hermite and Gauss-Laguerre quadratures. Int. J. Comput. Math. **87**(15), 3420–3429 (2010)
12. Choi, J.: Rapidly converging series for $\zeta(2n+1)$ from Fourier seires. Abst. Appl. Anal. 1–9 (2014). doi:10.1155/2014/457620
13. Srivastava, H.M.: Some simple algorithms for evaluations and representations of the Riemann zeta function at positive integer arguments. J. Math. Anal. Appl. **246**(2), 331–351 (2000)
14. Lima, F.M.S.: A simpler proof of a Katsurada's theorem and rapidly converging series for $\zeta(2n+1)$ and $\beta(2n)$. arXiv: 1203.5660v2
15. Dancs, M.J., He, T.X.: An Euler-type formula for $\zeta(2n+1)$. J. Num. Theory **118**(2), 192–199 (2006)
16. Dancs, M.J., He, T.X.: Numerical approximation to $\zeta(2n+1)$. J. Comput. Appl. Math. **196**(1), 150–154 (2006)
17. Dilcher, K.: Asymptotic behavior of Bernoulli, Euler, and generalized Bernoulli polynomials. J. Approx. Theory **49**(4), 321–330 (1987)
18. Lopez, J.L., Temme, N.M.: Large degree asymptotics of generalized Bernoulli and Euler polynomials. J. Math. Anal. Appl. **363**(1), 197–208 (2010)
19. Gibson, J.: The distribution of r-free numbers in arithmetic progressions. Int. J. Number Theory **10**(3), 559–563 (2014)
20. Harvey, D.: A multimodular algorithm for computing bernoulli numbers. Math. Comput. **79**(272), 2361–2370 (2010)
21. Borwein, J.M., Bradley, D.M., Cradall, R.E.: Computational strategies for the Riemann zeta function. J. Comput. Appl. Math. **121**(1), 247–296 (2000)

16

An improved adaptation of homotopy analysis method

Maasoomah Sadaf[1] · Ghazala Akram[1]

Abstract An improved adaptation of the well-known homotopy analysis method (HAM) is proposed to approximate the solutions of strongly nonlinear differential problems in terms of a rapidly convergent series. The proposed method involves simpler integrals and less computations than the standard HAM. The method is illustrated using different numerical examples. The comparative analysis confirms the applicability and efficiency of the proposed technique.

Keywords Homotopy analysis method · Improved adaptation · Series solution

Introduction

Liao [1] proposed an approximate analytical technique namely homotopy analysis method (HAM) built on the concept of homotopy for the solutions of nonlinear differential equations. The homotopy analysis is not only an efficient method to solve highly nonlinear differential equation problems but also allows great freedom to choose the initial approximation and is highly flexible in many respects so that it might overcome restrictions of perturbation techniques and other non-perturbation methods. The great freedom and flexibility of the HAM has inspired many mathematicians to attack HAM in search of better numerical techniques. Marinca and Herisanu [2] proposed optimal homotopy asymptotic method to investigate the solutions of nonlinear equations arising in heat transfer. Motsa et al. [3] introduced a new spectral-homotopy analysis method for solving a nonlinear second order BVP. Homotopy analysis method has been successfully applied to investigate the solutions of integral equations [4]. Abbasbandy and Shivanian [5] proposed predictor homotopy analysis method (PHAM) to predict the multiplicity of the solutions of some nonlinear boundary value problems. Shivanian et al. [6] used PHAM to study a case of boundary layer flows. They proved the existence of multiple solutions and also calculated the approximate solutions. Vosoughi et al. [7] also used PHAM to obtain two approximate solutions in a study of nonlinear reactive transport model. Abbasbandy et al. [8] studied the role of convergence-control parameter in homotopy analysis method. Motsa et al. [9] proposed an improved spectral homotopy analysis method for MHD flow in a semi-porous channel. Shaban et al. [10] proposed a Tau modification of the homotopy analysis method to study the magneto-hydrodynamic squeezing flow between two parallel disks with suction or injection. Shivanian and Abbasbandy [11] discussed PHAM for two points second order boundary value problems. Homotopy analysis method has been applied in a study of combined conduction–convection–radiation heat transfer [12]. Odibat and Bataineh constructed homotopy polynomials by introducing an adaptation of HAM [13]. In the present paper, an improved adaptation of homotopy analysis method is proposed for the numerical solution of differential equations.

Higher order boundary value problems are studied due to their mathematical importance and applications in different physical phenomena. Mathematical modeling of AFTI-F16 fighters involves ninth order differential

✉ Ghazala Akram
toghazala2003@yahoo.com

Maasoomah Sadaf
maasoomahsadaf@yahoo.com

1 Department of Mathematics, University of the Punjab, Lahore 54590, Pakistan

equation [14]. Ninth order boundary value problems also arise in the study of of astrophysics, hydrodynamic, and hydromagnetic stability [15, 16]. The study of hydrodynamic and hydromagnetic stability also involves eighth and tenth order boundary value problems [17]. The mathematical importance of boundary value problems of higher order motivates to study different mathematical techniques to obtain the solutions of these problems. Siddiqi and Twizell [18–21] presented the solutions of 6th, 8th, 10th, and 12th order boundary value problems using 6th, 8th, 10th, and 12th degree splines, respectively. Inc and Evans [22] used Adomian decomposition method to approximate solutions of eighth order boundary value problems. Siddiqi and Akram [23–27] presented the solutions of 5th, 6th, 8th, 10th, and 12th order boundary value problems using non-polynomial spline techniques. Hassan and Erturk [28] applied differential transformation method to obtain the solution of some linear and nonlinear higher order boundary value problems. Siddiqi and Iftikhar [29] used the variational iteration method to approximate the solution of seventh order boundary value problems in terms of a convergent series. Hakeemullah et al. [30] used the optimal homotopy asymptotic method (OHAM) for approximating the solution of modified Kawahara equations. The proposed method is numerically illustrated for the solutions of different higher order boundary value problems. Viswanadham and Ballem [31] used Galerkin method with septic B-splines to approximate the solution of tenth order boundary value problems.

Homotopy analysis method

Homotopy analysis method is an analytical technique which can be used to compute the solutions of linear and nonlinear differential equations. The solution is obtained in terms of a convergent series. Consider a nonlinear differential equation

$$N[y(x)] = 0, \quad x \in \Theta, \tag{1}$$

where N is a nonlinear operator, x is an independent variable, $y(x)$ is an unknown function, and Θ is the interval of domain. A homotopy $Y(x, p)$ can be constructed with an embedding parameter $p \in [0,1]$ by

$$(1-p)L[Y(x,p)-y_0(x)]-phH(x)N[Y(x,p]=0, \quad x\in\Theta, \tag{2}$$

where h is auxiliary parameter, $H(x)$ is an auxiliary function, and L is auxiliary linear operator. The homotopy analysis allows great freedom to select h, $H(x)$, and L. Also there is larger freedom to choose the initial approximation $y_0(x)$. For suitable selection, the unknown function $Y(x, p)$ can be determined. Moreover, the mth-order derivative of $y_0(x)$ with respect to the embedding parameter exists at $p=0$ for all positive integral values of m. This quantity is the so-called mth-order deformation derivative. Application of Taylor's theorem gives the series expansion of $Y(x, p)$ as

$$Y(x,p) = y_0(x) + \sum_{m=1}^{\infty} y_m(x)p^m, \tag{3}$$

where $y_m(x)$ is obtained by dividing the mth-order deformation derivative by $m!$. For suitably chosen $H(x)$, h, L, and $y_0(x)$, the series converges to $y(x)$ at $p = 1$. Moreover, $y_m(x), m = 1,2,3,\ldots$ can be calculated using the mth-order deformation equation

$$L[y_m(x) - \chi_m y_{m-1}(x)] - hH(x)R_m(\underline{y}_{m-1}) = 0, \quad x \in \Theta, \quad p \in [0,1], \tag{4}$$

where

$$R_m(\underline{y}_{m-1}) = \frac{1}{(m-1)!} \left. \frac{\partial^{m-1} N(Y(x,p))}{\partial p^{m-1}} \right|_{p=0},$$
$$\underline{y}_{m-1}(x) = \{y_0(x), y_1(x), \ldots, y_{m-1}(x)\}, \tag{5}$$
$$\chi_m = \begin{cases} 0, & m \le 1, \\ 1, & m > 1. \end{cases}$$

An improved adaptation of homotopy analysis method

Recently, Odibat and Bataineh [13] have presented an adaptation of homotopy analysis method for solving strongly nonlinear problems. The technique not only reduces the number of terms involved in each iteration but also overcomes the difficulty faced in solving complicated

Table 1 Numerical results for Example 1

x	Exact solution	Approximate solution	Absolute error
0.0	1.000000	1.000000	0.000000
0.1	0.814354	0.814354	3.648192×10^{-11}
0.2	0.654985	0.654985	6.893886×10^{-10}
0.3	0.518573	0.518573	2.896354×10^{-9}
0.4	0.402192	0.402192	6.203250×10^{-9}
0.5	0.303265	0.303265	8.573576×10^{-9}
0.6	0.219525	0.219525	8.182360×10^{-9}
0.7	0.148976	0.148976	5.222692×10^{-9}
0.8	0.089866	0.089866	1.870414×10^{-9}
0.9	0.040657	0.040657	1.950906×10^{-10}
1.0	0.000000	1.524790×10^{-15}	1.524790×10^{-15}

integrals. The adaptation is based on the assumption that the nonlinear operator N can be expressed as a power series in the dependent variable. The method can be easily implemented on equations of the form

$$N(y) = f(y)$$

but needs some extra calculation for equations of the form

$$N(y) = f(x, y).$$

In the present paper, the adaptation is improved to obliterate the need for extra calculations. The improved adaptation is based on the assumption that a power series expansion of the nonlinear operator N can be expressed as

$$N(y) = \sum_{i=0}^{\infty} c_i x^i, \tag{6}$$

where c_i's are real numbers. Using the fact that HAM yields the solution of Eq. (1) in terms of a power series as

$$Y(x, p) = \sum_{i=0}^{\infty} y_i p^i, \tag{7}$$

The modified homotopy $\hat{Y}(x, p)$ can be constructed with an embedding parameter $p \in [0, 1]$ by

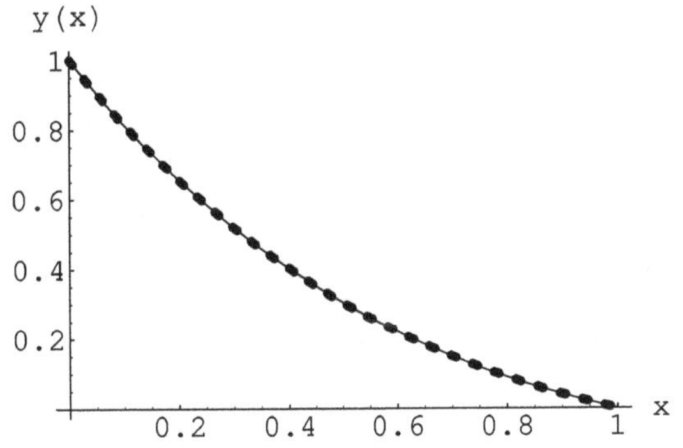

Fig. 1 Comparison of exact solution (*solid line*) and approximate solution (*dashed line*) for Example 1

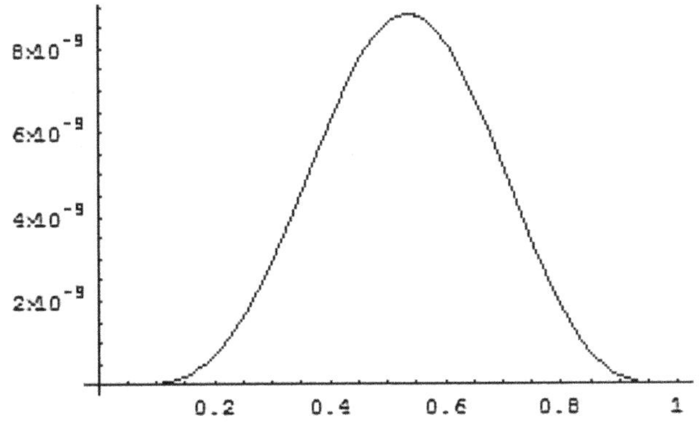

Fig. 2 Absolute errors for Example 1

$$(1-p)L\left[\sum_{i=0}^{\infty} y_i p^i - y_0(x)\right] - phH(x)\sum_{i=0}^{\infty} c_i x^i p^i = 0, \quad x \in \Theta, \tag{8}$$

where h is auxiliary parameter, $H(x)$ is an auxiliary function, and L is auxiliary linear operator. The first few modified higher order deformation equations are expressed as

$$\begin{cases} L[y_1(x)] = hH(x)c_0, \\ L[y_2(x)] = L[y_1(x)] + hH(x)c_1 y_0, \\ L[y_3(x)] = L[y_2(x)] + hH(x)(c_1 y_1 + c_2 y_0^2), \\ L[y_4(x)] = L[y_3(x)] + hH(x)(c_1 y_2 + 2c_2 y_0 y_1 + c_3 y_0^3). \end{cases}$$

In the next section, the proposed method is numerically illustrated using different linear and nonlinear higher order boundary value problems.

Numerical examples

Example 1 For $x \in [0, 1]$, the following ninth order nonlinear boundary value problem is considered

$$\left.\begin{array}{l} y^{(9)}(x) - y(x)y'(x) = e^{-2x}(2 + e^x(x-10) - 3x + x^2), \\ y(0) = 1, \quad y(1) = 0, \\ y'(0) = -2, \quad y'(1) = -e^{-1}, \\ y''(0) = 3, \quad y''(1) = 2e^{-1}, \\ y'''(0) = -4, \quad y'''(1) = -3e^{-1}, \\ y^{(4)}(0) = 5. \end{array}\right\} \tag{9}$$

The analytic solution of this differential system is

$$y(x) = (1-x)e^{-x}.$$

Table 2 Numerical results for Example 2

x	Exact solution	Approximate solution	Absolute error
0.0	1.000000	1.000000	0.000000
0.1	0.994654	0.994654	7.527856×10^{-11}
0.2	0.977122	0.977122	1.566206×10^{-9}
0.3	0.944901	0.944901	7.256635×10^{-9}
0.4	0.895095	0.895095	1.717163×10^{-8}
0.5	0.824361	0.824361	2.627787×10^{-8}
0.6	0.728848	0.728848	2.783580×10^{-8}
0.7	0.604126	0.604126	1.977577×10^{-8}
0.8	0.445108	0.445108	7.908433×10^{-9}
0.9	0.245960	0.245960	9.245274×10^{-10}
1.0	0.000000	$-1.421085 \times 10^{-14}$	1.421085×10^{-14}

Table 3 Comparison of maximum absolute error with DTM

DTM [28]	Proposed method
3.0×10^{-7}	2.8×10^{-8}

Table 4 Numerical results for Example 3

x	Exact solution	Approximate solution	Absolute error
0.0	0.000000	0.000000	0.000000
0.1	0.009983	0.009983	1.565661×10^{-11}
0.2	0.039734	0.039734	3.442625×10^{-10}
0.3	0.088656	0.088656	1.670801×10^{-9}
0.4	0.155767	0.155767	4.110871×10^{-9}
0.5	0.239713	0.239713	6.500192×10^{-9}
0.6	0.338785	0.338785	7.076563×10^{-9}
0.7	0.450952	0.450952	5.142951×10^{-9}
0.8	0.573885	0.573885	2.095353×10^{-9}
0.9	0.704994	0.704994	2.486530×10^{-10}
1.0	0.841471	0.841471	5.107026×10^{-15}

According to basic assumption of HAM, the initial approximation is chosen as

$$y_0(x) = \frac{1}{24e}(24e - 48ex + 36ex^2 - 16ex^3 + 5ex^4 + 660x^5 - 244ex^5 - 1788x^6 + 658ex^6 + 1620x^7 - 596ex^7 - 492x^8 + 181x^8). \tag{10}$$

The first, second, and third order deformation equations are obtained as

$$\begin{cases} L[y_1(x)] = 10hH(x), \\ L[y_2(x)] = L[y_1(x)] - 11hH(x)y_0, \\ L[y_3(x)] = L[y_2(x)] + hH(x)(-11y_1 + 6y_0^2). \end{cases}$$

Solving these differential equations for $y_1(x)$, $y_2(x)$ and $y_3(x)$, the third order approximation to the solution is calculated. Here, the auxiliary function $H(x)$ is taken as $H(x) = 1$ and the linear operator is chosen to be the homogeneous part of the nonlinear operator N.

Table 1 shows the approximate solution and corresponding absolute error values using the proposed method for $h = -0.2$. The comparison of the exact solution and the approximate solution is shown in Fig. 1 and the graph of absolute errors over the interval of domain is shown in Fig. 2.

Example 2 The following ninth order linear boundary value problem is considered, as

Table 5 Numerical results for Example 4

x	Exact solution	Approximate solution	Absolute error
0.0	0.000000	0.000000	0.000000
0.1	0.099465	0.099465	2.339568×10^{-8}
0.2	0.195424	0.195424	1.940541×10^{-7}
0.3	0.283470	0.283470	4.492431×10^{-7}
0.4	0.358038	0.358038	5.339237×10^{-7}
0.5	0.412180	0.412180	3.341755×10^{-7}
0.6	0.437309	0.437309	1.488507×10^{-8}
0.7	0.422888	0.422888	1.607205×10^{-7}
0.8	0.356087	0.356087	1.154905×10^{-7}
0.9	0.221364	0.221364	1.815175×10^{-8}
1.0	0.000000	$-7.105464 \times 10^{-15}$	7.105464×10^{-15}

Table 6 Comparison of maximum absolute errors

Inc and Evans [22]	Proposed method
1.83×10^{-4}	5.3×10^{-7}

$$\left.\begin{array}{l} y^{(9)}(x) - y(x) = -9e^x, \quad x \in [0, 1] \\ y(0) = 1, \quad y(1) = 0, \\ y'(0) = 0, \quad y'(1) = -e, \\ y''(0) = -1, \quad y''(1) = -2e, \\ y'''(0) = -2, \quad y'''(1) = -3e, \\ y^{(4)}(0) = -3. \end{array}\right\} \tag{11}$$

The analytic solution of this differential system is $y(x) = (1 - x)e^x$.

The initial approximation is calculated using the standard HAM, as

$$y_0(x) = 1 - \frac{x^2}{2} - \frac{x^3}{3} - \frac{x^4}{8} + \frac{1}{24}(-1012 + 372e)x^5 + \frac{1}{24}(2642 - 972e)x^6 + \frac{1}{24}(-2316 + 852e)x^7 + \frac{1}{24}(685 - 252e)x^8. \tag{12}$$

First, second, and third order deformation equations are obtained, as

$$\begin{cases} L[y_1(x)] = 8hH(x), \\ L[y_2(x)] = L[y_1(x)] + 9hH(x)y_0, \\ L[y_3(x)] = L[y_2(x)] + hH(x)(9y_1 + 5y_0^2). \end{cases}$$

The third order approximation to the solution is calculated by solving these differential equations for $y_1(x)$, $y_2(x)$ and $y_3(x)$. Here, the auxiliary function $H(x)$ is taken as $H(x) = 1$. For this choice of function, higher accuracy can be achieved without having to go to higher order of approximation. The linear operator is chosen to be the homogeneous part of the nonlinear operator N. The approximate solution and corresponding absolute error values, for $h = -0.2$, are summarized in Table 2.

Table 3 shows that the proposed method gives better results than differential transformation method [28].

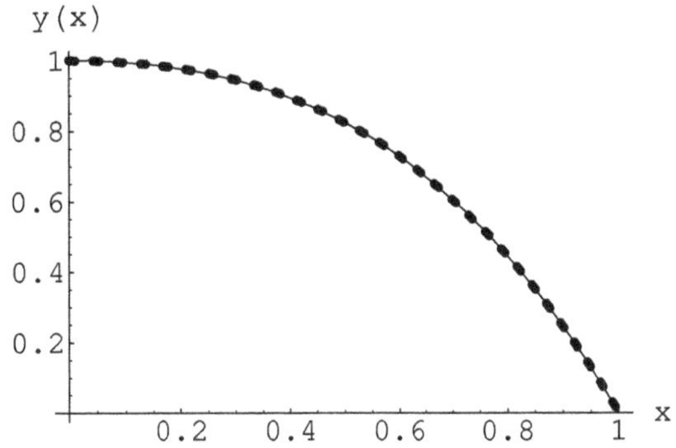

Fig. 3 Comparison of exact solution (*solid line*) and approximate solution (*dashed line*) for Example 2

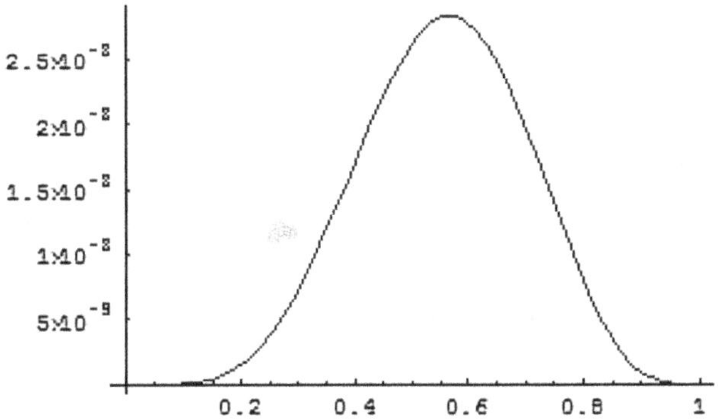

Fig. 4 Absolute errors for Example 2

Example 3 The following ninth order nonlinear boundary value problem is considered as

$$\left.\begin{aligned} &y^{(9)}(x)e^{y(x)} = 0, \qquad x \in [0,1],\\ &y(0) = 0, y(1) = \sin(1),\\ &y'(0) = 0, y'(1) = \cos(1) + \sin(1),\\ &y''(0) = 2, y''(1) = 2\cos(1) - \sin(1)+,\\ &y'''(0) = 0, y'''(1) = -\cos(1) - 3\sin(1),\\ &y^{(4)}(0) = -4. \end{aligned}\right\} \tag{13}$$

The analytic solution of this differential system is

$y(x) = x\sin(x)$.

The initial approximation is calculated according to the standard HAM, as

$$\begin{aligned} y_0(x) = x^2 - \frac{x^4}{6} &+ \frac{1}{6}x^6(264 + 231\cos(1) - 462\sin(1))\\ &+\frac{1}{6}x^8(59 + 59\cos(1) - 108\sin(1))\\ &+\frac{1}{6}x^5(-116 - 89\cos(1) + 195\sin(1))\\ &+\frac{1}{6}x^7(-212 - 201\cos(1) + 381\sin(1)). \end{aligned} \tag{14}$$

The third order approximation to the exact solution is calculated using the improved adaptation proposed in the present paper. The approximate solution values and corresponding absolute errors are shown in Table 4.

The auxiliary function $H(x)$ is taken as $H(x) = 1$. The linear operator is chosen to be the homogeneous part of the nonlinear operator N. Moreover, value of h is chosen as $h = 1$.

Example 4 The following eighth order linear boundary value problem is considered as

Table 7 Comparison of absolute errors for Example 5

x	Exact solution	Approximate solution	Present method	Viswanadham and Ballem [31]
0.1	0.904837	0.904837	6.916689×10^{-14}	6.735325×10^{-6}
0.2	0.818731	0.818731	1.122769×10^{-12}	4.410744×10^{-6}
0.3	0.740818	0.740818	3.966605×10^{-12}	3.629923×10^{-5}
0.4	0.670320	0.670320	6.939005×10^{-12}	4.839897×10^{-5}
0.5	0.606531	0.606531	7.532863×10^{-12}	4.929304×10^{-5}
0.6	0.548812	0.548812	5.340395×10^{-12}	3.945827×10^{-5}
0.7	0.496585	0.496585	2.327194×10^{-12}	9.834766×10^{-6}
0.8	0.449329	0.449329	4.971024×10^{-13}	1.996756×10^{-6}
0.9	0.406570	0.406570	2.053913×10^{-15}	5.066395×10^{-6}

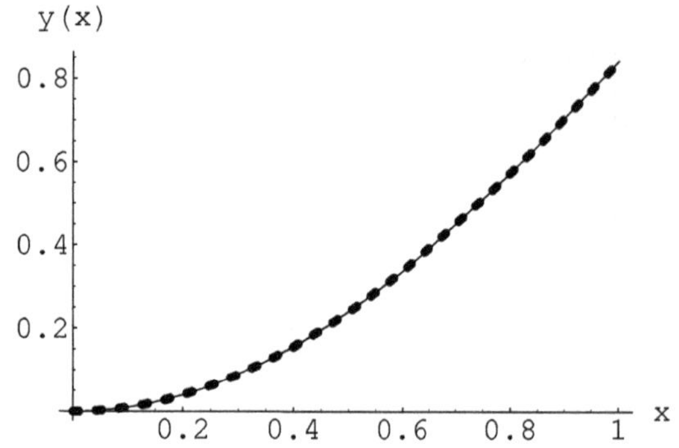

Fig. 5 Comparison of exact solution (*solid line*) and approximate solution (*dashed line*) for Example 3

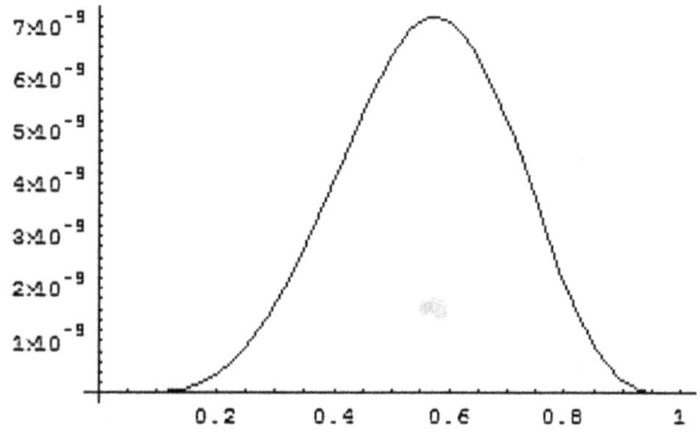

Fig. 6 Absolute errors for Example 3

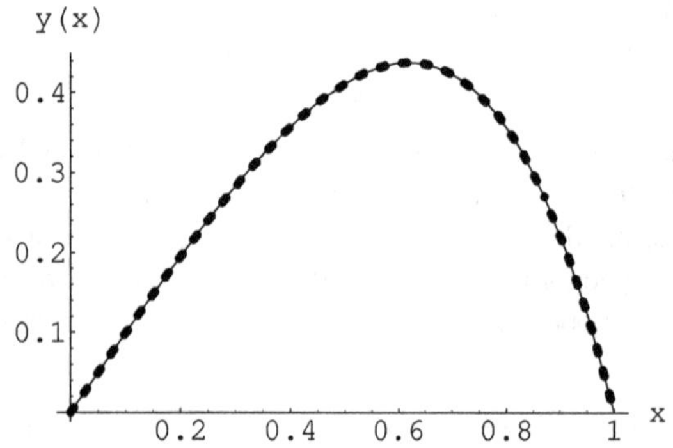

Fig. 7 Comparison of exact solution (*solid line*) and approximate solution (*dashed line*) for Example 4

$$\left.\begin{aligned}
&y^{(8)}(x) + xy(x) + e^x(48 + 15x + 2x^3) = 0, \quad x \in [0, 1],\\
&y(0) = 0, \quad y(1) = 0,\\
&y'(0) = 1, \quad y'(1) = -e,\\
&y''(0) = 0, \quad y''(1) = -4e,\\
&y'''(0) = -3, \quad y'''(1) = -9e.
\end{aligned}\right\} \tag{15}$$

The analytic solution of this differential system is

$y(x) = x(1 - x)e^x$.

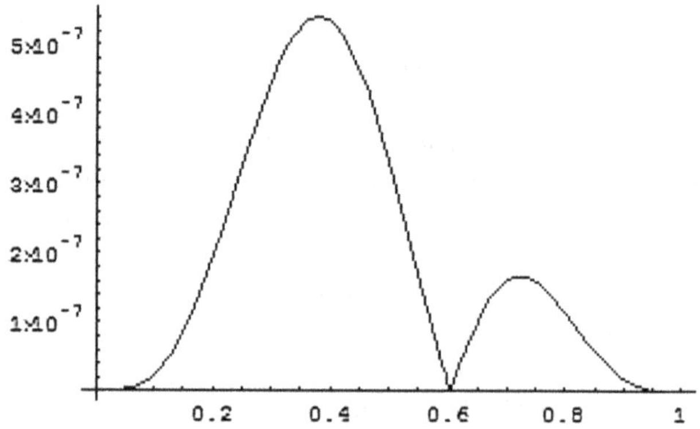

Fig. 8 Absolute errors for Example 4

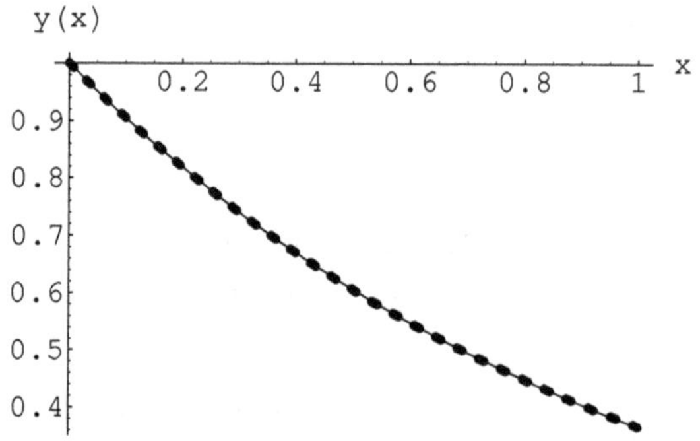

Fig. 9 Comparison of exact solution (*solid line*) and approximate solution (*dashed line*) for Example 5

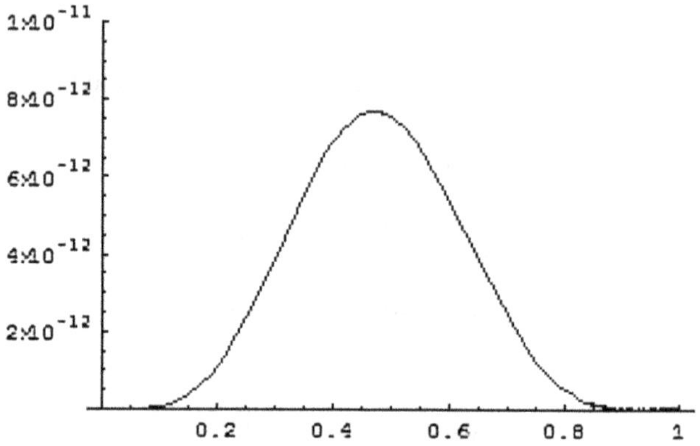

Fig. 10 Absolute errors for Example 5

The initial approximation is calculated according to the standard HAM, as

$$y_0(x) = x - \frac{x^3}{2} + \frac{1}{2}x^4(-36 + 13e) + \frac{1}{2}x^5(84 - 31e) + \frac{1}{2}x^6(-68 + 25e) + \frac{1}{2}x^7(19 - 7e).$$

Using the improved adaptation proposed in the present paper, the third order approximation to the exact solution is calculated. The approximate solution values and the corresponding absolute errors are shown in Table 5.

The auxiliary function $H(x)$ is taken as $H(x) = 1$. The linear operator is chosen to be the homogeneous part of the nonlinear operator N and the value of h is chosen as $h = -1$. Table 6 shows the comparison of the maximum absolute error using proposed method with that obtained by Inc and Evans [22].

Example 5 The following tenth order nonlinear boundary value problem is considered as

$$\left.\begin{aligned} &y^{(10)}(x) + e^{-x}y^2 = e^{-x} + e^{-3x}, \quad x \in [0,1],\\ &y(0) = 1, \quad y(1) = \frac{1}{e},\\ &y'(0) = -1, \quad y'(1) = -\frac{1}{e},\\ &y''(0) = 1, \quad y''(1) = \frac{1}{e},\\ &y'''(0) = -1, \quad y'''(1) = -\frac{1}{e},\\ &y^{(4)}(0) = 1, \quad y^{(4)}(1) = \frac{1}{e}. \end{aligned}\right\} \tag{16}$$

The analytic solution of this differential system is

$$y(x) = e^{-x}.$$

The initial approximation is calculated according to the standard HAM as

$$\begin{aligned} y_0(x) = 1 - x + \frac{x^2}{2} - \frac{x^3}{6} + \frac{x^4}{24} + \frac{(4645 - 1709e)x^5}{24e}\\ \times \frac{(-15804 + 5814e)x^6}{24e}\\ + \frac{(20610 - 7582e)x^7}{24e} + \frac{(-12148 + 4469e)x^8}{24e}\\ + \frac{(2721 - 1001e)x^9}{24e}. \end{aligned}$$

Using the improved adaptation proposed in the present paper, the third order approximation to the exact solution is calculated.

The auxiliary function $H(x)$ is taken as $H(x) = 1$. The linear operator is chosen to be the homogeneous part of the nonlinear operator N and the value of h is chosen as $h = -0.2$. Table 7 shows the approximate solution values and the comparison of the absolute errors using the proposed method with those obtained by Viswanadham and Ballem [31].

Figures 1, 2, 3, 4, 5, 6, 7, 8, 9, and 10 show the comparison of the approximate solutions to the exact solutions and the variation of the absolute errors over the interval of domain for Examples 1–5.

Conclusion

In this paper, an improved adaptation of the well-known homotopy analysis method is proposed for approximate solutions of the strongly nonlinear differential equations. The method has three major advantages over the traditional homotopy analysis method. First, it involves fewer terms in each iteration. Second, the integrals involved on each iteration step are easier to manipulate. At last, it obliterates the need for extra calculations which is the main advantage over the adaptation of homotopy analysis method proposed by Odibat and Bataineh [13]. The method is illustrated with the help of different numerical examples and the results are summarized in Tables 1, 2, 3, 4, 5, 6 and 7. Tables 3, 6 and 7 show the comparison of absolute errors using the proposed method with those obtained by other methods. The comparison of the results shows that the present method is an effective tool for determining the solutions of different linear and nonlinear problems.

References

1. Liao, S.J.: Proposed homotopy analysis techniques for the solution of nonlinear problems. Ph.D. dissertation, Shanghai Jiao Tong University (1992)
2. Marinca, V., Herisanu, N.: Application of optimal homotopy asymptotic method for solving nonlinear equations arising in heat transfer. Int. Commun. Heat Mass Transf. **35**, 710–715 (2008)
3. Motsa, S.S., Sibanda, P., Shateyi, S.: A new spectral-homotopy analysis method for solving a nonlinear second order BVP. Commun. Nonlinear Sci. Numer. Simul. **15**, 2293–2302 (2010)
4. Vosughi, H., Shivanian, E., Abbasbandy, S.: A new analytical technique to solve Volterra's integral equations. Math. Methods Appl. Sci. **34**(10), 1243–1253 (2011)
5. Abbasbandy, S., Shivanian, E.: Predictor homotopy analysis method and its application to some nonlinear problems. Commun. Nonlinear Sci. Numer. Simul. **16**, 2456–2468 (2011)
6. Shivanian, E., Alsulami, H.H., Alhuthali, M.S., Abbasbandy, S.: Predictor homotopy analysis method (PHAM) for nano boundary layer flows with nonlinear Navier boundary condition: Existence of four solutions. Filomat **28**(8), 1687–1697 (2014)
7. Vosoughi, H., Shivanian, E., Abbasbandy, S.: Unique and multiple PHAM series solutions of a class of nonlinear reactive transport model. Numer. Algorithms **61**(3), 515–524 (2012)
8. Abbasbandy, S., Shivanian, E., Vajravelu, K.: Mathematical properties of $\hbar$-curve in the frame work of the homotopy analysis method. Commun. Nonlinear Sci. Numer. Simul. **16**(11), 4268–4275 (2011)
9. Motsa, S.S., Shateyi, S., Marewo, G.T., Sibanda, P.: An improved spectral homotopy analysis method for MHD flow in a semi-porous channel. Numer. Algorithms **60**, 463–481 (2012)
10. Shaban, M., Shivanian, E., Abbasbandy, S.: Analyzing magneto-hydrodynamic squeezing flow between two parallel disks with suction or injection by a new hybrid method based on the Tau method and the homotopy analysis method. Eur. Phys. J. Plus **128**(11), 1–10 (2013)

11. Shivanian, E., Abbasbandy, S.: Predictor homotopy analysis method: two points second order boundary value problems. Nonlinear Anal.: Real World Appl. **15**, 89–99 (2014)
12. Soltani, L.A., Shivanian, E., Ezzati, R.: Convection-radiation heat transfer in solar heat exchangers filled with a porous medium: exact and shooting homotopy analysis solution. Appl. Therm. Eng. **103**, 537–542 (2016)
13. Odibat, Z., Bataineh, A.S.: An adaptation of homotopy analysismethod for reliable treatment of strongly nonlinear problems: construction of homotopy polynomials. Math. Methods Appl. Sci. **38**, 991–1000 (2015)
14. Lyshevski, S.E., Dunipace, K.R. (1997) Identification and tracking control of aircraft from real-time perspectives. In: Proceedings of the 1997 IEEE International Conference on Control Applications, Hartford, CT, pp. 499–504
15. Mohyud-Din, S.T., Yildirim, A.: Solution of tenth and ninth order boundary value problems by homotopy perturbation method. J. Korean Soc. Ind. Appl. Math. **14**(1), 17–27 (2010)
16. Mohyud-Din, S.T., Yildirim, A.: Solutions of tenth and ninth order boundary value problems by modified variational iteration method. Appl. Appl. Math. **5**(1), 11–25 (2010)
17. Chandrasekhar, S.: Hydrodynamic and Hydromagnetic Stability. The International Series of Monographs on Physics. Clarendon Press, Oxford (1961)
18. Siddiqi, S.S., Twizell, E.H.: Spline solution of linear sixth order boundary value problems. Int. J. Comput. Math. **60**(3), 295–304 (1996)
19. Siddiqi, S.S., Twizell, E.H.: Spline solution of linear eighth order boundary value problems. Comput. Methods Appl. Mech. Eng. **131**, 309–325 (1996)
20. Siddiqi, S.S., Twizell, E.H.: Spline solution of linear twelfth order boundary value problems. J. Comput. Appl. Math. **78**, 371–390 (1997)
21. Siddiqi, S.S., Twizell, E.H.: Spline solution of linear tenth order boundary value problems. Int. J. Comput. Math. **68**(3), 345–362 (1998)
22. Inc, M., Evans, D.J.: An efficient approach to approximate solutions of eighth order boundary value problems. Int. J. Comput. Math. **81**(6), 685–692 (2004)
23. Siddiqi, S.S., Akram, G.: Solution of fifth order boundary value problems using non-polynomial spline technique. Appl. Math. Comput. **175**, 1571–1581 (2006)
24. Siddiqi, S.S., Akram, G.: Solution of sixth order boundary value problems using non-polynomial spline technique. Appl. Math. Comput. **181**, 708–720 (2006)
25. Siddiqi, S.S., Akram, G.: Solution of eighth order boundary value problems using non-polynomial spline technique. Appl. Math. Comput. **84**, 347–368 (2007)
26. Siddiqi, S.S., Akram, G.: Solution of 10th order boundary value problems using non-polynomial spline technique. Appl. Math. Comput. **190**, 641–651 (2007)
27. Siddiqi, S.S., Akram, G.: Solution of 12th order boundary value problems using non-polynomial spline technique. Appl. Math. Comput. **199**, 559–571 (2008)
28. Hassan, I.H.A., Erturk, V.S.: Solutions of different types of the linear and non-linear higher order boundary value problems by differential transformation method. Eur. J. Pure Appl. Math. **2**(3), 426–447 (2009)
29. Siddiqi, S.S., Iftikhar, M.: Variational iteration method for the solution of seventh order boundary value problems using Hes polynomials. J. Assoc. Arab Univ. Basic Appl. Sci. **18**, 60–65 (2015)
30. Ullah, H., Nawaz, R., Islam, S., Idrees, M., Fiza, M.: The optimal homotopy asymptotic method with application to modified Kawahara equation. J. Assoc. Arab Univ. Basic Appl. Sci. **18**, 82–88 (2015)
31. Kasi Viswanadham, K.N.S., Ballem, S.: Numerical solution of tenth order boundary value problems by Galerkin method with septic B-splines. Int. J. Appl. Sci. Eng. **13**(3), 247–260 (2015)

17

Lower semicontinuity of solutions for order-perturbed parametric vector equilibrium problems

Shunyou Xia[1,2] · Shuwen Xiang[2] · Yanlong Yang[2] · Deping Xu[3]

Abstract The lower semicontinuity of the (weak) efficient solution mappings for parametric vector equilibrium problems under more weaker assumptions is established. Some examples are developed to illustrate our results are real generalization different from recent ones in the literature and to describe the essential conditions of the latest results in the references are not real essential.

Keywords Order-perturbed parametric vector equilibrium problems · Lower semicontinuity · Cone lower semicontinuity · Efficient solutions

Mathematics Subject Classification 49K40 · 90C31

✉ Shunyou Xia
xiashunyou@126.com

Shuwen Xiang
shwxiang@vip.163.com

Yanlong Yang
yylong1980@163.com

Deping Xu
xdpsc@163.com

1 School of Mathematics and Computer Science, Guizhou Education University, Guiyang 550018, China

2 College of Computer Science and Technology, Guizhou University, Guiyang 550025, China

3 College of Information Management, Chengdu University of Technology, Chengdu 610059, China

Introduction

Several classes of problems, including the vector variational inequality problem, the vector complementarity problem, the vector optimization problem and the vector saddle point problem, have been unified as a model of the vector equilibrium problem, which has been intensively studied in the literature (see [1–16]). One of the important topics in optimization theory is the stability analysis of the solution mappings for vector equilibrium problems. Stability may be understood as some types of lower or upper semicontinuity. Recently, the semicontinuity, especially the lower semicontinuity, of the solution mappings for parametric vector equilibrium problems has been intensively studied in various directions (see [1, 2, 4, 6, 10, 15] and references therein).

Anh and Khanh first obtained the semicontinuity of the solution mappings of parametric multivalued vector quasi-equilibrium problems (see [1]), and then obtained verifiable sufficient conditions for solution sets of general quasi-variational inclusion problems to have these semicontinuity-related properties and discussed in detail a traffic network problem as a sample for employing the main results in practical situations (see [2]), and latter established sufficient conditions for lower and Hausdorff lower semicontinuity, upper semicontinuity, and continuity of solution mappings of parametric quasi-equilibrium problems in topological vector spaces (see [3]).

Gong and Yao, by virtue of a density result and a scalarization technique, first discussed the lower semicontinuity of the set of efficient solutions to parametric vector equilibrium problems with monotone bifunctions (see [4]), and studied the continuity of the solution mapping to parametric weak vector equilibrium problems (see [5]), recently, established the lower semicontinuity of solutions to the parametric generalized strong vector equilibrium problem without the assumptions of monotonicity of the objective mapping and compactness of the constraint mapping (see [6]). Huang et al. used local existence results to establish the lower semicontinuity of solution mappings for parametric implicit vector equilibrium problems (see [7]). By using a new proof which is different from the ones of [4, 5], Chen et al. established the lower semicontinuity and continuity of the solution mappings to a parametric generalized vector equilibrium problem (see [9]). Li and Fang investigated the lower semicontinuity of the solutions mapping to parametric generalized Ky Fan inequality under a weaker assumption than C-strict monotonicity (see [10]). Recently, by using an idea of [4], Li et al. established the continuity of solution mappings to a parametric generalized strong vector equilibrium problem for set-valued mappings under an assumption which is different from the C-strict monotonicity (see [11]). Kimura and Yao discussed the semicontinuity of solution mappings of parametric vector quasi-equilibrium problems (see [13]). Cheng and Zhu obtained a lower semicontinuity result of the solution mapping to weak vector variational inequalities in finite-dimensional spaces by using the scalarization method (see [14]).

Zhang et al. obtained the lower semicontinuity of solution mappings for parametric vector equilibrium problems under the Höder-related assumptions [15]. Wangkeeree et al. extended the results in [15] to the case of set-valued mappings on parametric strong vector equilibrium problems (see [16]).

However, all these results are with respect to the fixed order relationship in the object space, that is, the cone partial order is not perturbed by the parameters. Motivated by the idea of variational domination structure, the main results in [15] will be obtained under more weaker conditions in this paper.

The organization of this work is as follows. In Sect. 2, we introduce the efficient solutions to parametric vector equilibrium problems with the cone partial order being perturbed by the parameters and recall some basic notions. In Sect. 3, we discuss the lower semicontinuity of the (weak) efficient solution mappings for parametric vector equilibrium problems in the case of weakly conditions. Some examples are given to illustrate that the assumptions of the main results in our work or in [15] are only sufficient for some special problems and indicate also that our outcomes are real extension from the corresponding ones in [15].

Preliminaries

Let X and Z be two metric spaces, and let Y be a metric vector space and C be a pointed closed convex cone in Y with nonempty interior int$C$; the zero element in Y is denoted by θ. Let A be a nonempty subset of X and F be a vector-valued mapping from $A \times A$ into Y. A vector equilibrium problem, in short (VEP), is described as:

Find $x \in A$ such that $F(x,y) \notin -C\backslash\{\theta\}$, for all $y \in A$

A point $x \in A$ is said to be an efficient solution to (VEP) if

$$F(x,y) \notin -C\backslash\{\theta\}, \text{ for all } y \in A$$

When the subset A of X and the function F are perturbed by the parameter $\lambda \in \Lambda$, where $\Lambda \subset Z$, a parametric vector equilibrium problem, in short (PVEP), is a problem as following:

$$\text{Find } x \in A(\lambda) \quad \text{such that } F(x,y,\lambda) \notin -C\backslash\{\theta\}, \text{ for all } y \in A(\lambda)$$

where $A : \Lambda \to 2^X\backslash\{\phi\}$ is a set-valued mapping, and $F{:}B \times B \times \Lambda \subset X \times X \times Z \to Y$ is a vector-valued mapping with $A(\Lambda) = \cup_{\lambda \in \Lambda} A(\lambda) \subset B$.

A point $x \in A(\lambda)$ is said to be an efficient solution to (PVEP) if

$$F(x,y,\lambda) \notin -C\backslash\{\theta\}, \text{ for all } y \in A(\lambda)$$

In this work, we consider (PVEP) with cone C being also perturbed by parameter $\lambda \in \Lambda$, described as follows:

$$\text{Find } x \in A(\lambda) \quad \text{such that } F(x,y,\lambda) \notin -C(\lambda)\backslash\{\theta\}, \text{ for all } y \in A(\lambda)$$

where, for each $\lambda \in \Lambda$, $C(\lambda)$ is a pointed closed convex cone in Y with nonempty interior, that is to say, $C{:}\Lambda \to 2^Y$ is a cone-valued mapping. In this case, we call the (PVEP) as parametric vector order-perturbed equilibrium problem, in short (*PVOPEP*).

A point $x \in A(\lambda)$ is called an efficient solution to (*PVOPEP*) if

$$F(x,y,\lambda) \notin -C(\lambda)\backslash\{\theta\}, \text{ for all } y \in A(\lambda)$$

The set of efficient solutions to (*PVOPEP*) is denoted by $S(\lambda)$, i.e.,

$$S(\lambda) := \{x \in A(\lambda) | F(x,y,\lambda) \notin -C(\lambda)\backslash\{\theta\}, \forall y \in A(\lambda)\}$$

It is easy to see that S is a set-valued mapping $S{:}\Lambda \to 2^X$. Throughout this work, we always assume that $S(\lambda) \neq \phi$; for all $\lambda \in \Lambda$. Next, we recall some basic definitions and their properties which will be needed in the following.

$B_X(\lambda, \delta)$ denotes the open ball with center λ and radius $\delta > 0$ in a metric space X, $d_X(\cdot,\cdot)$ denotes the distance in X, and the distance from x to the set $A \subset X$ is denoted by $d_X(x, A)$.

Definition 2.1 ([17]) A set-valued mapping S: $\Lambda \to 2^X$ is said to be

1. lower semicontinuous (*l.s.c.*) at $\lambda_0 \in \Lambda$ if for any open set V satisfying $V \cap S(\lambda_0) \neq \phi$, there exists $\delta > 0$ such that for every $\lambda \in B_\Lambda(\lambda_0, \delta), V \cap S(\lambda) \neq \phi$;

2. upper semicontinuous (*u.s.c.*) at $\lambda_0 \in \Lambda$ if for any open set V satisfying $S(\lambda_0) \subset V$, there exists $\delta > 0$ such that for every $\lambda \in B_\Lambda(\lambda_0, \delta), S(\lambda) \subset V$;
3. *l.s.c.*(resp.,*u.s.c.*) on Λ if it is *l.s.c.*(resp.,*u.s.c.*) at each $\lambda \in \Lambda$;
4. continuous on Λ if it is both *l.s.c.* and *u.s.c.* on Λ.

Proposition 2.1 ([18, 19])

1. $S:\Lambda \to 2^X$ is *l.s.c.* at $\lambda_0 \in \Lambda$ if and only if for any sequence $\{\lambda_n\} \subset \Lambda$ with $\lambda_n \to \lambda_0$ and any $x_0 \in S(\lambda_0)$, there exists $x_n \in S(\lambda_n)$ such that $x_n \to x_0$.
2. If S has compact values (i.e., $S(\lambda)$ is a compact set for each $\lambda \in \Lambda$), then S is *u.s.c.* at $\lambda_0 \in \Lambda$ if and only if for any sequence $\{\lambda_n\} \subset \Lambda$ with $\lambda_n \to \lambda_0$ and for any $x_n \in S(\lambda_n)$, there exist $x_0 \in S(\lambda_0)$ and a subsequence $\{x_{n_k}\}$ of $\{x_n\}$ such that $x_{n_k} \to x_0$.

Definition 2.2 A vector-valued mapping f: $X \to Y$ is called cone lower semicontinuity (*c.l.s.c.*) at $x_0 \in X$ if for each open set V of $f(x_0)$, there exists a neighborhood U of x_0, such that $f(x) \subset V + C$, for all $x \in U$, where $C \subset Y$ is a cone.

Definition 2.3 Let Λ be a topology space, Y be a topology vector space, and $C:\Lambda \to 2^Y$ be a cone-valued mapping, if for every $\lambda \in \Lambda$, $C(\lambda)$ is a closed convex pointed cone in Y. The closed unit ball with center θ in Y is denoted by $B_Y(\theta)$. We call the cone-valued mapping C is an upper semicontinuous cone-valued mapping, if for each $\lambda \in \Lambda$ and each open set Uin Y with $U \supset C(\lambda) \cap B_Y(\theta)$, there exists an open set V of λ such that $U \supset C(\lambda') \cap B_Y(\theta)$ for every $\lambda' \in V$.

The main results

In this section, we present the lower semicontinuity of the solution mapping to (*PVOPEP*).

Theorem 3.1. Suppose that the following conditions are satisfied:

① $A(\cdot)$ is continuous with compact values on Λ.
② $F(\cdot, \cdot, \cdot)$ is *c.l.s.c.* on $B \times B \times \Lambda$.
③ $C(\cdot)$ is an upper semicontinuous cone-valued mapping on Λ.
④ If $A(\lambda)\backslash S(\lambda) \neq \phi$ for each $\lambda \in \Lambda$, then for each $\lambda \in \Lambda$,for each $x \in A(\lambda)\backslash S(\lambda)$, there exist $y \in S(\lambda)$ and a positive function $M:\Lambda \to (0, +\infty)$ which is upper semicontinuous on Λ, such that

$$d_X(x,y) \leq M(\lambda) \cdot d_Y(F(x,y,\lambda), Y\backslash -\mathrm{int}C(\lambda))$$

Then, $S(\cdot)$ is *l.s.c.* on Λ.

Proof Suppose to the contrary that there exists $\lambda_0 \in \Lambda$ such that $S(\cdot)$is not *l.s.c.* at λ_0. Then, there exists a sequence $\{\lambda_n\} \subset \Lambda$ with $\lambda_n \to \lambda_0$ and $x_0 \in S(\lambda_0)$ with some open set V_1 of x_0, such that for any $x_n' \in S(\lambda_n)$, $x_n' \notin V_1$.

From $x_0 \in S(\lambda_0)$, we have $x_0 \in A(\lambda_0)$ and

$$F(x_0, y, \lambda_0) \notin -C(\lambda_0)\backslash\{\theta\}, \quad \forall y \in A(\lambda_0) \tag{1}$$

Since $A(\cdot)$ is *l.s.c.* at λ_0, there exists a sequence $\{x_n\} \subset A(\lambda_n)$ such that $x_n \in V_1$. Then, for the above open set V_1, there exists a positive integer N; such that $x_n \in V_1$, for all $n \geq N$. Obviously, we have $x_n \in A(\lambda_n)\backslash S(\lambda_n)$ for all $n \geq N$. For the sake of convenience, we consider n as still from one to infinity. By ④, for each $x_n \in A(\lambda_n)\backslash S(\lambda_n)$, there exist $y_n \in S(\lambda_n)$ and a positive function $M:\Lambda \to (0, +\infty)$ which is upper semicontinuous on Λ, such that

$$d_X(x_n, y_n) \leq M(\lambda_n) \cdot d_Y(F(x_n, y_n, \lambda_n), Y\backslash -\mathrm{int}C(\lambda_n)) \tag{2}$$

Since $y_n \in A(\lambda_n)$, it follows from the upper semicontinuity and compactness of $A(\cdot)$ at λ_0 that there exist $y_0 \in A(\lambda_0)$ and a subsequence $\{y_{n_i}\}$ of $\{y_n\}$ such that $y_{n_i} \to y_0$. In particular, from (2), we have

$$d_X(x_{n_i}, y_{n_i}) \leq M(\lambda_{n_i}) \cdot d_Y(F(x_{n_i}, y_{n_i}, \lambda_{n_i}), Y\backslash -\mathrm{int}C(\lambda_{n_i})) \tag{3}$$

Since the distance function $d(\cdot, \cdot)$ is continuous, $F(\cdot, \cdot, \cdot)$ is *c.l.s.c.*, $M(\cdot)$ is *u.s.c.*, and $C(\cdot)$ is an upper semicontinuous cone-valued mapping, then let $i \to +\infty$ on both sides of (3), we have

$$d_X(x_0, y_0) \leq M(\lambda_0) \cdot d_Y(F(x_0, y_0, \lambda_0), Y\backslash -\mathrm{int}C(\lambda_0)) \tag{4}$$

If $x_0 \neq y_0$, by (4), we can obtain

$$M(\lambda_0) \cdot d_Y(F(x_0, y_0, \lambda_0), Y\backslash -\mathrm{int}C(\lambda_0)) \geq d_X(x_0, y_0) > 0$$

From $M(\lambda_0) > 0$, we have

$$d_Y(F(x_0, y_0, \lambda_0), Y\backslash -\mathrm{int}C(\lambda_0)) > 0$$

Thus, we have

$$F(x_0, y_0, \lambda_0) \in -\mathrm{int}C(\lambda_0))$$

which contradicts (1) as we see by taking $y = y_0$. Therefore $x_0 = y_0$. This is impossible by the contradiction assumption. Thus, the proof is complete. □

Remark 3.1 The following examples indicate that the assumption ④ in Theorem 3.1 in our work (or the assumption (iii) of Theorem 3.1 in [15]) cannot be applied to the case when $A(\lambda)\backslash S(\lambda) = \phi$, for some $\lambda \in \Lambda$. So the assumption ④ of Theorem 3.1 in our paper (or the assumption (iii) of Theorem 3.1 in [15])is not essential.

Example 3.1 Let $X = Z = R, \Lambda_1 = [\frac{1}{4}, \frac{2}{3}]$, $\Lambda_2 = [2, 3]$, $A(\lambda) = B = [0, \quad 1]$, $Y = R^2$, $C(\lambda) = R_+^2$ and

$F(x, y, \lambda) = (\lambda(y-x), \lambda(y-x))$. It follows from a direct computation that

$S_1(\lambda) = \{0\}, \forall\lambda \in \Lambda_1$ and $S_2(\lambda) = \{0\}, \forall\lambda \in \Lambda_2$

For each $\lambda \in \Lambda_1 \cup \Lambda_2$, for every $x \in A(\lambda)\backslash S(\lambda) = (0, 1]$, taking $y = 0 \in S(\lambda)$; we have

$d_X(x, y) = x$ and $d_Y(F(x, y, \lambda), Y\backslash - \text{int}C(\lambda)) = \lambda x$

Obviously,

$d_X(x, y) \geq d_Y(F(x, y, \lambda), Y\backslash - \text{int}C(\lambda)), \forall\lambda \in \Lambda_1$

$d_X(x, y) \leq d_Y(F(x, y, \lambda), Y\backslash - \text{int}C(\lambda)), \forall\lambda \in \Lambda_2$

From above two inequalities, it easy to see that the assumption (iii) in Theorem 3.1. in [15] is violated when $\lambda \in \Lambda_1$. However, $S_1(\cdot)$ and $S_2(\cdot)$ are all continuous on Λ_1 and Λ_2, respectively. If we take M: $\Lambda_1 \to (0, +\infty)$, defined by $M(\lambda) = 4, \forall\lambda \in \Lambda_1 = \left[\frac{1}{4}, \frac{2}{3}\right]$, then

$d_X(x, y) \leq M(\lambda) \cdot d_Y(F(x, y, \lambda), Y\backslash - \text{int}C(\lambda)), \forall\lambda \in \Lambda_1$

Thus, the example 3.1 satisfies all the assumptions of Theorem 3.1. in our work, and then, it is obtained that $S(\cdot)$ is *l.s.c.* on Λ.Consequently, Theorem 3.1. in our work is real extension from Theorem 3.1. in [15]. □

Example 3.2 Let $X = Y = R$, $C(\lambda) = R_+$, $\Lambda = (0, 1]$, $A(\lambda) = B = [\lambda^2, 1+\lambda]$ and $F(x, y, \lambda) = \lambda(y-x)$. It follows from a direct computation that $S(\lambda) = \{\lambda^2\}$, $\forall\lambda \in \Lambda$. And it is easy to check that $S(\cdot)$ is continuous on Λ. For any $\lambda \in (0, 1]$, for each $x \in A(\lambda)\backslash S(\lambda) = (\lambda^2, 1+\lambda]$, taking the unique element $y = \lambda^2 \in S(\lambda)$, we have

$$\begin{aligned} x - \lambda^2 &= d_X(x, y) \geq d_Y(F(x, y, \lambda), Y\backslash - \text{int}C(\lambda)) \\ &= \lambda(x - \lambda^2) \end{aligned}$$

Obviously, the assumption (iii) of Theorem 3.1. in [15] is violated, but if taking M: $\Lambda \to (0, +\infty)$ as following:

$M(\lambda) = \frac{a}{\lambda}, \forall\lambda \in \Lambda$ where a is a constant number and $a \geq 1$.

Thus, the example 3.2 satisfies all the assumptions of Theorem 3.1. in our work, and it follows from Theorem 3.1. in our work that $S(\cdot)$ is *l.s.c.* on Λ. But the Theorem 3.1. in [15] is invalid.

Example 3.3 Let $X = Y = Z = R$, $C(\lambda) = R_+$, $\Lambda = [0, 1]$, $A(\lambda) = B = [1-\lambda^2, 1+\lambda^2]$ and $F(x, y, \lambda) = x(1+\lambda-y)$.

It follows from a direct computation that $S(\lambda) = [1-\lambda^2, 1+\lambda^2] = A(\lambda)$. For any $\lambda \in [0, 1]$, $A(\lambda)\backslash S(\lambda) = \phi$. Obviously, we cannot take any $x \in A(\lambda)\backslash S(\lambda)$. So the condition (iii) of Theorem 3.1 in [15] (or the assumption ④ in Theorem 3.1 in our work) cannot be applied. But it is easy to check that $S(\cdot)$ is continuous on Λ.

Our approach can also be applied to study the lower semicontinuity of the weak solution mappings. A point $x \in A(\lambda)$ is called a weak efficient solution to (PVOPEP) if

$F(x, y, \lambda) \notin -\text{int}C(\lambda), \forall y \in A(\lambda)$

The set of weak efficient solutions to (PVOPEP) is denoted by $S_W(\lambda)$, i.e.,

$S_W(\lambda) := \{x \in A(\lambda) | F(x, y, \lambda) \notin -\text{int}C(\lambda), \forall y \in A(\lambda)\}$

We can also obtain the following theorem on the lower semicontinuity of the weak efficient solution map to (PVOPEP) with a trivial adaptation of the proof.

Theorem 3.2 Suppose that the following conditions are satisfied:

① $A(\cdot)$ is continuous with compact values on Λ.
② $F(\cdot, \cdot, \cdot)$ is *c.l.s.c.* on $B \times B \times \Lambda$.
③ $C(\cdot)$ is an upper semicontinuous cone-valued mapping on Λ.
④ If $A(\lambda)\backslash S_W(\lambda) \neq \phi$; for each $\lambda \in \Lambda$, then for each $\lambda \in \Lambda$,for each $x \in A(\lambda)\backslash S_W(\lambda)$, there exist $y \in S_W(\lambda)$ and a positive function M:$\Lambda \to (0, +\infty)$ which is upper semicontinuous on Λ, such that

$d_X(x, y) \leq M(\lambda) \cdot d_Y(F(x, y, \lambda), Y\backslash - \text{int}C(\lambda))$

Then, $S_W(\cdot)$ is *l.s.c.* on Λ.

Acknowledgement This work is supported by NSFC(70661001; 11161008), DPFME P.R.China(20115201110002), NSFGP([2014] 2005; [2015]7298).

References

1. Anh, L.Q., Khanh, P.Q.: Semicontinuity of the solution set of parametric multivalued vector quasiequilibrium problems. J. Math. Anal. Appl. **294**, 699–711 (2004)
2. Anh, L.Q., Khanh, P.Q.: Semicontinuity of solution sets to parametric quasivariational inclusions with applications to traffic networks II. Lower semicontinuities applications. Set Valued Anal. **16**, 943–960 (2008)
3. Anh, L.Q., Khanh, P.Q.: Continuity of solution maps of parametric quasiequilibrium problems. J. Global Optim. **46**, 247–259 (2010)
4. Gong, X.H., Yao, J.C.: Lower semicontinuity of the set of efficient solutions for generalized systems. J. Optim. Theory Appl. **138**, 197–205 (2008)
5. Gong, X.H.: Continuity of the solution set to parametric weak vector equilibrium problems. J. Optim. Theory Appl. **139**, 35–46 (2008)
6. Han, Y., Gong, X.H.: Lower semicontinuity of solution mapping to parametric generalized strong vector equilibrium problems. Appl. Math. Lett. **28**, 38–41 (2014)
7. Huang, N.J., Li, J., Thompson, H.B.: Stability for parametric implicit vector equilibrium problems. Math. Comput. Model. **43**, 1267–1274 (2006)
8. Chen, B., Huang, N.J.: Continuity of the solution mapping to

parametric generalized vector equilibrium problems. J. Global Optim. **56**, 1515–1528 (2012)
9. Chen, C.R., Li, S.J., Teo, K.L.: Solution semicontinuity of parametric generalized vector equilibrium problems. J. Global Optim. **45**, 309–318 (2009)
10. Li, S.J., Fang, Z.M.: Lower semicontinuity of the solution mappings to a parametric generalized Ky Fan inequality. J. Optim. Theory Appl. **147**, 507–515 (2010)
11. Li, S.J., Liu, H.M., Zhang, Y., Fang, Z.M.: Continuity of the solution mappings to parametric generalized strong vector equilibrium problems. J. Global Optim. **55**, 597–610 (2013)
12. Li, X.B., Li, S.J.: Hölder continuity of perturbed solution set for convex optimization problems. Appl. Math. Comput. **232**, 908–918 (2014)
13. Kimura, K., Yao, J.C.: Sensitivity analysis of solution mappings of parametric vector quasi-equilibrium problems. J. Global Optim. **41**, 187–202 (2008)
14. Cheng, Y.H., Zhu, D.L.: Global stability results for the weak vector variational inequality. J. Global Optim. **32**, 543–550 (2005)
15. Zhang, W.Y., Fang, Z.M., Zhang, Y.: A note on the lower semicontinuity of efficient solutions for parametric vector equilibrium problems. Appl. Math. Lett. **26**, 469–472 (2013)
16. Wangkeeree, R., Wangkeeree, R., Preechasilp, P.: Continuity of the solution mappings to parametric generalized vector equilibrium problems. Appl. Math. Lett. **29**, 42–45 (2014)
17. Kien, B.T.: On the lower semicontinuity of optimal solution sets. Optimization **54**, 123–130 (2005)
18. Aubin, J.P., Ekeland, I.: Applied Nonlinear Analysis. Wiley, New York (1984)
19. Ferro, F.: A minimax theorem for vector-valued functions. J. Optim. Theory Appl. **60**, 19–31 (1989)

18

A mathematical–physical approach on regularity in hit-and-miss hypertopologies for fuzzy set multifunctions

Alina Gavriluţ[1] · Maricel Agop[2]

Abstract In this paper, an approach concerning hit-and-miss hypertopologies and especially regularity property viewed both as a continuity property in a hit-and-miss hypertopology (from a mathematical point of view) and also as a physical approximation property is intended.

Keywords Hit-and-miss hypertopologies · Hausdorff topology · Vietoris topology · Wijsman topology · Regularity · Approximations · Fractal theories · Non-differentiable physics · Scale relativity theory

Mathematics Subject Classification 28C15 · 49J53

Introduction

Nowadays, Hausdorff, Vietoris, Wijsman, Fell, Attouch-Wets, etc., hypertopologies are intensively studied due to their various applications in optimization, convex analysis, economics, image processing, sound analysis and synthesis (see Beer [7], Apreutesei [4], Hu and Papageorgiou [20], etc., concerning the Vietoris topology). Results involving the Hausdorff distance were obtained by Lorenzo and Maio [26] in melodic similarity, Lu et al. [27]—an approach to word image matching, etc. Recently, it was shown that using proximity, all hypertopologies known so far are of the type hit-and-miss, which led to the unification of all hypertopologies under one topology called *the Bombay Hypertopology* [29].

The idea of modeling at multiple scales the phenomena behavior has become a useful tool in pure mathematics, applied mathematics physics and so on. Fractals are multiscale objects, which often describe such phenomena better than traditional mathematical models do. That is why fractal-based techniques lie at the heart of these areas. Kunze et al. [21] and Wicks [42] developed hyperspace theories concerning the Hausdorff metric and the Vietoris topology, as a foundation for self-similarity and fractality. In fact, for many years, topological methods were used in many fields to study the chaotic nature in dynamical systems (see for instance Sharma and Nagar [40], Wang et al. [41], Goméz-Rueda et al. [18], Li [24], Liu et al. [25], Ma et al. [28], Fu and Xing [11], etc.). These phenomena seem to be collective (set-valued), emerging out of many segregated components, having collective dynamics of many units of individual systems. This arose the need of a topological study of such collective dynamics. Recent studies of dynamical systems, in engineering and physical sciences, have revealed that the underlying dynamics is set-valued (collective), and not of a normal, individual kind, as it was usually studied before.

Also, the reader can refer to Lewin et al. [22] and Brown [8] for interesting approaches of topology in psychology. We also mention different aspects concerning generalized fractals in hyperspaces endowed with Hausdorff, or, more generally, with Vietoris hypertopology (see Andres and Fi šer [2], Andres and Rypka [3], Banakh and Novosad [5], Kunze et al. [21]).

Since in some examples of fractals (like neural networks and the circulatory system), the uniform property of the

✉ Alina Gavriluţ
gavrilut@uaic.ro

Maricel Agop
m.agop@yahoo.com

1 Faculty of Mathematics, "Al.I. Cuza" University, Carol I Bd. 11, Iasi 700506, Romania

2 Department of Physics, Gheorghe Asachi Technical University of Iaşi, Iasi, Romania

Hausdorff topology is inappropriate, we could intend to choose a convenient topology on the set of values of the studied multifunctions. In this sense, Wijsman topology may be preferred instead of Hausdorff topology because Wijsman topology could describe better the pointwise properties of fractals.

On the other hand, recently, domain theory has been studied in theoretical computer science, as a mathematical theory of semantics of programming languages (Edalat [10], Gierz et al. [17], etc.). In this context, (hyper)-topological notions from Mathematical Analysis as well as measure theory, dynamical systems or fractality can be considered via domain theory, obtaining computational models. Namely, in denotational semantics and domain theory, power domains are domains of nondeterministic and concurrent computations. As it is well-known, domain theory was introduced by Scott in theoretical computer science as a mathematical theory of semantics of programming languages.

Together with the increasing interest in hypertopologies, non-additive set multifunctions theories developed. In this context, regularity is known as an important continuity property with respect to different topologies, but, at the same time, it can be interpreted as an approximation property. Using regularity, we can approximate "unknown" sets by other sets which we have more informations. Usually, from a mathematical perspective, this approximation is done from the left by closed sets, or more restrictive, by compact sets and/or from the right by open sets. As a mathematical direct application of regularity, the classical Lusin's theorem concerning the existence of continuous restrictions of measurable functions is very important and useful for discussing different kinds of approximation of measurable functions defined on special topological spaces and for numerous applications in the study of convergence of sequences of Sugeno and Choquet integrable functions (see Li et al. [23] for an interesting application of Lusin theorem), in the study of the approximation properties of neural networks, as the learning ability of a neural network is closely related to its approximating capabilities. Also, regular Borel measures are important tools in studies on the Kolmogorov fractal dimension (Barnsley [6], Mandelbrot [30], etc.). Lebesgue measure is a remarkable example of a regular measure.

The paper is organized as follows: in "Hit-and-miss hypertopologies: an overview" and "Regular set multifunctions" several remarkable hit-and-miss hypertopologies and their properties are listed from a mathematical perspective and regularity of set multifunctions is introduced in a unifying way with respect to these hypertopologies. In "Regularization by sets of functions of ε-approximation-type scale. Physical correspondences with hit-and-miss topologies" and "Conclusions", a physical perspective concerning regularity and fractality is provided.

Our unifying mathematical–physical point of view on fractality, hypertopologies and regularity was initiated in our recent works [12, 13, 15, 16].

Hit-and-miss hypertopologies: an overview

Hausdorff, Vietoris and Wijsman, etc., topologies are remarkable examples of the so-called hit-and-miss hypertopologies. Like some physical concepts, these hypertopologies, although are composed of two independent parts, upper and lower hypertopologies, they become consistent when seen together. For instance, in physical terms, the non-differentiability of the curve motion of the physical object involves the simultaneous definition at any point of the curve, of two differentials (left and right). Since we cannot favor one of the two differentials, the only solution is to consider them simultaneously through a complex differential. Its application, multiplied by dt, where t is an affine parameter, to the field of space coordinates implies complex speed fields.

We used the following (selected) references: Apreutesei [4], Beer [7], Gavriluţ and Apreutesei [14], Kunze et al. [21], Hu and Papageorgiou [20, Ch. 1], Precupanu et al. [38, Ch. 1], G. Apreutesei in Precupanu et al. [39, Ch. 8], Maio and Naimpally [29], etc.

We now briefly recall and list the definitions and main properties of the above-mentioned hypertopologies:

Vietoris topology

Let (X, τ) be a Hausdorff, topological space and $\mathcal{P}_0(X)$, the family of all nonvoid subsets of X. We consider

$M^- = \{C \in \mathcal{P}_0(X)/M \cap C \neq \emptyset\}$ (i.e., C *hits* M), $M^+ = \{C \in \mathcal{P}_0(X)/C \subseteq M\}$ (i.e., C *misses* cM), $S_{UV} = \{D^+/D \in \tau\}$ and $S_{LV} = \{D^-/D \in \tau\}$.

Vietoris topology $\widehat{\tau}_V$ on $\mathcal{P}_0(X)$ has as a subbase the class $S_{UV} \cup S_{LV}$ and it is the supremum $\widehat{\tau}_V = \widehat{\tau}_V^+ \cup \widehat{\tau}_V^-$ of the lower and upper Vietoris topologies:

$\widehat{\tau}_V^+$—*the upper Vietoris topology* ($\widehat{\tau}_V^-$—*the lower Vietoris topology*, respectively) is the topology which has as a subbase the class S_{UV} (S_{LV}, respectively).

For $U, V \in \tau$, define $\mathcal{B}_{U,V_1,V_2,\ldots,V_k} = U^+ \cap V_1^- \cap V_2^- \cap \cdots \cap V_k^-$.

The family $\mathcal{B}_{U,V_1,V_2,\ldots,V_k}$ of such subsets, where $U, V_1, V_2, \ldots, V_k \in \tau$, is a base for the topology $\widehat{\tau}_V$ and the family of subsets $\mathcal{B}_U = U^+$ ($\mathcal{B}_V = V^-$, respectively) is a base for $\widehat{\tau}_V^+$ ($\widehat{\tau}_V^-$, respectively).

In different continuity properties (regularity for instance), the following observation is used:

Remark 2.1 [39, Ch. 8] A net $(A_i)_{i\in I} \subset \mathcal{P}_0(X)$ is:

(i) $\hat{\tau}_V^-$-convergent to $A_0 \in \mathcal{P}_0(X)$ if for every $V \in \tau$, with $A_0 \cap V \neq \emptyset, \exists i_V \in I$ so that for every $i \in I, i \geq i_V$, we have $A_i \cap V \neq \emptyset$;
(ii) $\hat{\tau}_V^+$-convergent to $A_0 \in \mathcal{P}_0(X)$ if for every $V \in \tau$, with $A_0 \subset V, \exists i_V \in I$ so that for every $i \in I$, $i \geq i_V$, we have $A_i \subset V$.

In what follows, let (*X*, *d*) be a metric space. By $\mathcal{P}_f(X)$ we mean the family of closed, nonvoid sets of *X*, by $\mathcal{P}_{bf}(X)$ the family of bounded, closed, nonvoid sets of *X* and by $\mathcal{P}_k(X)$, the family of all nonvoid compact subsets of $X \cdot \tau_d$ denotes the topology induced by the metric *d*.

Wijsman topology

Wijsman topology τ_W on $\mathcal{P}_0(X)$ is the supremum of the *upper Wijsman topology* τ_W^+ and the *lower Wijsman topology* $\tau_W^- : \tau_W = \tau_W^+ \cup \tau_W^-$.

The family

$$\mathcal{F} = \{M \in \mathcal{P}_0(X); d(x, M) < \varepsilon\}_{\substack{x\in X\\ \varepsilon>0}} \cup \{M \in \mathcal{P}_0(X); d(x, M) > \varepsilon\}_{\substack{x\in X\\ \varepsilon>0}}$$

is a subbase for τ_W on $\mathcal{P}_0(X)$.

Let $M \in \mathcal{P}_0(X), \{x_1, x_2, \ldots, x_n\} \subset X, \varepsilon > 0$ be arbitrarily chosen.

τ_W^- (τ_W^+, respectively) is generated by the family $U_W^-(M, x_1, x_2, \ldots, x_n, \varepsilon) = \{N \in \mathcal{P}_0(X); d(x_i, N) < d(x_i, M) + \varepsilon$, for every $i = \overline{1,n}\}$ ($U_W^+(M, x_1, x_2, \ldots, x_n, \varepsilon) = \{N \in \mathcal{P}_0(X); d(x_i, M) < d(x_i, N) + \varepsilon$, for every $i = \overline{1,n}\}$, respectively).

Proposition 2.2 (Apreutesei, Ch. 8 in Precupanu et al. [39]) $\tau_W^- = \tau_V^-$.

Remark 2.3

(I) Suppose $\{M_i\}_{i\in I} \subset \mathcal{P}_0(X)$. The following statements are equivalent:

(i) $M_i \xrightarrow{\tau_W} M \in \mathcal{P}_0(X)$;
(ii) For every $x \in X$, $d(x, M_i) \xrightarrow{p} d(x, M)$ (pointwise convergence);
(iii) $M_i \xrightarrow{\tau_W^+} M$ and $M_i \xrightarrow{\tau_W^-} M$.

(II)

(i) $M_i \xrightarrow{\tau_W^+} M$ if and only if for every $x \in X$, $\liminf_i d(x, M_i) \geq d(x, M)$ (i.e., for every $0 < \varepsilon < \varepsilon'$ with $S(x, \varepsilon') \cap M = \emptyset$, there is $i_0 \in I$ so that for every $i \in I$, with $i \geq i_0$, we have $S(x, \varepsilon) \cap M_i = \emptyset$).
(ii) $M_i \xrightarrow{\tau_W^-} M$ if and only if for every $x \in X$, $\limsup_i d(x, M_i) \leq d(x, M)$ (i.e., for every $D \in \tau_d$ with $D \cap M \neq \emptyset$, there is $i_0 \in I$ so that for every $i \in I$, with $i \geq i_0$ we have $D \cap M_i \neq \emptyset$).

Remark 2.4 [14]

(i) If (*X*, *d*) is a complete, separable metric space, then $\mathcal{P}_f(X)$ with the Wijsman topology is a Polish space (Beer [7]). Moreover, the space $(\mathcal{P}_f(X), \tau_W)$ is Polish if and only if (*X*, *d*) is Polish.
(ii) $(\mathcal{P}_f(X), \tau_W)$ is a Tychonoff space. (*X*, *d*) is separable if and only if $\mathcal{P}_f(X)$ is either metrizable, first-countable or second-countable. The dependence of the Wijsman topology on the metric *d* is quite strong. Even if two metrics are uniformly equivalent, they may generate different Wijsman topologies. Necessary and sufficient conditions for two metrics to induce the same Wijsman topology have been found.

Hausdorff topology

In recent years, due to the development of computational graphics (for instance, in the automatic recognition of figures problems), it was necessary to measure accurately the matching, i.e., to calculate the distance between two sets of points. This led to the need to operate with an acceptable distance, which has to satisfy the first condition in the definition of a distance: the distance is zero if and only if the overlap is perfect. An appropriate metric in these issues is the Hausdorff metric on which we will refer in the following and which, roughly speaking, measures the degree of overlap of two compact sets.

Let $M, N \in \mathcal{P}_f(X)$. The *Hausdorff–Pompeiu pseudometric* h on $\mathcal{P}_f(X)$ is the "greatest" of all distances from any point in one of these two sets, to the nearest point from the other set, so, it is defined by

$$(*)\ h(M, N) = \max\{e(M, N), e(N, M)\},$$

where $e(M, N) = \sup_{x\in M} d(x, N)$ is *the excess* of *M* over *N* and $d(x, N) = \inf_{y\in N} d(x, y)$ is the distance from x to N (with respect to the metric *d*).

For instance, the Cantor set $\mathcal{C} \in \mathcal{P}_k(\mathbb{R})$ and its "steps" $I_n \in \mathcal{P}_k(\mathbb{R}), \forall n \in \mathbb{N}$ (Kunze et al. [21]).

The topology induced by the Hausdorff pseudometric *h* is called the *Hausdorff hypertopology* τ_H *on* $\mathcal{P}_f(X)$.

On $\mathcal{P}_{bf}(X)$, *h* becomes a veritable metric. If, in addition, X is complete, then the same is $\mathcal{P}_f(X)$ (Hu and Papageorgiou [20]).

We observe that $e(N,M) = h(M,N)$, for every $M,N \in \mathcal{P}_f(X)$, with $M \subseteq N$. Also, $e(M,N) \leq e(M,P)$, for every $M,N,P \in \mathcal{P}_f(X)$, with $P \subseteq N$ and

$e(M,P) \leq e(N,P)$, for every $M,N,P \in \mathcal{P}_f(X)$, with $M \subseteq N$.

Generally, even if $M,N \in \mathcal{P}_k(X)$, then $e(M,N) \neq e(N,M)$.

If $M \in \mathcal{P}_f(X)$ and $\varepsilon > 0$ is arbitrary, but fixed, we consider the *ε-dilatation of the set M*

$$S(M,\varepsilon) = \{x \in X; \exists m \in M, d(x,m) < \varepsilon\}$$
$$\left(= \bigcup_{m\in M} \{x \in X; m \in M, d(x,m) < \varepsilon\} \right).$$

Obviously, $M \subseteq S(M,\varepsilon)$.

Since $h(M,N) < \varepsilon$ iff $M \subset S(N,\varepsilon)$ and $N \subset S(M,\varepsilon)$, we have the following equivalent expression for $h(M, N)$:

$$(**)\ h(M,N) = \inf\{\varepsilon > 0, M \subset S(N,\varepsilon), N \subset S(M,\varepsilon)\}$$

($h(M, N)$ is the "smallest" $\varepsilon > 0$ which permits the ε-dilatation of M to cover N and the ε-dilatation of N to cover M).

In other words, $\tau_H = \tau_H^+ \cup \tau_H^-$, where τ_H^+ (*upper Hausdorff topology*), respectively, τ_H^- (*lower Hausdorff topology*) has as a base, the family $\{U^+(M,\varepsilon)\}_{\varepsilon>0}$, where $U^+(M,\varepsilon) = \{N \in \mathcal{P}_f(X); N \subset S(M,\varepsilon)\}$, respectively, the family $\{U^-(M,\varepsilon)\}_{\varepsilon>0}$, where $U^-(M,\varepsilon) = \{N \in \mathcal{P}_f(X); M \subset S(N,\varepsilon)\}$.

Another equivalent expression of the Hausdorff distance between two sets $M,N \in \mathcal{P}_f(X)$ is:

$$(***)\ h(M,N) = \sup\{|d(x,N) - d(x,M)|; x \in X\}.$$

And this highlights the uniform aspect of the Hausdorff topology: it is the topology on $\mathcal{P}_f(X)$ of uniform convergence on X of the distance functionals $x \mapsto d(x,M)$, with $M \in \mathcal{P}_f(X)$.

Hausdorff topology is invariant with respect to uniformly equivalent metrics (Apreutesei [4]).

In the following, we list some properties of the Hausdorff metric:

Proposition 2.5

(I)

(i) $h(M_1 \cup M_2, N_1 \cup N_2) \leq \max\{h(M_1,N_1), h(M_2,N_2)\}$, $\forall M_1,M_2,N_1,N_2 \in \mathcal{P}_f(X)$;

If X is a Banach space, then:

(II)

(i) $h(\alpha M, \alpha N) = |\alpha| h(M,N), \forall \alpha \in R, \forall M,N \in \mathcal{P}_f(X)$;

(ii) $h(M+P, N+P) \leq h(M,N), \forall M,N, P \in \mathcal{P}_f(X)$;

(ii′) $h(\sum_{i=1}^p M_i, \sum_{i=1}^p N_i) \leq \sum_{i=1}^p h(M_i,N_i), \forall M_i, N_i \in \mathcal{P}_f(X)$

(*where* $M+N = \{m+n; m \in M, n \in N\}$).

If, particularly, $X = \mathbb{R}$, and $a,b,c,d \in \mathbb{R}$, with $a<b, c<d$, then

$$h([a,b],[c,d]) = \max\{|a-c|, |b-d|\}.$$

Remark 2.6 [21] Hausdorff metric has some interesting characteristics:

(i) It is possible for a sequence of finite sets to converge to an uncountable set:

$$\forall n \geq 1, \text{ let be} M_n = \left\{0, \frac{1}{n}, \frac{2}{n}, \ldots, \frac{n-1}{n}, 1\right\}$$
$(\subset [0,1])$ (all of them are finite sets).

Since $h(M_n,[0,1]) = \frac{1}{2n}$, then $M_n \xrightarrow{h} [0,1]$ (in $\mathcal{P}_k(\mathbb{R})$), but [0, 1] is uncountable.

(ii) Adding or removing a single point often influences the Hausdorff distance between two (compact) sets: if $M = [0,1]$ and $N = [0,1] \cup \{x\}$, where $x \notin [0,1]$, then $h(M,N) = \max\{-x, x-1\}$ (so, it is a function of x).

(iii) $I_n \xrightarrow{h} \mathcal{C}$ and $J_n \xrightarrow{h} \mathcal{C}$ where $(I_n)_{n\in\mathbb{N}}$ (respectively, $(J_n)_{n\in\mathbb{N}}$) are the steps in the construction of Cantor set $\mathcal{C}$.

Remark 2.7 [4, 14], [38, Ch. 1], [39, Ch. 8]

(i) If the pointwise convergence of Wijsman convergence is replaced by uniform convergence (uniformly in x), then one obtains Hausdorff convergence induced by the Hausdorff pseudometric. Generally, Hausdorff topology τ_H is finer than Wijsman topology τ_W. Hausdorff and Wijsman topologies on $\mathcal{P}_f(X)$ coincide if and only if (X, d) is totally bounded.

(ii) If X is a real normed space, then Hausdorff topology, Vietoris topology and Wijsman topology are equivalent on the class of monotone sequences of subsets of $\mathcal{P}_k(X)$.

Remark 2.8 Hausdorff metric on $\mathcal{P}_k(X)$ is an essential tool in the study of fractals and their generalizations: hyperfractals, multifractals and superfractals—[2, 3, 21].

Barnsley [6] calls the space $(\mathcal{P}_k(X), h)$, *the life space of fractal*. Recently, Banakh and Novosad [5] proposed a fractal approach using Vietoris topology (in a more general setting than the one used for the Hausdorff topology).

Regular set multifunctions

Suppose that T is a locally compact, Hausdorff space, $\mathcal{C}$ a ring of subsets of T and X a real normed space space. Usually, it is assumed that $\mathcal{C}$ is $\mathcal{B}_0$ ($\mathcal{B}'_0$, respectively)—the Baire δ-ring (σ-ring, respectively) generated by compact sets, which are G_δ (i.e., countable intersections of open sets) or $\mathcal{C}$ is $\mathcal{B}$ ($\mathcal{B}'$, respectively)—the Borel δ-ring (σ-ring, respectively) generated by the compact sets of T.

$\mathcal{B}_0 \subset \mathcal{B} \subset \mathcal{B}'$, $\mathcal{B}_0 \subset \mathcal{B}'_0$. If T is metrisable or if it has a countable base, then any compact set $K \subset T$ is G_δ. In this case $\mathcal{B}_0 = \mathcal{B}$ (Dinculeanu [9, Ch. III, p. 187]) so $\mathcal{B}'_0 = \mathcal{B}'$.

By $\mathcal{K}$ we denote the family of all compact subsets of T and by $\mathcal{D}$ the family of all open subsets of T.

Regularity can be considered as a property of continuity with respect to a topology on $\mathcal{P}(T)$ (Dinculeanu [9, Ch. III, p. 197]):

For every $K \in \mathcal{K}$ and every $D \in \mathcal{D}$, with $K \subset D$, we denote $\mathcal{I}(K,D) = \{A \subset T / K \subset A \subset D\}$.

Since $\mathcal{I}(K,D) \cap \mathcal{I}(K',D') = \mathcal{I}(K \cup K', D \cap D')$, for every $\mathcal{I}(K,D), \mathcal{I}(K',D')$, the family $\{\mathcal{I}(\mathcal{K},\mathcal{D})\}_{\substack{K\in\mathcal{K}\\ D\in\mathcal{D}}}$ is a base of a topology $\tilde{\tau}$ on $\mathcal{P}(T)$. $\tilde{\tau}$ also denotes the topology induced on any subfamily $\mathcal{S} \subset \mathcal{P}(T)$ of subsets of T.

By $\tilde{\tau}_l$ ($\tilde{\tau}_r$, respectively) we denote the topology induced on $\{\mathcal{I}(\mathcal{K})\}_{K\in\mathcal{K}} = \{\{A \subset T / K \subset A\}\}_{K\in\mathcal{K}}$ ($\{\mathcal{I}(\mathcal{D})\}_{D\in\mathcal{D}} = \{\{A \subset T / A \subset D\}\}_{D\in\mathcal{D}}$, respectively) (Dinculeanu [9, Ch. III, p. 197–198]).

Definition 3.1 A class $\mathcal{F} \subset \mathcal{P}(T)$ is *dense* in $\mathcal{P}(T)$ with respect to the topology induced by $\tilde{\tau}$ if for every $K \in \mathcal{K}$ and every $D \in \mathcal{D}$, with $K \subset D$, there is $A \in \mathcal{C}$ such that $K \subset A \subset D$.

Since T is locally compact, the following statements can be easily verified (Dinculeanu [9, Ch. III, p. 197]):

Remark 3.2

(1) $\mathcal{B}_0, \mathcal{B}, \mathcal{B}'_0, \mathcal{B}'$ are dense in $\mathcal{P}(T)$ with respect to the topology induced by $\tilde{\tau}$.

(2)

(i) For every $A \in \mathcal{C}$, there exists $D \in \mathcal{D} \cap \mathcal{C}$ so that $A \subset D$.

(ii) If $\mathcal{C}$ is $\mathcal{B}$ or $\mathcal{B}'$, then for every $A \in \mathcal{C}$, there exist $K \in \mathcal{K} \cap \mathcal{C}$ and $D \in \mathcal{D} \cap \mathcal{C}$ so that $K \subset A \subset D$.

Let $\mu : \mathcal{C} \to \mathcal{P}_f(X)$ be an arbitrary set multifunction.

Definition 3.3 μ is said to be *monotone* or *fuzzy* (with respect to the inclusion of sets) if $\mu(A) \subseteq \mu(B)$, for every $A, B \in \mathcal{C}$ with $A \subseteq B$.

Example 3.4 (*of monotone set multifunctions*)

(i) Let $\mathcal{C}$ be a ring of subsets of an abstract space T, $m : \mathcal{C} \to \mathbb{R}_+$ a finitely additive set function and $\mu : \mathcal{C} \to \mathcal{P}_{bf}(\mathbb{R})$ the set multifunction defined for every $A \in \mathcal{C}$ by

$$\mu(A) = \begin{cases} [-m(A), m(A)], & \text{if } m(A) \leq 1 \\ [-m(A), 1], & \text{if } m(A) > 1 \end{cases}.$$

We easily observe that μ is monotone and $|\mu(A)| = m(A)$, for every $A \in \mathcal{C}$.

(ii) Let $\nu_1, \ldots, \nu_p : \mathcal{C} \to \mathbb{R}_+$, be p finitely additive set functions, where $\mathcal{C}$ is a ring of subsets of an abstract space T. We consider the set multifunction $\mu : \mathcal{C} \to \mathcal{P}_f(\mathbb{R})$, defined for every $A \in \mathcal{C}$ by

$$\mu(A) = \{\nu_1(A), \nu_2(A), \ldots, \nu_p(A)\}.$$

Then the set multifunction $\mu^\vee : \mathcal{C} \to \mathcal{P}_f(\mathbb{R})$, defined for every $A \in \mathcal{C}$ by:

$$\mu^\vee(A) = \overline{\bigcup_{\substack{B\subset A,\\ B\in\mathcal{C}}} \mu(B)}$$

is monotone.

In what follows, let $\mu : (\mathcal{C}, \tau_1) \to (\mathcal{P}_f(X), \tau_2)$ be a monotone set multifunction, where $\tau_1 \in \{\tilde{\tau}, \tilde{\tau}_l, \tilde{\tau}_r\}$ and $\tau_2 \in \{\tau_H, \tau_W, \tau_V\}$.

Let also be $\mathcal{B}_1 \in \{\{\mathcal{I}(\mathcal{K},\mathcal{D})\}_{\substack{K\in\mathcal{K}\\ D\in\mathcal{D}}}, \{\mathcal{I}(\mathcal{K})\}_{K\in\mathcal{K}}, \{\mathcal{I}(D)\}_{D\in\mathcal{D}}\}$, respectively, $\mathcal{B}_2$, bases for τ_1, respectively, τ_2 (as discussed in "Hit-and-miss hypertopologies: an overview").

$\tau_2 = \tau_2^+ \cup \tau_2^-$, where $\tau_2^+ \in \{\tau_H^+, \tau_V^+, \tau_W^+\}$ and $\tau_2^- \in \{\tau_H^-, \tau_V^-, \tau_W^-\}$.

In a unifying way,

Definition 3.5 A set $A \in \mathcal{C}$ is said to be (τ_2)-regular if $\mu : (\mathcal{C}, \tau_1) \to (\mathcal{P}_f(X), \tau_2)$ is continuous at A, that is, for every $\mathcal{V} \in \mathcal{B}_2$, with $\mu(A) \in \mathcal{V}$, there exists $\widetilde{\mathcal{V}} \in \mathcal{B}_1$ so that $\mu(\widetilde{\mathcal{V}} \cap \mathcal{C}) \subset \mathcal{V}$ (or, equivalently, for every $(A_i)_{i\in I}, A \subset \mathcal{C}$, with $A_i \xrightarrow{\tau_1} A$, it results $\mu(A_i) \xrightarrow{\tau_2} \mu(A)$).

When τ_1 is $\tilde{\tau}$, or $\tilde{\tau}_l$ or $\tilde{\tau}_r$, respectively, we get the notions of (τ_2)*-regularity* , $(\tau_2$-)R_l*-regularity* (inner regularity) or $(\tau_2$-)R_r*-regularity* (outer regularity).

Precisely, we have:

Proposition 3.6 *A is:*

(i) *regular iff for every $\mathcal{V} \in \mathcal{B}_2$, with $\mu(A) \in \mathcal{V}$, there exist $K \in \mathcal{K} \cap \mathcal{C}, K \subset A$ and $D \in \mathcal{D} \cap \mathcal{C}, D \supset A$ so that for every $B \in \mathcal{C}$, with $K \subset B \subset D$, we have $\mu(B) \in \mathcal{V}$;*

ii) *R_l-regular iff for every $\mathcal{V} \in \mathcal{B}_2$, with $\mu(A) \in \mathcal{V}$, there exists $K \in \mathcal{K} \cap \mathcal{C}, K \subset A$ so that for every $B \in \mathcal{C}$, with $K \subset B \subset A$, we have $\mu(B) \in \mathcal{V}$;*

(iii) *R_r-regular iff for every $\mathcal{V} \in \mathcal{B}_2$, with $\mu(A) \in \mathcal{V}$, there exists $D \in \mathcal{D} \cap \mathcal{C}, D \supset A$ so that for every $B \in \mathcal{C}$, with $A \subset B \subset D$, we have $\mu(B) \in \mathcal{V}$.*

Remark 3.7 Every $K \in \mathcal{K}$ is R_l-regular and every $D \in \mathcal{D}$ is R_r-regular.

The following results can be proved using the above definitions:

Proposition 3.8

(i) *A set A is (τ_2-)regular if and only if it is (τ_2-)R_l-regular and (τ_2-)R_r-regular.*

(ii) *A is (τ_2-)regular (R_l-regular or R_r-regular, respectively) if and only if it is (τ_2^+-)and (τ_2^--)regular (R_l-regular or R_r-regular, respectively).*

Theorem 3.9 *Suppose $\mu_1, \mu_2 : \mathcal{C} \to (\mathcal{P}_f(X), \tau_2)$ are two monotone set multifunctions.*

(i) If *μ_1, μ_2 are R_l-regular, then $\mu_1 = \mu_2$ on $\mathcal{C}$ if and only if $\mu_1 = \mu_2$ on $\mathcal{K} \cap \mathcal{C}$;*

(ii) If *μ_1, μ_2 are R_r-regular, then $\mu_1 = \mu_2$ on $\mathcal{C}$ if and only if $\mu_1 = \mu_2$ on $\mathcal{D} \cap \mathcal{C}$.*

Remark 3.10 For $\tau_2 = \tau_H, \tau_W$ or τ_V, respectively, we particularly get the notions of regularity as we defined and studied in [12–14]. For instance, if $\tau_2 = \tau_H$, then, by its monotonicity, μ is (in the sense of [12]):

(i) *regular* if for every $\varepsilon > 0$, there are $K \in \mathcal{K} \cap \mathcal{C}, K \subset A$ and $D \in \mathcal{D} \cap \mathcal{C}, D \supset A$ so that $h(\mu(A), \mu(B)) < \varepsilon$, for every $B \in \mathcal{C}$, with $K \subset B \subset D$.

ii) *R_l-regular* if for every $\varepsilon > 0$, there exists $K \in \mathcal{K} \cap \mathcal{C}, K \subset A$ so that $h(\mu(A), \mu(B)) = e(\mu(A), \mu(B)) < \varepsilon$, for every $B \in \mathcal{C}$, with $K \subset B \subset A$.

(iii) *R_r-regular* if for every $\varepsilon > 0$, there exists $D \in \mathcal{D} \cap \mathcal{C}, D \supset A$ such that $h(\mu(A), \mu(B)) = e(\mu(B), \mu(A)) < \varepsilon$, for every $B \in \mathcal{C}$, with $A \subset B \subset D$.

In fact, one may easily observe that (in τ_H):

(i) μ is regular iff for every $\varepsilon > 0$, there are $K \in \mathcal{K} \cap \mathcal{C}, K \subset A$ and $D \in \mathcal{D} \cap \mathcal{C}, D \supset A$ so that $e(\mu(D), \mu(K)) < \varepsilon$;

(ii) μ is R_l-regular iff for every $\varepsilon > 0$, there is $K \in \mathcal{K} \cap \mathcal{C}, K \subset A$ so that $e(\mu(A), \mu(K)) < \varepsilon$;

(iii) μ is R_r-regular iff for every $\varepsilon > 0$, there is $D \in \mathcal{D} \cap \mathcal{C}, D \supset A$ so that $e(\mu(D), \mu(A)) < \varepsilon$,

that is, in each case, we find an alternative expression of regularity as an approximation property.

Regularization by sets of functions of ε-approximation-type scale: physical correspondences with hit-and-miss topologies

In this section, analogously to our considerations from the previous section concerning regularity as an approximation property, we now study physical regularizations. Precisely, as we shall see, generally, the "reduction" of the complex dimensions to their real part requires the regularization by sets of functions of ε-approximation-type scale, while the "reduction" to their imaginary part requires regularization with "known" sets, that is, sets for which we have some informations.

We consider a fractal function $f(x)$, with $x \in [a, b]$ (for instance, one of the trajectory's equation) and the sequence of the variable x values:

$$x_a = x_0, x_1 = x_0 + \varepsilon, \ldots, x_k = x_0 + k\varepsilon, \ldots, x_n = x_0 + n\varepsilon = x_b. \tag{1}$$

By $f(x, \varepsilon)$, we denote the fractured line connecting the points $f(x_0), \ldots, f(x_k), \ldots, f(x_n)$.

This line will be considered as an approximation which is different from the one used before. We shall say that $f(x, \varepsilon)$ is an *ε-approximation scale.*

Now, we consider the $\bar{\varepsilon}$-approximation scale $f(x, \bar{\varepsilon})$ of the same function. When we study a fractal phenomenon by approximation, because $f(x)$ is similar almost everywhere, then, if ε and $\bar{\varepsilon}$ are small enough, the two approximations $f(x, \varepsilon)$ and $f(x, \bar{\varepsilon})$ must lead to the same results. If we compare the two cases, then to an infinitesimal increase $d\varepsilon$ of ε, it corresponds an increase $d\bar{\varepsilon}$ of $\bar{\varepsilon}$, if the scale is dilated.

In this case, $\frac{d\varepsilon}{\varepsilon} = \frac{d\bar{\varepsilon}}{\bar{\varepsilon}}$, i.e.,

$$\frac{d\varepsilon}{\varepsilon} = d\rho \tag{2}$$

is the ratio of the scale $\varepsilon + d\varepsilon$ and $d\varepsilon$ must be preserved.

Then, we can consider the infinitesimal transformation of the scale as

$$\varepsilon' = \varepsilon + d\varepsilon = \varepsilon + \varepsilon d\rho. \quad (3)$$

By such transformation, it results in the case of the function $f(x, \varepsilon)$:

$$f(x, \varepsilon') = f(x, \varepsilon + \varepsilon d\rho), \quad (4)$$

respectively, if we stop after the first approximation,

$$f(x, \varepsilon') = f(x, \varepsilon) + \frac{\partial f}{\partial \varepsilon}(\varepsilon' - \varepsilon), \quad (5)$$

i.e.,

$$f(x, \varepsilon') = f(x, \varepsilon) + \frac{\partial f}{\partial \varepsilon}\varepsilon d\rho. \quad (6)$$

We note that, for arbitrary, but fixed ε_0,

$$\frac{\partial \ln \frac{\varepsilon}{\varepsilon_0}}{\partial \varepsilon} = \frac{\partial(\ln \varepsilon - \ln \varepsilon_0)}{\partial \varepsilon} = \frac{1}{\varepsilon}, \quad (7)$$

so Eq. 5 becomes

$$f(x, \varepsilon') = f(x, \varepsilon) + \frac{\partial f(x, \varepsilon)}{\partial \ln \frac{\varepsilon}{\varepsilon_0}} d\rho. \quad (8)$$

Finally, we get

$$f(x, \varepsilon') = (1 + \frac{\partial}{\partial \ln \frac{\varepsilon}{\varepsilon_0}} d\rho) f(x, \varepsilon). \quad (9)$$

The operator

$$\widetilde{D} = \frac{\partial}{\partial \ln \frac{\varepsilon}{\varepsilon_0}} \quad (10)$$

is called the *dilatation operator*.

The above relation shows that the intrinsic variable of the resolution is not ε, but $\ln \frac{\varepsilon}{\varepsilon_0}$.

On the other hand, simultaneous invariance with respect to both space–time coordinates and the resolution scale induces general scale relativity theory (SRT) [32, 33]. These theories are more general than Einstein's general relativity theory, being invariant with respect to the generalized Poincaré group (standard Poincaré group and dilatation group) [32, 33].

Basically, we discuss various physical theories built on manifolds of fractal space–time and they all turn out to be reducible to one of the following classes:

(i) SRT [35, 36] and its possible extensions [34]. It is considered that the microparticles motion takes place on continuous but non-differentiable curves. In such context, regularization works using sets of functions of ε-approximation-type scale.

(ii) Transition in which to each point of the motion trajectory, a transfinite set is assigned (in particular, a Cantor-type set—see the El Naschie [34] $\varepsilon^{(\infty)}$ model of space–time), to mimic the continuous (the trans-physics). In such context, the regularization of "vague" sets by known sets works.

(iii) Fractal string theories containing simultaneously relativity and trans-physics [19, 37].

The reduction of the complex dimensions to their real part is equivalent to Scale Relativity-Type theories, while reducing them to the imaginary part of their complex dimensions generates trans-physics. In such context, the simultaneous regularization by sets of functions of ε-approximation-type scale and also by "known" sets works. The "reduction" of the complex dimensions to their real part requires the regularization by sets of functions of ε-approximation-type scale, while the "reduction" to their imaginary part requires regularization with "known" sets.

Dynamical systems behaviors are collective phenomena emerging out of many segregated components. Most of these systems are collective (that is, set-valued) dynamics of many units of individual systems, whence the need of a (hyper)fractal topological treatment of such collective dynamics.

We consider that the particle of a complex system moves on continuous, but non-differentiable curves (fractal curves). Once accepted such a hypothesis, some consequences of non-differentiability by SRT are evident [35, 36].

For instance, physical quantities that describe the complex system are fractal functions, i.e., functions depending both on spatial coordinates and time as well as on the scale resolution $\frac{\delta t}{\tau}$.

In classical physics, the physical quantities describing the dynamics of a complex system are continuous, but differentiable functions depending only on spatial coordinates and time.

Since [1, 30, 31, 35, 36], two representations are complementary: the formalism of the fractal hydrodynamics (at the continuum level), and the one of the Schrödinger-type theory (at the discontinuum level). Moreover, the chaoticity, either through turbulence in the fractal hydrodynamic approach, either through stochasticization in the Schrödinger-type approach, is generated only by the non-differentiability of the movement trajectories in a fractal space.

Conclusions

In this paper, we intend to present a unifying mathematical–physical perspective concerning the relationships, interpretations and similitudes existing among fractality, regularity and several hit-and-miss hypertopologies. We intend to continue the study of regularity in hypertopologies viewed in the context of domain theory (in correlation with [10, 17]). We are also interested in developing a neural network fractal theory using Wijsman topology (its pointwise character seems to characterize some properties better than the Hausdorff topology induced by the Hausdorff–Pompeiu metric (which has a uniform character).

Acknowledgments The authors are indebted to the Area editors and the unknown referees for their valuable remarks and suggestions in the improving of the paper.

References

1. Agop, M., Niculescu, O., Timofte, A., Bibire, L., Ghenadi, A.S., Nicuţă, A., Nejneru, C., Munceleanu, G.V.: Non-differentiable mechanical model and its implications. Int. J. Theor. Phys. **49**(7), 1489–1506 (2010)
2. Andres, J., Fišer, J.: Metric and topological multivalued fractals. Int. J. Bifur. Chaos Appl. Sci. Eng. **14**(4), 1277–1289 (2004)
3. Andres, J., Rypka, M.: Multivalued fractals and hyperfractals. Int. J. Bifur. Chaos Appl. Sci. Eng. **22**(1), 1250009 (2012), 27 pp
4. Apreutesei, G.: Families of subsets and the coincidence of hypertopologies. Ann. Alexandru Loan Cuza Univ. Math **XLIX**, 1–18 (2003)
5. Banakh, T., Novosad, N.: Micro and macro fractals generated by multi-valued dynamical systems (2013). arXiv:1304.7529v1 [math.GN]
6. Barnsley, M.: Fractals Everywhere. Academic Press, New York (1988)
7. Beer, G.: Topologies on Closed and Closed Convex Sets. Kluwer Academic Publishers, Dordrecht (1993)
8. Brown, S.: Memory and mathesis: for a topological approach to psychology. Theory Cultu. Soc. **29**(4–5), 137–164 (2012)
9. Dinculeanu, N.: Measure Theory and Real Functions (in Romanian), Ed. Did. şi Ped., Bucureşti (1964)
10. Edalat, A.: Dynamical systems. Measures and fractals via domain theory. Inf. Comput. **120**(1), 32–48 (1995)
11. Fu, H., Xing, Z.: Mixing properties of set-valued maps on hyperspaces via Furstenberg families. Chaos Solitons Fractals **45**(4), 439–443 (2012)
12. Gavriluţ, A.: Regularity and autocontinuity of set multifunctions. Fuzzy Sets Syst. **161**, 681–693 (2010)
13. Gavriluţ, A.: Continuity properties and Alexandroff theorem in Vietoris topology. Fuzzy Sets Syst. **194**, 76–89 (2012)
14. Gavriluţ, A., Apreutesei, G.: Wijsman topology and non-uniform aspects of non-additive set multifunctions. (submitted for publication)
15. Gavriluţ, A., Agop, M.: A Mathematical Approach in the Study of the Dynamics of Complex Systems. Ars Longa Publishing House (2013). (in Romanian)
16. Gavriluţ, A., Agop, M.: Approximation theorems for set multifunctions in Vietoris topology. Physical implications of regularity. Iran. J. Fuzzy syst. **12**(1), 27–42 (2015)
17. Gierz, G., Hofmann, K.H., Keimel, K., Lawson, J.D., Mislove, M.W., Scott, D.S.: Continuous lattices and domains. In: Encyclopedia of Mathematics and its Applications, vol. 93. Cambridge University Press, London (2003)
18. Gómez-Rueda, J.L., Illanes, A., Méndez, H.: Dynamic properties for the induced maps in the symmetric products. Chaos Solitons Fractals **45**(9–10), 1180–1187 (2012)
19. Hawking, S., Penrose, R.: The Nature of Space Time. Princeton University Press, Princeton (1996)
20. Hu, S., Papageorgiou, N.S.: Handbook of Multivalued Analysis, vol. I. Kluwer Acad. Publ, Dordrecht (1997)
21. Kunze, H., La Torre, D., Mendivil, F., Vrscay, E.R.: Fractal Based Methods in Analysis. Springer, Berlin (2012)
22. Lewin, K., Heider, G.M., Heider, F.: Principles of Topological Psychology. McGraw-Hill, New York (1936)
23. Li, J., Li, J., Yasuda, M.: Approximation of fuzzy neural networks by using Lusin's theorem, 86-92 (2007)
24. Li, R.: A note on stronger forms of sensitivity for dynamical systems. Chaos Solitons Fractals **45**(6), 753–758 (2012)
25. Liu, L., Wang, Y., Wei, G.: Topological entropy of continuous functions on topological spaces. Chaos Solitons Fractals **39**(1), 417–427 (2009)
26. di Lorenzo, P., di Maio, G.: The Hausdorff metric in the melody space: a new approach to melodic similarity. In: The 9th International Conference on Music Perception and Cognition, Alma Mater Studiorum University of Bologna (2006)
27. Lu, Y., Tan, C. L., Huang, W., Fan, L.: An approach to word image matching based on weighted Hausdorff distance. In: Proceedings of Sixth International Conference on Document Analysis and Recognition, Seattle, 2001, pp. 921–925 (2001)
28. Ma, X., Hou, B., Liao, G.: Chaos in hyperspace system. Chaos Solitons Fractals **40**(2), 653–660 (2009)
29. di Maio, G., Naimpally, S.: Hit-and-far-miss hypertopologies. Mat. Vesn. **60**, 59–78 (2008)
30. Mandelbrot, B.B.: The Fractal Geometry of Nature. W.H. Freiman, New York (1983)
31. Munceleanu, G.V., Păun, V.P., Casian-Botez, I., Agop, M.: The microscopic-macroscopic scale transformation through a chaos scenario in the fractal space–time theory. Int. J. Bifurc. Chaos **21**, 603–618 (2011)
32. El-Nabulsi, A.R.: Fractional derivatives generalization of Einstein's field equations. Indian J. Phys. **87**, 195–200 (2013)
33. El-Nabulsi, A.R.: New astrophysical aspects from Yukawa fractional potential correction to the gravitational potential in D dimensions. Indian J. Phys. **86**, 763–768 (2012)
34. El Naschie, M.S., Rösler, O.E., Prigogine, I. (eds.): Quantum Mechanics, Diffusion and Chaotic Fractals. Elsevier, Oxford (1995)
35. Nottale, L.: Fractal Space–Time and Microphysics: Towards Theory of Scale Relativity. World Scientific, Singapore (1993)
36. Nottale, L.: Scale Relativity and Fractal Space–Time, A New Approach to Unifying Relativity and Quantum Mechanics. Imperial College Press, London (2011)
37. Penrose, R.: The Road to Reality: A Complete Guide to the Laws of the Universe. Jonathan Cape, London (2004)
38. Precupanu, A., Croitoru, A., Godet-Thobie, Ch.: Set-valued Integrals (in Romanian), Iaşi, in progress
39. Precupanu, A., Precupanu, T., Turinici, M., Apreutesei Dumitriu, N., Stamate, C., Satco, B.R., Văideanu, C., Apreutesei, G., Rusu, D., Gavriluţ, A.C., Apetrii, M.: Modern Directions in Multivalued Analysis and Optimization Theory, Venus Publishing House, Iaşi (2006) (in Romanian)
40. Sharma, P., Nagar, A.: Topological dynamics on hyperspaces. Appl. Gen. Topol. **11**(1), 1–19 (2010)
41. Wang, Y., Wei, G., Campbell, W.H., Bourquin, S.: A framework of induced hyperspace dynamical systems equipped with the hit-or-miss topology. Chaos Solitons Fractals **41**(4), 1708–1717 (2009)
42. Wicks, K.R.: Fractals and Hyperspaces. Springer, Berlin (1991)

19

Approximations to the distribution of sum of independent non-identically gamma random variables

H. Murakami[1]

Abstract Calculating the sum of independent non-identically distributed random variables is necessary in the scientific field. Computing the probability of the corresponding significance point is important in cases that have a finite sum of random variables. However, it is difficult to evaluate this probability when the number of random variables increases. Under these circumstances, consideration of a more accurate approximation of the distribution function is extremely important. A saddlepoint approximation is performed using upper probabilities from the distribution of the sum of independent non-identically gamma random variables under finite sample sizes. In this study, we compared the results from a saddlepoint approximation to those from normal and moment-based approximations to identify the most appropriate method to use for the distribution function.

Keywords Independent and non-identically distributed · Saddlepoint approximation · Sum of gamma random variables

Introduction

The distribution of the sum of independent identically distributed gamma random variables is well known. However, within the scientific field, it is necessary to know the distribution of the sum of independent non-identically distributed (i.n.i.d.) gamma random variables. For example, it would be necessary to know this distribution for calculating total waiting times where component times are assumed to be independent exponential or gamma random variables. In addition, engineers calculate total excess water flow into a dam as the sum of i.n.i.d. gamma random variables. To calculate the exact probability distribution of the sum of i.n.i.d. gamma random variables, the probability of all possible elements consistent with the sum must be computed. Mathai [12] derived the distribution of the sum of i.n.i.d. gamma random variables by converting the moment-generating function. Additionally, Moschopoulos [13] calculated the distribution of the sum of i.n.i.d. gamma random variables using a simple recursive relation approach. For the detail of the gamma distribution family, we refer the reader to Khodabin and Ahmadabadi [9]. However, Mathai [12] and Moschopoulos [13] derived the density of the sum of i.n.i.d. gamma random variables with infinite summation. This method of computation is intractable in practice, especially in cases in which there is an increase in the number of random variables. An exact calculation is feasible by applying the standard inversion formula to the characteristic function in computer algebra systems, such as Mathematica. However, in these calculations, the probability is estimated with an approximation method. Approximation methods are widely used and have been studied extensively. From a practical view, approximations are typically precise and straightforward to implement in various statistical software programs. Hence, obtaining a more accurate approximation for evaluating the density or the distribution function of i.n.i.d. random variables remains an important area of debate in statistics. In this study, we describe the use of approximation methods to calculate the distribution of the sum of i.n.i.d. gamma random variables in Sect. "A Saddlepoint approximation to the distribution of sum of i.n.i.d. gamma random variables". Furthermore, we

✉ H. Murakami
murakami@gug.math.chuo-u.ac.jp

1 Department of Mathematical Information Science, Tokyo University of Science, Tokyo, Japan

discuss the derivation of the order of errors of suggested approximation for the given distribution. For the approximation presented in this paper, we used the saddlepoint formula employed previously by Daniels [2, 3] and developed by Lugannani [11]. The saddlepoint approximation can be obtained for any statistic or random variable that contains a cumulant generating function. Additionally, the saddlepoint generates accurate probabilities in the tail of distribution. Saddlepoint approximations have been used with great success by several researchers. Excellent discussions of their applications to a range of distributional problems are found in the following studies: Jensen [8], Huzurbazar [7], Kolassa [10], and Butler [1]. Recently, Eisinga et al. [4] discussed the use of the saddlepoint approximation for the sum of i.n.i.d. binomial random variables. Additionally, Murakami [14] and Nadarajah [15] considered the use of the saddlepoint approximation for the sum of i.n.i.d. uniform and beta random variables, respectively. In Sect. "Numerical results", we discuss the results obtained from using the saddlepoint approximation. In Sect. "Concluding remarks", we summarize our conclusions.

A Saddlepoint approximation to the distribution of sum of i.n.i.d. gamma random variables

In this section, we discuss the use of the saddlepoint approximation of the sum of independent non-identically gamma random variables. We assumed that $X_1, \ldots, X_n$ are independent random variables, with shape, $\alpha_i > 0$, and scale parameters, $\beta_i > 0$, for $i = 1, \ldots, n$. Next, we let $S_n = X_1 + X_2 + \cdots + X_n$. The moment-generating function of S_n is

$$M_n(s) = \prod_{i=1}^{n} (1 - \beta_i s)^{-\alpha_i}.$$

It is important to note that Mathai [12] derived the density function of the sum of i.n.i.d. gamma random variables by converting its moment-generating function as follows:

$$\begin{aligned} f_{S_n}(x) = & \left\{ \prod_{i=1}^{n} \beta_i^{\alpha_i} \Gamma(\rho) \right\}^{-1} x^{\rho-1} \exp\left(-\frac{x}{\beta_1}\right) \\ & \times \sum_{r_2=0}^{\infty} \cdots \sum_{r_n=0}^{\infty} \left\{ (\alpha_2)_{r_2} \cdots (\alpha_n)_{r_n} \right\} \\ & \times \left[\left(\frac{1}{\beta_1} - \frac{1}{\beta_2} \right) x \right]^{r_2} \cdots \left[\left(\frac{1}{\beta_1} - \frac{1}{\beta_n} \right) x \right]^{r_n} \frac{1}{r_2! \cdots r_n! (\rho)_\rho}, \\ & x > 0, \end{aligned}$$

where $\rho = \alpha_1 + \cdots + \alpha_n$ and $(y)_z$ denote the Pochhammer symbol. In addition, Moschopoulos [13] obtained the density function of the sum of i.n.i.d. gamma random variables using the following simple recursive relation approach:

$$f_{S_n}(x) = \frac{C}{\Gamma(\rho + k) \beta_1^{\rho+k}} \sum_{k=0}^{\infty} \delta_k x^{\rho+k-1} \exp\left(-\frac{x}{\beta_1}\right), \quad x > 0,$$

where

$$C = \prod_{i=1}^{n} \left(\frac{\beta_1}{\beta_i} \right)^{\alpha_i},$$

$$\delta_{k+1} = \frac{1}{k+1} \sum_{i=1}^{k+1} i \omega_i \delta_{k+1-i}, \quad k = 0, 1, 2, \ldots,$$

$$\omega_k = \frac{1}{k} \sum_{i=1}^{n} \alpha_i (1 - \frac{\beta_1}{\beta_i})^k, \quad k = 1, 2, \ldots,$$

with $\delta_0 = 1$. It is difficult to evaluate the exact density of S_n with increasing n.

Herein, we consider an approximation to the distribution of S_n. The cumulant generating function of S_n is

$$\kappa_n(s) = -\sum_{i=1}^{n} \alpha_i \log(1 - \beta_i s).$$

Using the cumulant generating function, the mean, μ, and variance, σ^2, of S_n are given below:

$$\mu = \sum_{i=1}^{n} \alpha_i \beta_i \quad \text{and} \quad \sigma^2 = \sum_{i=1}^{n} \alpha_i \beta_i^2.$$

According to Daniels [3], the saddlepoint approximation of the density function of S_n is as follows:

$$f_s(v) = f_*(v) \{1 + O(n^{-1})\},$$

where

$$f_*(v) = \{2\pi \kappa_n''(\hat{s})\}^{-\frac{1}{2}} \exp\{\kappa_n(\hat{s}) - \hat{s}v\},$$

$$\kappa_n'(s) = \sum_{i=1}^{n} \frac{\alpha_i \beta_i}{1 - \beta_i s}, \quad \kappa_n''(s) = \sum_{i=1}^{n} \frac{\alpha_i \beta_i^2}{(1 - \beta_i s)^2},$$

and $\hat{s}$ is the root of $\kappa_n'(s) = v$ which is readily solved numerically by the Newton–Raphson algorithm.

Several approaches have been used to further minimize the error of the saddlepoint approximation [5]. For example, one method uses a higher order approximation by including adjustments for the third and fourth cumulants [3]. A higher order saddlepoint approximation uses the following correction term:

$$f_s(v) = f_*(v) \left\{ 1 + \frac{1}{8} \frac{\kappa_n^{(4)}(\hat{s})}{\kappa_n''(\hat{s})^2} - \frac{5}{24} \frac{\kappa_n^{(3)}(\hat{s})^2}{\kappa_n''(\hat{s})^3} + O(n^{-2}) \right\}, \tag{1}$$

where

$$\kappa_n^{(3)}(s) = \sum_{i=1}^{n} \frac{2\alpha_i \beta_i^3}{(1 - \beta_i s)^3}, \quad \kappa_n^{(4)}(s) = \sum_{i=1}^{n} \frac{6\alpha_i \beta_i^4}{(1 - \beta_i s)^4}.$$

The approximate tail probabilities of S_n are determined by numerically integrating Eq. (1).

An alternative approach is to use the Lugannani and Rice [11] for the continuous tail probability approximation as follows:

$$\Pr(S_n < v) \approx \Phi(\hat{w}) - \phi(\hat{w})\left(\frac{1}{\hat{u}} - \frac{1}{\hat{w}}\right),$$

where $\phi(\cdot)$ is the standard normal density function, $\Phi(\cdot)$ is the corresponding cumulative distribution function, and

$$\hat{w} = \sqrt{2(\hat{s}v - \kappa_n(\hat{s}))}\mathrm{sgn}(\hat{s}), \quad \hat{u} = \hat{s}\sqrt{\kappa_n''(\hat{s})},$$

where $\mathrm{sgn}(\hat{s}) = \pm 1, 0$ if $\hat{s}$ is positive, negative, or zero.

Numerical results

In this section, we investigated the upper probability using the saddlepoint approximations to S_n. In this study, we focused on the Lugannani–Rice formula. Note that Mathai [12] obtained a normal approximation with $n \to \infty$. Moschopoulos [13] derived the density of S_n with infinite summation as follows:

$$f_{S_n}(x) = \frac{C}{\Gamma(\rho+k)\beta_1^{\rho+k}} \sum_{k=0}^{\infty} \delta_k x^{\rho+k-1} \exp\left(-\frac{x}{\beta_1}\right), \quad x > 0.$$

We used a finite number and truncated the infinite series to meet an acceptable precision as published by Moschopoulos [13]. This equation is listed as follows:

$$\begin{aligned} f_{S_n}(x) &= \frac{C}{\Gamma(\rho+k)\beta_1^{\rho+k}} \sum_{k=0}^{\infty} \delta_k x^{\rho+k-1} \exp\left(-\frac{x}{\beta_1}\right) \\ &= \frac{C}{\Gamma(\rho)\beta_1^{\rho}} x^{\rho-1} \exp\left(-\frac{x}{\beta_1}\right) \sum_{k=0}^{\infty} \left(\frac{\delta_k}{(\rho)_k}\right)\left(\frac{x}{\beta_1}\right)^k \\ &\le \frac{C}{\Gamma(\rho)\beta_1^{\rho}} x^{\rho-1} \exp\left(-\frac{x}{\beta_1}\right) \sum_{k=0}^{\infty} \frac{1}{k!}\left(\frac{bx}{\beta_1}\right)^k \\ &= \frac{C}{\Gamma(\rho)\beta_1^{\rho}} x^{\rho-1} \exp\left(-\frac{x(1-b)}{\beta_1}\right), \end{aligned}$$

where $(\rho)_k = \rho(\rho+1)\cdots(\rho+k-1)$, $(\rho)_0 = 1$ and $b = \max_{2\le i\le n}(1 - \beta_1/\beta_i)$. To bound the truncation error with the sum of the first $\ell + 1$, we used the following equation:

$$\begin{aligned} E(w) &= \int_0^w \frac{C}{\Gamma(\rho)\beta_1^{\rho}} x^{\rho-1} \exp\left(-\frac{x(1-b)}{\beta_1}\right) dx \\ &\quad - \int_0^w \frac{C}{\Gamma(\rho+k)\beta_1^{\rho+k}} \sum_{k=0}^{\ell} \delta_k x^{\rho+k-1} \exp\left(-\frac{x}{\beta_1}\right) dx. \end{aligned}$$

In addition, we used another approximation method for the distribution of S_n, a moment-based approximation proposed by Ha and Provost [6]. The distribution of S_n is approximated by the polynomial adjusted $\tilde{f}_k(v)$ such that

$$\tilde{f}_k(v) = \psi(v) \sum_{\ell=0}^{k} \xi_\ell v^\ell,$$

where

$$\begin{pmatrix} \xi_0 \\ \xi_1 \\ \vdots \\ \xi_k \end{pmatrix} = \begin{pmatrix} m(0) & m(1) & \cdots & m(k-1) & m(k) \\ m(1) & m(2) & \cdots & m(k) & m(k+1) \\ \vdots & \vdots & \ddots & \vdots & \vdots \\ m(k) & m(k+1) & \cdots & m(2k-1) & m(2k) \end{pmatrix}^{-1} \begin{pmatrix} 1 \\ \mathrm{E(M)} \\ \vdots \\ \mathrm{E(M^k)} \end{pmatrix}$$

and $m(k)$ and $\mathrm{E(M^k)}$ denote the kth moment of the adjusted distribution $\psi(v)$ and the kth moment of S_n, respectively.

Herein, we consider the approximation adjusted with the skew-normal distribution as follows:

$$\psi_*(v) = \frac{2}{\sigma}\phi\left(\frac{v-\ell}{g}\right)\Phi\left(\frac{\lambda(v-\ell)}{g}\right),$$

where

$$\ell = \mu - \delta g\sqrt{\frac{2}{\pi}}, \quad g^2 = \sigma^2\left(1 - \frac{2\delta^2}{\pi}\right)^{-1},$$

$$\zeta = \frac{\mathrm{E(M_p^3)} - 3\mu\sigma^2 - \mu^3}{\sigma^3}$$

$$\lambda = \frac{\delta}{\sqrt{1-\delta^2}}, \quad |\delta| = \mathrm{Sign}(\zeta)\sqrt{\frac{\pi}{2}\frac{|\xi|^{2/3}}{|\xi|^{2/3} + \left(\frac{4-\pi}{2}\right)^{2/3}}},$$

$$\xi = \min(0.99, |\zeta|).$$

Then,

$$m(k) = \frac{\mathrm{d}^k}{\mathrm{d}t^k}\left\{2\exp\left(\ell t + \frac{g^2t^2}{2}\right)\Phi\left(\frac{\lambda g t}{\sqrt{1+\lambda^2}}\right)\right\}\Big|_{t=0}.$$

Note that we obtained $\xi_0 = 1, \xi_1 = \xi_2 = 0$ for $k = 2$. Afterwards, the moment-based approximation with skew-normal polynomial was as follows:

$$\tilde{f}_2(v) = \psi_*(v) \sum_{\ell=0}^{2} \xi_\ell v^\ell = \psi_*(v)(1 + 0 + 0) = \psi_*(v).$$

An important step for the proposed method is to determine the optimal degrees for the polynomials. We followed the selection rule, which is based on the integrated squared differences between density approximations as previously published by Ha and Provost [6].

For this study, the following notations were utilized: exact probability of S_n, E_P, (as proposed by Moschopoulos

Table 1 Numerical results for α % significance level for case 1 ($n = 5$)

v	E_P	A_L	A_N	A_M	r.e. A_L	r.e. A_N	r.e. A_M
Case A							
18.4956	0.6000	0.6002	0.5504	0.5997	0.0003	0.0827	0.0005
20.3703	0.7000	0.7002	0.6615	0.6975	0.0002	0.0550	0.0036
22.7281	0.8000	0.8001	0.7826	0.7974	0.0002	0.0218	0.0033
26.3012	0.9000	0.9001	0.9088	0.8998	0.0001	0.0098	0.0003
29.5275	0.9500	0.9500	0.9665	0.9510	0.0000	0.0174	0.0011
32.5345	0.9750	0.9750	0.9892	0.9758	0.0000	0.0145	0.0009
36.2813	0.9900	0.9900	0.9980	0.9901	0.0000	0.0081	0.0001
Case B							
726.059	0.6000	0.6000	0.5824	0.6005	0.0000	0.0293	0.0009
753.042	0.7000	0.7000	0.6872	0.7002	0.0000	0.0183	0.0003
785.493	0.8000	0.8000	0.7952	0.7998	0.0000	0.0060	0.0002
832.073	0.9000	0.9000	0.9045	0.8996	0.0000	0.0050	0.0004
871.945	0.9500	0.9500	0.9574	0.9498	0.0000	0.0078	0.0002
907.574	0.9750	0.9750	0.9817	0.9750	0.0000	0.0069	0.0000
950.242	0.9900	0.9900	0.9944	0.9901	0.0000	0.0044	0.0001
Case C							
10.1143	0.6000	0.6004	0.4883	0.5785	0.0006	0.1861	0.0365
12.2272	0.7000	0.7001	0.6035	0.6937	0.0001	0.1379	0.0532
15.1163	0.8000	0.7999	0.7457	0.8106	0.0002	0.0679	0.0224
19.9089	0.9000	0.8998	0.9070	0.9148	0.0003	0.0077	0.0297
24.5886	0.9500	0.9498	0.9755	0.9537	0.0002	0.0268	0.0107
29.1965	0.9750	0.9749	0.9954	0.9715	0.0001	0.0209	0.0129
35.2135	0.9900	0.9899	0.9997	0.9862	0.0001	0.0098	0.0047
Case D							
61.4894	0.6000	0.6006	0.5605	0.5998	0.0009	0.0659	0.0003
66.2173	0.7000	0.7006	0.6694	0.6993	0.0008	0.0437	0.0010
72.0867	0.8000	0.8005	0.7863	0.7994	0.0006	0.0171	0.0007
80.8566	0.9000	0.9004	0.9073	0.9003	0.0004	0.0081	0.0003
88.6810	0.9500	0.9502	0.9639	0.9504	0.0002	0.0147	0.0003
95.9162	0.9750	0.9751	0.9873	0.9751	0.0001	0.0126	0.0001
104.878	0.9900	0.9901	0.9973	0.9899	0.0001	0.0073	0.0001
Case E							
0.326492	0.6000	0.6088	0.4596	0.5533	0.0146	0.2339	0.0778
0.426459	0.7000	0.7104	0.5691	0.6850	0.0149	0.1870	0.0215
0.570043	0.8000	0.8120	0.7155	0.8228	0.0150	0.1056	0.0285
0.821751	0.9000	0.9107	0.8967	0.9321	0.0119	0.0037	0.0356
1.080503	0.9500	0.9569	0.9759	0.9545	0.0073	0.0273	0.0048
1.345989	0.9750	0.9787	0.9966	0.9638	0.0038	0.0222	0.0115
1.707191	0.9900	0.9913	0.9999	0.9832	0.0013	0.0100	0.0069
Case F							
259.5839	0.6000	0.6000	0.5772	0.6005	0.0001	0.0380	0.0008
272.0152	0.7000	0.7000	0.6832	0.7002	0.0000	0.0240	0.0003
287.0790	0.8000	0.8000	0.7935	0.7998	0.0000	0.0082	0.0002
308.9076	0.9000	0.9000	0.9055	0.8997	0.0000	0.0061	0.0003
327.7724	0.9500	0.9500	0.9592	0.9499	0.0000	0.0097	0.0001
344.7615	0.9750	0.9750	0.9833	0.9751	0.0000	0.0085	0.0001
365.2610	0.9900	0.9900	0.9952	0.9901	0.0000	0.0053	0.0001

Table 2 Numerical results for α % significance level for case 2 ($n = 10$)

v	E_P	A_L	A_N	A_M	r.e. A_L	r.e. A_N	r.e. A_M
Case A							
26.3644	0.6000	0.6002	0.5579	0.5997	0.0003	0.0702	0.0006
28.5636	0.7000	0.7002	0.6677	0.6992	0.0003	0.0462	0.0012
31.3020	0.8000	0.8002	0.7858	0.7995	0.0002	0.0177	0.0006
35.4041	0.9000	0.9001	0.8909	0.9004	0.0001	0.0090	0.0004
39.0681	0.9500	0.9501	0.9647	0.9503	0.0001	0.0155	0.0004
42.4555	0.9750	0.9750	0.9878	0.9750	0.0000	0.0132	0.0000
46.6462	0.9900	0.9900	0.9975	0.9898	0.0000	0.0075	0.0002
Case B							
1096.671	0.6000	0.6000	0.5855	0.6003	0.0000	0.0242	0.0005
1128.481	0.7000	0.7000	0.6895	0.7002	0.0000	0.0150	0.0003
1166.561	0.8000	0.8000	0.7961	0.7999	0.0000	0.0049	0.0001
1220.920	0.9000	0.9000	0.9038	0.8998	0.0000	0.0042	0.0002
1267.189	0.9500	0.9500	0.9562	0.9499	0.0000	0.0066	0.0001
1308.347	0.9750	0.9750	0.9807	0.9750	0.0000	0.0059	0.0000
1357.419	0.9900	0.9900	0.9938	0.9901	0.0000	0.0038	0.0001
Case C							
3.62148	0.6000	0.6186	0.4116	0.4838	0.0310	0.3140	0.1936
4.71105	0.7000	0.7107	0.5012	0.6281	0.0152	0.2840	0.1027
6.54256	0.8000	0.8050	0.6494	0.8154	0.0062	0.1882	0.0193
10.1893	0.9000	0.9011	0.8732	0.9645	0.0013	0.0298	0.0716
14.2326	0.9500	0.9500	0.9763	0.9552	0.0000	0.0277	0.0055
18.5259	0.9750	0.9747	0.9980	0.9638	0.0003	0.0236	0.0115
24.4566	0.9900	0.9898	1.0000	0.9829	0.0002	0.0101	0.0170
Case D							
65.5973	0.6000	0.6007	0.5653	0.6000	0.0012	0.0578	0.0000
69.7384	0.7000	0.7007	0.6732	0.6995	0.0010	0.0038	0.0007
74.8464	0.8000	0.8007	0.7880	0.7995	0.0008	0.0151	0.0006
82.4243	0.9000	0.9005	0.9065	0.9001	0.0006	0.0073	0.0001
89.1426	0.9500	0.9503	0.9625	0.9503	0.0003	0.0132	0.0003
95.3289	0.9750	0.9752	0.9863	0.9751	0.0002	0.0116	0.0001
102.966	0.9900	0.9901	0.9968	0.9899	0.0001	0.0069	0.0001
Case E							
3.24261	0.6000	0.6023	0.5000	0.5879	0.0038	0.1666	0.0201
3.83360	0.7000	0.7019	0.6131	0.6957	0.0028	0.1241	0.0061
4.63308	0.8000	0.8013	0.7506	0.8058	0.0017	0.0618	0.0072
5.95032	0.9000	0.9007	0.9061	0.9091	0.0008	0.0067	0.0102
7.23386	0.9500	0.9504	0.9739	0.9524	0.0004	0.0251	0.0025
8.49852	0.9750	0.9752	0.9947	0.9723	0.0002	0.0202	0.0028
10.1529	0.9900	0.9901	0.9996	0.9871	0.0001	0.0097	0.0029
Case F							
325.9534	0.6000	0.6000	0.5823	0.6005	0.0000	0.0295	0.0009
338.0179	0.7000	0.7000	0.6871	0.7002	0.0000	0.0184	0.0003
352.5290	0.8000	0.8000	0.7952	0.7998	0.0000	0.0060	0.0003
373.3628	0.9000	0.9000	0.9046	0.8996	0.0000	0.0051	0.0005
391.1991	0.9500	0.9500	0.9575	0.9498	0.0000	0.0078	0.0002
407.1389	0.9750	0.9750	0.9818	0.0000	0.0000	0.0069	0.0000
426.2288	0.9900	0.9900	0.9944	0.9901	0.0000	0.0044	0.0001

Table 3 Numerical results for α % significance level for case 3 ($n = 15$)

v	E_P	A_L	A_N	A_M	r.e. A_L	r.e. A_N	r.e. A_M
Case A							
38.7943	0.6000	0.6002	0.5679	0.6001	0.0003	0.0535	0.0002
41.1704	0.7000	0.7002	0.6757	0.6996	0.0002	0.0347	0.0006
44.0891	0.8000	0.8001	0.7897	0.7995	0.0002	0.0128	0.0007
48.3914	0.9000	0.9001	0.9068	0.9000	0.0001	0.0075	0.0000
52.1756	0.9500	0.9501	0.9620	0.9503	0.0001	0.0127	0.0003
55.6331	0.9750	0.9750	0.9857	0.9752	0.0000	0.0110	0.0002
59.8647	0.9900	0.9900	0.9965	0.9900	0.0000	0.0066	0.0000
Case B							
1588.181	0.6000	0.6000	0.5880	0.6002	0.0000	0.0200	0.0003
1625.798	0.7000	0.7000	0.6914	0.7002	0.0000	0.0123	0.0002
1670.661	0.8000	0.8000	0.7990	0.8000	0.0000	0.0039	0.0000
1734.395	0.9000	0.9000	0.9032	0.8999	0.0000	0.0036	0.0001
1788.379	0.9500	0.9500	0.9553	0.9499	0.0000	0.0055	0.0001
1836.205	0.9750	0.9750	0.9798	0.9750	0.0000	0.0050	0.0000
1893.005	0.9900	0.9900	0.9932	0.9900	0.0000	0.0033	0.0000
Case C							
368.9250	0.6000	0.6004	0.5773	0.6005	0.0007	0.0379	0.0008
385.2496	0.7000	0.7006	0.6977	0.7001	0.0008	0.0242	0.0002
405.0313	0.8000	0.8008	0.7931	0.7997	0.0010	0.0086	0.0004
433.7166	0.9000	0.9012	0.9052	0.8996	0.0013	0.0057	0.0004
458.5522	0.9500	0.9516	0.9590	0.9499	0.0017	0.0095	0.0001
480.9795	0.9750	0.9772	0.9833	0.9751	0.0023	0.0085	0.0001
508.1583	0.9900	0.9925	0.9952	0.9901	0.0025	0.0053	0.0001
Case D							
236.3144	0.6000	0.6001	0.5777	0.6005	0.0002	0.0372	0.0008
245.9966	0.7000	0.7001	0.6834	0.7002	0.0002	0.0237	0.0003
257.7243	0.8000	0.8001	0.7933	0.7998	0.0001	0.0083	0.0002
274.7178	0.9000	0.9001	0.9052	0.8997	0.0001	0.0058	0.0003
289.4113	0.9500	0.9500	0.9589	0.9499	0.0000	0.0094	0.0001
302.6551	0.9750	0.9750	0.9831	0.9751	0.0000	0.0084	0.0001
318.6544	0.9900	0.9900	0.9952	0.9901	0.0000	0.0052	0.0001
Case E							
1.78598	0.6000	0.6156	0.4898	0.5948	0.0260	0.1836	0.0253
2.10679	0.7000	0.7173	0.5966	0.6977	0.0247	0.1478	0.0032
2.55103	0.8000	0.8174	0.7318	0.8139	0.0218	0.0852	0.0174
3.31351	0.9000	0.9131	0.8962	0.9176	0.0146	0.0043	0.0196
4.09487	0.9500	0.9576	0.9724	0.9525	0.0080	0.0236	0.0027
4.90124	0.9750	0.9787	0.9953	0.9675	0.0038	0.0208	0.0077
6.00860	0.9900	0.9912	0.9998	0.9846	0.0012	0.0099	0.0055
Case F							
624.837	0.6000	0.6000	0.5866	0.6003	0.0000	0.0223	0.0004
642.419	0.7000	0.7000	0.6903	0.7002	0.0000	0.0138	0.0003
663.431	0.8000	0.8000	0.7965	0.0000	0.0000	0.0044	0.0000
693.358	0.9000	0.9000	0.9036	0.8998	0.0000	0.0040	0.0002
718.771	0.9500	0.9500	0.9558	0.9499	0.0000	0.0061	0.0001
741.331	0.9750	0.9750	0.9803	0.9750	0.0000	0.0055	0.0000
768.173	0.9900	0.9900	0.9935	0.9900	0.0000	0.0036	0.0000

[13]); normal approximation, A_N; saddlepoint approximation with Lugannani–Rice formula, A_L; moment-based approximation with skew-normal polynomial, A_M; and the relative error of approximations, r.e. (Tables 1, 2, 3). We used different values for $\vec{\alpha} = (\alpha_1, \alpha_2, \ldots, \alpha_n)$ and $\vec{\beta} = (\beta_1, \beta_2, \ldots, \beta_n)$. These values were grouped into cases 1–3. Herein, we assumed that α_i and β_i for $n = 5, 10$ and 15 as follows:

Case 1: α_i and β_i were simulated from Case A: Uniform distribution with interval [0, 3] independently as

$$\alpha_i = (1.04022, 1.52149, 2.96165, 0.77156, 1.93264)$$

$$\beta_i = (2.93353, 2.60821, 2.49735, 1.57684, 1.05720)$$

Case B: Poisson distribution with parameter $\lambda = 10$ independently as

$$\alpha_i = (9, 10, 18, 8, 11)$$
$$\beta_i = (17, 14, 13, 10, 9)$$

Case C: Lognormal distribution with parameters location $\mu = 0$ and scale $\sigma = 2$ independently as

$$\alpha_i = (0.05459, 0.87723, 0.98562, 1.37783, 6.40726)$$

$$\beta_i = (1.68872, 0.39881, 0.25645, 6.14009, 0.18285)$$

Case D: Gamma distribution with parameters shape $\gamma = 2.0$ and scale $\xi = 1.5$ independently as

$$\alpha_i = (1.96560, 1.89408, 3.00261, 4.28812, 3.01364)$$

$$\beta_i = (6.99957, 5.68468, 3.15081, 3.49359, 3.32123)$$

Case E: Exponential distribution with parameter $\lambda = 2.0$ independently as

$$\alpha_i = (0.52959, 0.33946, 0.00643, 0.67897, 0.21986)$$

$$\beta_i = (0.01120, 0.06997, 0.09169, 0.32160, 0.52149)$$

Case F: Binomial distribution with parameter $N = 20$, $p = 0.3$ independently as

$$\alpha_i = (5, 6, 11, 5, 7)$$
$$\beta_i = (10, 8, 8, 6, 5)$$

Case 2: α_i and β_i were simulated from Case A: Uniform [0, 3] independently as

$$\alpha_i = (0.22417, 1.14752, 0.50906, 1.98942, 2.72316, 2.50722, 1.28708, 0.52985, 2.61593, 2.06543)$$

$$\beta_i = (2.85308, 0.91297, 2.36745, 0.57299, 2.95146, 2.01277, 2.77988, 0.36263, 0.10206, 1.98484)$$

Case B: Poisson distribution with parameter $\lambda = 10$ independently as

$$\alpha_i = (6, 9, 7, 11, 14, 13, 9, 7, 14, 11)$$
$$\beta_i = (15, 8, 12, 7, 17, 11, 15, 6, 5, 11)$$

Case C: Lognormal distribution with parameters location $\mu = 0$ and scale $\sigma = 2$ independently as

$$\alpha_i = (0.11560, 10.5703, 0.18732, 0.05774, 9.51800, 0.32369, 0.01573, 0.38308, 0.15837, 0.50578)$$

$$\beta_i = (0.57765, 0.00716, 0.04465, 0.01232, 0.13829, 0.69592, 0.62148, 7.70751, 0.11445, 0.04548)$$

Case D: Gamma distribution with parameters shape $\gamma = 2.0$ and scale $\xi = 1.5$ independently as

$$\alpha_i = (2.17916, 1.10074, 1.40375, 0.55393, 3.18918, 3.27868, 5.79357, 0.74198, 2.54858, 1.52722)$$

$$\beta_i = (4.51414, 3.89484, 2.66041, 3.88386, 1.83366, 2.45540, 2.29499, 2.67110, 1.79620, 6.17920)$$

Case E: Exponential distribution with parameter $\lambda = 2.0$ independently as

$$\alpha_i = (0.63848, 0.49992, 0.68366, 0.20362, 0.39683, 0.03837, 1.13507, 0.31168, 0.24129, 0.29315)$$

$$\beta_i = (1.54349, 0.15572, 1.81201, 0.21383, 0.25284, 0.31853, 0.27480, 0.62515, 0.89060, 0.21336)$$

Case F: Binomial distribution with parameter $N = 20$, $p = 0.3$ independently as

$$\alpha_i = (5, 5, 5, 7, 6, 9, 3, 6, 7, 6)$$
$$\beta_i = (3, 7, 2, 7, 6, 6, 6, 5, 4, 7)$$

Case 3: α_i and β_i were simulated from Case A: Uniform distribution with interval [0, 3] independently as

$$\alpha_i = (0.58008, 2.22637, 2.51611, 1.12297, 2.29383, 1.18906, 1.57483, 0.31849, 2.53235, 2.27937, 2.91811, 2.57865, 0.79358, 1.86520, 2.46924)$$

$$\beta_i = (0.28821, 1.91593, 2.49916, 0.86430, 0.84279, 0.11141, 1.67239, 2.36912, 2.71671, 0.77938, 1.87762, 0.17339, 0.96113, 0.79465, 1.37481)$$

Case B: Poisson distribution with parameter $\lambda = 10$ independently as

$$\alpha_i = (14, 8, 7, 6, 11, 8, 9, 10, 14, 9, 10, 7, 13, 12, 15)$$

$$\beta_i = (11, 8, 12, 10, 16, 11, 4, 7, 5, 10, 8, 10, 18, 6, 14)$$

Case C: Lognormal distribution with parameter location $\mu = 0$ and scale $\sigma = 2$ independently as

$$\alpha_i = (0.65787, 2.10546, 27.7337, 5.30692, 1.32748, 6.60756, 0.06394, 0.44348, 0.62371, 2.24283, 2.82459, 9.36521, 0.82459, 0.27752, 0.78127)$$

$$\beta_i = (0.35243, 0.03793, 9.39041, 0.64510, 5.61762, 1.26038, 33.0086, 2.74834, 0.82037, 9.86410, 9.98734, 0.85253, 18.6350, 0.06063, 0.15637)$$

Case D: Gamma distribution with parameter shape $\gamma = 2.0$ and scale $\xi = 1.5$ independently as

$$\alpha_i = (6.89214, 8.05464, 1.88477, 7.43300, 3.20878, 3.90603, 0.939177, 1.5469, 12.2503, 5.63549, 1.49472, 3.16031, 1.32145, 1.69085, 0.815398)$$

$$\beta_i = (2.62198, 1.1429, 4.84865, 1.37041, 7.08359, 5.63902, 3.86697, 4.96188, 3.69634, 7.9203, 6.06043, 4.08442, 0.501449, 3.76872, 9.79386)$$

Case E: Exponential distribution with parameter $\lambda = 2.0$ independently as

$$\alpha_i = (0.32981, 0.46440, 0.08766, 0.29351, 0.16135, 1.34496, 0.17082, 0.05857, 0.34622, 0.11345, 0.31936, 0.23438, 0.21816, 0.22887, 0.06095)$$

$$\beta_i = (0.05570, 0.88621, 0.78224, 0.42277, 1.07697, 0.23754, 0.43072, 0.20499, 0.24145, 0.38523, 0.13308, 0.19507, 1.63801, 0.15344, 0.11302)$$

Case F: Binomial distribution with parameter $N = 20$, $p = 0.3$ independently as

$$\alpha_i = (6, 5, 8, 6, 7, 3, 7, 9, 6, 8, 6, 7, 7, 7, 8)$$
$$\beta_i = (9, 4, 4, 6, 4, 7, 6, 7, 7, 6, 7, 7, 3, 7, 8)$$

The results listed in Tables 1, 2 and 3 indicate that the A_M approximation was more suitable than the normal approximation for the distribution of S_n. In support of this, we observed that the A_L approximation was more accurate than the A_M approximation in all cases tested. Therefore, we suggest estimating the probability using the A_L approximation in cases with large n.

Concluding remarks

In this paper, we considered both the saddlepoint and moment-based approximations on the distribution of the sum of i.n.i.d. gamma random variables. Use of the saddlepoint approximation was an accurate method for calculating distribution. From our results, we determined that the precision of the saddlepoint approximation was superior to both the normal and moment-based approximations.

Acknowledgments The author would like to thank the editor and the referee for their valuable comments and suggestions. The author appreciates that this research is supported by the Grant-in-Aid for Young Scientists B of JSPS, KAKENHI Number 26730025.

References

1. Butler, R.W.: Saddlepoint Approximations with Applications. Cambridge University Press, Cambridge (2007)
2. Daniels, H.E.: Saddlepoint approximations in statistics. Ann. Math. Stat. **25**, 631–650 (1954)
3. Daniels, H.E.: Tail probability approximations. Int. Stat. Rev. **55**, 37–48 (1987)
4. Eisinga, R., Grotenhuis, M.T., Pelzer, B.: Saddlepoint approximation for the sum of independent non-identically distributed binomial random variables. Stat. Neerl. **67**, 190–201 (2013)
5. Gillespie, C.S., Renshaw, E.: An improved saddlepoint approximation. Math. Biosci. **208**, 359–374 (2007)
6. Ha, H.-T., Provost, S.B.: A viable alternative to resorting to statistical tables. Commun. Stat. Simul. Comput. **36**, 1135–1151 (2007)
7. Huzurbazar, S.: Practical saddlepoint approximations. Am. Stat. **53**, 225–232 (1999)
8. Jensen, J.L.: Saddlepoint Approximations. Oxford University Press, Oxford (1995)
9. Khodabin, M., Ahmadabadi, A.: Some properties of generalized gamma distribution. Math. Sci. **4**, 9–28 (2010)
10. Kolassa, J.E.: Series Approximation Methods in Statistics. Springer-Verlag, New York (2006)
11. Lugannani, R., Rice, S.O.: Saddlepoint approximation for the distribution of the sum of independent random variables. Adv. Appl. Probab. **12**, 475–490 (1980)
12. Mathai, A.M.: Storage capacity of a dam with gamma type inputs. Ann. Inst. Stat. Math. **34**, 591–597 (1982)
13. Moschopoulos, P.G.: The distribution of the sum of independent gamma random variables. Ann. Inst. Stat. Math. **37**, 541–544 (1985)
14. Murakami, H.: A saddlepoint approximation to the distribution of the sum of independent non-identically uniform random variables. Stat. Neerl. **68**, 267–275 (2014)
15. Nadarajah, S., Jiang, X., Chu, J.: A saddlepoint approximation to the distribution of the sum of independent non-identically beta random variables. Stat. Neerl. **69**, 102–114 (2015)

20

Module character inner amenability of Banach algebras

H. Sadeghi[1] · M. Lashkarizadeh Bami[2]

Abstract In the present paper, we introduce the notion of module (ϕ, φ)-inner amenability and module character inner amenability for a Banach algebra A which is a Banach module over another Banach algebra $\mathfrak{A}$ with compatible actions. We characterize module (ϕ, φ)-inner amenability and prove some hereditary properties.

Keywords Module (ϕ, φ)-inner amenability · Module character inner amenability · Banach $\mathfrak{A}$-bimodule

Mathematics Subject Classification 46H25

Introduction and preliminaries

Lau [10] introduced a wide class of Banach algebras, called F-algebras, and studied the notion of left amenability for these algebras. In [12], Nasr-Isfahani introduced the concept of inner amenability for Lau algebras. A Lau algebra A was said to be inner amenable if there exists a topological inner invariant mean on the W^*-algebra A^*, that is, a positive linear functional m of norm 1 on A^*, such that $m(f.a) = m(a.f)$ for all $f \in A^*$ and all $a \in P_1(A) = \{a \in A : \|a\| = 1\}$ (or equivalently, for all $a \in A$). Commutative Lau algebras, such as the Fourier algebra $A(G)$ of a locally compact group G, are examples of inner amenable algebras. In addition, the group algebra $L^1(G)$ of any locally compact group G is inner amenable.

Recently, Jabbari et al. [8] have introduced the notion of φ-inner amenability for a Banach algebra A, where $\varphi \in \Delta(A)$, the character space of A. A Banach algebra A was said to be φ-inner amenable if there exists a $m \in A^{**}$ satisfying $m(\varphi) = 1$ and $m(f.a) = m(a.f)(a \in A, f \in A^*)$. A is said to be character inner amenable if and only if A is φ-inner amenable for every $\varphi \in \Delta(A)$.

In [6], Ebrahimi Vishki and Khoddami have investigated the character inner amenability for certain products of Banach algebras consist of projective tensor product $A\widehat{\otimes}B$, Lau product $A \times_\theta B$, where $\theta \in \Delta(B)$ and the module extension $A \oplus X$. For instance, they showed that the projective tensor product $A\widehat{\otimes}B$ is character inner amenable if and only if both A and B are character inner amenable.

Let $\mathfrak{A}$ and A be Banach algebras, such that A be a Banach $\mathfrak{A}$-bimodule with compatible actions

$$\alpha.(ab) = (\alpha.a)b, (ab).\alpha = a(b.\alpha), \alpha.(\beta.a) = (\alpha\beta).a, (a.\beta).\alpha = a.(\beta\alpha),$$

for all $a, b \in A$ and $\alpha \in \mathfrak{A}$.

Let X be a Banach A-bimodule and a Banach $\mathfrak{A}$-bimodule with compatible left actions defined by

$$\alpha.(a.x) = (\alpha.a).x, a.(\alpha.x) = (a.\alpha).x, (\alpha.x).a = \alpha.(x.a)(a \in A, \alpha \in \mathfrak{A}, \mathfrak{x} \in \mathfrak{X}),$$

and similar for the right or two-sided actions. Then, we say that X is a Banach A-$\mathfrak{A}$-module.

✉ H. Sadeghi
h.sadeghi@iaufr.ac.ir

M. Lashkarizadeh Bami
lashkari@sci.ui.ac.ir

[1] Department of Mathematics, Fereydan Branch, Islamic Azad University, Isfahan, Iran

[2] Department of Mathematics, Faculty of Science, University of Isfahan, P.O. Box 81745-163, Isfahan, Iran

Let $A\widehat{\otimes}A$ be the projective tensor product of A and A which is a Banach A-bimodule and a Banach $\mathfrak{A}$-bimodule by the following actions:

$$\alpha.(a\otimes b)=(\alpha.a)\otimes b, c.(a\otimes b)=(ca)\otimes b(\alpha\in\mathfrak{A}, \mathfrak{a},\mathfrak{b},\mathfrak{c}\in\mathfrak{A}),$$

similarly for the right actions. Let $I_{A\widehat{\otimes}A}$ be the closed ideal of $A\widehat{\otimes}A$ generated by elements of the form:

$$\{a.\alpha\otimes b-a\otimes\alpha.b|\alpha\in\mathfrak{A}, \mathfrak{a},\mathfrak{b}\in\mathfrak{A}\}. \tag{1}$$

Consider the map $\omega_A\in\mathcal{L}(A\widehat{\otimes}A,A)$ defined by $\omega_A(a\otimes b)=ab$ and extended by linearity and continuity. Let J_A be the closed ideal of A generated by

$$\omega(I_{A\widehat{\otimes}A})=\{(a.\alpha)b-a(\alpha.b) \quad |a,b\in A,\alpha\in\mathfrak{A}\}. \tag{2}$$

Then, the module projective tensor product $A\widehat{\otimes}_{\mathfrak{A}}A$, which is $(A\widehat{\otimes}A)/I_{A\widehat{\otimes}A}$ by [16], and the quotient Banach algebra A/J_A are both Banach A-bimodules and Banach $\mathfrak{A}$-bimodules. In addition, A/J_A is A-$\mathfrak{A}$-module with compatible actions when A acts on A/J_A canonically.

Let A and $\mathfrak{A}$ be Banach algebras, such that A is a Banach $\mathfrak{A}$-bimodule with compatible actions. Let $\varphi\in\Delta(\mathfrak{A})\cup\{\}$ and consider the set $\Omega_{A,\varphi}$ of linear continuous maps $\phi:A\longrightarrow\mathfrak{A}$, such that

$$\phi(ab)=\phi(a)\phi(b), \phi(\alpha.a)=\phi(a.\alpha)=\varphi(\alpha)\phi(a) \\ (a,b\in A,\alpha\in\mathfrak{A}). \tag{3}$$

The concept of module (ϕ,φ)-amenability and module character amenability for Banach algebra A, where $\varphi\in\Delta(\mathfrak{A})$ and $\phi\in\Omega_A$ were introduced by Bodaghi and Amini in [4].

Our aim in this paper is to introduce and study module (ϕ,φ)-inner amenability and module character inner amenability of Banach algebras. We characterize (ϕ,φ)-inner amenability and prove some hereditary properties. Moreover, we investigate that module (ϕ,φ)-inner amenability for certain class of Banach algebras consists of projective tensor product $A\widehat{\otimes}B$, $A\oplus_\infty B$, and $A\oplus_p B$, the l^p-direct sum of A and B, where $1\leq p<\infty$.

Characterization and hereditary properties

We commence this section with the following definition.

Definition 2.1 Let A be a Banach $\mathfrak{A}$-bimodule and let $\varphi\in\Delta(\mathfrak{A})$ and $\phi\in\Omega_A$. Then, A is called module (ϕ,φ)-inner amenable if there exists $m\in A^{**}$, such that $m(\varphi\circ\phi)=1$, $m(f.a)=m(a.f)$ and $m(\alpha.f)=m(f.\alpha)$ for all $a\in A, f\in A^*$ and $\alpha\in\mathfrak{A}$. A Banach $\mathfrak{A}$-bimodule A is called module character inner amenable if it is module (ϕ,φ)-inner amenable for each $\varphi\in\Delta(\mathfrak{A})$ and $\phi\in\Omega_A$.

We note that if $\mathfrak{A}=\mathbb{C}$ and φ is the identity map, then the module (ϕ,φ)-inner amenability and module character inner amenability coincide with ϕ-inner amenability and character inner amenability (see [8] and [6]).

The next theorem characterizes module (ϕ,φ)-inner amenability of Banach algebras that is analogue of Proposition 2.1 of [5] on module (ϕ,φ)-amenable Banach algebras.

Theorem 2.2 *Let A be a Banach $\mathfrak{A}$-bimodule and let $\varphi\in\Delta(\mathfrak{A})$ and $\phi\in\Omega_A$. Then, the following statements are equivalent:*

(i) *A is module (ϕ,φ)-inner amenable;*
(ii) *There exists a bounded net $(a_i)_i$ in A such that $\|aa_i-a_ia\|\longrightarrow 0$, $\|\alpha.a_i-a_i.\alpha\|\longrightarrow 0(a\in A,\alpha\in\mathfrak{A})$ and $\varphi\circ\phi(a_i)=1$ for all i;*
(iii) *There exists a bounded net $(a_i)_i$ in A such that $\|aa_i-a_ia\|\longrightarrow 0$, $\|\alpha.a_i-a_i.\alpha\|\longrightarrow 0(a\in A,\alpha\in\mathfrak{A})$ and $\varphi\circ\phi(a_i)\longrightarrow 1$.*

Proof (iii) $\Rightarrow$(i) Assume that a net $(a_i)_i$ exists. Let m be a w^*-cluster point of the net $(a_i)_i$ in A^{**}. Then, $\langle m,\varphi\circ\phi\rangle=\lim_i\langle\varphi\circ\phi,a_i\rangle=1$. For every $a\in A$ and $f\in A^*$, we have

$$\begin{aligned}\langle m,f.a\rangle&=\lim_i\langle f.a,a_i\rangle=\lim_i\langle f,aa_i\rangle\\&=\lim_i\langle f,aa_i-a_ia\rangle+\lim_i\langle f,a_ia\rangle\\&=\lim_i\langle a.f,a_i\rangle=\langle m,a.f\rangle,\end{aligned}$$

and similarly, we have $\langle m,f.\alpha\rangle=\langle m,\alpha.f\rangle(\alpha\in\mathfrak{A})$. Therefore, A is module (ϕ,φ)-inner amenable.

(i)$\Rightarrow$ (ii) Suppose that A is module (ϕ,φ)-inner amenable. Then, there exists $m\in A^{**}$ such that $m(\varphi\circ\phi)=1$, $m(f.a)=m(a.f)$, and $m(\alpha.f)=m(f.\alpha)$ for all $a\in A, f\in A^*$ and $\alpha\in\mathfrak{A}$. Choose a net $(u_\beta)_\beta$ in A with $u_\beta\longrightarrow m$ in the w^*-topology of A^{**} and $\|u_\beta\|\leq\|m\|$ for all β. Since $\langle\varphi\circ\phi,u_\beta\rangle\longrightarrow\langle\varphi\circ\phi,m\rangle=1$, passing to a subnet and replacing u_β by $(1/\varphi\circ\phi(u_\beta))u_\beta$, we may assume that $\varphi\circ\phi(u_\beta)=1$ and $\|u_\beta\|\leq\|m\|+1$ for all β. Consider the product space A^A endowed with the product of norm topological. Define a linear map $T:A\longrightarrow A^A$ by $T(b)=(ab-ba+\alpha.b-b.\alpha)_{a\in A}$, for all $b\in A$ and $\alpha\in\mathfrak{A}$. Let

$$B=\{b\in A:\|b\|\leq\|m\|+1 \text{ and } \varphi\circ\phi(\mathrm{b})=1\}\subseteq \mathrm{A}.$$

Clearly, B is convex and so $T(B)$ is a convex subset of A^A. For every $f\in A^*$, we have

$$\begin{aligned}
&\langle f, au_\beta - u_\beta a + \alpha.u_\beta - u_\beta.\alpha\rangle \\
&= \langle f, au_\beta\rangle - \langle f, u_\beta a\rangle + \langle f, \alpha.u_\beta\rangle + \langle f, u_\beta.\alpha\rangle \\
&= \langle f.a, u_\beta\rangle - \langle a.f, u_\beta\rangle + \langle f.\alpha, u_\beta\rangle + \langle \alpha.f, u_\beta\rangle \\
&\to \langle m, f.a\rangle - \langle m, a.f\rangle + \langle m, f.\alpha\rangle - \langle m, \alpha.f\rangle \\
&= 0.
\end{aligned}$$

This product of weak topologies coincide with topology on A^A (see Theorem 4.3 of [17]). By Mazur's theorem, $0 \in \overline{T(B)}^w = \overline{T(B)}^{\|.\|}$. Therefore, there exists a bounded net $(a_i)_i$ in A, such that $\varphi \circ \phi(a_i) = 1$ and

$$\|aa_i - a_i a\| \longrightarrow 0, \|\alpha.a_i - a_i.\alpha\| \longrightarrow 0 (a \in A, \alpha \in \mathfrak{A}).$$

(ii)$\Rightarrow$ (iii) It is clear. □

Definition 2.3 We say that the Banach algebra $\mathfrak{A}$ acts trivially on A from the left (right) if there is a multiplicative linear functional f on $\mathfrak{A}$, such that $\alpha.a = f(\alpha)a$ (resp. $a.\alpha = f(\alpha)a$) for all $\alpha \in \mathfrak{A}$ and $a \in A$.

For the proof of the following result, we refer to Lemma 3.13 of [1].

Lemma 2.4 *Let $\mathfrak{A}$ acts on A trivially from the left or right and A/J_A has a right bounded approximate identity, then for each $\alpha \in \mathfrak{A}$ and $a \in A$ we have $f(\alpha)a - a.\alpha \in J_A$.*

Let $\varphi \in \Delta(\mathfrak{A})$ and $\phi \in \Omega_A$. Clearly, $\phi\big((a.\alpha)b - a(\alpha.b)\big) = 0(\alpha \in \mathfrak{A}, \mathfrak{a}, \mathfrak{b} \in \mathfrak{A})$, and hence, $\phi = 0$ on J_A and $\tilde{\phi} : A/J_A \longrightarrow \mathfrak{A}$ given by $\tilde{\phi}(a + J_A) = \phi(a)$ is well defined. Then, $\tilde{\phi} \in \Omega_{A/J_A}$.

Proposition 2.5 *Let A be a Banach $\mathfrak{A}$-bimodule, and let $\mathfrak{A}$ acts on A trivially from the left and A/J_A has a bounded approximate identity. Then A/J_A is $\varphi \circ \tilde{\phi}$-inner amenable for every $\varphi \in \Delta(\mathfrak{A})$ and $\phi \in \Omega_A$.*

Proof Let $(e_\alpha + J_A)_\alpha$ be a bounded approximate identity of A/J_A and let $\varphi \in \Delta(\mathfrak{A})$ and $\phi \in \Omega_A$. Then, $\varphi \circ \tilde{\phi}(e_\alpha + J_A) \longrightarrow 1$. Clearly

$$\|(a + J_A)(e_\alpha + J_A) - (e_\alpha + J_A)(a + J_A)\| \longrightarrow 0$$

for all $a + J_A \in A/J_A$. By Proposition 2.2 of [6], A/J_A is $\varphi \circ \tilde{\phi}$-inner amenable. □

Note that in the above proposition, both left and right actions of $\mathfrak{A}$ on A/J_A are trivial, by Lemma 2.4. Therefore, for every $\alpha \in \mathfrak{A}$, we have

$$\begin{aligned}
&\|\alpha.(e_\alpha + J_A) - (e_\alpha + J_A).\alpha\| \\
&= \|\alpha.e_\alpha + J_A - e_\alpha.\alpha + J_A\| \\
&= \|f(\alpha)(e_\alpha + J_A) - f(\alpha)(e_\alpha + J_A)\| = 0.
\end{aligned}$$

Then, the net $(e_\alpha + J_A)_\alpha$ satisfies condition (iii) of Theorem 2.2, and hence, A/J_A is module $(\tilde{\phi}, \varphi)$-inner amenable.

Remark 2.6 A inverse semigroup is a discrete semigroup S, such that for each $s \in S$, there is a unique element $s^* \in S$ with $ss^*s = s$ and $s^*ss^* = s^*$. An element $e \in S$ is called an idempotent if $e^2 = e^* = e$. The set of idempotent elements of S is denoted by E_S. Define the relation $\leq$ on E_S by $e \leq d \Leftrightarrow ed = e(e, d \in E_S)$. Then, E_S is a commutative subsemigroup of S, and $l^1(E_S)$ may be regarded as a subalgebra of $l^1(S)$.

Let s be an inverse semigroup with the set of idempotents E_S. We let $l^1(E_S)$ acts on $l^1(S)$ by multiplication from the right and trivially from the left, that is

$$\delta_e.\delta_s = \delta_s\delta_s.\delta_e = \delta_{se} = \delta_s * \delta_e(e \in E_S, s \in S).$$

By these actions, $l_1(S)$ becomes a Banach $l_1(E_S)$-module. In this case

$$J_{l^1(S)} = \{\delta_{set} - \delta_{st} | e \in E_S, s, t \in S\}.$$

We consider an equivalence relation on S as follows $s \approx t \Leftrightarrow \delta_s - \delta_t \in J_{l^1(S)}(s, t \in S)$. For inverse semigroup S, the quotient semigroup $S/\approx$ is discrete group and so $l^1(S/\approx)$ has an identity (see [2, 13]). Indeed, $S/\approx$ is homomorphic to the maximal group homomorphic image G_S of S (see [11, 14]). It is also shown in Theorem 3.3 of [15] that $l^1(S)/J_{l^1(S)} \cong l^1(S/\approx) = l^1(G_S)$ is a commutative $l^1(E_S)$-bimodule with the following actions:

$$\delta_e.\delta_{[s]} = \delta_{[s]}, \delta_{[s]}.\delta_e = \delta_{[se]}(s \in S, e \in E_S),$$

where $[s]$ denotes the equivalence class of s in G_S.

It is shown in [4] that the maps φ and ϕ satisfying (3) exist for $l^1(S)$.

Example 2.7 Let S be an inverse semigroup with the set of idempotents E_S. Consider $l^1(S)$ as a Banach module over $l^1(E_S)$ with the trivial left action and natural right action. Then, by Proposition 2.5, $l^1(G_S)$ is $\varphi \circ \tilde{\phi}$-inner amenable (module $(\tilde{\phi}, \varphi)$-inner amenable) for all $\varphi \in \Delta(l^1(E_S))$ and $\phi \in \Omega_{l^1(S)}$.

Example 2.8 Let A be a commutative Banach algebra and commutative $\mathfrak{A}$-bimodule (i.e., $\alpha.a = a.\alpha(a \in A, \alpha \in \mathfrak{A})$). Let $\varphi \in \Delta(\mathfrak{A}), \phi \in \Omega_{\mathfrak{A}}$ and let $a \in A$ be such that $\varphi \circ \phi(a) = 1$. put $m = \widehat{a}$. Then, $m(\varphi \circ \phi) = \widehat{a}(\varphi \circ \phi) = \varphi \circ \phi(a) = 1$ and clearly, $m(f.a) = m(a.f)$ and $m(\alpha.f) = m(f.\alpha)$ for all $a \in A, \alpha \in \mathfrak{A}$. Therefore, A is module (ϕ, φ)-inner amenable. In particular, if S is a commutative inverse semigroup, then $l^1(S)$ is commutative and commutative $l^1(E_S)$-bimodule. Therefore, $l^1(S)$ is module (ϕ, φ)-inner amenable for all $\varphi \in \Delta(l^1(E_S))$ and $\phi \in \Omega_{l^1(S)}$.

Example 2.9 Let $S = (\mathbb{N}, \wedge)$ be the inverse semigroup of positive integers with the minimum operation. Let $A = l^1(S), \mathfrak{A} = \mathrm{l}(\mathfrak{E}_{\mathfrak{S}})$ and $\mathfrak{A}$ acts on A by the following actions:

$$\delta_e.\delta_s = \delta_s.\delta_e = \delta_{se} \quad (e \in E_S, s \in S).$$

A is module amenable (see page 42 of [3]). By Theorem 2.1 of [4], A is module character amenable. Let $\varphi \in \Delta(\mathfrak{A})$ and $\phi \in \Omega_A$. Then, there exist $m \in A^{**}$, such that $m(f.a) = \varphi \circ \phi(a)m(f), m(f.\alpha) = \varphi(\alpha)m(f)$ and $m(\varphi \circ \phi) = 1$ for every $f \in A^*, a \in A$ and $\alpha \in \mathfrak{A}$. Let $(a_\alpha)_\alpha$ be a net in A converging to m in the w^*-topology of A^{**}. Since A is commutative and commutative $\mathfrak{A}$-bimodule, for every $a \in A, f \in A^*$

$$\begin{aligned} m(f.a) &= \lim_\alpha \langle f.a, a_\alpha \rangle = \lim_\alpha \langle f, aa_\alpha \rangle \\ &= \lim_\alpha \langle f, a_\alpha a \rangle = \lim_\alpha \langle a.f, a_\alpha \rangle \\ &= m(a.f). \end{aligned}$$

Similarly, for every $\alpha \in \mathfrak{A}$ and $f \in A^*$, we have

$$m(f.\alpha) = m(\alpha.f).$$

Thus, A is module (ϕ, φ)-inner amenable. Therefore, A is module character inner amenable.

The proof of the following proposition is adapted from that of Proposition 2.3 of [4].

Proposition 2.10 *Let A and B be Banach $\mathfrak{A}$-bimodules and let h be an $\mathfrak{A}$-module homomorphism with dense range. If $\phi \in \Omega_B, \varphi \in \Delta(\mathfrak{A})$ and A is module $(\phi \circ h, \varphi)$-inner amenable, then B is module (ϕ, φ)-inner amenable.*

Proof Let $m \in A^{**}$ be such that $m(\varphi \circ (\phi \circ h)) = 1, m(f.a) = m(a.f)$ and $m(f.\alpha) = m(\alpha.f)$ for all $a \in A, f \in A^*$ and $\alpha \in \mathfrak{A}$. Define $m_B \in B^{**}$ by $m_B(g) = m(g \circ h)(g \in B^*)$. We show that $m_B(g.b) = m_B(b.g)(b \in B)$. For see this let $b \in B$ be such that $h(a) = b$. One can easily check that $(g.h(a)) \circ h = (g \circ h).a$ and $(h(a).g) \circ h = a.(g \circ h)$. Hence for every $g \in B^*$

$$\begin{aligned} m_B(g.b) &= m_B(g.h(a)) = m((g.h(a)) \circ h) \\ &= m((g \circ h).a) = m(a.(g \circ h)) \\ &= m((h(a).g) \circ h) = m_B(h(a).g) \\ &= m_B(b.g). \end{aligned}$$

By density of the range of h and the continuity of h, we conclude that $m_B(g.b) = m_B(b.g)(b \in B)$. In addition, for every $\alpha \in \mathfrak{A}$, we have

$$\begin{aligned} m_B(g.\alpha) &= m((g.\alpha) \circ h) = m((g \circ h).\alpha) \\ &= m(\alpha.(g \circ h)) = m((\alpha.g) \circ h)) \\ &= m_B(\alpha.g). \end{aligned}$$

Furthermore, $m_B(\varphi \circ \phi) = m((\varphi \circ \phi) \circ h) = m(\varphi \circ (\phi \circ h)) = 1$. Therefore, B is module (ϕ, φ)-inner amenable. □

Corollary 2.11 *Let A and B be Banach $\mathfrak{A}$-bimodules and let h be an $\mathfrak{A}$-module homomorphism with dense range. Then the module character inner amenability of A implies the module character inner amenability of B. In particular, if A is module character inner amenable, then so is A/J_A.*

The proof idea of the following result is taken from the proof of Lemma 2.6 of [4].

Proposition 2.12 Let A be a Banach $\mathfrak{A}$-bimodule and I be a closed ideal and $\mathfrak{A}$-submodule of A, and let $\varphi \in \Delta(\mathfrak{A})$ and $\phi \in \Omega_A$ be such that $\phi|_I \neq 0$. If A is module (ϕ, φ)-inner amenable, then I is module $(\phi|_I, \varphi)$-inner amenable.

Proof Let $m \in A^{**}$ satisfy $m(\varphi \circ \phi) = 1$, $m(f.a) = m(a.f)$ and $m(\alpha.f) = m(f.\alpha)$ for all $a \in A, f \in A^*$ and $\alpha \in \mathfrak{A}$. By a similar argument as in the proof of Lemma 3.1 of [9], one can define a bounded linear functional n on I^* by, $n(g) = m(f)$ for all $g \in I^*$, where f is an arbitrary element of A^* extending g. Now, for every $g \in I^*, a \in I$ and $\alpha \in \mathfrak{A}$, we have

$$n(g.a) = m(f.a) = m(a.f) = n(a.g),$$

and

$$n(g.\alpha) = m(f.\alpha) = m(\alpha.f) = n(\alpha.g).$$

In addition, $n(\varphi \circ \phi|_I) = m(\varphi \circ \phi) = 1$. Therefore, I is module $(\phi|_I, \varphi)$-inner amenable. □

We need to recall the following remark from [4] to give the next result:

Remark 2.13 Let $\mathfrak{A}$ be a Banach algebra and $\mathfrak{A}$ be the unitization of $\mathfrak{A}$ which is $\mathfrak{A} = \mathfrak{A} \oplus \mathbb{C}$ is a unital Banach algebra which contains $\mathfrak{A}$ as a closed ideal. Let A be a Banach $\mathfrak{A}$-module. Then, A is a Banach $\mathfrak{A}$-bimodule with the following module actions:

$$(\alpha, \lambda).a = \alpha.a + \lambda a, a.(\alpha, \lambda) = a.\alpha + \lambda a(\lambda \in \mathbb{C}, \alpha \in \mathfrak{A}, a \in \mathfrak{A}).$$

Let $A^\sharp = (A \oplus \mathfrak{A}, \cdot)$, where the multiplication $\cdot$ is defined through

$$(a, u) \cdot (b, v) = (ab + a.v + u.b, uv)(a, b \in A, u, v \in \mathfrak{A}).$$

Then, with the actions defined by

$$u.(a, v) = (u.a, uv), (a, v).u = (a.u, vu)(a \in A, u, v \in \mathfrak{A}),$$

$A^\sharp$ is a unital $\mathfrak{A}$-module Banach algebra with the identity $e_{A^\sharp} = (0, e_{\mathfrak{A}})$, where $e_{\mathfrak{A}} = (0, 1)$. Now, suppose that $\phi \in \Omega_A$ and $\varphi^\#$ is the extension of φ on $\mathfrak{A}$ defined by $\varphi^\#(\alpha, \lambda) = \varphi(\alpha) + \lambda(a \in A, \alpha \in \mathfrak{A}, \lambda \in \mathbb{C}$. If $u = (\alpha, \lambda) \in \mathfrak{A}$, it is easy to see that

$$\phi(a.u) = \phi(u.a) = \varphi^{\#}(u)\phi(a)(a \in A). \tag{4}$$

Define $\phi^{\sharp} : A^{\sharp} \longrightarrow \mathfrak{A}$ by

$$\phi^{\sharp}(a,u) = (\phi(a), \varphi^{\#}(u))(a \in A, u \in \mathfrak{A}). \tag{5}$$

Using (4), one can show that $\phi^{\sharp}$ is multiplicative and

$$\phi^{\sharp}(u.(a,v)) = \phi^{\sharp}((a,v).u) = \varphi^{\#}(u)\phi^{\sharp}(a,v)(a \in A, u, v \in \mathfrak{A}).$$

Therefore, $\phi^{\sharp}$ is an extension of ϕ, such that $\phi^{\sharp}(0,u) = \varphi^{\#}(u)$ is the extension $h_0 = \tilde{0}$ of the zero function given by (5).

The proof of the following theorem is inspired by the proof of Proposition 2.7 of [4].

Proposition 2.14 *Let A be a Banach $\mathfrak{A}$-bimodule, and let $\varphi \in \Delta(\mathfrak{A})$ and $\phi \in \Omega_A$. Then, A is module (ϕ, φ)-inner amenable if and only if $A^{\sharp}$ is module $(\phi^{\sharp}, \varphi^{\#})$-inner amenable.*

Proof Let $A^{\sharp}$ be module $(\phi^{\sharp}, \varphi^{\#})$-inner amenable. Since the image of $\phi^{\sharp}|_A$ is included in $\mathfrak{A}$, by 2.12, we conclude that A is module (ϕ, φ)-inner amenable.

Conversely, suppose that A is module (ϕ, φ)-inner amenable. Then, there exists a $m \in A^{**}$, such that $m(\varphi \circ \phi) = 1$, $m(f.a) = m(a.f)$, and $m(\alpha.f) = m(f.\alpha)$ for all $a \in A, f \in A^*$ and $\alpha \in \mathfrak{A}$. By Remark 2.13, we may identify the dual space $(A^{\sharp})^*$ with $A^* \oplus \mathbb{C}h_0$, where $h_0|_A = 0$ and $h_0(e_{A^{\sharp}}) = 1$. Define $n \in (A^{\sharp})^{**}$ by $n(f) = m(f)(f \in A^*)$ and $n(h_0) = 0$. Since A is an ideal and $\mathfrak{A}$-submodule of $A^{\sharp}$, it follows that $h_0.a = 0$ and $h_0.\alpha = 0$ for all $a \in A$ and $\alpha \in \mathfrak{A}$. A simple computation shows that

$$n\big((f + \lambda h_0).(a + \lambda' e_{A^{\sharp}})\big) = n\big((a + \lambda' e_{A^{\sharp}}).(f + \lambda h_0)\big),$$

and

$$n\big((f + \lambda h_0).u\big) = n\big(u.(f + \lambda h_0)\big),$$

for all $f \in A^*, a \in A, u \in \mathfrak{A}$ and $\lambda, \lambda' \in \mathbb{C}$. For $f \in A^*$, consider the map $\bar{f} : A^{\sharp} \longrightarrow \mathbb{C}$ defined by $\bar{f}(a,u) = f(a) + \tilde{\varphi}(u)(a \in A, u \in \mathfrak{A})$. Thus, $\varphi^{\#} \circ \phi^{\sharp} = \overline{\varphi \circ \phi}$, and hence

$$n(\varphi^{\#} \circ \phi^{\sharp}) = n(\overline{\varphi \circ \phi}) = n(\varphi \circ \phi + \lambda h_0) = m(\varphi \circ \phi) = 1.$$

Therefore, $A^{\sharp}$ is module $(\phi^{\sharp}, \varphi^{\#})$-inner amenable.

Module inner amenability of certain Banach algebras

Let $A \widehat{\otimes} B$ be the projective tensor product of two Banach algebras A and B. For every $f \in A^*$ and $g \in B^*$, let $f \otimes g$ denote the element of $(A \widehat{\otimes} B)^*$ satisfying, $f \otimes g(a \otimes b) = f(a)g(b)(a \in A, b \in B)$. In addition, note that $A \widehat{\otimes} B$ is a Banach $\mathfrak{A} \widehat{\otimes} \mathfrak{A}$-bimodule with the following actions:

$$(\alpha \otimes \beta).(a \otimes b) = (\alpha.a) \otimes (\beta.b)(a \in A, b \in B, \alpha, \beta \in \mathfrak{A}),$$

and similarly for right action. For $\varphi_1, \varphi_2 \in \Delta(\mathfrak{A}), \psi \in \Omega_{\mathfrak{A}}(= \Omega_{\mathfrak{A},\varphi})$ and $\phi \in \Omega_A(= \Omega_{A,\varphi_2})$, define $(\phi \otimes \psi) : A \widehat{\otimes} B \rightarrow \mathfrak{A} \widehat{\otimes} \mathfrak{A}$ by $(\phi \otimes \psi)(a \otimes b) = \phi(a) \otimes \psi(b)(a \in A, b \in B)$. Clearly, $\phi \otimes \psi \in \Omega_{A \widehat{\otimes} B}(= \Omega_{A \widehat{\otimes} B, \varphi_1 \otimes \varphi_2})$ and $\varphi_1 \otimes \varphi_2 \in \Delta(\mathfrak{A} \widehat{\otimes} \mathfrak{A})$. In addition, if $\overline{\varphi} \in \Delta(\mathfrak{A} \widehat{\otimes} \mathfrak{A})$, then $\overline{\varphi} = \varphi_1 \otimes \varphi_2$, where $\varphi_1, \varphi_2 \in \Delta(\mathfrak{A})$ (see [4]).

The technique of proof of the following theorem (one side) is similar to that of Theorem 2.8 of [4].

Theorem 3.1 *Let A and B be Banach $\mathfrak{A}$-bimodules, and let $\varphi_1, \varphi_2 \in \Delta(\mathfrak{A})$ and $\phi \in \Omega_A, \psi \in \Omega_B$. Then $A \widehat{\otimes} B$ is module $(\phi \otimes \psi, \varphi_1 \otimes \varphi_2)$-inner amenable (as $\mathfrak{A} \widehat{\otimes} \mathfrak{A}$ module) if and only if A is module (ϕ, φ_1)-inner amenable and B is module (ψ, φ_2)-inner amenable.*

Proof Suppose that $A \widehat{\otimes} B$ is module $(\phi \otimes \psi, \varphi_1 \otimes \varphi_2)$-inner amenable. Then, there exists $m \in (A \widehat{\otimes} B)^{**}$, such that $m\big((\varphi_1 \otimes \varphi_2) \circ (\phi \otimes \psi)\big) = 1$ and

$$\begin{aligned} m\big(f \otimes g.(a \otimes b)\big) &= m\big((a \otimes b).f \otimes g\big), m\big(f \otimes g.(\alpha \otimes \beta)\big) \\ &= m\big((\alpha \otimes \beta).f \otimes g\big), \end{aligned}$$

for all $a \otimes b \in A \widehat{\otimes} B, f \in A^*, g \in B^*$ and $\alpha \otimes \beta \in \mathfrak{A} \widehat{\otimes} \mathfrak{A}$. Define $m_A : A^* \rightarrow \mathbb{C}$ by $m_A(f) = m\big(f \otimes (\varphi_2 \circ \psi)\big)(f \in A^*)$. Therefore, $m_A(\varphi_1 \circ \phi) = m\big((\varphi_1 \circ \phi) \otimes (\varphi_2 \circ \psi)\big) = m\big((\varphi_1 \otimes \varphi_2) \circ (\phi \otimes \psi)\big) = 1$. Choose $b_0 \in A$, such that $\varphi_2 \circ \psi(b_0) = 1$. Therefore, for every $a \in A$ and $f \in A^*$, we have

$$\begin{aligned} m_A(f.a) &= m\big((f.a) \otimes (\varphi_2 \circ \psi)\big) \\ &= m\big((f.a) \otimes (\varphi_2 \circ \psi).b_0\big) \\ &= m\big(f \otimes (\varphi_2 \circ \psi).(a \otimes b_0)\big) \\ &= m\big((a \otimes b_0).f \otimes (\varphi_2 \circ \psi)\big) \\ &= m\big((a.f) \otimes b_0.(\varphi_2 \circ \psi)\big) \\ &= m_A(a.f). \end{aligned}$$

Similarly, for every $\alpha \in \mathfrak{A}$, if we take $\beta \in \mathfrak{A}$, such that $\varphi_2(\beta) = 1$, then

$$\begin{aligned} m_A(f.\alpha) &= m\big((f.\alpha) \otimes (\varphi_2 \circ \psi)\big) \\ &= m\big((f.\alpha) \otimes (\varphi_2 \circ \psi).\beta\big) \\ &= m\big(f \otimes (\varphi_2 \circ \psi).(\alpha \otimes \beta)\big) \\ &= m\big((\alpha \otimes \beta).f \otimes (\varphi_2 \circ \psi)\big) \\ &= m\big((\alpha.f) \otimes \beta.(\varphi_2 \circ \psi)\big) \\ &= m\big((\alpha.f) \otimes (\varphi_2 \circ \psi)\big) \\ &= m_A(\alpha.f), \end{aligned}$$

for all $f \in A^*$. Therefore, A is module (φ, ϕ_1)-inner amenable. Similarly, one can prove that B is module (ψ, ϕ_2)-inner amenable.

For the converse, let A is module (ϕ, φ_1)-inner amenable and B is module (ψ, φ_2)-inner amenable. Then, by Theorem 2.2, there exist bounded nets $(a_i)_i$ in A and $(b_j)_j$ in B with bounds M_1 and M_2, respectively, such that

$$\varphi_1 \circ \phi(a_i) = 1, \|aa_i - a_i a\| \longrightarrow 0, \|\alpha.a_i - a_i.\alpha\| \longrightarrow 0 (a \in A, \alpha \in \mathfrak{A}),$$

and

$$\varphi_2 \circ \psi(b_j) = 1, \|bb_j - b_j b\| \longrightarrow 0, \|\alpha.b_j - b_j.\alpha\| \longrightarrow 0 (b \in B, \alpha \in \mathfrak{A}).$$

Consider the bounded net $(a_i \otimes b_j)_{(i,j)}$ in $A \widehat{\otimes} B$. Therefore, $((\varphi_1 \otimes \varphi_2) \circ (\phi \otimes \psi))(a_i \otimes b_j) = \varphi_1 \circ \phi(a_i)\varphi_2 \circ \psi(b_j) = 1$. Let $\mathfrak{F} = \sum_{l=}^{\mathfrak{N}} \alpha_l \otimes \beta_l \in \mathfrak{A} \widehat{\otimes} \mathfrak{A}$, then

$$\begin{aligned} &\|\mathfrak{F}.(a_i \otimes b_j) - (a_i \otimes b_j).\mathfrak{F}\| \\ &= \| \sum_{l=1}^{N} [(\alpha_l.a_i - a_i.\alpha_l) \otimes \beta_l.b_j + a_i.\alpha_l \otimes (\beta_l.b_j - b_j.\beta_l)] \| \\ &\leq \sum_{l=1}^{N} M_2 \|\beta_l\| \|\alpha_l.a_i - a_i.\alpha_l\| + \sum_{l=1}^{N} M_1 \|\alpha_l\| \|\beta_l.b_j - b_j.\beta_l\| \longrightarrow 0. \end{aligned}$$

Now, let $\mathfrak{G} \in \mathfrak{A} \widehat{\otimes} \mathfrak{A}$, so there exist sequences $(\alpha_l)_l \subseteq \mathfrak{A}$ and $(\beta_l)_l \subseteq \mathfrak{A}$, such that $\mathfrak{G} = \sum_{l=}^{\infty} \alpha_l \otimes \beta_l$ with $\sum_{l=1}^{\infty} \|\alpha_l\| \|\beta_l\| < \infty$. By the same argument as in the proof of the Theorem 3.1 of [6], for every $\mathfrak{G} \in \mathfrak{A} \widehat{\otimes} \mathfrak{A}$, one can show that $\|\mathfrak{G}.(a_i \otimes b_j) - (a_i \otimes b_j).\mathfrak{G}\| \longrightarrow 0$. Similarly, we may show that $\|G(a_i \otimes b_j) - (a_i \otimes b_j)G\| \longrightarrow 0$ for all $G \in A \widehat{\otimes} B$. Therefore, Proposition 2.2 implies that $A \widehat{\otimes} B$ is module $(\phi \otimes \psi, \varphi_1 \otimes \varphi_2)$-inner amenable.

Let A and B be Banach algebras, it is well known that $A \oplus_\infty B$ and $A \oplus_p B$, the l^p-direct sum of A and B, are Banach algebras with respect to the canonical multiplication defined by

$$(a, b)(c, d) := (ac, bd) \quad (a, c \in A, b, d \in B),$$

and norms $\|(a, b)\| = \max\{\|a\|, \|b\|\}$ and $\|(a, b)\| = (\|a\|^p + \|b\|^p)^{\frac{1}{p}} (a \in A, b \in B)$. Furthermore, if A and B are two Banach $\mathfrak{A}$-bimodules, then $A \oplus_\infty B$ and $A \oplus_p B$ are Banach $\mathfrak{A}$-bimodules under the module actions

$$\alpha.(a, b) = (\alpha.a, \alpha.b), \quad (a, b).\alpha = (a.\alpha, b.\alpha) \quad (a \in A, b \in B, \alpha \in \mathfrak{A}).$$

Before stating the next theorem, we note that if for every $\varphi \in \Delta(\mathfrak{A}), \phi \in \Omega_{\mathfrak{A}}$ and $\psi \in \Omega_B$, we define $(0, \psi) : A \oplus_p B \to \mathfrak{A}$ and $(\phi, 0) : A \oplus_p B \to \mathfrak{A}$ by

$$(0, \psi)(a, b) = \psi(b), (\phi, 0)(a, b) = \phi(a)(a \in A, b \in B),$$

where $1 \leq p \leq \infty$, then $(0, \psi)$ and $(\phi, 0) \in \Omega_{A \oplus_p B} (= \Omega_{A \oplus_p B, \varphi})$.

Theorem 3.2 *Let A and B be two $\mathfrak{A}$-bimodule Banach algebras, $\varphi \in \Delta(\mathfrak{A}), \phi \in \Omega_A, \psi \in \Omega_B$ and $1 \leq p \leq \infty$. Then the following statements are valid:*

(i) *$A \oplus_p B$ is module $((\phi, 0), \varphi)$-inner amenable if and only if A is module (ϕ, φ)-inner amenable.*
(ii) *$A \oplus_p B$ is module $((0, \psi), \varphi)$-inner amenable if and only if B is module (ψ, φ)-inner amenable.*

Proof (i) Assume that $A \oplus_p B$ is module $((\phi, 0), \varphi)$-inner amenable. By Theorem 2.2, there exists a net $(a_i, b_i)_i$ in $A \oplus_p B$, such that $\varphi \circ (\phi, 0)(a_i, b_i) \longrightarrow 1$ and

$$\|(a, b).(a_i, b_i) - (a_i, b_i).(a, b)\| \longrightarrow 0, \|\alpha.(a_i, b_i) - (a_i, b_i).\alpha\| \longrightarrow 0,$$

for all $(a, b) \in A \oplus_p B$ and $\alpha \in \mathfrak{A}$. Consider the bounded net $(a_i)_i$ in A. One can easily show that $\|aa_i - a_i a\| \longrightarrow 0$ and $\|\alpha.a_i - a_i.\alpha\| \longrightarrow 0$ for all $a \in A, \alpha \in \mathfrak{A}$. In addition, it is clear that $\varphi \circ \phi(a_i) \longrightarrow 1$. Therefore, Theorem 2.2 implies that A is module (ϕ, φ)-inner amenable.

Conversely, suppose that A is module (ϕ, φ)-inner amenable. Then, there exists a bounded net $(a_i)_i$ in A, such that $\varphi \circ \phi(a_i) \longrightarrow 1$, $\|aa_i - a_i a\| \longrightarrow 0$ and $\|\alpha.a_i - a_i.\alpha\| \longrightarrow 0$ for all $a \in A, \alpha \in \mathfrak{A}$. Clearly, the bounded net $(a_i, 0)_i \subset A \oplus_p B$ satisfies in the condition (iii) of Theorem 2.2. Therefore, $A \oplus_p B$ is module $((\phi, 0), \varphi)$-inner amenable.

Similarly, we can prove (ii). □

Corollary 3.3 *Let A and B be two $\mathfrak{A}$-bimodule Banach algebras and $1 \leq p \leq \infty$. Then $A \oplus_p B$ is module $((\phi, 0), \varphi)$-inner amenable and module $((0, \psi), \varphi)$-inner amenable for every $\varphi \in \Delta(\mathfrak{A}), \phi \in \Omega_A$* and *$\psi \in \Omega_B$ if and only if both A and B are module character inner amenable.*

We note that for two Banach algebras A and B, a direct verification shows that

$$\Delta(A \oplus_p B) = (\Delta(A) \times \{0\}) \cup (\{0\} \times \Delta(B)), 1 \leq p \leq \infty.$$

Now, if we take $\mathfrak{A} = \mathbb{C}$ and φ is the identity map in the above corollary, then we obtain that $A \oplus_p B$ is character inner amenable if and only if both A and B are character inner amenable. Therefore, the above corollary generalizes Proposition 4.2 of [6].

Let A be a Banach algebra and X be a Banach A-bimodule. The l^1-direct sum of A and X, denoted by $A \oplus_1 X$, with the product defined by

$$(a,x)(a',x') = (aa', a.x' + x.a')(a,a' \in A, x, x' \in X),$$

is a Banach algebra that is called the module extension Banach algebra of A and X.

If A is $\mathfrak{A}$-bimodule and X is a Banach A-$\mathfrak{A}$-module, then $A \oplus_1 X$ is Banach $\mathfrak{A}$-bimodules under the module actions:

$$\alpha.(a,x) = (\alpha.a, \alpha.x), (a,x).\alpha = (a.\alpha, x.\alpha) \quad (a \in A, x \in X, \alpha \in \mathfrak{A}). \tag{6}$$

Let A and B be Banach algebras and let X be a Banach A, B-module; that is, a left A-module and a right B-module satisfying $\|axb\| \le \|a\|\|x\|\|b\|, (a \in A, b \in B, x \in X)$. The corresponding triangular Banach algebra

$$\tau = \left\{ \begin{pmatrix} a & x \\ 0 & b \end{pmatrix} : a \in A, x \in X, b \in B \right\},$$

is equipped with the usual 2×2-matrix operations and the norm

$$\| \begin{pmatrix} a & x \\ 0 & b \end{pmatrix} \| = \|a\| + \|x\| + \|b\|.$$

This Banach algebra were introduced by Forrest and Marcoux in [7]. Note that τ can be identified with the module extension $(A \oplus_1 B) \oplus_1 X$, in which X is considered as a $A \oplus_1 B$-module under the operations:

$$(a,b).x = ax, x.(a,b) = xb(a \in A, b \in B, x \in X).$$

Furthermore, if A and B are two Banach $\mathfrak{A}$-bimodules and X is a Banach $A \oplus_1 B$-$\mathfrak{A}$-module, then τ is Banach $\mathfrak{A}$-bimodules under the module actions defined as (6).

Let $\phi \in \Omega_A$ and define $\tilde{\phi} : A \oplus_1 X \longrightarrow \mathfrak{A}$ by $\tilde{\phi}(a,x) = \phi(a)(a \in A, x \in X)$. Then, $\tilde{\phi} \in \Omega_{A \oplus_1 X}$.

Using Theorem 2.2, we can routinely prove the following proposition and so we omit its proof.

Proposition 3.4 *Let A be $\mathfrak{A}$-bimodule and X be a Banach A-$\mathfrak{A}$-module and let $\varphi \in \Delta(\mathfrak{A})$ and $\phi \in \Omega_A$. Then $A \oplus_1 X$ is module $(\tilde{\phi}, \varphi)$-inner amenable if and only if there exists a bounded net $(a_i, x_i)_i$ in $A \oplus_1 X$* satisfying

(i) $\varphi \circ \phi(a_i) \longrightarrow 1$ *and* $\|aa_i - a_i a\| \longrightarrow 0, \|\alpha.a_i - a_i.\alpha\| \longrightarrow 0$ and $\|\alpha.x_i - x_i.\alpha\| \longrightarrow 0$ for all $a \in A$ and $\alpha \in \mathfrak{A}$,

(ii) $\|x.a_i - a_i.x\| \longrightarrow 0$ *for all $x \in X$, and*

(iii) $\|a.x_i - x_i.a\| \longrightarrow 0$ *for all $a \in A$.*

Corollary 3.5 *Let A be $\mathfrak{A}$-bimodule and X be a Banach A-$\mathfrak{A}$-module and let $\varphi \in \Delta(\mathfrak{A})$ and $\phi \in \Omega_A$. If $A \oplus_1 X$ is module $(\tilde{\phi}, \varphi)$-inner amenable, then A is module (ϕ, φ)-inner amenable.*

Corollary 3.6 *Let Aand B be $\mathfrak{A}$-bimodules and X be a Banach $A \oplus_1 B$-$\mathfrak{A}$-module. If τ is module character inner amenable, then so are A and B.*

Proof Suppose that τ is module character inner amenable. By Corollary 3.5, $A \oplus_1 B$ is module character inner amenable. Therefore, corollary 3.3 implies that A and B are module character inner amenable. □

Acknowledgements We are thankful to the referees for their valuable suggestions and comments.

References

1. Amini, M., Bodaghi, A., Babaee, R.: Module derivations into iterated duals of Banach algebras , Proc. Rom. Aca., Series A, 12, 277–284 (2011)
2. Amini, M., Bodaghi, A., Ebrahimi Bagha, D.: Module amenability of the second dual and module topological center of semigroup algebras. Semigroup Forum **80**, 302–312 (2010)
3. Bami, M. L., Valaei, M., Amini, A.: Module amenability and weak module amenability of Banach algebras, U.P.B. Sci. Bull., Series A, 76, 35–44 (2014)
4. Bodaghi, A., Amini, M.: Module character amenability of Banach algebras. Arch. Math. **99**, 353–365 (2012)
5. Bodaghi, A., Ebrahimi, H., Lashkarizadeh Bami, M., Nemati, M.: Module mean for Banach algebras, U.P.B. Sci. Bull., Series A, 78, 21–30 (2016)
6. Ebrahimi Vishki, H.R., Khoddami, A.R.: Character inner amenability of certain Banach algebras. Colloq. Math. **122**, 225–232 (2011)
7. Forrest, B.E., Marcoux, L.W.: Derivations on triangular Banach algebras. Indiana Univ. Math. J. **45**, 441–462 (1996)
8. Jabbari, A., Mehdi Abad, T., Zaman Abadi, M.: On φ-inner amenable Banach algebras. Colloq. Math. **122**, 1–10 (2011)
9. Kaniuth, E., Lau, A., Pym, J.: On φ-amenability of Banach algebras. Math. Proc. Camp. Phil. Soc. **144**, 85–96 (2008)
10. Lau, A.T.: Analysis on a class of Banach algebras with applications to harmonic analysis on locally compact groups and semigroups. Fund. Math. **118**, 161–175 (1983)
11. Mun, W.D.: A class of irreducible matrix representation of an arbitrary inverse semigroup. Proc. Glasgow Math. Assoc **5**, 41–48 (1961)
12. Nasr-Isfahani, R.: Inner amenability of Lau algebras. Arch. Math. (Brno) **37**, 45–55 (2001)
13. Pourmahmood Aghababa, H.: (Super) module amenability, module topological centre and semigroup algebras. Semigroup Forum **81**, 344–356 (2010)
14. Pourmahmood Aghababa, H.: H.: A note on two equivalence relations on inverse semigroups. Semigroup Forum **48**, 200–202 (2012)
15. Rezavand, R., Amini, M., Sattari, M.H., Ebrahim Bagha, D.: Module Arens regularity for semigroup algebras. Semigroup Forum **77**, 300–305 (2008)
16. Rieffel, M.A.: Induced Banach representations of Banach algebras and locally compact groups. J. Funct. Anal. **1**, 443–491 (1967)
17. Schaefer, H.H.: Topological Vector Space, Springer-Verlag (1979)

21

A survey on Christoffel–Darboux type identities of Legendre, Laguerre and Hermite polynomials

Asghar Arzhang[1]

Abstract In this paper, we construct some new Christoffel–Darboux type identities for Legendre, Laguerre and Hermite polynomials. We obtain these types of identities for the derivatives of these polynomials.

Keywords Christoffel–Darboux identity · Cauchy kernel · Legendre polynomials · Laguerre polynomials · Hermite polynomials

Mathematical Subject Classification 33D45 · 33D50 · 45E05

Introduction

In [1], we have simplified the fraction

$$P_n(x,y) = \frac{P_n(y) - P_n(x)}{y - x}, \tag{1}$$

in terms of $P_i(x)$,$P_j(y)$ where $P_n \in \{T_n, U_n, V_n, W_n\}$. Also, for every kind of Chebyshev polynomials, we have obtained the expanded form of the fraction

$$P_{n,s}(x,y) = \frac{P_n^{(s)}(y) - P_n^{(s)}(x)}{y - x}, \quad P_n \in \{T_n, U_n, V_n, W_n\}. \tag{2}$$

in terms of $P_i(x)$, $P_j(y)$ where $P_n^{(s)}(x)$ is the sth derivative of $P_n(x)$.

✉ Asghar Arzhang
arzhang@kiau.ac.ir

1 Department of Mathematics, Karaj Branch, Islamic Azad University, Karaj, Iran

In this paper, we expand the fraction (1) where $P_n(x)$ is Legendre, Laguerre and Hermite polynomials.

Christoffel–Darboux type identities of Legendre, Laguerre and Hermite polynomials

Theorem 2.1 *Let $\{P_n(x)\}_{n=0}^{\infty}$ be a sequence of orthogonal polynomials with respect to the weight function $w(x)$ on interval $[a, b]$ then*

$$P_{n+1}(x,y) = \frac{P_{n+1}(y) - P_{n+1}(x)}{y - x} = \sum_{i=0}^{n}\sum_{j=0}^{n-i} A_{i,j}^{n+1} P_i(x)P_j(y)$$
$$= \sum_{i=0}^{n}\sum_{j=0}^{i} A_{n-i,j}^{n+1} P_{n-i}(x)P_j(y), \tag{3}$$

where

$$A_{i,j}^{n+1} = \frac{1}{\gamma_i\gamma_j}\sum_{k=i+j+1}^{n+1}\sum_{v=i}^{k-j-1} C_{n+1,k}\, B_{v,i}\, B_{k-v-1,j}$$
$$= \frac{1}{\gamma_i\gamma_j}\sum_{k=0}^{n-i-j}\sum_{v=0}^{k} C_{n+1,k+i+j+1}\, B_{v+i,i}\, B_{k+j-v,j} \tag{4}$$

$$\gamma_i = \int_a^b P_n^2(x) w(x)\, dx,$$

$$B_{m,n} = \int_a^b x^m P_n(x) w(x)\, dx,$$

and $C_{n+1,k}$ is the coefficient of x^k in $P_{n+1}(x)$.

Proof $P_n(x)$ is orthogonal to every polynomial of degree less than n. So, if $i+j > n$ then $A_{i,j}^{n+1} = 0$. If $i+j \leq n$ then use orthogonality and expanded form of $P_n(x)$ to obtain the result. □

Corollary 2.1 *If the interval* $[a, b]$ *is symmetric about the origin and* $P_n(-x) = (-1)^n P_n(x)$ *then for even* $n+i+j$, $A^n_{i,j} = 0$.

If the linearization formula of $P_n(x)$ is available then we can compute $A_{i,j}$ coefficients in Eq. (3) by using one sum instead of using double sum in Eq. (4).

Christoffel–Darboux type identities of Hermite polynomials

Theorem 3.1 *Let* $H_n(x)$ *be Hermite polynomial of degree n then*

$$H_n(x,y) = \frac{H_n(y) - H_n(x)}{y-x} = \sum_{i=0}^{n-1}\sum_{j=0}^{n-i-1} A^n_{i,j} H_i(x) H_j(y),$$

where

$$A^n_{i,j} = \left(\frac{1-(-1)^{i+j+n}}{2}\right) \times \left\{ \frac{1}{2^i i!}\sum_{k=0}^{j}(-1)^{\left(\frac{3n-3j+6k+i+1}{2}\right)} \frac{2^{\left(\frac{n-j+i-1}{2}\right)}}{k!}\binom{n}{j-k} \times \Gamma\left(\frac{n-j+i+2k+1}{2}\right) + \frac{1}{2^j j!}\sum_{k=0}^{i}(-1)^{\left(\frac{3n-3i+6k+j+1}{2}\right)} \times \frac{2^{\left(\frac{n-i+j-1}{2}\right)}}{k!}\binom{n}{i-k} \times \Gamma\left(\frac{n-i+j+2k+1}{2}\right)\right\}. \tag{5}$$

Proof First, we prove that

$$H_{m,n}(x,y) = P.V. \int_{-\infty}^{\infty}\int_{-\infty}^{\infty} \frac{H_m(x)H_n(y)e^{-x^2}e^{-y^2}}{y-x}\,dy dx = 2^{\frac{m+n-1}{2}}(-1)^{n+1}\sin\left(\frac{m+n}{2}\pi\right)\Gamma\left(\frac{m+n+1}{2}\right)\pi, \quad m,n = 0,1,2,\ldots. \tag{6}$$

From [5], use the Hilbert transform of $H_n(y)e^{-y^2}$ to obtain

$$P.V. \int_{-\infty}^{\infty} \frac{H_n(y)e^{-y^2}}{y-x}\,\mathrm{d}y = (2\pi)^{n+1}\sqrt{\pi}(-1)^{n+1} \times \int_0^{\infty} f^n e^{-\pi^2 f^2}\sin\left(2\pi f x + \frac{n\pi}{2}\right)\mathrm{d}f$$

So

$$H_{m,n}(x,y) = P.V. \int_{-\infty}^{\infty}\int_{-\infty}^{\infty} \frac{H_m(x)H_n(y)e^{-x^2}e^{-y^2}}{y-x}\,\mathrm{d}y\mathrm{d}x = (2\pi)^{n+1}\sqrt{\pi}(-1)^{n+1}\int_0^{\infty}\int_{-\infty}^{\infty} f^n H_m(x)e^{-x^2}e^{-\pi^2 f^2} \times \sin\left(2\pi f x + \frac{n\pi}{2}\right)\mathrm{d}x\mathrm{d}f. \tag{7}$$

On the other hand, we have

$$\int_{-\infty}^{\infty} e^{-x^2}H_m(x)\sin\left(2\pi f x + \frac{n\pi}{2}\right)\mathrm{d}x = \begin{cases} 0, & m+n \text{ is even,} \\ 2\int_0^{\infty} e^{-x^2}H_m(x)\sin\left(2\pi f x + \frac{n\pi}{2}\right)\mathrm{d}x, & m+n \text{ is odd,}\end{cases} \tag{8}$$

So, if $m+n$ is even then $H_{m,n}(x,y) = 0$. If $m+n$ is odd, then use relation (8) and integration by parts and Rodrigue's formula of Hermite polynomials to obtain

$$\int_0^{\infty} e^{-x^2}H_m(x)\sin\left(2\pi f x + \frac{n\pi}{2}\right)\mathrm{d}x = (2\pi f)^m \sin\left(\frac{m+n}{2}\pi\right)\int_0^{\infty} e^{-x^2}\cos(2\pi f x)\,\mathrm{d}x. \tag{9}$$

From the relations (7), (9) by using change of the variable $\pi f = y$, we obtain

$$\begin{aligned} H_{m,n}(x,y) &= 2^{m+n+2}(-1)^{n+1}\sqrt{\pi}\sin\left(\frac{m+n}{2}\pi\right) \times \int_0^{\infty} y^{m+n}e^{-y^2}\,\mathrm{d}y\int_0^{\infty} e^{-x^2}\cos(2xy)\,\mathrm{d}x \\ &= 2^{m+n+2}(-1)^{n+1}\sqrt{\pi}\sin\left(\frac{m+n}{2}\pi\right) \times \int_0^{\infty} y^{m+n}e^{-y^2}\left(\frac{\sqrt{\pi}}{2}e^{-y^2}\right)\mathrm{d}y \\ &= 2^{\frac{m+n-1}{2}}(-1)^{n+1}\sin\left(\frac{m+n}{2}\pi\right)\Gamma\left(\frac{m+n+1}{2}\right)\pi \end{aligned} \tag{10}$$

So, for odd $m+n$, we have

$$H_{m,n}(x,y) = 2^{\frac{m+n-1}{2}}(-1)^{\frac{m+3n+1}{2}}\Gamma\left(\frac{m+n+1}{2}\right)\pi \tag{11}$$

The famous linearization formula of Hermite polynomials is [2]

$$H_m(x)H_n(x) = 2^m m!\sum_{k=0}^{m}\frac{1}{2^k k!}\binom{n}{m-k}H_{n-m+2k}(x), \quad m \leq n. \tag{12}$$

By using the relations (11) and (12), we can obtain $A^n_{i,j}$ in relation (5). □

Corollary 3.1 *The* $A^n_{i,j}$ *coefficients in relation* (5) *can be computed as follows:*

$$A^n_{i,j} = \frac{n!}{2^{i+j}i!j!\pi}\sum_{k=i+j+1}^{n}\sum_{v=i}^{k-j-1}(-1)^{\frac{n-k}{2}}\left(\frac{1+(-1)^{n+k}}{2}\right) \times \left(\frac{1+(-1)^{v+i}}{2}\right)\left(\frac{1+(-1)^{k+j-v-1}}{2}\right) \times \frac{2^k v!(k-v-1)!}{k!(v-i)!(k-v-j-1)!\left(\frac{n-k}{2}\right)!}\Gamma\left(\frac{v-i+1}{2}\right)\Gamma\left(\frac{k-v-j}{2}\right).$$

Now, we can obtain Christoffel–Darboux type identities for the derivatives of Hermite polynomials.

Corollary 3.2 *Let*

$$H_n^{(s)}(x,y) = \frac{H_n^{(s)}(y) - H_n^{(s)}(x)}{y-x} = \sum_{i=0}^{n-s-1}\sum_{j=0}^{n-s-i-1} A_{i,j}^{n,s} H_i(x)H_j(y), \quad s = 0\ldots n, \tag{13}$$

where

$$\begin{aligned} A_{i,j}^{n,s} = {} & \frac{2^s n!}{(n-s)!}\left(\frac{1-(-1)^{n-s+i+j}}{2}\right) \\ & \times \left\{ \frac{1}{2^i i!}\sum_{k=0}^{j}(-1)^{\left(\frac{3n-3s-3j+6k+i+1}{2}\right)} \frac{2^{\left(\frac{n-s-j+i-1}{2}\right)}}{k!} \right. \\ & \times \binom{n-s}{j-k}\Gamma\left(\frac{n-s-j+i+2k+1}{2}\right) \\ & + \frac{1}{2^j j!}\sum_{k=0}^{i}(-1)^{\left(\frac{3n-3s-3i+6k+j+1}{2}\right)} \frac{2^{\left(\frac{n-s-i+j-1}{2}\right)}}{k!} \\ & \left. \times \binom{n-s}{i-k}\Gamma\left(\frac{n-s-i+j+2k+1}{2}\right) \right\}. \end{aligned} \tag{14}$$

Christoffel–Darboux type identities of Legendre polynomials

Theorem 4.1 *Let $P_n(x)$ be Legendre polynomial of degree n* then

$$P_n(x,y) = \frac{P_n(y) - P_n(x)}{y-x} = \sum_{i=0}^{n-1}\sum_{j=0}^{n-i-1} A_{i,j}^{n} P_i(x)P_j(y), \tag{15}$$

where

$$\begin{aligned} A_{i,j}^{n} = {} & -\frac{1}{2}(2i+1)(2j+1) \\ & \sum_{k=0}^{\min(i,j)} \frac{1+(-1)^{i+j+n-1}}{i+j+n-2k+1}\left\{\frac{B_{i,j}^{k}}{i+j-n-2k} \right. \\ & \left. - \frac{B_{\min(i,j),n}^{k}}{\min(i,j)-\max(i,j)+n-2k}\right\}, \end{aligned}$$

$$B_{i,j}^{k} = \frac{B_{i-k}B_kB_{j-k}}{B_{i+j-k}}\left(\frac{2i+2j-4k+1}{2i+2j-2k+1}\right),$$

$$B_k = \frac{1\cdot3\cdot5\cdots(2k-1)}{k!}, \quad k = 1,2,3,\ldots.$$

$$B_0 = 1. \tag{16}$$

Proof Legendre function of the second kind is defined by

$$Q_n(x) = -\frac{1}{2}P.V.\int_{-1}^{1}\frac{P_n(y)}{y-x}\,\mathrm{d}y,$$

and

$$P.V.\int_{-1}^{1}P_m(x)Q_n(x)\,\mathrm{d}x = \frac{1+(-1)^{m+n}}{(m-n)(m+n+1)}, \quad m \neq n. \tag{17}$$

Therefore

$$\begin{aligned} A_{i,j}^{n} &= \frac{1}{4}(2i+1)(2j+1)\int_{-1}^{1}\int_{-1}^{1}\frac{P_n(y)-P_n(x)}{y-x}P_i(x)P_j(y)\,\mathrm{d}y\,\mathrm{d}x \\ &= -\frac{1}{2}(2i+1)(2j+1)\int_{-1}^{1}P_i(x)\Big(P_j(x)Q_n(x) - P_n(x)Q_j(x)\Big)\,\mathrm{d}x. \end{aligned} \tag{18}$$

The following famous linearization formula of Legendre polynomials is Neumann-Adams formula [2]:

$$P_m(x)P_n(x) = \sum_{k=0}^{m}\frac{B_{m-k}B_kB_{n-k}}{B_{m+n-k}}\left(\frac{2m+2n-4k+1}{2m+2n-2k+1}\right) P_{m+n-2k}(x), \quad m \leq n,$$

$$B_k = \frac{1\cdot3\cdot5\cdots(2k-1)}{k!}, \quad k = 1,2,3,\ldots.$$

$$B_0 = 1. \tag{19}$$

Now, use the relations (17), (18) and (19) to obtain the result. □

Corollary 4.1 *The $A_{i,j}^{n}$ coefficients in relation* (15) *can be computed as follows*:

$$\begin{aligned} A_{i,j}^{n} = {} & \frac{(2i+1)(2j+1)}{2^{n+i+j+2}}\sum_{k=i+j+1}^{n}\sum_{v=i}^{k-j-1}(-1)^{\frac{n-k}{2}}\left(\frac{1+(-1)^{n+k}}{2}\right) \\ & \times \left(\frac{1+(-1)^{v+i}}{2}\right)\left(\frac{1+(-1)^{k+j-v-1}}{2}\right) \\ & \times \frac{(n+k)!v!(k-v-1)!}{k!(v-i)!(k-v-j-1)!\left(\frac{n-k}{2}\right)!\left(\frac{n+k}{2}\right)!} \frac{\Gamma\left(\frac{v-i+1}{2}\right)\Gamma\left(\frac{k-v-j}{2}\right)}{\Gamma\left(\frac{v+i+3}{2}\right)\Gamma\left(\frac{k-v+j+2}{2}\right)} \end{aligned}$$

Now, we can obtain Christoffel–Darboux type identities for the derivatives of Legendre polynomials.

From [4], for the case $\gamma = 0$, we can derive

$$P_n^{(s)}(x) = \sum_{k=0}^{n-s} a_k^s P_k(x), \quad s = 0\ldots n, \tag{20}$$

where

$$\begin{aligned} a_k^s = {} & \left(\frac{1+(-1)^{n+k+s}}{2}\right)\frac{2k+1}{2^{s-2}(s-1)!}\frac{(n+k+s-1)!}{(n+k+s+2)!} \\ & \times \frac{\left(\frac{n-k+s}{2}-1\right)!}{\left(\frac{n-k-s}{2}\right)!}\frac{\left(\frac{n+k-s}{2}+1\right)!}{\left(\frac{n+k+s}{2}-1\right)!}. \end{aligned} \tag{21}$$

Corollary 4.2

$$\frac{P_n^{(s)}(y)-P_n^{(s)}(x)}{y-x}=\sum_{i=0}^{n-s-1}\sum_{j=0}^{n-s-i-1}A_{i,j}^{n,s}P_i(x)P_j(y),\quad s=0\ldots n, \tag{22}$$

where

$$\begin{aligned}A_{i,j}^{n,s}=&-\frac{1}{2}(2i+1)(2j+1)\sum_{k=i+j+1}^{n-s}\sum_{v=0}^{\min(i,j)}\left(\frac{1+(-1)^{n+k+s}}{2}\right)\\&\times\frac{2k+1}{2^{s-2}(s-1)!}\frac{(n+k+s-1)!}{(n+k+s+2)!}\\&\times\frac{(\frac{n-k+s}{2}-1)!}{(\frac{n-k-s}{2})!}\frac{(\frac{n+k-s}{2}+1)!}{(\frac{n+k+s}{2}-1)!}\frac{1+(-1)^{i+j+k-1}}{i+j+k-2v+1}\\&\times\left\{\frac{B_{i,j}^{v}}{i+j-k-2v}-\frac{B_{\min(i,j),k}^{v}}{\min(i,j)-\max(i,j)+k-2v}\right\},\\B_{i,j}^{v}=&\frac{B_{i-v}B_vB_{j-v}}{B_{i+j-v}}\left(\frac{2i+2j-4v+1}{2i+2j-2v+1}\right),\\B_v=&\frac{1\cdot3\cdot5\cdots(2v-1)}{v!},\quad v=1,2,3,\ldots.\\B_0=&1.\end{aligned} \tag{23}$$

Christoffel–Darboux type identities of Laguerre polynomials

The famous linearization formula of associated laguerre polynomials is Feldheim formula [6]

$$L_m^{\alpha}L_n^{\beta}(x)=\sum_{k=0}^{m+n}\sum_{v=0}^{k}(-1)^{m+n+k}\binom{k}{v}\binom{m+\alpha}{n-k+v}\binom{n+\beta}{m-v}\times L_k^{\alpha+\beta}(x).$$

In spite of Hermite and Legendre polynomials, the linearization formula of Laguerre polynomials is presented by double summation. The coefficients $A_{i,j}^n$ of Hermite and Legendre polynomials are obtained from (5) and (16) by one summation, and in the following relations, the $A_{i,j}^n$ coefficients of Laguerre polynomials are given by double summation.

Corollary 5.1 *Let $L_n^m(x)$ be associated laguerre polynomials of degree n then*

$$L_{n+1}^m(x,y)=\frac{L_{n+1}^m(y)-L_{n+1}^m(x)}{y-x}=\sum_{i=0}^{n}\sum_{j=0}^{n-i}A_{i,j}^{n+1}L_i^m(x)L_j^m(y), \tag{24}$$

where

$$A_{i,j}^{n+1}=\sum_{k=i+j+1}^{n+1}\sum_{v=i}^{k-j-1}(-1)^{i+j+k}\frac{i!j!}{k!}\times\frac{(m+v)!(m+k-v-1)!}{(m+i)!(m+j)!}\binom{m+n+1}{n-k+1}\binom{v}{i}\binom{k-v-1}{j} \tag{25}$$

The related formula for Laguerre polynomials of degree n is

$$L_{n+1}(x,y)=\frac{L_{n+1}(y)-L_{n+1}(x)}{y-x}=\sum_{i=0}^{n}\sum_{j=0}^{n-i}A_{i,j}^{n+1}L_i(x)L_j(y), \tag{26}$$

where

$$A_{i,j}^{n+1}=\sum_{k=i+j+1}^{n+1}\sum_{v=i}^{k-j-1}(-1)^{i+j+k}\frac{\binom{n+1}{k}\binom{v}{i}\binom{k-v-1}{j}}{(v+1)\binom{k}{v+1}} \tag{27}$$

From [3], we have

$$\frac{\mathrm{d}^s}{\mathrm{d}x^s}L_n^m(x)=(-1)^s\sum_{k=0}^{n-s}\binom{n-k-1}{s-1}L_k^m(x),\quad s=0\ldots n.$$

Let

$$L_n^{m,s}(x)=\frac{d^s}{\mathrm{d}x^s}L_n^m(x),$$

then

$$\frac{L_n^{m,s}(y)-L_n^{m,s}(x)}{y-x}=\sum_{i=0}^{n-s-1}\sum_{j=0}^{n-s-i-1}A_{i,j}^{n,s}L_i^m(x)L_j^m(y),\quad s=0\ldots n,$$

where

$$A_{i,j}^{n,s}=\sum_{k=i+j+1}^{n-s}\sum_{k'=i+j+1}^{k}\sum_{v=i}^{k'-j-1}(-1)^{i+j+s+k'}\frac{i!j!}{k'!}\frac{(m+v)!(m+k'-v-1)!}{(m+i)!(m+j)!}\times\binom{n-k-1}{s-1}\binom{m+k}{k-k'}\binom{v}{i}\binom{k'-v-1}{j}.$$

Conclusion

In this paper, we obtained some new Christoffel–Darboux type identities for Legendre, Laguerre and Hermite polynomials. We also obtained these types of identities for the derivatives of these polynomials. These formulas are good theoretically and the correctness of the obtained formulas are checked by Maple 17, and Some of these formulas are not efficient numerically.

Acknowledgments This work has been funded and supported by Islamic Azad University, Karaj Branch, and the author is thankful to it.

References

1. Arzhang, A.: A survey on Christoffel–Darboux-type identities of Chebyshev polynomials. Integral Transforms Spec. Funct. **24**(10), 840–849 (2013). doi:10.1080/10652469.2012.762711
2. Andrews, G.E., Roy, R.: Special Functions. Cambridge University Press, Richard Askey, Cambridge (1999)
3. Doha, E.H.: On the connection coefficients and recurrence relations arising from expansions in series of Laguerre polynomials. J. Phys. A **36**, 5449–5462 (2003)
4. Karageorghis, A., Phillips, T.N.: On the coefficients of differentiated expansions of ultraspherical polynomials. Appl. Numer. Math. **9**, 133–141 (1992)
5. Poularikas, A.D.: The Transforms and Applications Handbook. CRC Press LLC, Boca Raton (2000)
6. Popov, B.S., Srivastava, H.M.: Linearization of a product of two polynomials of different orthogonal systems. Facta Univ. Ser. Math. Inform. **18**, 1–8 (2003)

22

A one-step implicit iterative process for a finite family of I-nonexpansive mappings in Kohlenbach hyperbolic spaces

Birol Gunduz[1] · **Sezgin Akbulut**[2]

Abstract The purpose of the present paper is threefold. First, to give definition of I-nonexpansive mappings in a Kohlenbach hyperbolic space. Second, to define a new one-step implicit iterative process. Finally, to establish strong and Δ-convergence theorems for this iterative process in a Kohlenbach hyperbolic space. In addition, an example is provided to validate our result. Our results extend some existing results.

Keywords Kohlenbach hyperbolic space · I-nonexpansive map · Common fixed point · Implicit iterative process · Strong convergence · Δ-convergence

Mathematics Subject Classification 47H09 · 47H10 · 49M05

Introduction

In nonlinear functional analysis, one of the most productive tools is the fixed point theory, which has numerous applications in many quantitative disciplines such as biology, chemistry, computer science, and additionally in many branches of engineering. So, the metric fixed point theory has been investigated extensively in the past two decades by numerous mathematicians. Takahashi [1] introduced the concept of convexity in a metric space (X, d) as follows.

A convex structure in a metric space (X, d) is a mapping $W : X \times X \times [0, 1] \to X$ satisfying, for all $x, y, u \in X$ and all $\lambda \in [0, 1]$,

$$d(u, W(x, y, \lambda)) \leq \lambda d(u, x) + (1 - \lambda)d(u, y).$$

A metric space together with a convex structure is called a convex metric space. A nonempty subset C of X is said to be convex if $W(x, y, \lambda) \in C$ for all $(x, y, \lambda) \in C \times C \times [0, 1]$.

Recently, Kohlenbach [2] enriched the concept of convex metric space by defining hyperbolic space.

A hyperbolic space [2] is a triple (X, d, W), where (X, d) is a metric space and $W : X^2 \times [0, 1] \to X$ is such that

W1. $d(u, W(x, y, \alpha)) \leq \alpha d(u, x) + (1 - \alpha)d(u, y)$
W2. $d(W(x, y, \alpha), W(x, y, \beta)) = |\alpha - \beta| d(x, y)$
W3. $W(x, y, \alpha) = W(y, x, (1 - \alpha))$
W4. $d(W(x, z, \alpha), W(y, w, \alpha)) \leq \alpha d(x, y) + (1 - \alpha)d(z, w)$

for all $x, y, z, w \in X$ and $\alpha, \beta \in [0, 1]$. If (X, d, W) satisfies only (W1), then it coincides with the convex metric space introduced by Takahashi [1]. All normed spaces and their subsets are the examples of hyperbolic spaces as well as convex metric spaces. It is remarked that every CAT(0) and Banach spaces are very special cases of hyperbolic space.

From now on, $\mathbb{N}$ denotes the set of natural numbers and $J = \{1, 2, \ldots, N\}$, the set of first N natural numbers. Denote by $F(T)$, the set of fixed points of T, that is, $F(T) = \{x \in K : Tx = x\}$ and by $F := \cap_{i=1}^{N}(F(T_i) \cap F(I_i))$, the set of common fixed points of two families $\{T_i : i \in J\}$ and $\{I_i : i \in J\}$. In what follows, we fix $x_0 \in K$ as a starting point of a process unless stated otherwise, and take $\{\alpha_n\}, \{\beta_n\}, \{\gamma_n\}$ sequences in $(0, 1)$.

✉ Birol Gunduz
birolgndz@gmail.com

Sezgin Akbulut
sezginakbulut@atauni.edu.tr

1 Department of Mathematics, Faculty of Science and Art, Erzincan University, 24000 Erzincan, Turkey

2 Present Address: Department of Mathematics, Faculty of Science, Ataturk University, 25240 Erzurum, Turkey

Mann iterative process for fixed point of a mapping T is as follows:

$$x_n = \alpha_n x_{n-1} + (1-\alpha_n)Tx_{n-1}, \quad n \in \mathbb{N}. \tag{1.1}$$

In [3], Shahzad defined I-nonexpansivity of a mapping T in a Banach space. Now, we give metric version of I-nonexpansivity of a mapping T.

Let K be a nonempty subset of a metric space (X, d) and T, I be two selfmaps on K. T is said to be nonexpansive if $d(Tx, Ty) \le d(x, y)$ for all $x, y \in K$, and T is said to be I-nonexpansive [3] if $d(Tx, Ty) \le d(Ix, Iy)$ for all $x, y \in K$.

In [4], Rhoades and Temir established the weak convergence of the sequence of the Mann iterates to a common fixed point of T and I by considering the map T to be I-nonexpansive. More precisely, they proved the following theorem.

Theorem 1 [4] *Let K be a closed convex bounded subset of uniformly convex Banach space X, which satisfies Opial's condition, and let T, I self-mappings of K with T be an I-nonexpansive mapping, I a nonexpansive on K. Then, for $x_0 \in K$,* the sequence $\{x_n\}$ of Mann iterates converges weakly to a common fixed point of $F(T) \cap F(I)$.

Temir and Gul [5] obtained weak convergence theorems of fixed points for I-nonexpansive mappings and I-asymptotically quasi-nonexpansive mappings in Hilbert space. Later on, some authors [6–8] studied convergence theorems for generalization of the class of I-nonexpansive mappings.

Concerning the common fixed points of the finite family $\{T_i : i \in J\}$, Xu and Ori [9] introduced the following implicit iterative process:

$$x_n = \alpha_n x_{n-1} + (1-\alpha_n)T_n x_n, \quad n \in \mathbb{N} \tag{1.2}$$

where $T_n = T_{n(\mathrm{mod}\, N)}$.

Zhao et al. [10] introduced the following implicit iterative process for the same purpose.

$$x_n = \alpha_n x_{n-1} + \beta_n T_n x_{n-1} + \gamma_n T_n x_n, \quad n \in \mathbb{N} \tag{1.3}$$

where $T_n = T_{n(\mathrm{mod}\, N)}$.

Khan [11] generalized results of Zhao et al. [10] for two finite families of nonexpansive mappings. Plubtieng et al. [12] defined an implicit iterative process for two finite families of nonexpansive mappings $\{T_i : i \in J\}$ and $\{S_i : i \in J\}$ as follows:

$$\begin{aligned} x_n &= \alpha_n x_{n-1} + (1-\alpha_n)T_n y_n, \\ y_n &= \beta_n x_n + (1-\beta_n)S_n x_n, \quad n \in \mathbb{N} \end{aligned} \tag{1.4}$$

where $T_n = T_{n(\mathrm{mod}\, N)}$ and $S_n = S_{n(\mathrm{mod}\, N)}$.

Khan et al. [13] studied implicit iteration (1.4) for two finite families of nonexpansive mappings in a hyperbolic space as follows:

$$\begin{aligned} x_n &= W(x_{n-1}, T_n y_n, \alpha_n), \\ y_n &= W(x_n, S_n x_n, \beta_n), \quad n \in \mathbb{N} \end{aligned} \tag{1.5}$$

Later on, some authors discussed the convergence of the iterative process in hyperbolic spaces (see, for example, [14–17]).

Motivated by the above facts, in this paper we define a new algorithm as follows:

Let K be a nonempty closed convex subset of a convex metric space X, $\{T_i : i \in J\}$ be a finite family of I_i-nonexpansive mappings and $\{I_i : i \in J\}$ be a finite family of nonexpansive mappings. Then $\{x_n\}$ is defined as:

$$x_n = W\left(T_n x_n, W\left(I_n x_{n-1}, x_{n-1}, \frac{\beta_n}{1-\alpha_n}\right), \alpha_n\right) \tag{1.6}$$

where $T_n = T_{n(\mathrm{mod}\, N)}$ and $I_n = I_{n(\mathrm{mod}\, N)}$, $0 < a \le \alpha_n$, $\beta_n \le b < 1$ and satisfy $\alpha_n + \beta_n < 1$.

Obviously, (1.6) is equivalent to $x_n = \alpha_n x_{n-1} + \beta_n I_n x_{n-1} + (1-\alpha_n-\beta_n)T_n x_n$ in the Banach space setting. Thus, iteration process (1.6) is more general than the iteration process (1.1)–(1.5) and iteration process of Khan [11].

Using process (1.6), we prove some Δ and strong convergence theorems for approximating common fixed points of a finite family of I-nonexpansive mappings and a finite family of nonexpansive mappings in a uniformly convex hyperbolic space. These results improve and extend the corresponding results of Rhoades and Temir [4], Soltuz [18], Chidume and Shahzad [19], Zhao et al. [10], Khan [11]. Our results also improve the corresponding results of Plubtieng et al. [12] and Khan et al. [13] being computationally simpler.

Preliminaries

Let K be a nonempty closed convex subset of a convex metric space X, $\{T_i : i \in J\}$ be a finite family of I_i-nonexpansive mappings and $\{I_i : i \in J\}$ be a finite family of nonexpansive mappings. Let $\{x_n\}$ be defined by (1.6). Define a mapping $B_1 : K \to K$ by $B_1 x = W(T_1 x, W(I_1 x_0, x_0, \frac{\beta_1}{1-\alpha_1}), \alpha_1)$ for all $x \in K$. Existence of x_1 is guaranteed if B_1 has a fixed point. Now for any $x, y \in K$, we have

$$\begin{aligned} d(B_1 x, B_1 y) &\le \alpha_1 d(T_1 x, T_1 y) \\ &\quad + (1-\alpha_n) d\left(W\left(I_1 x_0, x_0, \frac{\beta_1}{1-\alpha_1}\right), W\left(I_1 x_0, x_0, \frac{\beta_1}{1-\alpha_1}\right)\right) \\ &\le \alpha_1 d(Ix, Iy) \le \alpha_1 d(x, y). \end{aligned}$$

Since $\alpha_1 < 1$, B_1 is a contraction. By Banach contraction principle, B_1 has a unique fixed point. Thus, the existence of x_1 is established. Similarly, the existence of $x_2, x_3, \dots$ is established. Thus, the iteration process (1.6) is well defined.

A hyperbolic space (X, d, W) is said to be uniformly convex [20] if for all $u, x, y \in X$, $r > 0$ and $\varepsilon \in (0, 2]$, there exists a $\delta \in (0, 1]$ such that

$$\left.\begin{array}{l} d(x,u) \leq r \\ d(y,u) \leq r \\ d(x,y) \geq \varepsilon r \end{array}\right\} \Rightarrow d\left(W\left(x, y, \frac{1}{2}\right), u\right) \leq (1-\delta) r.$$

A map $\eta : (0, \infty) \times (0, 2] \to (0, 1]$ which provides such a $\delta = \eta(r, \varepsilon)$ for given $r > 0$ and $\varepsilon \in (0, 2]$ is called modulus of uniform convexity. We call η monotone if it decreases with r (for a fixed ε).

Let $\{x_n\}$ be a bounded sequence in a hyperbolic space X. For $x \in X$, define a continuous functional $r(., \{x_n\}) : X \to [0, \infty)$ by $r(x, \{x_n\}) = \limsup_{n\to\infty} d(x, x_n)$. The asymptotic radius $\rho = r(\{x_n\})$ of $\{x_n\}$ is given by $\rho = \inf\{r(x, \{x_n\}) : x \in X\}$. The asymptotic center of a bounded sequence $\{x_n\}$ with respect to a subset K of X is defined as follows:

$$A_K(\{x_n\}) = \{x \in X : r(x, \{x_n\}) \leq r(y, \{x_n\}) \quad \text{for any } y \in K\}.$$

The set of all asymptotic centers of $\{x_n\}$ is denoted by $A(\{x_n\})$.

It has been shown in [21] that bounded sequences have unique asymptotic center with respect to closed convex subsets in a complete and uniformly convex hyperbolic space with monotone modulus of uniform convexity.

A sequence $\{x_n\}$ in X is said to Δ-converge to $x \in X$ if x is the unique asymptotic center of $\{u_n\}$ for every subsequence $\{u_n\}$ of $\{x_n\}$ [22]. In this case, we write Δ-$\lim_n x_n = x$.

Recall that Δ-convergence coincides with weak convergence in Banach spaces with Opial's property [23].

A sequence $\{x_n\}$ in a metric space X is said to be Fej ér monotone with respect to K (a subset of X) if $d(x_{n+1}, p) \leq d(x_n, p)$ for all $p \in K$ and $n \geq 1$. A map $T : K \to K$ is semi-compact if any bounded sequence $\{x_n\}$ satisfying $d(x_n, Tx_n) \to 0$ as $n \to \infty$ has a convergent subsequence.

Two mappings $T, S : K \to K$ with $F \neq \emptyset$ are said to satisfy the Condition (A′) [24] if there exists a nondecreasing function $f : [0, \infty) \to [0, \infty)$ with $f(0) = 0$, $f(t) > 0$ for all $t \in (0, \infty)$ such that

$$\text{either } d(x, Tx) \geq f(d(x, F)) \quad \text{or} \quad d(x, Sx) \geq f(d(x, F))$$

for all $x \in K$, where $d(x, F) = \inf\{d(x, p) : p \in F\}$.

Let $\{T_i : i \in J\}$ be a finite family of nonexpansive mappings of K with nonempty fixed points set F. Then $\{T_i : i \in J\}$ is said to satisfy the Condition (B) on K [19] if there exists a nondecreasing function $f : [0, \infty) \to [0, \infty)$ with $f(0) = 0$, $f(t) > 0$ for all $t \in (0, \infty)$ such that

$$\max_{i\in J} d(x, T_i x) \geq f(d(x, F))$$

for all $x \in K$.

Khan [11] modified Condition (B) for two finite families of mappings as follows. Let $\{T_i : i \in J\}$ and $\{I_i : i \in J\}$ be two finite families of mappings of K with nonempty fixed points set F. These families are said to satisfy Condition (B′) on K if there exists a nondecreasing function $f : [0, \infty) \to [0, \infty)$ with $f(0) = 0$, $f(t) > 0$ for all $t \in (0, \infty)$ such that

$$\text{either} \max_{i\in J} d(x, T_i x) \geq f(d(x, F)) \quad \text{or} \quad \max_{i\in J} d(x, I_i x) \geq f(d(x, F))$$

for all $x \in K$. The Condition (B′) reduces to the Condition (A′) when $T_1 = T_2 = \cdots = T_N = T$ and $S_1 = S_2 = \cdots = S_N = S$, and to the Condition (B) when $S_i = T_i$ for all $i \in J$.

For the development of our main results, some key results are listed below.

Lemma 1 [13] *Let (X, d, W) be a uniformly convex hyperbolic space with monotone modulus of uniform convexity η. Let $x \in X$ and $\{\alpha_n\}$ be a sequence in $[b, c]$ for some $b, c \in (0, 1)$. If $\{x_n\}$ and $\{y_n\}$ are sequences in* X such that $\limsup_{n\to\infty} d(x_n, x) \leq r$, $\limsup_{n\to\infty} d(y_n, x) \leq r$ *and $\lim_{n\to\infty} d(W(x_n, y_n, \alpha_n), x) = r$ for some $r \geq 0$, then* $\lim_{n\to\infty} d(x_n, y_n) = 0$.

Lemma 2 [13] *Let K be a nonempty closed convex subset of a uniformly convex hyperbolic space and $\{x_n\}$ a bounded sequence in K such that $A(\{x_n\}) = \{y\}$ and $r(\{x_n\}) = \rho$. If $\{y_m\}$ is another sequence in K such that $\lim_{m\to\infty} r(y_m, \{x_n\}) = \rho$, then $\lim_{m\to\infty} y_m = y$.*

Lemma 3 [25] *Let K be a nonempty closed subset of a complete metric space (X, d) and $\{x_n\}$ be Fejér monotone with respect to K.* Then *$\{x_n\}$ converges to some $p \in K$ if and only if* $\lim_{n\to\infty} d(x_n, K) = 0$.

Main results

Lemma 4 *Let K be a closed and convex subset of a convex metric space X. Let $\{T_i : i \in J\}$ be a finite family of I_i-nonexpansive mappings and $\{I_i : i \in J\}$ be a finite family of nonexpansive mappings on K such that $F \neq \emptyset$. Then the sequence $\{x_n\}$ defined by (1.6) is Fej ér monotone with respect to F.*

Proof Let $p \in F$. It follows from (1.6) that

$$\begin{aligned}d(x_n,p) &= d\left(W\left(T_nx_n, W\left(I_nx_{n-1},x_{n-1},\frac{\beta_n}{1-\alpha_n}\right),\alpha_n\right),p\right)\\ &\leq \alpha_n d(T_nx_n,p)+(1-\alpha_n)d\left(W\left(I_nx_{n-1},x_{n-1},\frac{\beta_n}{1-\alpha_n}\right),p\right)\\ &\leq \alpha_n d(I_nx_n,p)+\beta_n d(I_nx_{n-1},p)+(1-\alpha_n-\beta_n)d(x_{n-1},p)\\ &\leq \alpha_n d(x_n,p)+\beta_n d(x_{n-1},p)+(1-\alpha_n-\beta_n)d(x_{n-1},p)\end{aligned}$$

and this implies that

$$d(x_n,p)\leq d(x_{n-1},p).$$

Hence, $\{x_n\}$ is Fejér monotone with respect to F. □

Lemma 5 *Let K be a closed and convex subset of a uniformly convex hyperbolic space X with monotone modulus of uniform convexity η. Let $\{T_i : i \in J\}$ be a finite family of I_i-nonexpansive mappings and $\{I_i : i \in J\}$ be a finite family of nonexpansive mappings on K such that $F \neq \emptyset$. Then for the sequence $\{x_n\}$ in (1.6), we have*

$$\lim_{n\to\infty} d(x_n,T_kx_n)=\lim_{n\to\infty} d(x_n,I_kx_n)=0 \quad v \text{ for each } k\in J.$$

Proof Let $p \in F$. Then $\lim_{n\to\infty} d(x_n,p)$ exists by above lemma. Suppose that $\lim_{n\to\infty} d(x_n,p)=c$. Then

$$\lim_{n\to\infty} d(x_n,p)=\lim_{n\to\infty} d\left(W\left(T_nx_n, W\left(I_nx_{n-1},x_{n-1},\frac{\beta_n}{1-\alpha_n}\right),\alpha_n\right),p\right)=c. \tag{3.1}$$

Since T_n is a I_n-nonexpansive mapping for all n, we have $d(T_nx_n,p)\leq d(I_nx_n,p)\leq d(x_n,p)$. Taking lim sup on both sides of this inequality, we obtain

$$\limsup_{n\to\infty} d(T_nx_n,p)\leq \limsup_{n\to\infty} d(x_n,p)=c. \tag{3.2}$$

Now

$$\begin{aligned}&d\left(W\left(I_nx_{n-1},x_{n-1},\frac{\beta_n}{1-\alpha_n}\right),p\right)\\ &\leq \left(\frac{\beta_n}{1-\alpha_n}\right)d(I_nx_{n-1},p)+\left(1-\frac{\beta_n}{1-\alpha_n}\right)d(x_{n-1},p)\\ &\leq \left(\frac{\beta_n}{1-\alpha_n}\right)d(x_{n-1},p)+\left(1-\frac{\beta_n}{1-\alpha_n}\right)d(x_{n-1},p)\\ &\leq d(x_{n-1},p)\end{aligned}$$

implies that

$$\limsup_{n\to\infty} d\left(W\left(I_nx_{n-1},x_{n-1},\frac{\beta_n}{1-\alpha_n}\right),p\right)\leq c. \tag{3.3}$$

Using (3.1)–(3.3) and Lemma 1, we get

$$\lim_{n\to\infty} d\left(T_nx_n, W\left(I_nx_{n-1},x_{n-1},\frac{\beta_n}{1-\alpha_n}\right)\right)=0. \tag{3.4}$$

Observe that

$$\begin{aligned}d(x_n,T_nx_n) &= d\left(W\left(T_nx_n, W\left(I_nx_{n-1},x_{n-1},\frac{\beta_n}{1-\alpha_n}\right),\alpha_n\right),T_nx_n\right)\\ &\leq \alpha_n d(T_nx_n,T_nx_n)+(1-\alpha_n)\\ &\quad\times d\left(W\left(I_nx_{n-1},x_{n-1},\frac{\beta_n}{1-\alpha_n}\right),T_nx_n\right).\end{aligned}$$

Taking lim sup on both sides in the above inequality and using (3.4), we have

$$\lim_{n\to\infty} d(x_n,T_nx_n)=0. \tag{3.5}$$

Next,

$$\begin{aligned}d(x_n,p) &= d\left(W\left(T_nx_n, W\left(I_nx_{n-1},x_{n-1},\frac{\beta_n}{1-\alpha_n}\right),\alpha_n\right),p\right)\\ &\leq \alpha_n d(T_nx_n,p)+(1-\alpha_n)d\left(W\left(I_nx_{n-1},x_{n-1},\frac{\beta_n}{1-\alpha_n}\right),p\right)\\ &\leq \alpha_n d(I_nx_n,p)+(1-\alpha_n)d\left(W\left(I_nx_{n-1},x_{n-1},\frac{\beta_n}{1-\alpha_n}\right),p\right)\end{aligned}$$

yields that

$$d(x_n,p)\leq d\left(W\left(I_nx_{n-1},x_{n-1},\frac{\beta_n}{1-\alpha_n}\right),p\right).$$

Thus, we get

$$c\leq \liminf_{n\to\infty} d\left(W\left(I_nx_{n-1},x_{n-1},\frac{\beta_n}{1-\alpha_n}\right),p\right). \tag{3.6}$$

Combining (3.3) and (3.6), we get

$$\lim_{n\to\infty} d\left(W\left(I_nx_{n-1},x_{n-1},\frac{\beta_n}{1-\alpha_n}\right),p\right)=c. \tag{3.7}$$

Taking lim sup on both sides of $d(I_nx_{n-1},p)\leq d(x_{n-1},p)$, we obtain

$$\limsup_{n\to\infty} d(I_nx_{n-1},p)\leq \limsup_{n\to\infty} d(x_{n-1},p)=c. \tag{3.8}$$

Now using (3.7), (3.8) and Lemma 1, we obtain

$$\lim_{n\to\infty} d(x_{n-1},I_nx_{n-1})=0. \tag{3.9}$$

Consider

$$\begin{aligned}d(x_n,x_{n-1}) &= d\left(W\left(T_nx_n, W\left(I_nx_{n-1},x_{n-1},\frac{\beta_n}{1-\alpha_n}\right),\alpha_n\right),x_{n-1}\right)\\ &\leq \alpha_n d(T_nx_n,x_{n-1})+(1-\alpha_n)\\ &\quad\times d\left(W\left(I_nx_{n-1},x_{n-1},\frac{\beta_n}{1-\alpha_n}\right),x_{n-1}\right)\\ &\leq \alpha_n\{d(T_nx_n,x_n)+d(x_n,x_{n-1})\}\\ &\quad+(1-\alpha_n)d\left(W\left(I_nx_{n-1},x_{n-1},\frac{\beta_n}{1-\alpha_n}\right),x_{n-1}\right)\\ &\leq \alpha_n\{d(T_nx_n,x_n)+d(x_n,x_{n-1})\}\\ &\quad+(1-\alpha_n)\left[\frac{\beta_n}{1-\alpha_n}d(I_nx_{n-1},x_{n-1})\right.\\ &\quad\left.+\left(1-\frac{\beta_n}{1-\alpha_n}\right)d(x_{n-1},x_{n-1})\right].\end{aligned}$$

By this inequality, we have

$$d(x_n, x_{n-1}) \le \frac{\alpha_n}{1-\alpha_n} d(T_n x_n, x_n) + \frac{\beta_n}{1-\alpha_n} d(I_n x_{n-1}, x_{n-1}).$$

Taking limsup on both the sides in the above inequality and then using (3.5) and (3.9), we have

$$\lim_{n\to\infty} d(x_n, x_{n-1}) = 0$$

and so

$$\lim_{n\to\infty} d(x_n, x_{n+k}) = 0, \quad \text{for all } k \in J. \tag{3.10}$$

We have also,

$$\begin{aligned} d(x_n, I_n x_n) &\le d(x_n, x_{n-1}) + d(x_{n-1}, I_n x_{n-1}) + d(I_n x_{n-1}, I_n x_n) \\ &\le d(x_n, x_{n-1}) + d(x_{n-1}, I_n x_{n-1}) + d(x_{n-1}, x_n) \\ &= 2d(x_n, x_{n-1}) + d(x_{n-1}, I_n x_{n-1}). \end{aligned}$$

Thus, from (3.9) and (3.10), we obtain

$$\lim_{n\to\infty} d(x_n, I_n x_n) = 0. \tag{3.11}$$

Finally, note that

$$\begin{aligned} d(x_n, I_{n+k} x_n) &\le d(x_n, x_{n+k}) + d(x_{n+k}, I_{n+k} x_{n+k}) + d(I_{n+k} x_{n+k}, I_{n+k} x_n) \\ &\le d(x_n, x_{n+k}) + d(x_{n+k}, I_{n+k} x_{n+k}) + d(x_{n+k}, x_n) \\ &\le 2d(x_n, x_{n+k}) + d(x_{n+k}, I_{n+k} x_{n+k}). \end{aligned}$$

Taking lim on both sides of the above inequality, we have

$$\lim_{n\to\infty} d(x_n, I_{n+k} x_n) = 0 \quad \text{for each } k \in J.$$

Since for each $k \in J$, the sequence $\{d(x_n, I_k x_n)\}$ is a subsequence of $\bigcup_{i=1}^N \{d(x_n, I_{n+k} x_n)\}$ and $\lim_{n\to\infty} d(x_n, I_{n+k} x_n) = 0$ for each $k \in J$; therefore,

$$\lim_{n\to\infty} d(x_n, I_k x_n) = 0 \quad \text{for each } k \in J.$$

Similarly, we have

$$\lim_{n\to\infty} d(x_n, T_{n+k} x_n) = 0 \quad \text{for each } k \in J,$$

and hence

$$\lim_{n\to\infty} d(x_n, T_k x_n) = 0 \quad \text{for each } k \in J.$$

□

Firstly, we state a result concerning Δ-convergence for algorithm (1.6). The method of proof closely follows [13, Theorem 3.1].

Theorem 2 *Let K be a nonempty closed convex subset of a complete uniformly convex hyperbolic space X with monotone modulus of uniform convexity η. Let $\{T_i : i \in J\}$ be a finite family of I_i-nonexpansive mappings and $\{I_i : i \in J\}$ be a finite family of nonexpansive mappings on K such that $F \neq \emptyset$. Then the sequence $\{x_n\}$ defined in* (1.6), *Δ-converges to a common fixed point of $\{T_i : i \in J\}$ and $\{I_i : i \in J\}$.*

Proof It follows from Lemma 4 that $\{x_n\}$ is bounded. Therefore, $\{x_n\}$ has a unique asymptotic center, that is, $A(\{x_n\}) = \{x\}$. Assume that $\{u_n\}$ is any subsequence of $\{x_n\}$ such that $A(\{u_n\}) = \{u\}$. Then by Lemma 5, we have $\lim_{n\to\infty} d(u_n, T_k u_n) = \lim_{n\to\infty} d(u_n, I_k u_n) = 0$ for each $k = 1, 2, \ldots, N$. Now we prove that u is the common fixed point of $\{T_i : i \in J\}$ and $\{I_i : i \in J\}$.

Define a sequence $\{v_m\}$ in K by $v_m = T_m u$, where $T_m = T_{m(\text{mod}\, N)}$. Clearly,

$$\begin{aligned} d(v_m, u_n) &\le d(T_m u, T_m u_n) + d(T_m u_n, T_{m-1} u_n) + \cdots + d(T u_n, u_n) \\ &\le d(u, u_n) + \sum_{i=1}^{m-1} d(u_n, T_i u_n). \end{aligned}$$

Thus, we have

$$r(v_m, \{u_n\}) = \limsup_{n\to\infty} d(v_m, u_n) \le \limsup_{n\to\infty} d(u, u_n) = r(u, \{u_n\}).$$

This implies that $|r(v_m, \{u_n\}) - r(u, \{u_n\})| \to 0$ as $m \to \infty$. By Lemma 2, we obtain $T_{m(\text{mod}\, N)} u = u$, which implies that u is the common fixed point of $\{T_i : i \in J\}$. Similarly, we can show that u is the common fixed point of $\{I_i : i \in J\}$. Therefore, $u \in F$. Moreover, $\lim_{n\to\infty} d(x_n, u)$ exists by Lemma 4.

Assume $x \neq u$. By the uniqueness of asymptotic centers,

$$\begin{aligned} \limsup_{n\to\infty} d(u_n, u) &< \limsup_{n\to\infty} d(u_n, x) \\ &\le \limsup_{n\to\infty} d(x_n, x) \\ &< \limsup_{n\to\infty} d(x_n, u) \\ &= \limsup_{n\to\infty} d(u_n, u), \end{aligned}$$

a contradiction. Thus, $x = u$. Since $\{u_n\}$ is an arbitrary subsequence of $\{x_n\}$; therefore, $A(\{u_n\}) = \{u\}$ for all subsequences $\{u_n\}$ of $\{x_n\}$. This proves that $\{x_n\}$ Δ-converges to a common fixed point of $\{T_i : i \in J\}$ and $\{I_i : i \in J\}$. □

Secondly, we give strong convergence theorems of algorithm (1.6).

Theorem 3 *Let K be a nonempty closed convex subset of a complete uniformly convex hyperbolic space X with monotone modulus of uniform convexity η. Let $\{T_i : i \in J\}$ be a finite family of I_i-nonexpansive mappings and $\{I_i : i \in J\}$ be a finite family of nonexpansive mappings on K such that $F \neq \emptyset$. Then the sequence $\{x_n\}$ defined in* (1.6) *converges strongly to $p \in F$ if and only if $\lim_{n\to\infty} d(x_n, F) = 0$.*

Proof It follows from Lemma 4 that $\{x_n\}$ is Fejér monotone with respect to F and $\lim_{n\to\infty} d(x_n, F)$ exists. Now applying the Lemma 3, we obtain the result. □

Theorem 4 *Let K be a nonempty closed convex subset of a complete uniformly convex hyperbolic space X with monotone modulus of uniform convexity η. Let $\{T_i : i \in J\}$ be a finite family of I_i-nonexpansive mappings and $\{I_i : i \in J\}$ be a finite family of nonexpansive mappings on K such that $F \neq \emptyset$. Suppose that $\{T_i : i \in J\}$ and $\{I_i : i \in J\}$ satisfy condition (B'). Then the sequence $\{x_n\}$ defined in* (1.6) *converges strongly to $p \in F$.*

Proof By Lemma 4, $\lim_{n\to\infty} d(x_n, F)$ exists for all $p \in F$. In addition, by Lemma 5, $\lim_{n\to\infty} d(x_n, T_k x_n) = \lim_{n\to\infty} d(x_n, I_k x_n) = 0$ for each $k \in J$. It follows from condition (B′) that $\lim_{n\to\infty} f(d(x_n, F)) = 0$. Since f is nondecreasing with $f(0) = 0$, we have $\lim_{n\to\infty} d(x_n, F) = 0$. Therefore, Theorem 3 implies that $\{x_n\}$ converges strongly to a point p in F. □

Note that the Condition (B′) is weaker than both the compactness of K and the semi-compactness of the non-expansive mappings $\{T_i : i \in J\}$ and $\{I_i : i \in J\}$ (see [26]); therefore, we already have the following result.

Theorem 5 *Let K be a nonempty closed convex subset of a complete uniformly convex hyperbolic space X with monotone modulus of uniform convexity η. Let $\{T_i : i \in J\}$ be a finite family of I_i-nonexpansive mappings and $\{I_i : i \in J\}$ be a finite family of nonexpansive mappings on K such that $F \neq \emptyset$. Suppose that either K is compact or one of the map in $\{T_i : i \in J\}$ and $\{I_i : i \in J\}$ is semi-compact. Then the sequence $\{x_n\}$ defined in* (1.6) *converges strongly to $p \in F$.*

Proof Use Lemma 5 and the line of action given in the proof of Theorem 3 in [11]. □

By taking $T_i = T$, $I_i = I$ for all $i \in J$ in the Theorem 2–5, we get the following corollary, yet is new in itself.

Theorem 6 *Let K be a nonempty closed convex subset of a complete uniformly convex hyperbolic space X with monotone modulus of uniform convexity η. Let T be a I-nonexpansive mapping and I be a nonexpansive mapping on K such that $F = F(T) \cap F(I) \neq \emptyset$. Let the sequence $\{x_n\}$ defined by*

$$x_n = W\left(Tx_n, W\left(Ix_{n-1}, x_{n-1}, \frac{\beta_n}{1-\alpha_n}\right), \alpha_n\right) \tag{3.12}$$

(i) *Then $\{x_n\}$ Δ-converges to a common fixed point of T and I.*
(ii) *Then $\{x_n\}$ converges strongly to $p \in F$ if and only if $\lim_{n\to\infty} d(x_n, F) = 0$.*
(iii) *If T and I satisfy condition (A'), then $\{x_n\}$converges strongly to a point in F.*
(iv) *If either K is compact or one of the mappins in TandIis semi-compact, then $\{x_n\}$ converges strongly to $p \in F$.*

To testify our above theorem, we give the following numerical example.

Example 1 Let $X = (-\infty, +\infty)$ with the usual norm $\|\,\|$ and $K = [0, 1]$. In this case, (1.6) reduces to $x_n = \alpha_n x_{n-1} + \beta_n I x_{n-1} + (1 - \alpha_n - \beta_n) T x_n$. Define $T, I : K \to K$ as $Tx = \frac{2x+1}{4}$ and $Ix = 1 - x$. Then one can see that T is a I-non-expansive mapping and I is a nonexpansive mapping on K with common fixed point set $\{\frac{1}{2}\}$. Also *K is compact and* T,I are semi-compact. *Set* $\alpha_n = \frac{1}{n+1}$ and $\beta_n = \frac{1}{n+2}$. *Thus, all assumptions of above theorem are satisfied.* Now we show that $\{x_n\}$ converges strongly to $\frac{1}{2}$. By taking $n = 1$ and $x_0 \in K$, we get $\alpha_1 = \frac{1}{2}, \beta_1 = \frac{1}{3}, 1 - \alpha_1 - \beta_1 = \frac{1}{6}$ and find x_1 from $x_1 = \alpha_1 x_0 + \beta_1 I x_0 + (1 - \alpha_1 - \beta_1) T x_1$. Similarly, $x_2, x_3, \ldots, x_n, \ldots$. We obtain the first ten terms of $\{x_n\}$ as in following table for initial value $x_0 = 0$, $x_0 = 0.2$, $x_0 = 0.7$ and $x_0 = 1$, respectively. From the table below, we see that the sequence $\{x_n\}$ converges strongly to $\frac{1}{2}$. This means that above theorem is applicable.

Iteration no. n	$x_0 = 0$ x_n	$x_0 = 0.2$ x_n	$x_0 = 0.7$ x_n	$x_0 = 1$ x_n
1	0.4090909091	0.4454545455	0.5363636364	0.5909090909
2	0.4904306220	0.4942583733	0.5038277512	0.5095693780
3	0.4993400429	0.4996040258	0.5002639829	0.5006599571
4	0.4999678070	0.4999806842	0.5000128772	0.5000321930
5	0.4999988293	0.4999992976	0.5000004682	0.5000011707
6	0.4999999670	0.4999999803	0.5000000132	0.5000000330
7	0.4999999993	0.4999999995	0.5000000002	0.5000000007
8	0.5000000000	0.5000000000	0.5000000000	0.5000000000
9	0.5000000000	0.5000000000	0.5000000000	0.5000000000
10	0.5000000000	0.5000000000	0.5000000000	0.5000000000

Remark 1

(1) Our results generalize the corresponding results in [10, 11] for more general class of I-nonexpansive maps in the general setup of hyperbolic spaces.
(2) Theorem 2 gives an analogue of Khan's weak convergence result [11] for a finite family of I_i-nonexpansive maps on unbounded domain in a uniformly convex hyperbolic space X.
(3) In view of simplicity of the iterative process (1.6) as compared with (1.4), our results improve and generalize the results of Khan et al. [13].

References

1. Takahashi, W.: A convexity in metric spaces and nonexpansive mappings. Kodai Math. Semin. Rep. **22**, 142–149 (1970)
2. Kohlenbach, U.: Some logical metatheorems with applications in functional analysis. Trans. Am. Math. Soc. **357**, 89–128 (2005)
3. Shahzad, N.: Generalized I-nonexpansive maps and best approximations in Banach spaces. Demonstr. Math. **37**(3), 597–600 (2004)
4. Rhoades, B.E., Temir, S.: Convergence theorem for I-nonexpansive mapping. Int. J. Math. Math. Sci. **2006**(63435) (2006)
5. Temir, S., Gul, O.: Convergence theorem for I-asymptotically quasi-nonexpansive mapping in Hilbert space. J. Math. Anal. Appl. **329**, 759–765 (2007)
6. Temir, S.: On the convergence theorems of implicit iteration process for a finite family of I-asymptotically nonexpansive mappings. J. Comput. Appl. Math. **225**(2), 398–405 (2009)
7. Gu, F.: Some convergence theorems of non-implicit iteration process with errors for a finite families of I-asymptotically nonexpansive mappings. Appl. Math. Comput. **216**, 161–172 (2010)
8. Mukhamedov, F., Saburov, M.: Weak and strong convergence of an implicit iteration process for an asymptotically quasi-I-nonexpansive mapping in Banach space. Fixed Point Theory Appl **2010**(719631). doi:10.1155/2010/719631
9. Xu, H.K., Ori, R.G.: An implicit iteration process for nonexpansive mappings. Numer. Funct. Anal. Optim. **22**, 767–773 (2001)
10. Zhao, J., He, S., Su, Y.: Weak and strong convergence theorems for nonexpansive mappings in Banach spaces. Fixed Point Theory Appl. **2008**(751383). doi:10.1155/2008/751383
11. Khan, S.H.: A modified iterative process for common fixed points of two finite families of nonexpansive mappings. Anal. Şt. Univ. Ovidius Constanta **19**(1), 161–174 (2011)
12. Plubtieng, S., Wangkeeree, R., Punpaeng, R.: On the convergence of modified Noor iterations with errors for asymptotically nonexpansive mappings. J. Math. Anal. Appl. **322**(2), 1018–1029 (2006)
13. Khan, A.R., Fukhar-ud-din, H., Khan, M.A.A.: An implicit algorithm for two finite families of nonexpansive maps in hyperbolic spaces. Fixed Point Theory Appl. **2012**, 54 (2012)
14. Gunduz, B., Akbulut, S.: Strong and -convergence theorems in hyperbolic spaces. Miskolc Math. Notes **14**(3), 915–925 (2013)
15. Akbulut, S., Gunduz, B.: Strong and -convergence of a faster iteration process in hyperbolic space. Commun. Korean Math. Soc. **30**(3), 209–219 (2015)
16. Sahin, A., Basarir, M.: Some convergence results for modified SP-iteration scheme in hyperbolic spaces. Fixed Point Theory Appl. **2014**, 133 (2014)
17. Chang, S.S., Wang, G., Wang, L., Tang, Y.K., Ma, Z.L.: Δ-convergence theorems for multi-valued nonexpansive mappings in hyperbolic spaces. Appl. Math. Comput. **249**, 535–540 (2014)
18. Soltuz, S.M.: The backward Mann iteration. Octogon Math. Mag. **9**(2), 797–800 (2001)
19. Chidume, C.E., Shahzad, N.: Strong convergence of an implicit iteration process for a finite family of nonexpansive mappings. Nonlinear Anal. **65**, 1149–1156 (2005)
20. Shimizu, T., Takahashi, W.: Fixed points of multivalued mappings in certain convex metric spaces. Topol. Methods Nonlinear Anal. **8**, 197–203 (1996)
21. Leuştean, L.: Nonexpansive iterations in uniformly convex W-hyperbolic spaces. Contemp. Math. **513**, 193–210 (2010)
22. Kirk, W., Panyanak, B.: A concept of convergence in geodesic spaces. Nonlinear Anal. **68**, 3689–3696 (2008)
23. Kuczumow, T.: An almost convergence and its applications. Ann. Univ. Mariae Curie-Sklodowska Sect. A **32**, 79–88 (1978)
24. Fukhar-ud-din, H., Khan, S.H.: Convergence of iterates with errors of asymptotically quasi-nonexpansive mappings and applications. J. Math. Anal. Appl. **328**, 821–829 (2007)
25. Bauschke, H.H., Combettes, P.L.: Convex Analysis and Monotone Operator Theory in Hilbert Spaces. Springer, New York (2011)
26. Senter, H.F., Dotson, W.G.: Approximatig fixed points of nonexpansive mappings. Proc. Am. Math. Soc. **44**(2), 375–380 (1974)

23

Suzuki-type fixed point results in *b*-metric spaces

Jamal Rezaei Roshan[1] · Nawab Hussain[2] · Shaban Sedghi[1] · Nabi Shobkolaei[3]

Abstract An ingenious approach to generalize Banach contraction principle was adopted by Suzuki in his seminal papers (Proc Am Math Soc 136:1861–1869, 2008, Nonlinear Anal Theory Methods Appl 71:5313–5317, 2009). In this paper we prove certain common fixed point results for generalized Suzuki contractions in the set-up of b-metric spaces, where the b-metric function is not necessarily continuous. Finally, some examples are presented to verify the effectiveness and applicability of our main results.

Keywords Fixed point · b-metric space · Suzuki contarction

Mathematics Subject Classification 47H10 · 54H25

✉ Jamal Rezaei Roshan
Jmlroshan@gmail.com

Nawab Hussain
nhusain@kau.edu.sa

Shaban Sedghi
sedghi_gh@yahoo.com

Nabi Shobkolaei
nabi_shobe@yahoo.com

1 Department of Mathematics, Qaemshahr Branch, Islamic Azad University, Qaemshahr, Iran

2 Department of Mathematics, King Abdulaziz University, P.O. Box 80203, Jeddah 21589, Saudi Arabia

3 Department of Mathematics, Babol Branch, Islamic Azad University, Babol, Iran

Introduction

There are a lot of generalizations of Banach fixed point principle in the literature. In 2008 Suzuki introduced an interesting generalization of Banach fixed point principle. This interesting fixed-point result is as follows.

Theorem 1 [26] *Let (X, d) be a complete metric space, and let T be a mapping on X. Define a non-increasing function θ from [0, 1) into (1/2, 1] by*

$$\theta(r)=\begin{cases}1, & 0\le r\le \dfrac{\sqrt{5}-1}{2}\\ \dfrac{1-r}{r^2}, & \dfrac{\sqrt{5}-1}{2}\le r\le \dfrac{1}{\sqrt{2}}\\ \dfrac{1}{1+r}, & \dfrac{1}{\sqrt{2}}\le r<1.\end{cases}$$

Assume that there exists $r\in[0,1)$, such that

$$\theta(r)\mathrm{d}(x,Tx)\le \mathrm{d}(x,y)\Longrightarrow \mathrm{d}(Tx,Ty)\le r\mathrm{d}(x,y),$$

for all $x,y\in X$, then there exists a unique fixed-point z of T. Moreover, $\lim_{n\to\infty}T^nx=z$ for all $x\in X$.

Suzuki proved also the following version of Nemytckii fixed point theorem.

Theorem 2 *Let (X, d) be a compact metric space. Let $T:X\to X$ be a selfmap, satisfying for all $x,y\in Xx\neq y$ the condition*

$$\frac{1}{2}\mathrm{d}(x,Tx)\le \mathrm{d}(x,y)\Longrightarrow \mathrm{d}(Tx,Ty)<\mathrm{d}(x,y)$$

Then T has a unique fixed point in X.

This theorem was also generalized in [6].

In addition to the above results, Kikkawa and Suzuki [11] provided a Kannan-type version of the theorems mentioned above. In [21], a Chatterjea-type version is provided, whereas Popescu [20] obtained a Cirić-type version. Recently, Kikkawa and Suzuki also provided multivalued versions in [12, 13].

Very recently, Hussain et al. in [8] have extended Suzuki's Theorems 1 and 2, as well as Popescu's results from [20] to the case of metric-type spaces and cone metric-type spaces.

Czerwik in [5] introduced the concept of b-metric space. Since then, several papers deal with fixed point theory for single-valued and multivalued operators in b-metric spaces (see also [1–5, 7–10, 14–17, 19, 24, 25]). Pacurar [19] proved results on sequences of almost contractions and fixed points in b-metric spaces. Recently, Hussain and Shah [9] obtained results on KKM mappings in cone b-metric spaces. Khamsi ([14, 15]) also showed that each cone metric space has a b-metric structure.

The aim of this paper is to present some common fixed point results for two mappings under generalized contractive condition in b-metric space, where the b-metric function is not necessarily continuous. Because many of the authors in their works have used the b-metric spaces in which the b-metric function is assumed to be continuous. From this point of view the results obtained in this paper generalize and extend several ones obtained earlier concerning b-metric space.

Consistent with [5] and [25, p. 264], the following definition and results will be needed in the sequel.

Definition 1 [5] Let X be a (nonempty) set and $b \geq 1$ be a given real number. A function $d : X \times X \to R^+$ is a b-metric spaces iff, for all $x, y, z \in X$, the following condition are satisfied:

(b1) $\mathrm{d}(x,y) = 0$ iff $x = y$,
(b2) $\mathrm{d}(x,y) = \mathrm{d}(y,x)$,
(b3) $\mathrm{d}(x,z) \leq b[\mathrm{d}(x,y) + \mathrm{d}(y,z)]$.

The pair (X, d) is called a b-metric space.

It should be noted that, the class of b-metric spaces is effectively larger than that of metric spaces, since a b-metric is a metric only if $b = 1$.

We present an example which shows that a b-metric on X need not be a metric on X. (see also [25, p. 264]):

Example 1 [22] Let (X, d) be a metric space, and $\rho(x,y) = (d(x,y))^p$, where $p > 1$ is a real number. Then ρ is a b-metric with $b = 2^{p-1}$.

However, if (X, d) is a metric space, then (X, ρ) is not necessarily a metric space.

For example, if $X = \mathbb{R}$ is the set of real numbers and $d(x,y) = |x - y|$ is the usual Euclidean metric, then $\rho(x,y) = (x-y)^2$ is a b-metric on $\mathbb{R}$ with $b = 2$, but is not a metric on $\mathbb{R}$.

Before stating and proving our results, we present some definition and proposition in b-metric space. We recall first the notions of convergence and completeness in a b-metric space.

Definition 2 [3] Let (X, d) be a b-metric space. Then a sequence $\{x_n\}$ in X is called:

(a) convergent if and only if there exists $x \in X$ such that $d(x_n, x) \to 0$ as $n \to \infty$. In this case, we write $\lim_{n\to\infty} x_n = x$.
(b) Cauchy if and only if $d(x_n, x_m) \to 0$ as $n, m \to \infty$.

Proposition 1 (see remark 2.1 in [3]) *In a b-metric space (X, d) the following assertions hold*:

(i) *a convergent sequence has a unique limit,*
(ii) *each convergent sequence is Cauchy,*

Definition 3 [3] The b-metric space (X, d) is complete if every Cauchy sequence in X converges.

It should be noted that, in general a b-metric function $d(x, y)$ for $b > 1$ is not jointly continuous in all two of its variables. Now we present an example of a b-metric which is not continuous.

Example 2 (see Example 3 in [8]) Let $X = \mathbb{N} \cup \{\infty\}$ and let $D : X \times X \to \mathbb{R}^+$ be defined by,

$$D(m,n) = \begin{cases} 0 & \text{if } m = n, \\ \left|\dfrac{1}{m} - \dfrac{1}{n}\right|, & \text{if one of } m,n \text{ is even and the other is even or } \infty, \\ 5 & \text{if one of } m, n \text{ is odd and other is odd (and } m \neq n) \text{ or } \infty, \\ 2 & \text{otherwise.} \end{cases}$$

Then it is easy to see that for all $m,n,p \in X$, we have

$$D(m,p) \leq \frac{5}{2}(D(m,n) + D(n,p)).$$

Thus, (X, D) is a b-metric space with $b = \frac{5}{2}$. In [8], it is proved that $D(x, y)$ is not a continuous function.

Since in general a b-metric is not continuous, we need the following simple lemma about the b-convergent sequences.

Lemma 1 [22] *Let (X, d) be a b-metric space with $b \geq 1$, and suppose that $\{x_n\}$ and $\{y_n\}$b-converge to x, y, respectively. Then, we have*

$$\frac{1}{b^2}\mathrm{d}(x,y) \leq \liminf_{n\to\infty} \mathrm{d}(x_n, y_n) \leq \limsup_{n\to\infty} \mathrm{d}(x_n, y_n) \leq b^2 \mathrm{d}(x,y).$$

In particular, if $x = y$, then $\lim_{n\to\infty} \mathrm{d}(x_n, y_n) = 0$. Moreover, for each $z \in X$ we have

$$\frac{1}{b}\mathrm{d}(x,z) \leq \liminf_{n\to\infty} \mathrm{d}(x_n, z) \leq \limsup_{n\to\infty} \mathrm{d}(x_n, z) \leq b\mathrm{d}(x,z).$$

Main result

We start our work by proving the following crucial Theorem.

Theorem 3 *Let (X, d) be a complete b-metric space. Let $T, S : X \longrightarrow X$ be two self-maps and $\theta : [0,1) \longrightarrow (\frac{1}{2}, 1]$ be defined by*

$$\theta(r) = \begin{cases} 1, & 0 \leq r \leq \frac{\sqrt{5}-1}{2} \\ \frac{1-r}{r^2}, & \frac{\sqrt{5}-1}{2} \leq r \leq \frac{1}{\sqrt{2}} \\ \frac{1}{1+r}, & \frac{1}{\sqrt{2}} \leq r < 1. \end{cases} \tag{1}$$

Suppose there exists $r \in [0,1)$ such that for each $x, y \in X$, the following condition is satisfied

$$\frac{1}{b}\theta(r)\min\{d(x,Tx), d(x,Sx)\} \leq d(x,y) \Longrightarrow$$
$$\max\left\{\begin{array}{l} d(Sx,Sy), d(Tx,Ty), \\ d(Sx,Ty), d(Sy,Tx) \end{array}\right\} \leq \frac{r}{b^2} d(x,y). \tag{2}$$

Then T, S have a unique common fixed point $z \in X$.

Proof At first we show that if z is a fixed point of S or T, then z is a common fixed point of T and S. Let z be a fixed point of T that is $Tz = z$ then we show that $Sz = z$. From

$$0 = \frac{1}{b}\theta(r)\min\{d(z,Tz); d(z,Sz)\} \leq d(z,Tz),$$

it follows

$$d(Sz,z) \leq \max\left\{\begin{array}{l} d(Sz,STz), d(Tz,T^2z), \\ d(Sz,T^2z), d(STz,Tz) \end{array}\right\}$$
$$\leq \frac{r}{b^2} d(z,Tz) = 0,$$

thus $Sz = z$. Therefore it is enough to show that T have a fixed point. Putting $y = Sx$ in (2)

$$\frac{1}{b}\theta(r)\min\{d(x,Tx), d(x,Sx)\} \leq d(x,Sx),$$

it follows

$$\max\left\{\begin{array}{l} d(Sx,S^2x), d(Tx,TSx), \\ d(Sx,TSx), d(S^2x,Tx) \end{array}\right\} \leq \frac{r}{b^2} d(x,Sx), \tag{3}$$

for every $x \in X$. Hence,

$$d(Sx,TSx) \leq \frac{r}{b^2} d(x,Sx). \tag{4}$$

Now, putting $y = Tx$ in (2)

$$\frac{1}{b}\theta(r)\min\{d(x,Tx), d(x,Sx)\} \leq d(x,Tx),$$

it follows

$$\max\left\{\begin{array}{l} d(Sx,STx), d(Tx,T^2x), \\ d(Sx,T^2x), d(STx,Tx) \end{array}\right\} \leq \frac{r}{b^2} d(x,Tx), \tag{5}$$

for every $x \in X$. Hence,

$$d(Tx,T^2x) \leq \frac{r}{b^2} d(x,Tx), \tag{6}$$

and

$$d(STx,Tx) \leq \frac{r}{b^2} d(x,Tx). \tag{7}$$

Let $x_0 \in X$ be arbitrary and form the sequence $\{x_n\}$ by, $x_{2n+1} = Sx_{2n}$ and $Tx_{2n+1} = x_{2n+2}$ for $n \in \mathbb{N} \cup \{0\}$. We show that $\{x_n\}$ is a Cauchy sequence.

By (4), we have

$$\begin{aligned} d(x_{2n+1},x_{2n+2}) &= d(Sx_{2n}, TSx_{2n}) \\ &\leq \frac{r}{b^2} d(x_{2n}, Sx_{2n}) = \frac{r}{b^2} d(x_{2n}, x_{2n+1}). \end{aligned} \tag{8}$$

By (7), we have

$$\begin{aligned} d(x_{2n+1},x_{2n}) &= d(STx_{2n-1}, Tx_{2n-1}) \\ &\leq \frac{r}{b^2} d(x_{2n-1}, Tx_{2n-1}) = \frac{r}{b^2} d(x_{2n-1}, x_{2n}). \end{aligned}$$

Therefore,

$$\begin{aligned} d(x_n,x_{n+1}) &\leq \frac{r}{b^2} d(x_{n-1},x_n) \\ &\leq \frac{r^2}{b^4} d(x_{n-2},x_{n-1}) \\ &\vdots \\ &\leq \frac{r^n}{b^{2n}} d(x_0,x_1). \end{aligned}$$

Also, by definition of b-metric spaces for all $m \geq n$, we have

$$\begin{aligned} d(x_n,x_m) &\leq bd(x_n,x_{n+1}) + b^2 d(x_{n+1},x_{n+2}) + \cdots + b^{m-n-1} d(x_{m-1},x_m) \\ &\leq b\frac{r^n}{b^{2n}} d(x_0,x_1) + b^2 \frac{r^{n+1}}{b^{2n+2}} d(x_0,x_1) + \cdots + b^{m-n-1}\frac{r^{m-1}}{b^{2m-2}} d(x_0,x_1) \\ &= \frac{r^n}{b^{2n-1}} d(x_0,x_1) + \frac{r^{n+1}}{b^{2n}} d(x_0,x_1) + \cdots + \frac{r^{m-1}}{b^{m+n-1}} d(x_0,x_1) \\ &= \frac{r^n}{b^{2n-1}}\left(1 + \frac{r}{b} + \cdots + \frac{r^{m-n-1}}{b^{m-n}}\right) d(x_0,x_1) \\ &\leq \frac{r^n}{b^{2n-1}}\left(1 + \frac{r}{b} + \cdots + (\frac{r}{b})^{m-n-1}\right) d(x_0,x_1) \\ &\leq \frac{r^n}{b^{2n-1}} d(x_0,x_1)\left(\frac{1}{1-\frac{r}{b}}\right) \longrightarrow 0 \text{ as } n \rightarrow \infty. \end{aligned}$$

So, we have

$$\lim_{n,m\rightarrow\infty} d(x_n,x_m) = 0.$$

Hence, $\{x_n\}$ is a Cauchy sequence. Since X is complete, we conclude $\{x_n\}$ converges to z for some $z \in X$. That is

$$\lim_{n\rightarrow\infty} Sx_{2n} = \lim_{n\rightarrow\infty} x_{2n+1} = z,$$

and

$$\lim_{n\to\infty} Tx_{2n+1} = \lim_{n\to\infty} x_{2n+2} = z.$$

Let us prove now that

$$d(z, Tx) \le r d(z, x),$$

holds for each $x \neq z$. Since $d(x_{2n}, Sx_{2n}) \longrightarrow 0$, and by Lemma 1

$$\frac{1}{b} d(z, x) \le \limsup_{n\longrightarrow\infty} d(x_{2n}, x),$$

thus $\limsup_{n\longrightarrow\infty} d(x_{2n}, x) > 0$, it follows that there exists a $x_{2n_k} \in X$ such that

$$\frac{1}{b}\theta(r)\min\{d(x_{2n_k}, Sx_{2n_k}), d(x_{2n_k}, Tx_{2n_k})\} \le d(x_{2n_k}, x).$$

Assumption (2) implies that for such x_{2n_k}

$$\begin{aligned} d(Sx_{2n_k}, Tx) &\le \max\left\{ \begin{array}{l} d(Sx_{2n_k}, Sx), d(Tx_{2n_k}, Tx) \\ d(Sx_{2n_k}, Tx), d(Sx, Tx_{2n_k}) \end{array} \right\} \\ &\le \frac{r}{b^2} d(x_{2n_k}, x), \end{aligned}$$

hence by Lemma 1

$$\begin{aligned} \frac{1}{b} d(z, Tx) &\le \limsup_{n\to\infty} d(Sx_{2n_k}, Tx) \le \frac{r}{b^2} \limsup_{n\to\infty} d(x_{2n_k}, x) \\ &\le \frac{r}{b} d(z, x), \end{aligned}$$

thus for each $x \neq z$ we get that

$$d(z, Tx) \le r d(z, x). \tag{9}$$

We will prove that

$$d(T^n z, z) \le d(Tz, z), \tag{10}$$

for each $n \in \mathbb{N}$. For $n = 1$ this relation is obvious. Suppose that it holds for some $m \in \mathbb{N}$. If $T^m z = z$ then $T^{m+1} z = Tz$ it follows that the above inequality is true. If $T^m z \neq z$, we can apply (9) and the induction hypothesis, we get that

$$\begin{aligned} d(z, T^{m+1} z) &\le r d(z, T^m z) \\ &\le r d(Tz, z) \le d(Tz, z), \end{aligned}$$

and (10) is proved by induction.

In order to prove that $Tz = z$. We consider two possible cases.

Case I. $0 \le r < \frac{1}{\sqrt{2}}$ (and hence $\theta(r) \le \frac{1-r}{r^2}$). We will prove first that

$$d(T^n z, Tz) \le \frac{r}{b} d(Tz, z) \tag{11}$$

for each $n \in \mathbb{N}$. For $n = 1$ it is obvious. For $n = 2$ it follows from (6). Suppose that (11) holds for some $n > 2$. Since

$$\begin{aligned} d(Tz, z) &\le b d(z, T^n z) + b d(T^n z, Tz) \\ &\le b d(z, T^n z) + r d(z, Tz), \end{aligned}$$

hence $(1 - r) d(z, Tz) \le b d(z, T^n z)$. It follows [using (6) with $x = T^{n-1} z$] that

$$\begin{aligned} \frac{1}{b}\theta(r)\min\{d(ST^n z, T^n z), d(T^n z, T^{n+1} z)\} &\le \frac{1-r}{br^2} d(T^n z, T^{n+1} z) \\ &\le \frac{1-r}{br^n} d(T^n z, T^{n+1} z) \\ &\le \frac{1-r}{br^n} \cdot \frac{r^n}{b^{2n}} d(z, Tz) \\ &= \frac{1-r}{b^{2n+1}} d(z, Tz) \\ &\le \frac{1}{b^{2n}} d(z, T^n z) \\ &\le d(z, T^n z). \end{aligned}$$

Assumptions (2) and (10) imply that

$$\begin{aligned} &\max\{d(ST^n z, Sz), d(ST^n z, Tz), \\ &d(T^{n+1} z, Tz), d(Sz, T^{n+1} z)\} \le \frac{r}{b^2} d(z, T^n z) \\ &\le \frac{r}{b^2} d(z, Tz) \le \frac{r}{b} d(z, Tz). \end{aligned}$$

Thus

$$d(T^{n+1} z, Tz) \le \frac{r}{b} d(Tz, z). \tag{12}$$

So relation (11) is proved by induction.

Now $Tz \neq z$ and (11) imply that $T^n z \neq z$ for each $n \in \mathbb{N}$. Hence, (9) implies that

$$\begin{aligned} d(z, T^{n+1} z) &\le r d(z, T^n z) \\ &\le r^2 d(z, T^{n-1} z) \\ &\vdots \\ &\le r^n d(z, Tz). \end{aligned}$$

Hence $\lim_{n\to\infty} d(z, T^{n+1} z) = 0$. On the other hand using Lemma 1, we have

$$\frac{1}{b} d(z, \liminf_{n\longrightarrow\infty} T^{n+1} z) \le \liminf_{n\longrightarrow\infty} d(z, T^{n+1} z) = 0,$$

so

$$d(z, \liminf_{n\longrightarrow\infty} T^{n+1} z) = 0.$$

Similarly,

$$d(z, \limsup_{n\longrightarrow\infty} T^{n+1} z) = 0,$$

therefore $d(z, \lim_{n\longrightarrow\infty} T^{n+1} z) = 0$.

Thus $T^{n+1} z \longrightarrow z$ and, using Lemma 1 in (12), we have

$$\frac{1}{b} d(z, Tz) \le \limsup_{n\longrightarrow\infty} d(T^{n+1} z, Tz) \le \frac{r}{b} d(z, Tz),$$

which implies that $d(z, Tz) = 0$, a contradiction.

Case II. $\frac{1}{\sqrt{2}} \le r < 1$ (and so $\theta(r) = \frac{1}{1+r}$). We will prove that there exists a subsequence $\{x_{n_k}\}$ of $\{x_n\}$ such that

$$\frac{1}{b(1+r)}\min\{d(x_{n_k}, Sx_{n_k}), d(x_{n_k}, Tx_{n_k}) \le d(x_{n_k}, z) \qquad (13)$$

holds for each $k \in \mathbb{N}$. Suppose the contrary

$$\frac{1}{b(1+r)}d(x_n, Tx_n) \ge \frac{1}{b(1+r)}\min\{d(x_n, Sx_n), d(x_n, Tx_n)\} > d(x_n, z),$$

and

$$\frac{1}{b(1+r)}d(x_n, Sx_n) \ge \frac{1}{b(1+r)}\min\{d(x_n, Sx_n), d(x_n, Tx_n)\} > d(x_n, z),$$

holds for each $n \in \mathbb{N}$. Now if n is odd then

$$\frac{1}{b(1+r)}d(x_{2n+1}, Tx_{2n+1}) \ge \frac{1}{b(1+r)}\min\{d(x_{2n+1}, Sx_{2n+1}), d(x_{2n+1}, Tx_{2n+1}) > d(x_{2n+1}, z),$$

if n is even then

$$\frac{1}{b(1+r)}d(x_{2n}, Sx_{2n}) \ge \frac{1}{b(1+r)}\min\{d(x_{2n}, Sx_{2n}), d(x_{2n}, Tx_{2n}) > d(x_{2n}, z),$$

holds for each $n \in \mathbb{N}$. Then from (8) we have

$$\begin{aligned} d(x_{2n}, x_{2n+1}) &\le bd(x_{2n}, z) + bd(x_{2n+1}, z) \\ &< \frac{b}{b(1+r)}d(x_{2n}, Sx_{2n}) + \frac{b}{b(1+r)}d(x_{2n+1}, Tx_{2n+1}) \\ &= \frac{1}{1+r}d(x_{2n}, x_{2n+1}) + \frac{1}{1+r}d(x_{2n+1}, x_{2n+2}) \\ &\le \frac{1}{1+r}d(x_{2n}, x_{2n+1}) + \frac{r}{b^2(1+r)}d(x_{2n}, x_{2n+1}) \\ &\le \frac{1}{1+r}d(x_{2n}, x_{2n+1}) + \frac{r}{1+r}d(x_{2n}, x_{2n+1}) \\ &= d(x_{2n}, x_{2n+1}), \end{aligned}$$

which is impossible. Hence one of the following inequalities is satisfied for each $n \in \mathbb{N}$:

$$\frac{1}{b}\theta(r)\min\{d(x_{2n}, Sx_{2n}), d(x_{2n}, Tx_{2n}) \le d(x_{2n}, z).$$

or

$$\frac{1}{b}\theta(r)\min\{d(x_{2n+1}, Sx_{2n+1}), d(x_{2n+1}, Tx_{2n+1}) \le d(x_{2n+1}, z).$$

In other words, there is a subsequence $\{x_{n_k}\}$ of $\{x_n\}$ such that (13) holds for each $k \in \mathbb{N}$. Hence assumption (2) implies that

$$d(Sx_{2n}, Tz) \le \max\left\{\begin{matrix} d(Sx_{2n}, Sz), d(Tx_{2n}, Tz), \\ d(Tz, Sx_{2n}), d(Sz, Tx_{2n}) \end{matrix}\right\} \le \frac{r}{b^2}d(x_{2n}, z).$$

or

$$d(Tx_{2n+1}, Tz) \le \max\left\{\begin{matrix} d(Sx_{2n+1}, Sz), d(Tx_{2n+1}, Tz), \\ d(Tz, Sx_{2n+1}), d(Sz, Tx_{2n+1}) \end{matrix}\right\} \le \frac{r}{b^2}d(x_{2n+1}, z).$$

By Lemma 1, we get

$$\frac{1}{b}d(z, Tz) \le \limsup_{n\to\infty} d(Sx_{2n}, Tz) \le \frac{r}{b^2}\limsup_{n\to\infty} d(x_{2n}, z) \le \frac{r}{b}d(z, z) = 0,$$

or

$$\frac{1}{b}d(z, Tz) \le \limsup_{n\to\infty} d(Tx_{2n+1}, Tz) \le \frac{r}{b^2}\limsup_{n\to\infty} d(x_{2n+1}, z) \le \frac{r}{b}d(z, z) = 0,$$

hence $d(z, Tz) \le 0$, which is possible only if $Tz = z$.

Thus, we have proved that z is a fixed point of T. The uniqueness of the common fixed point follows easily from (2). Indeed, if z, z' are two common fixed points of T,

$$\frac{1}{b}\theta(r)\min\{d(z, Tz), d(z, Sz)\} \le d(z, z'),$$

then (2) implies that

$$d(z, z') = \max\{d(Sz, Sz'), d(Tz, Tz'), d(Sz, Tz'), d(Sz', Tz)\} \le \frac{r}{b^2}d(z, z'),$$

which is possible only if $z = z'$. This proves that z is a unique common fixed point of T and S. □

According to Theorem 3 we get the following result.

Corollary 1 *Let (X, d) be a complete b-metric space, and let T be a mapping on X. Define a non-increasing function θ from [0, 1) into (1/2, 1] by (1).*

Suppose there exists $r \in [0, 1)$ such that for each $x, y \in X$, the following condition is satisfied

$$\frac{1}{b}\theta(r)d(x, Tx) \le d(x, y) \Longrightarrow d(Tx, Ty) \le \frac{r}{b^2}d(x, y),$$

then there exists a unique fixed-point z of T. Moreover, $\lim_{n\to\infty} T^n x = z$ for all $x \in X$.

Proof It is enough set $S = T$ in the Theorem 3 then the desired result is obtained. □

Remark 1 Note that for $b = 1$, Corollary 1 reduces to Theorem.

Corollary 2 *Let (X, d) be a complete b-metric space, and $f,S,T : X \longrightarrow X$ be three self-maps and $\theta : [0,1) \longrightarrow (\frac{1}{2},1]$ be defined by* (1).

Suppose there exists $r \in [0,1)$ such that for each $x,y \in X$, the following condition is satisfied

$$\frac{1}{b}\theta(r)\min\{d(x,fTx),d(x,fSx)\} \le d(x,y) \Longrightarrow$$
$$\max\left\{\begin{array}{l} d(fSx,fSy),d(fTx,fTy), \\ d(fSx,fTy),d(fSy,fTx) \end{array}\right\} \le \frac{r}{b^2}d(x,y).$$

Also, if f is one to one, $fS = Sf$ and $fT = Tf$, then we have f, T and *S have a unique common fixed point $z \in X$.*

Proof By Theorem 3, fT, fS have a unique common fixed point $z \in X$. That is $fSz = fTz = z$, since f is one to one it follows that $Sz = Tz$. From

$$0 = \frac{1}{b}\theta(r)\min\{d(z,fTz),d(z,fSz)\} \le d(z,Tz),$$

it follows that

$$\begin{aligned} d(z,Tz) &\le \max\left\{\begin{array}{l} d(fSz,fSTz),d(fTz,fT^2z), \\ d(fSz,fT^2z),d(fSTz,fTz) \end{array}\right\} \\ &= \max\left\{\begin{array}{l} d(fSz,SfTz),d(fTz,TfTz), \\ d(fSz,TfTz),d(SfTz,fTz) \end{array}\right\} \\ &= \max\left\{\begin{array}{l} d(z,Sz),d(z,Tz), \\ d(z,Tz),d(Sz,z) \end{array}\right\} \\ &\le \frac{r}{b^2}d(z,Tz), \end{aligned}$$

it follows that $Tz = Sz = z$, hence $fz = fTz = z$. □

Corollary 3 Let (X,d) be a complete metric space, and $f,S,T : X \longrightarrow X$ be three self-maps and $\theta : [0,1) \longrightarrow (\frac{1}{2},1]$ be defined by (1).

Suppose there exists $r \in [0,1)$ such that for each $x,y \in X$, the following condition is satisfied

$$\theta(r)\min\{d(x,fTx),d(x,fSx)\} \le d(x,y) \Longrightarrow$$
$$\max\left\{\begin{array}{l} d(fSx,fSy),d(fTx,fTy), \\ d(fSx,fTy),d(fSy,fTx) \end{array}\right\} \le rd(x,y).$$

Also, if f is one to one, $fS = Sf$ and $fT = Tf$, then we have f, T and S have a unique common fixed point $z \in X$.

Proof It is enough to set $b = 1$ in the Corollary 2 then the desired result is obtained. □

Now, in order to support the useability of our results, let us introduce the following examples.

Let $X = [0,\infty)$. Define $d : X \times X \to \mathbb{R}^+$ by

$$d(x,y) = \begin{cases} 0, & x = y \\ (x+y)^2, & x \ne y. \end{cases}$$

for all $x,y \in X$. Then (X,d) is a complete b-metric space for $b = 2$. Define two maps $T,S : X \to X$ by

$$\begin{aligned} T(x) &= \ln(1+\frac{1}{4\sqrt{2}}x), \\ S(x) &= \ln(1+\frac{1}{8\sqrt{2}}x) \end{aligned}$$

for $x \in X$. Then for each $x,y \in X$ we have

$$\begin{aligned} \frac{1}{2}\theta(r)\min\{d(x,Tx),d(x,Sx)\} &= \frac{1}{4}\min\left\{\begin{array}{l} (x+\ln(1+\frac{1}{4\sqrt{2}}x))^2, \\ (x+\ln(1+\frac{1}{8\sqrt{2}}x))^2 \end{array}\right\} \\ &= \frac{1}{4}(x+\ln(1+\frac{1}{8\sqrt{2}}x))^2 \\ &\le \frac{1}{4}(x+\frac{1}{8\sqrt{2}}x)^2 = \frac{1}{4}(1+\frac{1}{8\sqrt{2}})^2x^2 \\ &\le x^2 \le (x+y)^2 = d(x,y). \end{aligned}$$

On the other hand, we have

$$\begin{aligned} \max\left\{\begin{array}{l} d(Sx,Sy),d(Tx,Ty), \\ d(Sx,Ty),d(Sy,Tx) \end{array}\right\} &= \max\left\{\begin{array}{l} (\ln(1+\frac{1}{8\sqrt{2}}x)+\ln(1+\frac{1}{8\sqrt{2}}y))^2, \\ (\ln(1+\frac{1}{4\sqrt{2}}x)+\ln(1+\frac{1}{4\sqrt{2}}y))^2, \\ (\ln(1+\frac{1}{8\sqrt{2}}x)+\ln(1+\frac{1}{4\sqrt{2}}y))^2, \\ (\ln(1+\frac{1}{4\sqrt{2}}x)+\ln(1+\frac{1}{8\sqrt{2}}y))^2 \end{array}\right\} \\ &\le \max\left\{\begin{array}{l} (\frac{1}{8\sqrt{2}}x+\frac{1}{8\sqrt{2}}y)^2, \\ (\frac{1}{4\sqrt{2}}x+\frac{1}{4\sqrt{2}}y)^2, \\ (\frac{1}{8\sqrt{2}}x+\frac{1}{4\sqrt{2}}y)^2, \\ (\frac{1}{4\sqrt{2}}x+\frac{1}{8\sqrt{2}}y)^2 \end{array}\right\} \\ &\le (\frac{1}{4\sqrt{2}}x+\frac{1}{4\sqrt{2}}y)^2 = [\frac{1}{4\sqrt{2}}(x+y)]^2 \\ &\le \frac{1}{4}\cdot\frac{1}{2\sqrt{2}}(x+y)^2 = \frac{r}{b^2}d(x,y). \end{aligned}$$

Thus T and S satisfy all the hypotheses of Theorem 3 and hence T and S have a unique common fixed point. Indeed, $r = \frac{1}{2\sqrt{2}} < \frac{\sqrt{5}-1}{2}$, $\theta(r) = \frac{1}{2}$ and 0 is the unique common fixed point of T and S.

Inspired by [8, Example 4] and [26, Example 1], we present the following example:

Example 3 Let $X = \{(0,0),(10,12),(12,10),(40,42),(42,40)\} \subset \mathbb{R}^2$. Define $d : X \times X \to \mathbb{R}^+$ by

$$d((x_1,y_1),(x_2,y_2)) = (x_1-x_2)^2+(y_1-y_2)^2,$$

for all $x = (x_1,y_1), y = (x_2,y_2) \in X$. Then (X, d) is a complete b-metric space for $b = 2$. Define two maps $T,S : X \to X$ by

$$\begin{cases} T(0,0) = T(10,12) = T(12,10) = (0,0), \\ T(40,42) = (10,12), \\ T(42,40) = (12,10). \end{cases}$$

$$\begin{cases} S(0,0) = S(10,12) = S(12,10) = (0,0), \\ S(40,42) = (12,10), \\ S(42,40) = (10,12). \end{cases}$$

Then for each $x, y \in X$, if

$$\frac{1}{2}\frac{\sqrt{2}}{\sqrt{2}+1}\min\{d(x,Tx), d(x,Sx)\} \leq d(x,y),$$

this implies that

$$\max\left\{\begin{array}{l} d(Sx,Sy), d(Tx,Ty), \\ d(Sx,Ty), d(Sy,Tx) \end{array}\right\} \leq \frac{r}{4}d(x,y).$$

Because,

i) $\frac{1}{2}\frac{\sqrt{2}}{\sqrt{2}+1}\min\{d((0,0),T(0,0)), d((0,0),S(0,0))\} \leq d((0,0),y), \quad \forall y \in X.$

ii) $\frac{1}{2}\frac{\sqrt{2}}{\sqrt{2}+1}\min\{d((10,12),T(10,12)), d((10,12),S(10,12))\} \leq d((10,12),y), \quad \forall y = (0,0),(40,42),(42,40).$

iii) $\frac{1}{2}\frac{\sqrt{2}}{\sqrt{2}+1}\min\{d((12,10),T(12,10)), d((12,10),S(12,10))\} \leq d((12,10),y), \quad \forall y = (0,0),(40,42),(42,40).$

iv) $\frac{1}{2}\frac{\sqrt{2}}{\sqrt{2}+1}\min\{d((40,42),T(40,42)), d((40,42),S(40,42))\} \leq d((40,42),y), \quad \forall y = (0,0),(10,12),(12,10).$

v) $\frac{1}{2}\frac{\sqrt{2}}{\sqrt{2}+1}\min\{d((42,40),T(42,40)), d((42,40),S(42,40))\} \leq d((42,40),y), \quad \forall y = (0,0),(10,12),(12,10).$

On the other hand, in all of the cases we have

$$\max\left\{\begin{array}{l} d(Sx,Sy), d(Tx,Ty), \\ d(Sx,Ty), d(Sy,Tx) \end{array}\right\} \leq \frac{1}{4\sqrt{2}}d(x,y).$$

Thus T satisfy all the hypotheses of Theorem 3 and hence T has a unique fixed point. Indeed, $r = \frac{1}{\sqrt{2}}$, $\theta(r) = \frac{\sqrt{2}}{\sqrt{2}+1}$ and (0, 0) is the unique common fixed point of T and S.

But,

$$d(T(40,42), T(42,40)) \leq \frac{r}{4}d((40,42),(42,40)),$$

that is $2^2 + 2^2 \leq \frac{r}{4}(2^2 + 2^2)$ this implies that $r \geq 4$. It is contradiction. This proves that Theorem 1 is not applicable to T.

Authors' contributions All authors contributed equally and significantly in writing this paper. All authors read and approved the final manuscript.

Compliance with ethical standards

Conflict of interest The authors declare that they have no conflict of interest.

References

1. Akkouchi, M.: Common fixed point theorems for two selfmappings of a b-metric space under an implicit relation. Hacet. J. Math. Stat. **40**(6), 805–810 (2011)
2. Aydi, H., Bota, M., Karapinar, E., Mitrović, S.: A fixed point theorem for set-valued quasi-contractions in b-metric spaces. Fixed Point Theory Appl. **2012**, 88 (2012)
3. Boriceanu, M., Bota, M., Petrusel, A.: Multivalued fractals in b-metric spaces. Cent. Eur. J. Math. **8**(2), 367–377 (2010)
4. Bota, M., Molnar, A., Varga, C.: On Ekeland's variational principle in b-metric spaces. Fixed Point Theory **12**(2), 21–28 (2011)
5. Czerwik, S.: Contraction mappings in b-metric spaces. Acta Math. et Inform. Univ. **1**, 5–11 (1993)
6. Dorić, D., Kadelburg, Z., Radenović, S.: Edelstein-Suzuki-type fixed point results in metric spaces. Nonlinear Anal Theory Methods Appl. **75**, 1927–1932 (2012)
7. Ly, D., Hieu, T.: Suzuki-type fixed point theorems for two maps on metric-type spaces. J. Nonlinear Anal. Optim. **4**(2), 17–29 (2013)
8. Hussain, N., Dorić, D., Kadelburg, Z., Radenović, S.: Suzuki-type fixed point results in metric type spaces. Fixed Point Theory Appl. (2012). doi:10.1186/1687-1812-2012-126
9. Hussain, N., Shah, M.H.: KKM mappings in cone b-metric spaces. Comput. Math. Appl. **62**, 1677–1684 (2011)
10. Hussain, N., Saadati, R., Agarwal, R.P.: On the topology and wt-distance on metric type spaces. Fixed Point Theory Appl. **2014**, 88 (2014)
11. Kikkawa, M., Suzuki, T., Some similarity between contractions and Kannan mappings, Fixed Point Theory Appl., vol. 2008, Article ID 649749, pp. 8 (2008)
12. Kikkawa, M., Suzuki, T.: Some notes on Fixed point theorems with constants. Kyushu Inst. Technol. Pure Appl. Math. Bull. Kyushu Inst. Technol. **56**, 11–18 (2009)
13. Kikkawa, M., Suzuki, T.: hree fixed point theorems for generalized contractions with constants in complete metric spaces. Nonlinear Anal. Theory Methods Appl. **69**(9), 2942–2949 (2008)
14. Khamsi, M. A., Remarks on cone metric spaces and fixed point theorems of contractive mappings, Fixed Point Theory Appl., Article ID 315398, p. 7. (2010), doi:10.1155/2010/315398
15. Khamsi, M.A., Hussain, N.: KKM mappings in metric type spaces. Nonlinear Anal. **73**(9), 3123–3129 (2010)
16. Latif, A., Parvaneh, V., Salimi, P., El-Mazrooei, A.E.: Various Suzuki type theorems in b-metric spaces. J. Nonlinear Sci. Appl. **8**, 363–377 (2015)
17. Olatinwo, M.O.: Some results on multi-valued weakly jungck mappings in b-metric space. Cent. Eur. J. Math. **6**(4), 610–621 (2008)
18. Mustafa, Z., Roshan, J.R., Parvaneh, V., Kadelburg, Z.: Some common fixed point results in ordered partial b-metric spaces. J. Inequal. Appl. **2013**, 562 (2013)
19. Pacurar, M.: Sequences of almost contractions and fixed points in b- metric spaces. Analele Universitatii de Vest, Timisoara Seria Matematica Informatica XLVIII **3**, 125–137 (2010)
20. Popescu, O.: Two fixed point theorems for generalized contractions with constants in complete metric space. Cent. Eur. J. Math. **7**(3), 529–538 (2009)
21. Popescu, O.: Fixed point theorem in metric spaces. Bull. Transilvania Univ. Brasüov **150**, 479–482 (2008)
22. Roshan, J.R., Parvaneh, V., Altun, I.: Some coincidence point

results in ordered b-metric spaces and applications in a system of integral equations. Appl. Math. Comput. **226**, 725–737 (2014)
23. Roshan, J.R., Parvaneh, V., Sedghi, S., Shobkolaei, N., Shatanawi, W.: Common fixed points of almost generalized $(\psi, \varphi)_s$-contractive mappings in ordered b-metric spaces. Fixed Point Theory Appl. **2013**, 159 (2013)
24. Shobe, N. Sedghi, S. Roshan, J.R and Hussain, N., Suzuki-type fixed point results in metric-like spaces, J. Funct. Spaces Appl., Vol., 2013, Article ID 143686 (2013)
25. Singh, S.L., Prasad, B.: Some coincidence theorems and stability of iterative proceders. Comput. Math. Appl. **55**, 2512–2520 (2008)
26. Suzuki, T.: A generalized Banach contraction principle that characterizes metric completeness. Proc. Am. Math. Soc. **136**, 1861–1869 (2008)
27. Suzuki, T.: A new type of fixed point theorem in metric spaces. Nonlinear Anal. Theory Methods Appl. **71**, 5313–5317 (2009)
28. Parvaneh, V., Roshan, J.R., Radenović, S.: Existence of tripled coincidence points in ordered b-metric spaces and an application to a system of integral equations. Fixed Point Theory Appl. **2013**, 130 (2013). doi:10.1186/1687-1812-2013-130

24

Three variable symmetric identities involving Carlitz-type q-Euler polynomials

Dae San Kim · Taekyun Kim

Abstract In this paper, we derive several identities of symmetry in three variables related to Carlitz-type q-Euler polynomials and alternating q-power sums. These and most of their identities are new, since there have been results only about identities of symmetry in two variables. The derivations of identities are based on the fermionic p-adic q-integral expressions of the generating functions for the Carlitz-type q-Euler polynomials.

Keywords Carlitz-type q-Euler polynomial · Fermionic p-adic q-integral

Mathematics Subject Classification 05A19 · 11B65 · 11B68

Introduction

Let p be a prime number with $p \equiv 1 \pmod 2$. Throughout this paper, $\mathbb{Z}_p$, $\mathbb{Q}_p$ and $\mathbb{C}_p$ will, respectively, denote the ring of p-adic integers, the field of p-adic rational numbers and the completion of algebraic closure of $\mathbb{Q}_p$. Let $|\cdot|_p$ be the normalized p-adic absolute value with $|p|_p = \frac{1}{p}$ and let q be an indeterminate in $\mathbb{C}_p$ with $|1-q|_p < p^{-\frac{1}{p-1}}$. For a continuous function $f : \mathbb{Z}_p \to \mathbb{C}_p$, the fermionic p-adic q-integral on $\mathbb{Z}_p$ is defined by Kim to be [9–17]

$$I_{-q}(f) = \int_{\mathbb{Z}_p} f(x) d\mu_{-q}(x) = \lim_{N\to\infty} \frac{1}{[p^N]_{-q}} \sum_{x=0}^{p^N-1} f(x)(-q)^x, \tag{1}$$

where $[x]_{-q} = \frac{1-(-q)^x}{1+q}$ and $[x]_q = \frac{1-q^x}{1-q}$.

From (1), we have

$$qI_{-q}(f_1) + I_{-q}(f) = [2]_q f(0), \tag{2}$$

where $f_1(x) = f(x+1)$.

In general, one derives that

$$q^n I_{-q}(f_n) + (-1)^{n-1} I_{-q}(f) = [2]_q \sum_{l=0}^{n-1} (-1)^{n-1-l} q^l f(l), \tag{3}$$

where $f_n(x) = f(x+n)$, $(n \geq 1)$.

So, for $n \equiv 1 \pmod 2$,

$$q^n I_{-q}(f_n) + I_{-q}(f) = [2]_q \sum_{l=0}^{n-1} (-1)^l q^l f(l); \tag{4}$$

for $n \equiv 0 \pmod 2$,

$$q^n I_{-q}(f_n) - I_{-q}(f) = -[2]_q \sum_{l=0}^{n-1} (-1)^l q^l f(l). \tag{5}$$

In particular, for $q = 1$, we have

$$I_{-1}(f) = \int_{\mathbb{Z}_p} f(x) d\mu_{-1}(x) = \lim_{N\to\infty} \sum_{x=0}^{p^N-1} f(x)(-1)^x, \tag{6}$$

and

D. S. Kim
Department of Mathematics, Sogang University, Seoul 121-742, Republic of Korea
e-mail: dskim@sogang.ac.kr

T. Kim (✉)
Department of Mathematics, Kwangwoon University, Seoul 139-701, Republic of Korea
e-mail: tkkim@kw.ac.kr

$$I_{-1}(f_1)+I_{-1}(f)=2f(0). \tag{7}$$

As is well known, the ordinary Euler polynomials are defined by the generating function to be

$$\frac{2}{e^t+1}e^{xt}=\sum_{n=0}^{\infty}E_n(x)\frac{t^n}{n!} \tag{8}$$

(see [1–24]). When $x=0$, $E_n=E_n(0)$ are called the Euler numbers.

From (5) and (6), we can easily derive

$$\int_{\mathbb{Z}_p}e^{(x+y)t}d\mu_{-1}(y)=\frac{2}{e^t+1}e^{xt}=\sum_{n=0}^{\infty}E_n(x)\frac{t^n}{n!}. \tag{9}$$

By (7), we get

$$E_0=1,\quad (E+1)^n+E_n=2\delta_{0,n}, \tag{10}$$

[11, 16, 17] with the usual convention about replacing E^n by E_n.

From (9), we note that

$$\begin{aligned}E_n(x)&=\int_{\mathbb{Z}_p}(x+y)^n d\mu_{-1}(y)=\sum_{l=0}^{n}\binom{n}{l}x^{n-l}\int_{\mathbb{Z}_p}y^l d\mu_{-1}(y)\\&=\sum_{l=0}^{n}\binom{n}{l}x^{n-l}E_l.\end{aligned} \tag{11}$$

In light of (10), the Carlitz-type q-Euler numbers are given by

$$q(q\mathcal{E}_q+1)^n+\mathcal{E}_{n,q}=[2]_q\delta_{n,0},\quad (n\geq 0), \tag{12}$$

with the usual convention about replacing $\mathcal{E}_q^n$ by $\mathcal{E}_{n,q}$ (see [10, 12, 15]).

The q-Euler polynomials are defined by (see [10, 16])

$$\begin{aligned}\mathcal{E}_{n,q}(x)&=\left(q^x\mathcal{E}_q+[x]_q\right)^n\\&=\sum_{l=0}^{n}\binom{n}{l}q^{lx}\mathcal{E}_{l,q}[x]_q^{n-l}.\end{aligned} \tag{13}$$

From (2), we derive

$$\begin{aligned}\sum_{n=0}^{\infty}\mathcal{E}_{n,q}\frac{t^n}{n!}&=\int_{\mathbb{Z}_p}e^{[y]_qt}d\mu_{-q}(y)\\&=[2]_q\sum_{m=0}^{\infty}(-1)^m q^m e^{[m]_qt},\end{aligned} \tag{14}$$

and

$$\begin{aligned}\sum_{n=0}^{\infty}\mathcal{E}_{n,q}(x)\frac{t^n}{n!}&=\int_{\mathbb{Z}_p}e^{[x+y]_qt}d\mu_{-q}(y)\\&=[2]_q\sum_{m=0}^{\infty}(-1)^m q^m e^{[m+x]_qt}.\end{aligned} \tag{15}$$

By (13) and (15), we get

$$\begin{aligned}\mathcal{E}_{n,q}(x+y)&=\sum_{l=0}^{n}\binom{n}{l}q^{lx}\mathcal{E}_{l,q}(y)[x]_q^{n-l}\\&=\sum_{l=0}^{n}\binom{n}{l}q^{(n-l)x}\mathcal{E}_{n-l,q}(y)[x]_q^{l}.\end{aligned} \tag{16}$$

Explicit expressions for Carlitz-type q-Euler numbers can be obtained, for example, from their generating functions :

$$\mathcal{E}_{n,q}(x)=\frac{[2]_q}{(1-q)^n}\sum_{l=0}^{n}\binom{n}{l}(-1)^l q^{lx}\frac{1}{1+q^{l+1}}, \tag{17}$$

and

$$\mathcal{E}_{n,q}=\frac{[2]_q}{(1-q)^n}\sum_{l=0}^{n}\binom{n}{l}(-1)^l\frac{1}{1+q^{l+1}}. \tag{18}$$

In [10, 14, 15], Kim introduced the polynomials $\mathcal{E}_{n,q}^{(h,k)}(x)$ in terms of the following multiple fermionic p-adic q-integral on $\mathbb{Z}_p$:

$$\begin{aligned}\mathcal{E}_{n,q}^{(h,k)}(x)=&\int_{\mathbb{Z}_p}\cdots\int_{\mathbb{Z}_p}q^{\sum_{l=1}^{k}(h-l)y_l}\\&\times[x+y_1+\cdots+y_k]_q^n d\mu_{-q}(y_1)\cdots d\mu_{-q}(y_k).\end{aligned} \tag{19}$$

In particular, if $k=1$, $\mathcal{E}_{n,q}^{(h,1)}(x)$ will be simply denoted by $\mathcal{E}_{n,q}^{(h)}(x)$ so that

$$\mathcal{E}_{n,q}^{(h)}(x)=\int_{\mathbb{Z}_p}q^{(h-1)y}[x+y]_q^n d\mu_{-q}(y). \tag{20}$$

One can derive the following explicit expression of $\mathcal{E}_{n,q}^{(h,k)}(x)$:

$$\mathcal{E}_{n,q}^{(h,k)}(x)=\frac{[2]_q^k}{(1-q)^n}\sum_{l=0}^{n}\binom{n}{l}(-1)^{n-l}q^{lx}\frac{1}{(-q^{l+h};q^{-1})_k}, \tag{21}$$

where $(x:q)_n=(1-x)(1-xq)\cdots(1-xq^{n-1})$ (see [10, 14, 15]).

The following simple facts will be used over and over again :

$$[a+b]_q=[a]_q+q^a[b]_q. \tag{22}$$

From (22), one can show

$$[a+b+c]_q=[a]_q+q^a[b]_q+q^{a+b}[c]_q, \tag{23}$$

$$[ab]_q=[a]_q[b]_{q^a}.$$

In this paper, we give several identities of symmetry in three variables related to Carlitz-type q-Euler polynomials and alternating q-power sums which are derived from the triple fermionic p-adic q-integral on $\mathbb{Z}_p$. These and most of their identities are new, since there have been results only about identities of symmetry in two variables.

Symmetric identities of q-Euler polynomials

First, we will consider the following triple integral which is obviously invariant under any permutations of w_1, w_2, w_3. This simple observation is the philosophy that underlies this paper.

Let

$$I = \int_{\mathbb{Z}_p^3} e^{[w_2w_3x_1+w_1w_3x_2+w_1w_2x_3+w_1w_2w_3(y_1+y_2+y_3)]_q t} \\ \times d\mu_{-q^{w_2w_3}}(x_1)d\mu_{-q^{w_1w_3}}(x_2)d\mu_{-q^{w_1w_2}}(x_3). \tag{24}$$

It is not difficult to show that

$$[w_2w_3x_1 + w_1w_3x_2 + w_1w_2x_3 + w_1w_2w_3(y_1 + y_2 + y_3)]_q \\ = [w_2w_3]_q[x_1 + w_1y_1]_{q^{w_2w_3}} + q^{w_2w_3(x_1+w_1y_1)}[w_1w_3]_q[x_2 + w_2y_2]_{q^{w_1w_3}} \\ + q^{w_2w_3(x_1+w_1y_1)+w_1w_3(x_2+w_2y_2)}[w_1w_2]_q[x_3 + w_3y_3]_{q^{w_1w_2}}. \tag{25}$$

So the integrand of I is

$$e^{[w_2w_3x_1+w_1w_3x_2+w_1w_2x_3+w_1w_2w_3(y_1+y_2+y_3)]_q t} \\ = e^{[w_2w_3]_q[x_1+w_1y_1]_{q^{w_2w_3}} t} e^{q^{w_2w_3(x_1+w_1y_1)}[w_1w_3]_q[x_2+w_2y_2]_{q^{w_1w_3}} t} \\ \times e^{q^{w_2w_3(x_1+w_1y_1)+w_1w_3(x_2+w_2y_2)}[w_1w_2]_q[x_3+w_3y_3]_{q^{w_1w_2}} t} \\ = \sum_{n=0}^{\infty}\Bigg\{\sum_{k+l+m=n}\binom{n}{k,l,m}[w_2w_3]_q^k[w_1w_3]_q^l[w_1w_2]_q^m q^{w_1w_2w_3(l+m)y_1} \\ \times q^{w_1w_2w_3my_2}q^{w_2w_3(l+m)x_1}[x_1 + w_1y_1]_{q^{w_2w_3}}^k q^{w_1w_3mx_2}[x_2 + w_2y_2]_{q^{w_1w_3}}^l \\ \times [x_3 + w_3y_3]_{q^{w_1w_2}}^m\Bigg\}\frac{t^n}{n!}. \tag{26}$$

Thus the integral in (24) is

$$I = \sum_{n=0}^{\infty}\Bigg\{\sum_{k+l+m=n}\binom{n}{k,l,m}[w_2w_3]_q^k[w_1w_3]_q^l[w_1w_2]_q^m q^{w_1w_2w_3(l+m)y_1} \\ \times q^{w_1w_2w_3my_2}\int_{\mathbb{Z}_p} q^{w_2w_3(l+m)x_1}[x_1 + w_1y_1]_{q^{w_2w_3}}^k d\mu_{-q^{w_2w_3}}(x_1) \\ \times \int_{\mathbb{Z}_p} q^{w_1w_3mx_2}[x_2 + w_2y_2]_{q^{w_1w_3}}^l d\mu_{-q^{w_1w_3}}(x_2) \\ \times \int_{\mathbb{Z}_p}[x_3 + w_3y_3]_{q^{w_1w_2}}^m d\mu_{-q^{w_1w_2}}(x_3)\Bigg\}\frac{t^n}{n!} \\ = \sum_{n=0}^{\infty}\Bigg\{\sum_{k+l+m=n}\binom{n}{k,l,m}[w_2w_3]_q^k[w_1w_3]_q^l[w_1w_2]_q^m q^{w_1w_2w_3(l+m)y_1} \\ \times q^{w_1w_2w_3my_2}\mathcal{E}_{k,q^{w_2w_3}}^{(l+m+1)}(w_1y_1)\mathcal{E}_{l,q^{w_1w_3}}^{(m+1)}(w_2y_2)\mathcal{E}_{m,q^{w_1w_2}}(w_3y_3)\Bigg\}\frac{t^n}{n!}. \tag{27}$$

Thus, we get the following theorem.

Theorem 1 *Let w_1, w_2, w_3 be any positive integers, n any nonnegative integer. Then the following expression is invariant under any permutations of w_1, w_2, w_3, so that it gives us six symmetries:*

$$\sum_{k+l+m=n}\binom{n}{k,l,m}[w_2w_3]_q^k[w_1w_3]_q^l[w_1w_2]_q^m q^{w_1w_2w_3(l+m)y_1} \\ \times q^{w_1w_2w_3my_2}\mathcal{E}_{k,q^{w_2w_3}}^{(l+m+1)}(w_1y_1)\mathcal{E}_{l,q^{w_1w_3}}^{(m+1)}(w_2y_2)\mathcal{E}_{m,q^{w_1w_2}}(w_3y_3) \\ = \sum_{k+l+m=n}\binom{n}{k,l,m}[w_1w_3]_q^k[w_2w_3]_q^l[w_1w_2]_q^m q^{w_1w_2w_3(l+m)y_1} \\ \times q^{w_1w_2w_3my_2}\mathcal{E}_{k,q^{w_1w_3}}^{(l+m+1)}(w_2y_1)\mathcal{E}_{l,q^{w_2w_3}}^{(m+1)}(w_1y_2)\mathcal{E}_{m,q^{w_1w_2}}(w_3y_3) \\ = \sum_{k+l+m=n}\binom{n}{k,l,m}[w_1w_3]_q^k[w_1w_2]_q^l[w_2w_3]_q^m q^{w_1w_2w_3(l+m)y_1} \\ \times q^{w_1w_2w_3my_2}\mathcal{E}_{k,q^{w_1w_3}}^{(l+m+1)}(w_2y_1)\mathcal{E}_{l,q^{w_1w_2}}^{(m+1)}(w_3y_2)\mathcal{E}_{m,q^{w_2w_3}}(w_1y_3) \\ = \sum_{k+l+m=n}\binom{n}{k,l,m}[w_2w_3]_q^k[w_1w_2]_q^l[w_1w_3]_q^m q^{w_1w_2w_3(l+m)y_1} \\ \times q^{w_1w_2w_3my_2}\mathcal{E}_{k,q^{w_2w_3}}^{(l+m+1)}(w_1y_1)\mathcal{E}_{l,q^{w_1w_2}}^{(m+1)}(w_3y_2)\mathcal{E}_{m,q^{w_1w_3}}(w_2y_3) \\ = \sum_{k+l+m=n}\binom{n}{k,l,m}[w_1w_2]_q^k[w_2w_3]_q^l[w_1w_3]_q^m q^{w_1w_2w_3(l+m)y_1} \\ \times q^{w_1w_2w_3my_2}\mathcal{E}_{k,q^{w_1w_2}}^{(l+m+1)}(w_3y_1)\mathcal{E}_{l,q^{w_2w_3}}^{(m+1)}(w_1y_2)\mathcal{E}_{m,q^{w_1w_3}}(w_2y_3) \\ = \sum_{k+l+m=n}\binom{n}{k,l,m}[w_1w_2]_q^k[w_1w_3]_q^l[w_2w_3]_q^m q^{w_1w_2w_3(l+m)y_1} \\ \times q^{w_1w_2w_3my_2}\mathcal{E}_{k,q^{w_1w_2}}^{(l+m+1)}(w_3y_1)\mathcal{E}_{l,q^{w_1w_3}}^{(m+1)}(w_2y_2)\mathcal{E}_{m,q^{w_2w_3}}(w_1y_3).$$

We define, for nonnegative integers n, m, w, $K_{n,m}(w|q)$ as

$$K_{n,m}(w|q) = \sum_{i=0}^{w}(-1)^i q^{ni}[i]_q^m. \tag{28}$$

In particular, for $w = 0$, or $m = 0$, we have

$$K_{n,m}(0|q) = \begin{cases} 1, & \text{if } m = 0 \\ 0, & \text{if } m > 0, \end{cases} \tag{29}$$

and

$$K_{n,0}(w|q) = [w+1]_{-q^n}. \tag{30}$$

We now apply the formula of (3) as follows :

$$q^n I_{-q}(f_n) + (-1)^{n-1}I_{-q}(f) = [2]_q\sum_{l=0}^{n-1}(-1)^{n-1-l}q^l f(l), \tag{31}$$

with

$f(x) = e^{[w_1w_2x]_q t}$, "$q = q^{w_1w_2}$", $n = w_3$.

From (35), we have

$$
\begin{aligned}
&(q^{w_1w_2})^{w_3}\int_{\mathbb{Z}_p} e^{[w_1w_2(x+w_3)]_q t} d\mu_{-q^{w_1w_2}}(x) + (-1)^{w_3-1}\int_{\mathbb{Z}_p} e^{[w_1w_2x]_q t} d\mu_{-q^{w_1w_2}}(x)\\
&= [2]_{q^{w_1w_2}} \sum_{l=0}^{w_3-1} (-1)^{w_3-1-l} q^{w_1w_2l} e^{[w_1w_2l]_q t}\\
&= (-1)^{w_3-1}[2]_{q^{w_1w_2}} \sum_{m=0}^{\infty} K_{1,m}(w_3-1|q^{w_1w_2}) \frac{\left([w_1w_2]_q t\right)^m}{m!}\\
&= \sum_{m=0}^{\infty}\Big\{[w_1w_2]_q^m\\
&\times \int_{\mathbb{Z}_p}\left(q^{w_1w_2w_3}[x_1+w_3]_{q^{w_1w_2}}^m + (-1)^{w_3-1}[x]_{q^{w_1w_2}}^m\right) d\mu_{-q^{w_1w_2}}(x)\Big\}\frac{t^m}{m!}.
\end{aligned} \tag{32}
$$

Thus, by (32), we obtain the following lemma.

Lemma 2 *Let w_1, w_2 be any positive integers.*

(i) *For $w_3 \equiv 1$ (mod 2), we have*

$$
\begin{aligned}
&q^{w_1w_2w_3}\int_{\mathbb{Z}_p} e^{[w_1w_2(x+w_3)]_q t} d\mu_{-q^{w_1w_2}}(x) + \int_{\mathbb{Z}_p} e^{[w_1w_2x]_q t} d\mu_{-q^{w_1w_2}}(x)\\
&= \sum_{m=0}^{\infty}\left\{[w_1w_2]_q^m \int_{\mathbb{Z}_p}\left(q^{w_1w_2w_3}[x+w_3]_{q^{w_1w_2}}^m + [x]_{q^{w_1w_2}}^m\right) d\mu_{-q^{w_1w_2}}(x)\right\}\frac{t^m}{m!}\\
&= [2]_{q^{w_1w_2}} \sum_{i=0}^{w_3-1} (-1)^i q^{w_1w_2i} e^{[w_1w_2i]_q t}\\
&= [2]_{q^{w_1w_2}} \sum_{m=0}^{\infty} K_{1,m}(w_3-1|q^{w_1w_2}) \frac{\left([w_1w_2]_q t\right)^m}{m!}.
\end{aligned}
$$

(ii) *For $w_3 \equiv 0$ (mod 2), we have*

$$
\begin{aligned}
&q^{w_1w_2w_3}\int_{\mathbb{Z}_p} e^{[w_1w_2(x+w_3)]_q t} d\mu_{-q^{w_1w_2}}(x) - \int_{\mathbb{Z}_p} e^{[w_1w_2x]_q t} d\mu_{-q^{w_1w_2}}(x)\\
&= \sum_{m=0}^{\infty}\left\{[w_1w_2]_q^m \int_{\mathbb{Z}_p}\left(q^{w_1w_2w_3}[x+w_3]_{q^{w_1w_2}}^m - [x]_{q^{w_1w_2}}^m\right) d\mu_{-q^{w_1w_2}}(x)\right\}\frac{t^m}{m!}\\
&= -[2]_{q^{w_1w_2}} \sum_{i=0}^{w_3-1} (-1)^i q^{w_1w_2i} e^{[w_1w_2i]_q t}\\
&= -[2]_{q^{w_1w_2}} \sum_{m=0}^{\infty} K_{1,m}(w_3-1|q^{w_1w_2}) \frac{\left([w_1w_2]_q t\right)^m}{m!}.
\end{aligned}
$$

Consider the following sum of triple integrals.

$$
\begin{aligned}
I_1 &= q^{w_1w_2w_3}\int_{\mathbb{Z}_p^3} e^{[w_2w_3x_1+w_1w_3x_2+w_1w_2x_3+w_1w_2w_3(y_1+y_2+1)]_q t}\\
&\times d\mu_{-q^{w_2w_3}}(x_1) d\mu_{-q^{w_1w_3}}(x_2) d\mu_{-q^{w_1w_2}}(x_3)\\
&+ \int_{\mathbb{Z}_p^3} e^{[w_2w_3x_1+w_1w_3x_2+w_1w_2x_3+w_1w_2w_3(y_1+y_2)]_q t}\\
&\times d\mu_{-q^{w_2w_3}}(x_1) d\mu_{-q^{w_1w_3}}(x_2) d\mu_{-q^{w_1w_2}}(x_3),
\end{aligned} \tag{33}
$$

which is obviously invariant under any permutations of w_1, w_2, w_3.

For simplicity, we put

$$
a = a(x_1) = q^{w_2w_3(x_1+w_1y_1)}, \quad b = b(x_2) = q^{w_1w_3(x_2+w_2y_2)}. \tag{34}
$$

Then, from (33), we note that

$$
\begin{aligned}
I_1 &= \sum_{k,l=0}^{\infty}[w_2w_3]_q^k[w_1w_3]_q^l \frac{t^{k+l}}{k!l!}\int_{\mathbb{Z}_p^2} a^l[x_1+w_1y_1]_{q^{w_2w_3}}^k [x_2+w_2y_2]_{q^{w_1w_3}}^l\\
&\times \left\{\sum_{m=0}^{\infty}\frac{[w_1w_2]_q^m (abt)^m}{m!}\right.\\
&\times \left.\int_{\mathbb{Z}_p}\left(q^{w_1w_2w_3}[x_3+w_3]_{q^{w_1w_2}}^m + [x_3]_{q^{w_1w_2}}^m\right) d\mu_{-q^{w_1w_2}}(x_3)\right\}\\
&\times d\mu_{-q^{w_2w_3}}(x_1) d\mu_{-q^{w_1w_3}}(x_2).
\end{aligned} \tag{35}
$$

Assume now that $w_3 \equiv 1$ (mod 2). Then, by (i) of Lemma 2, we get

$$
\begin{aligned}
I_1 =& [2]_{q^{w_1w_2}} \sum_{k,l,m=0}^{\infty} [w_2w_3]_q^k [w_1w_3]_q^l [w_1w_2]_q^m \frac{t^{k+l+m}}{k!l!m!}\\
&\times K_{1,m}(w_3-1|q^{w_1w_2}) \int_{\mathbb{Z}_p} a^{l+m}[x_1+w_1y_1]_{q^{w_2w_3}}^k d\mu_{-q^{w_2w_3}}(x_1)\\
&\times \int_{\mathbb{Z}_p} b^m [x_2+w_2y_2]_{q^{w_1w_3}}^l d\mu_{-q^{w_1w_3}}(x_2).
\end{aligned} \tag{36}
$$

Recovering $a = q^{w_2w_3(x_1+w_1y_1)}$, $b = q^{w_1w_3(x_2+w_2y_2)}$, I_1 can be rewritten as

$$
\begin{aligned}
I_1 =& [2]_{q^{w_1w_2}} \sum_{k,l,m=0}^{\infty} [w_2w_3]_q^k [w_1w_3]_q^l [w_1w_2]_q^m \frac{t^{k+l+m}}{k!l!m!}\\
&\times K_{1,m}(w_3-1|q^{w_1w_2}) q^{w_1w_2w_3(l+m)y_1} q^{w_1w_2w_3my_2}\\
&\times \int_{\mathbb{Z}_p} q^{w_2w_3(l+m)x_1}[x_1+w_1y_1]_{q^{w_2w_3}}^k d\mu_{-q^{w_2w_3}}(x_1)\\
&\times \int_{\mathbb{Z}_p} q^{w_1w_3mx_2}[x_2+w_2y_2]_{q^{w_1w_3}}^l d\mu_{-q^{w_1w_3}}(x_2)\\
=& \sum_{n=0}^{\infty}\left\{[2]_{q^{w_1w_2}} \sum_{k+l+m=n}\binom{n}{k,l,m}[w_2w_3]_q^k[w_1w_3]_q^l[w_1w_2]_q^m\right.\\
&\times K_{1,m}(w_3-1|q^{w_1w_2}) q^{w_1w_2w_3(l+m)y_1} q^{w_1w_2w_3my_2}\\
&\times \left.\mathcal{E}_{k,q^{w_2w_3}}^{(l+m+1)}(w_1y_1)\mathcal{E}_{l,q^{w_1w_3}}^{(m+1)}(w_2y_2)\right\}\frac{t^n}{n!}.
\end{aligned} \tag{37}
$$

As the expression in (37) is invariant under any permutations of w_1, w_2, w_3 and it is equal to (37) provided that $w_3 \equiv 1$ (mod 2), we see that the expression in the curly bracket of (37) is invariant under any permutations of w_1, w_2, w_3, when $w_1 \equiv w_2 \equiv w_3 \equiv 1$ (mod 2). Instead of the sum of triple integrals in (33), we now consider their difference, namely,

$$q^{w_1w_2w_3}\int_{\mathbb{Z}_p^3} e^{[w_2w_3x_1+w_1w_3x_2+w_1w_2x_3+w_1w_2w_3(y_1+y_2+1)]_q t}$$
$$\times d\mu_{-q^{w_2w_3}}(x_1)d\mu_{-q^{w_1w_3}}(x_2)d\mu_{-q^{w_1w_2}}(x_3)$$
$$-\int_{\mathbb{Z}_p^3} e^{[w_2w_3x_1+w_1w_3x_2+w_1w_2x_3+w_1w_2w_3(y_1+y_2)]_q t} \tag{38}$$
$$\times d\mu_{-q^{w_2w_3}}(x_1)d\mu_{-q^{w_1w_3}}(x_2)d\mu_{-q^{w_1w_2}}(x_3),$$

which is invariant under any permutations of w_1, w_2, w_3. Proceeding analogously to the above and using (ii) of Lemma 2, we see that (38) is equal to the negative of the expression in (37), provided that $w_3 \equiv 0 \pmod 2$. Thus, we see that the expression in curly bracket of (37) is invariant under any permutations of w_1, w_2, w_3, when $w_1 \equiv w_2 \equiv w_3 \equiv 0 \pmod 2$.

Thus, we obtain the following theorem.

Theorem 3 *Let w_1, w_2, w_3 be positive integers satisfying either $w_1 \equiv w_2 \equiv w_3 \equiv 1$ (mod 2) or $w_1 \equiv w_2 \equiv w_3 \equiv 0$ (mod 2). Then, for any nonnegative integer n, the following expressions*

$$[2]_{q^{w_{\sigma(1)}w_{\sigma(2)}}}\sum_{k+l+m=n}\binom{n}{k,l,m}\mathcal{E}^{(l+m+1)}_{k,q^{w_{\sigma(2)}w_{\sigma(3)}}}\left(w_{\sigma(1)}y_1\right)$$
$$\times\mathcal{E}^{(m+1)}_{l,q^{w_{\sigma(1)}w_{\sigma(3)}}}\left(w_{\sigma(2)}y_2\right)K_{1,m}\left(w_{\sigma(3)}-1|q^{w_{\sigma(1)}w_{\sigma(2)}}\right)$$
$$\times q^{w_1w_2w_3(l+m)y_1}q^{w_1w_2w_3my_2}\left[w_{\sigma(2)}w_{\sigma(3)}\right]_q^k$$
$$\times\left[w_{\sigma(1)}w_{\sigma(3)}\right]_q^l\left[w_{\sigma(1)}w_{\sigma(2)}\right]_q^m$$

are all the same for any $\sigma \in S_3$.

Remark 1 We can obtain many interesting identities by letting $w_3 = 1$ or $w_2 = w_3 = 1$, in view of (29). However, writing those down requires much space and so we omit it.

With the same

$$a = q^{w_2w_3(x_1+w_1y_1)},\quad b = q^{w_1w_3(x_2+w_2y_2)},$$

I_1 can be written as

$$I_1=\int_{\mathbb{Z}_p^2} e^{[w_2w_3]_q[x_1+w_1y_1]_{q^{w_2w_3}}t} e^{[w_1w_3]_q[x_2+w_2y_2]_{q^{w_1w_3}}at}$$
$$\times d\mu_{-q^{w_2w_3}}(x_1)d\mu_{-q^{w_1w_3}}(x_2)$$
$$\times\int_{\mathbb{Z}_p}\left(q^{w_1w_2w_3}e^{[w_1w_2(x_3+w_3)]_q abt}+e^{[w_1w_2x_3]_q abt}\right)d\mu_{-q^{w_1w_2}}(x_3). \tag{39}$$

Let $w_3 \equiv 1 \pmod 2$. Then, from (i) of Lemma 2, the inner integral in (39) is

$$[2]_{q^{w_1w_2}}\sum_{i=0}^{w_3-1}(-1)^i q^{w_1w_2i}e^{[w_1w_2i]_q abt}. \tag{40}$$

So (40) is equal to

$$I_1=[2]_{q^{w_1w_2}}\sum_{i=0}^{w_3-1}(-1)^i q^{w_1w_2i}\int_{\mathbb{Z}_p^2} e^{[w_2w_3]_q[x_1+w_1y_1]_{q^{w_2w_3}}t}$$
$$\times e^{[w_1w_3]_q\left[x_2+w_2y_2+\frac{w_2}{w_3}i\right]_{q^{w_1w_3}}at}d\mu_{-q^{w_2w_3}}(x_1)d\mu_{-q^{w_1w_3}}(x_2)$$
$$=[2]_{q^{w_1w_2}}\sum_{i=0}^{w_3-1}(-1)^i q^{w_1w_2i}\sum_{k,l=0}\frac{t^{k+l}}{k!l!}q^{w_1w_2w_3ly_1}[w_2w_3]_q^k[w_1w_3]_q^l$$
$$\times\int_{\mathbb{Z}_p} q^{w_2w_3lx_1}[x_1+w_1y_1]^k_{q^{w_2w_3}}d\mu_{-q^{w_2w_3}}(x_1)$$
$$\times\int_{\mathbb{Z}_p}\left[x_2+w_2y_2+\frac{w_2}{w_3}i\right]^l_{q^{w_1w_3}}d\mu_{-q^{w_1w_3}}(x_2)$$
$$=\sum_{n=0}^{\infty}\left\{[2]_{q^{w_1w_2}}\sum_{k=0}^{n}\binom{n}{k}\mathcal{E}^{(n-k+1)}_{k,q^{w_2w_3}}(w_1y_1)q^{w_1w_2w_3(n-k)y_1}[w_2w_3]_q^k[w_1w_3]_q^{n-k}\right.$$
$$\left.\times\sum_{i=0}^{w_3-1}(-1)^i q^{w_1w_2i}\mathcal{E}_{l,q^{w_1w_3}}\left(w_2y_2+\frac{w_2}{w_3}i\right)\right\}\frac{t^n}{n!}. \tag{41}$$

Recalling that I_1 is invariant under any permutations of w_1, w_2, w_3 and it is equal to (41) for $w_3 \equiv 1 \pmod 2$, we see that the expression in the curly bracket of (41) is invariant under any permutations of w_1, w_2, w_3, when $w_1 \equiv w_2 \equiv w_3 \equiv 1 \pmod 2$. Also, starting from (39), using (ii) of Lemma 2, and proceeding analogous to the above, we see that the expression in the curly bracket of (41) is also invariant under any permutations of w_1, w_2, w_3, when $w_1 \equiv w_2 \equiv w_3 \equiv 0 \pmod 2$. Thus, we have the following theorem.

Theorem 4 *Let w_1, w_2, w_3 be positive integers satisfying either $w_1 \equiv w_2 \equiv w_3 \equiv 1$ (mod 2) or $w_1 \equiv w_2 \equiv w_3 \equiv 0$ (mod 2). Then, for any nonnegative integer n, the following expressions*

$$[2]_{q^{w_{\sigma(1)}w_{\sigma(2)}}}\sum_{k=0}^{n}\binom{n}{k}\mathcal{E}^{(n-k+1)}_{k,q^{w_{\sigma(2)}w_{\sigma(3)}}}\left(w_{\sigma(1)}y_1\right)q^{w_1w_2w_3(n-k)y_1}\left[w_{\sigma(2)}w_{\sigma(3)}\right]_q^k$$
$$\times\left[w_{\sigma(1)}w_{\sigma(3)}\right]_q^{n-k}\sum_{i=0}^{w_{\sigma(3)}-1}(-1)^i q^{w_{\sigma(1)}w_{\sigma(2)}i}\mathcal{E}_{l,q^{w_{\sigma(1)}w_{\sigma(3)}}}\left(w_{\sigma(2)}y_2+\frac{w_{\sigma(2)}}{w_{\sigma(3)}}i\right)$$

are all the same for any $\sigma \in S_3$.

Remark 2 In view of (29), by specializing $w_3 = 1$ or $w_2 = w_3 = 1$, we can obtain many interesting identities. However, we will omit those, as this requires much space.

Acknowledgments This work was supported by the National Research Foundation of Korea (NRF) grant funded by the Korea government (MOE) (No.2012R1A1A2003786).

References

1. Araci, S., Bagdasaryan, A., Ozel, C., Srivastava, H.M.: New Symmetric Identities Involving q-Zeta Type Functions. Appl. Math. Inf. Sci. **8**(6), 2803–2808 (2014). MR 3228678
2. Bayad, A., Chikhi, J.: Non linear recurrences for Apostol-Bernoulli-Euler numbers of higher order. Adv. Stud. Contemp. Math. (Kyungshang) **22**(1), 1–6 (2012). MR 2931600
3. Can, M., Cenkci, M., Kurt, V., Simsek, Y.: Twisted Dedekind type sums associated with Barnes' type multiple Frobenius-Euler l-functions. Adv. Stud. Contemp. Math. (Kyungshang) **18**(2), 135–160 (2009). MR 2508979 (2010a:11072)
4. Cangul, I.N., Kurt, V., Ozden, H., Simsek, Y.: On the higher-order w-q-Genocchi numbers. Adv. Stud. Contemp. Math. (Kyungshang) **19**(1), 39–57 (2009). MR 2542124 (2011b:05010)
5. Dan, D., Yang, J.: Some identities related to the Apostol-Euler and Apostol-Bernoulli polynomials. Adv. Stud. Contemp. Math. (Kyungshang) **19**(1), 39–57 (2009)
6. Kim, D.S., Kim, T., Lee, S.H., Seo, J.J.: Symmetric Identities of the q-Euler Polynomials. Adv. Stud. Theor. Phys. **7**(24), 1149–1155 (2013)
7. Kim, D.S., Kim, T., Rim, S.-H., Seo, J.-J.: A note on symmetric properties of the multiple q-Euler zeta functions and higher-order q-Euler polynomials. Appl. Math. Sci. **8**(32), 1585–1591 (2014). MR 3200156
8. Kim, D.S., Lee, N., Na, J., Park, K.H.: Identities of symmetry for higher-order Euler polynomials in three variables (I). Adv. Stud. Contemp. Math. (Kyungshang) **221**(1), 51–74 (2012). MR 2931605
9. Kim, T.: q-Volkenborn integration. Russ. J. Math. Phys. **9**(3), 288–299 (2002). MR 1965383 (2004f:11138)
10. Kim, T.: q-Euler numbers and polynomials associated with p-adic q-integrals. J. Nonlinear Math. Phys. **14**(1), 15–27 (2007). MR 2287831 (2007k:11202)
11. Kim, T.: Symmetry p-adic invariant integral on $\mathbb{Z}_p$ for Bernoulli and Euler polynomials. J. Differ. Equ. Appl. **14**(12), 1267–1277 (2008). MR 2462529 (2009i:11023)
12. Kim, T.: Some identities on the q-Euler polynomials of higher order and q-Stirling numbers by the fermionic p-adic integral on $\mathbb{Z}_p$. Russ. J. Math. Phys. **16**(4), 484–491 (2009). MR 2587805 (2011e:33045)
13. Kim, T.: Symmetry of power sum polynomials and multivariate fermionic p-adic invariant integral on $\mathbb{Z}_p$. Russ. J. Math. Phys. **16**(1), 93–96 (2009). MR 2486809 (2010c:11028)
14. Kim, T.: Barnes-type multiple q-zeta functions and q-Euler polynomials, J. Phys. A **43**(25), 255201, 11. MR 2653033 (2011f:11106) (2010)
15. Kim, T.: New approach to q-Euler polynomials of higher order. Russ. J. Math. Phys. **17**(2), 218–225 (2010). MR 2660920 (2011f:33026)
16. Kim, T.: A study on the q-Euler numbers and the fermionic q-integral of the product of several type q-Bernstein polynomials on $\mathbb{Z}_p$. Adv. Stud. Contemp. Math. (Kyungshang) **23**(1), 5–11 (2013). MR 3059313
17. Kim, T., Kim, D.S., Dolgy, D.V., Rim, S.H.: Some identities on the Euler numbers arising from Euler basis polynomials. Ars Combin. **109**, 433–446 (2013). MR 3087232
18. Luo, Q.-M.: q-analogues of some results for the Apostol-Euler polynomials. Adv. Stud. Contemp. Math. (Kyungshang) **20**(1), 103–113 (2010). MR 2597996 (2011e:05031)
19. Rim, S.-H., Jeong, J., Jin, J.-H.: On the twisted q-Euler numbers and polynomials with weight 0. J. Comput. Anal. Appl. **15**(2), 374–380 (2013). MR 3075614
20. Ryoo, C.S.: A note on the weighted q-Euler numbers and polynomials. Adv. Stud. Contemp. Math. (Kyungshang) **21**(1), 47–54 (2011). MR 2984863
21. Ryoo, C.S.: An identity of the symmetry for the second kind q-Euler polynomials. J. Comput. Anal. Appl. **15**(2), 294–299 (2013). MR 3075604
22. Ryoo, C.S., Song, H., Agarwal, R.P.: On the roots of the q-analogue of Euler-Barnes' polynomials. Adv. Stud. Contemp. Math. (Kyungshang) **9**(2), 153–163 (2004). MR 2090118 (2005d:11024)
23. Şen, E.: Theorems on Apostol-Euler polynomials of higher order arising from Euler basis. Adv. Stud. Contemp. Math. (Kyungshang) **23**(2), 337–345 (2013). MR 3088764
24. Simsek, Y.: Interpolation functions of the Eulerian type polynomials and numbers. Adv. Stud. Contemp. Math. (Kyungshang) **23**(2), 301–307 (2013). MR 3088760

25

Analytical solutions for stochastic differential equations via Martingale processes

Rahman Farnoosh[1] · Hamidreza Rezazadeh[1] · Amirhossein Sobhani[1] · Maryam Behboudi[2]

Abstract In this paper, we propose some analytical solutions of stochastic differential equations related to Martingale processes. In the first resolution, the answers of some stochastic differential equations are connected to other stochastic equations just with diffusion part (or drift free). The second suitable method is to convert stochastic differential equations into ordinary ones that it is tried to omit diffusion part of stochastic equation by applying Martingale processes. Finally, solution focuses on change of variable method that can be utilized about stochastic differential equations which are as function of Martingale processes like Wiener process, exponential Martingale process and differentiable processes.

Keywords Martingale process · Itô formula · Change of variable · Differentiable process · Analytical solution

✉ Rahman Farnoosh
rfarnoosh@iust.ac.ir

Hamidreza Rezazadeh
hr_rezazadeh@MathDep.iust.ac.ir

Amirhossein Sobhani
a_sobhani@MathDep.iust.ac.ir

Maryam Behboudi
m.behboudi@srbiau.ac.ir

1 School of Mathematics, Iran University of Science and Technology, 16844 Narmak, Tehran, Iran

2 Department of Statistics, Science and Research Branch, Islamic Azad University, Tehran, Iran

Introduction

The purpose of this article is to put forward some analytical and numerical solutions to solve the Itô stochastic differential equation (SDE):

$$\begin{cases} dX(t) = \mathcal{A}(X(t), t)dt + \mathcal{B}(X(t), t)dW_t, \\ X(0) = X_0, \end{cases} \tag{1}$$

where $W(t)$ is a Wiener process and triple $(\Omega, \mathcal{F}, \mathbb{P})$ is a probability space under some conditions and special relations between drift and volatility.

Both the drift vector $\mathcal{A} : \mathbb{R} \times [0, T] \longrightarrow \mathbb{R}$ and the diffusion matrix $a := \mathcal{B}\mathcal{B}^T : \mathbb{R} \times [0, T] \longrightarrow \mathbb{R}$ are considered Borel measurable and locally bounded functions. It is assumed that X_0 is a non-random vector. As usual, $\mathcal{A}$ and $\mathcal{B}$ are globally Lipschitz in $\mathbb{R}$ that is:

$$|\mathcal{A}(X,t) - \mathcal{A}(Y,t)| + |\mathcal{B}(X,t) - \mathcal{B}(Y,t)| \leq D|X - Y|,$$
$$X, Y \in \mathbb{R} \quad \text{and} \quad t \in [0, T],$$

and result in the linear growth condition:

$$|\mathcal{A}(X,t)| + |\mathcal{B}(X,t)| \leq C(1 + |X|).$$

These conditions guarantee (see [1, 2]) the Eq. (1) has a unique t-continuous solution adapted to the filtration $\mathcal{F}_t t \geq 0$ generated by $W(t)$ and

$$E\left[\int_0^T |X(s)|^2 \, ds\right] < \infty. \tag{2}$$

It is generally accepted that, analytical solutions of partial and ordinary differential equations are so important particularly in physics and engineering, whereas most of them do not have an exact solution and even a limited number of these equations, (e.g., in classical form), have implicit solutions. Analytical methods and solutions, especially in stochastic differential equations, could be excessive fundamental in some cases therefore we draw to take a comparison and analyze computation error between them and different numerical methods. Numerous numerical methods can be applied to solve stochastic differential equations like Monte Carlo simulation method, finite elements and finite differences [2, 3]. On the other hand, due to the importance of Martingale processes and finding their representation according to Martingale representation theorem, it is struggled to express arbitrary stochastic processes as a function of Martingale processes and found numerical methods so as to solve drift-free SDEs [4].

In this paper, we resolve to represent analytical methods for stochastic differential equations, specially reputed and famous equations in pricing and investment rate models, based on Martingale processes with various examples about them which we have found in a couple of papers like [2, 5–7]. There are two main reasons for this approach. Firstly, the each solutions of these kind of equations are Martingale processes or analytic function of Martingale Processes. Thus, due to drift-free property, it will be caused computational error less than numerical computations with existing classic methods. Secondly, for each Martingale process (especially differentiable process), there exists a spectral expansion of two-dimensional Hermite polynomials with constant coefficients [8]. Therefore, it could be made higher the strong order of convergence with increasing the number of polynomials in this expansion. Equations are just obtained with diffusion part or drift free, by making Martingale process from other process. This method can be done by Itô product formula on initial process and an appropriate Martingale process. Another suitable method to convert SDEs into ODEs that we try is to omit the diffusion part of the stochastic equation.

This article is organized as follows. In Sect. 2, it is verified the making of Martingales processes by exponential Martingale process. In Sect. 3, we solve equations as a function of Martingales with prominent analytical solution, by applying change of appropriate variables method on drift-free SDEs. In Sect. 5, some analytical and numerical examples of expressed methods are demonstrated. Finally, the conclusions and remarks are brought in last section.

Change of measure and Martingale process

In this section under some conditions, we intend to make a Martingale process from a random one in $\mathbb{L}^2(\mathbb{R} \times [0,T])$, where T is called maturity time. The exponential Martingale process associated with $\lambda(t)$ is defined as follows:

$$\mathcal{Z}_t^\lambda = \exp\left(\int_0^t \lambda(s)\,\mathrm{d}W_s - \frac{1}{2}\int_0^t \lambda^2(s)\,\mathrm{d}s\right). \tag{3}$$

It can be indicated by Itô formula that $\mathcal{Z}_t^\lambda$ is a Martingale due to the drift-free property:

$$\mathrm{d}\mathcal{Z}_t^\lambda = \lambda \mathcal{Z}_t^\lambda \mathrm{d}W_t, \quad \mathcal{Z}_t^\lambda(0) = 1. \tag{4}$$

Theorem 1 *Suppose that stochastic processes X_t verify in differential equation:*

$$\mathrm{d}X_t = \mu(X_t,t)\mathrm{d}t + \sigma(X_t,t)\mathrm{d}W_t, \tag{5}$$

and let $\lambda(t) := -\mu(X_t,t)/\sigma(X_t,t)$. Therefore, $X\mathcal{Z}_t^\lambda$ is a Martingale process.

Proof With attention to real function $\lambda(t)$, we have:

$$\begin{cases} \mathrm{d}X = \mu(X,t)\mathrm{d}t + \sigma(X,t)\mathrm{d}W_t = -\lambda(t)\sigma(X,t)\mathrm{d}t + \sigma(X,t)\mathrm{d}W_t, \\ \mathrm{d}\mathcal{Z}_t^\lambda = \mathcal{Z}_t^\lambda \lambda \mathrm{d}W_t. \end{cases}$$

By utilizing Itô product formula, we get:

$$\begin{aligned} \mathrm{d}(X\mathcal{Z}_t^\lambda) &= X\mathrm{d}(\mathcal{Z}_t^\lambda) + \mathcal{Z}_t^\lambda \mathrm{d}X + \mathrm{d}X\mathrm{d}(\mathcal{Z}_t^\lambda) \\ &= \lambda X \mathcal{Z}_t^\lambda \mathrm{d}W_t + \mu(X,t)\mathcal{Z}_t^\lambda \mathrm{d}t + \sigma(X,t)\mathcal{Z}_t^\lambda \mathrm{d}W_t \\ &\quad + \lambda\sigma(X,t)\mathcal{Z}_t^\lambda \mathrm{d}t. \end{aligned}$$

According to theorem assumption, we obtain:

$$\mathrm{d}(X\mathcal{Z}_t^\lambda) = \mathcal{Z}_t^\lambda (X\lambda + \sigma(X,t))\mathrm{d}W_t. \tag{6}$$

It emphasizes that $X\mathcal{Z}_t^\lambda$ is a P-Martingale. □

Therefore, $\lambda(t) = \frac{-\mu(X,t)}{\sigma(X,t)}$ is the sufficient condition for following SDEs equivalence:

$$\mathrm{d}X = \mu(X,t)\mathrm{d}t + \sigma(X,t)\mathrm{d}W_t \Leftrightarrow \mathrm{d}(X\mathcal{Z}_t^\lambda) = \mathcal{Z}_t^\lambda (X\lambda(t) + \sigma(X,t))\mathrm{d}W_t. \tag{7}$$

Consequently, by solving the obtained equation in Eq. (6), we obtain the following result when $\mathcal{Z}_0^\lambda = 1$:

$$X\mathcal{Z}_t^\lambda = \int_0^t \mathcal{Z}_t^\lambda (X\lambda(s) + \sigma(X,t))\,\mathrm{d}W_t + X_0. \tag{8}$$

By taking mathematical expectation from both sides of Eq. (8):

$$E^P[X\mathcal{Z}_t^\lambda] = X_0 \Rightarrow E^P[X] = X_0(\mathcal{Z}_t^\lambda)^{-1}. \tag{9}$$

In addition, to compute the variance of this stochastic process:

$$E^P[(X\mathcal{Z}_t^\lambda)^2] = X_0^2 + E\left[\int_0^t (\mathcal{Z}_s^\lambda)^2 (X\lambda(s) + \sigma(X,t))^2 \, ds\right]$$

(by Itǒ isometry)

$$= X_0^2 + \int_0^t (\mathcal{Z}_s^\lambda)^2 E\Big(\big[(X\lambda(s) + \sigma(X,t))^2\big]\Big) \, ds.$$

$$\text{var } (X\mathcal{Z}_t^\lambda) = (\mathcal{Z}_t^\lambda)^2 \text{ var } (X)$$

$$= \int_0^t (\mathcal{Z}_s^\lambda)^2 E\Big(\big[(X\lambda(s) + \sigma(X,t))^2\big]\Big) \, ds. \quad (10)$$

Applying (6) and using numerical approximation by EM method, we have:

$$\Delta X_i \mathcal{Z}_{t_i}^\lambda = \mathcal{Z}_{t_i}^\lambda (X_i \lambda(t_i) + \sigma_i) \Delta W_i.$$

$$X_{t_{i+1}} \mathcal{Z}_{t_{i+1}}^\lambda = X_{t_i} \mathcal{Z}_{t_i}^\lambda + \mathcal{Z}_{t_i}^\lambda (X_{t_i} \lambda(t_i) + \sigma_i) \Delta W_i.$$

$$X_{t_{i+1}} = (\mathcal{Z}_{t_{i+1}}^\lambda)^{-1} \mathcal{Z}_{t_i}^\lambda (X_{t_i} + (X_{t_i} \lambda(t_i) + \sigma_i) \Delta W_i).$$

Direct calculations would lead to the conclusion that:

$$R_{t_i} = (\mathcal{Z}_{t_{i+1}}^\lambda)^{-1} \mathcal{Z}_{t_i}^\lambda = \exp\left(-\int_{t_i}^{t_{i+1}} \lambda(s) dW_s + \frac{1}{2}\int_{t_i}^{t_{i+1}} |\lambda^2(s)| ds\right).$$

So the following Milstein recursive method is inferred as a good numerical method to find $X(t_{i+1})$:

$$X_{t_{i+1}} = R_{t_i}(X_{t_i} + (X_{t_i}\lambda(t_i) + \sigma_i)\Delta W_i) + \frac{1}{2} R_{t_i}^2 \lambda(t_i)(X_{t_i}\lambda(t_i) + \sigma_i)(\Delta^2 W_i - \Delta t_i). \quad (11)$$

In example 1, we compare this method with usual Milstein method in the case that a stochastic differential equation contains drift and volatility both parts and indicate that this method could be better in some cases.

Change of variable method

This section intends to analyze the change of variable method like [9], to get explicitly the solution of arbitrary SDE:

$$dX = \mathcal{A}(X,t)dt + \mathcal{B}(X,t)dW_t, \quad X(0) = x.$$

By finding appropriate variables $u(Y) = X$ and their conditions so that Y is the answer of a well-known SDEs related to Martingale processes.

$$dY = f(X,t)dt + g(X,t)dW_t, \quad y(0) = y.$$

For more explanation and different conditions under which they are possible, we could see [5, 10]. Now we consider following various cases.

Case 1 Consider the following SDE:

$$dY = a(t)dt + b(t)dW_t. \quad (12)$$

Applying Itô formula for $u(Y) = X$, to (12), we get:

$$\begin{cases} u'(a(t)) + \frac{1}{2}u''b^2(t) = \mathcal{A}(u(Y),t), \\ u'b(t) = \mathcal{B}(u(Y),t). \end{cases} \quad (13)$$

Thus, it concludes that:

$$\frac{a(t)}{b(t)}\mathcal{B} + \frac{1}{2}\mathcal{B}\mathcal{B}' = \mathcal{A} \Rightarrow \frac{\mathcal{A}}{\mathcal{B}} - \frac{1}{2}\mathcal{B}' = \frac{a(t)}{b(t)}. \quad (14)$$

Finally, the equation $\frac{\partial}{\partial Y}\left(\frac{\mathcal{A}}{\mathcal{B}} - \frac{1}{2}\mathcal{B}'\right) = 0$ is necessary condition to solve an equation via change of variable in (12) $\left(\mathcal{B}' = \frac{\partial \mathcal{B}}{\partial X}\right)$.

Case 2 Consider the exponential Martingale process SDE (3):

$$\begin{cases} dY = \lambda(t)YdW_t, \\ Y(0) = Y_0. \end{cases} \quad (15)$$

Applying Itô formula for $u(Y) = X$, to (15), we acquire:

$$\begin{cases} u'\lambda Y = \mathcal{B}(u,t) = \lambda(t)Y\hat{\mathcal{B}}(u) \quad \text{or} \quad u' = \hat{\mathcal{B}}(u), \\ \frac{1}{2}u''\lambda^2 Y^2 = \mathcal{A}(u,t). \end{cases} \quad (16)$$

So from the last equality, we have $\frac{\mathcal{B}'}{\lambda(t)} - \frac{2\mathcal{A}}{\mathcal{B}} = \lambda(t)$. Therefore, $\frac{\partial}{\partial u}\left(\mathcal{B}'_u - \frac{2\lambda(t)\mathcal{A}}{\mathcal{B}}\right) = 0$ is necessary condition to solve SDE, with this change of variable.

Case 3 Consider the well-known equation:

$$\begin{cases} dY = a(t)Ydt + b(t)YdW_t, \\ Y(0) = Y_0. \end{cases} \quad (17)$$

Which is Black–Scholes equation with exact solution

$$Y_0 = \exp\left(\int_0^t b(s)dW_s + \int_0^t \left(a(s) - \frac{1}{2}b^2(s)\right) ds\right).$$

Applying Itô formula for $u(Y) = X$, to (17), we get:

$$\begin{cases} u'a(t)Y + \frac{1}{2}u''b^2(t)Y^2 = \mathcal{A}(u,t), \\ u'Yb(t) = \mathcal{B}(u,t) = b(t)Y\hat{\mathcal{B}}(u). \end{cases} \quad (18)$$

For this reason, $u' = \hat{\mathcal{B}}(u)$ and we have:

$$\frac{a(t)}{b(t)} = \frac{\mathcal{A}}{\mathcal{B}} - \frac{1}{2}(\mathcal{B}'_u - b(t)) = \gamma(u,t). \quad (19)$$

It means that $\frac{\partial}{\partial u}\gamma(u,t) = 0$, is a necessary condition to solve the initial stochastic differential equation by this change of variable.

Case 4 Another appropriate and prominent case is as follows:

$$\begin{cases} dY_t = f(Y_t,t)dt + c(t)Y_t dW_t, \\ Y(0) = Y_0. \end{cases} \quad (20)$$

This kind of equations, applying Itô formula on $X_t = Y_t \mathcal{Z}_t^c(t)^{-1}$, is converted to a ordinary differential equations.

Theorem 2 *The stochastic differential equations in* (20) *given by continuous functions $f : \mathbb{R} \times \mathbb{R} \to \mathbb{R}$ and $C : \mathbb{R} \to \mathbb{R}$ can be written as:*

$$d(Y_t(\mathcal{Z}_t^c(t))^{-1}) = (\mathcal{Z}_t^c(t))^{-1} f(Y_t, t)dt, \tag{21}$$

where $\mathcal{Z}_t^c(t)$ is an exponential Martingale process.

(See Oksendal [1], Chapter 5, Exercise 17]). To be more precise, using change of variable $V = X(\mathcal{Z}_t^{c(t)})^{-1}$, it is enough to solve

$$\begin{cases} X_t' = (\mathcal{Z}_t^{c(t)})^{-1} f(X_t \mathcal{Z}_t^{c(t)}), \\ X(0) = X_0. \end{cases} \tag{22}$$

Applying Itô formula for $u(Y) = M_t$, in (20) we get:

$$dM_t = M_t' dY + \frac{1}{2} M_t''(dY)^2.$$

$$\begin{cases} f(Y,t)M_t' + \frac{1}{2} M_t'' c^2(t) Y^2 = \mathcal{A}(M_t, t), & (1) \\ c(t) Y M_t' = \mathcal{B}(M_t, t), \ u(Y_0) = M_0. & (2) \end{cases} \tag{23}$$

According to (23), we have $\mathcal{B}(M_t, t) = c(t)\hat{\mathcal{B}}(M_t)$. Besides, if the new stochastic differential equation is related to a Martingale process, we have $\mathcal{A}(M_t, t) = 0$ and:

$$f(Y,t) = -\frac{c^2(t)Y}{2}(\hat{\mathcal{B}}(M_t)' - 1). \tag{24}$$

Again, applying Itô formula for $\phi(M_t) = V_t$ to Martingale equation contributes to

$$dM_t = \mathcal{B}(M_t, t)dW_t = c(t)\hat{\mathcal{B}}(M_t)dW_t,$$

we can achieve to a novel group of stochastic differential equation that its solution is as a function of a Martingale process.

Examples

Example 1 Consider the following SDE

$$\begin{cases} dX = (a(t)\sqrt{X})dt + (b(t)\sqrt{X})dW_t, \\ X(0) = X_0. \end{cases} \tag{25}$$

from (9), we can get immediately $E[X] = X_0(\mathcal{Z}_t^\lambda)^{-1}$ such that $\lambda = \frac{a(t)}{b(t)}$. The graphs of various numerical solutions of this example by Milstein method, proposed formula (11) that is drift free and Taylor method of order 2 introduced as exact solution.

Example 2 Consider the following SDE that is named Black–Scholes equation.

$$dX = \mu(t)Xdt + \sigma(t)XdW_t.$$

Using (6), we have:

$$\begin{aligned} d(X\mathcal{Z}_t^\lambda) &= \mathcal{Z}_t^\lambda(X\lambda + \sigma(t))dW_t = \mathcal{Z}_t^\lambda(X\lambda + X\sigma(t))dW_t \\ &= X\mathcal{Z}_t^\lambda(\lambda + \sigma(t))dW_t. \end{aligned}$$

From this equality we could conclude that $X\mathcal{Z}_t^\lambda$, is the exponential Martingale $\mathcal{Z}_t^{\lambda+\sigma}$. Finally, $X = (\mathcal{Z}_t^\lambda)^{-1}\mathcal{Z}_t^{\lambda+\sigma} = \exp\left(\int_0^t \sigma(t)\, dW_s + \int_0^t(\mu(t) - \sigma^2)\, ds\right)$. This is the exact solution of Black–Scholes equation.

Example 3 Consider the following stochastic model

$$\begin{cases} dX = \frac{3}{4}t^2X^2dt + tX^{3/2}dW_t, \\ X(0) = 0. \end{cases}$$

It can be checked that for this equation the necessary condition holds for this equation. According to (13), we have $u'b(t) = tu^{3/2}$. Since u is just a function of Y, we should get $b(t) = t$, $u = \frac{4}{Y^2}$ and $\frac{a(t)}{b(t)} = 0$ (or $a(t) = 0$). Thus, $dY = tdW_t$ and $Y = \int_0^t sdW_s + Y(0)$, and ultimately $X = u(Y) = 4\left(\int_0^t sdW_s + Y(0)\right)^{-2}$, is the exact solution (Fig. 1).

Example 4 Consider the following SDE model

$$\begin{cases} dX = \frac{1}{2}(c^2(t)rX^{2r-1} - c^2(t)X^r)dt + c(t)X^r dW_t, \ (r \neq -1) \\ X(0) = 0. \end{cases}$$

First of all, we check the necessary condition in case 2:

$$\mathcal{B}_u' - \frac{2\mathcal{A}}{\mathcal{B}} = c(t)ru^{r-1} - \frac{c^2(t)ru^{2r-1} - c^2(t)u^r}{c(t)u^r} = c(t) = \lambda(t).$$

Utilizing the first equation in Eq. (16), $u'\lambda(t)Y = c(t)u^r$. Hence, $\ln Y = \frac{u^{-r+1}}{-r+1}$, that $r \neq -1$, $Y(0) = 1$ and $u(1) = 0$. Therefore, the exact solution is as follows:

$$X = u(Y) = \left((1-r)\left(\int_0^t c(s), dW_t - \frac{1}{2}\int_0^t c^2(s)ds\right)\right)^{\frac{1}{1-r}}.$$

In a particular case, if $r = \frac{1}{2}$, we reach the following model:

$$dX = \left(\frac{c^2(t)}{4} - c^2(t)\sqrt{X}\right)dt + \left(c(t)\sqrt{X}\right)dW_t,$$

$$X = \frac{1}{4}\left(\int_0^t c(t)dW_t - \frac{1}{2}\int_0^t c^2(s)ds\right)^2.$$

Example 5 Consider the following SDE model:

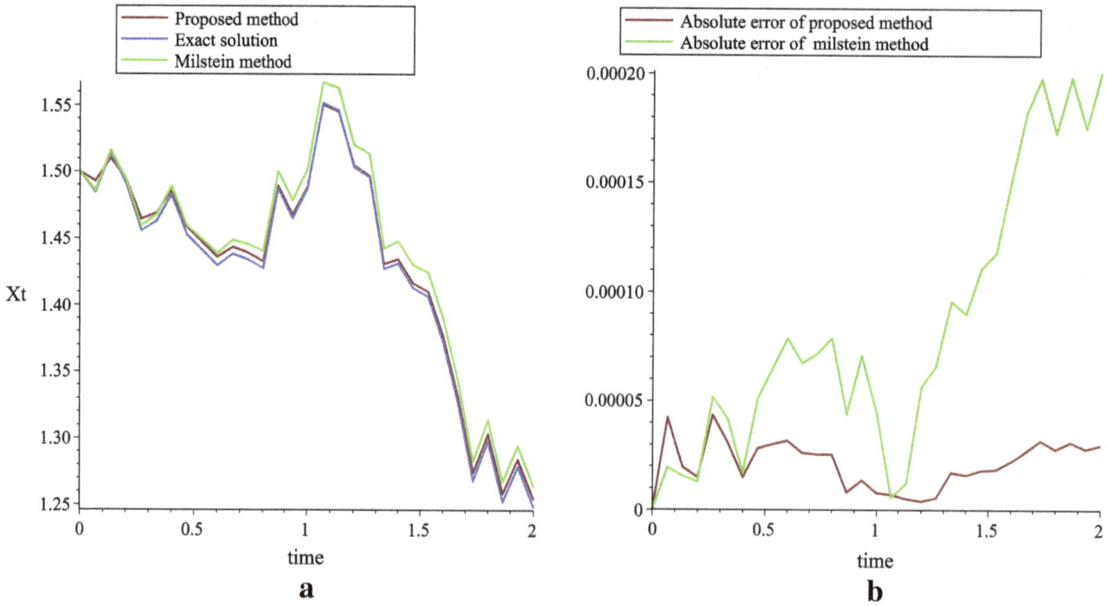

Fig. 1 **a** The graphs of the numerical solutions of Example 1 by Milstein method, proposed formula (11) and Taylor method of order 2 introduced as exact solution. In this example, it is considered that drift, volatility and initial condition respectively are: $a(t) = -0.2, b(t) = -1.0, X_0 = 1.5$, maturity time $T = 2$ and number of points $N = 30$. **b** The graphs of absolute error

$$\begin{cases} dX = X^3 dt + X^2 dW_t, \\ X(0) = 1. \end{cases} \quad (26)$$

First of all, we check the necessary condition in Case 3:

$$\gamma(u, t) = u - \frac{1}{2}(2u - b(t)) = \frac{a(t)}{b(t)} = \frac{b(t)}{2}.$$

From (18), we should have $u'b(t)Y = u^2$. Therefore, if $b(t) = 1$, we can get immediately $u = \frac{-1}{\ln Y}$ and $a(t) = \frac{b^2(t)}{2} = \frac{1}{2}$, so that Y is the solution of following equation.

$$\begin{cases} dY = \frac{1}{2} Y dt + Y dW_t, \\ Y(0) = \frac{1}{e}. \end{cases}$$

Therefore, according to geometric Brownian motion process, the exact solution is determined $Y = \frac{1}{e}\exp\left(\int_0^t dW_t\right) = e^{W(t)-1}$, and finally exact solution is equal to $X = \frac{1}{1-W(t)}$.

Example 6 Consider the stochastic model as follows:

$$\begin{cases} d\mathcal{Z}_t = \frac{-\mathcal{Z}_t^2}{2} - (\ln 2)\mathcal{Z}_t dt + (\ln 2 + \mathcal{Z}_t) dW_t, \\ \mathcal{Z}_t(0) = 0. \end{cases} \quad (27)$$

First, by applying Girsanov theorem so that $W_t^Q = W_t + \frac{(\ln 2)^2}{2} t$, we reach the following equation:

$$\begin{cases} d\mathcal{Z}_t = -\frac{(\ln 2)^2}{2} + \frac{-\mathcal{Z}_t^2}{2} - (\ln 2)\mathcal{Z}_t dt + (\ln 2 + \mathcal{Z}_t) dW_t^Q, \\ \mathcal{Z}_t(0) = 0. \end{cases} \quad (28)$$

Applying Itô formula for $X_t = e^{\mathcal{Z}_t}$, to the last equation, we obtain the following drift-free stochastic equation:

$$\begin{cases} dX_t = X_t \ln(2X_t) dW_t^Q, \\ X_t(0) = 1. \end{cases} \quad (29)$$

according to (23), we have $Yu' = u\ln(2u)$. Consequently, $Y = \frac{\ln(2X)}{2}$, $X = u(Y) = \frac{1}{2}e^{2Y}$.

From (24), we have $f = -Y^2$ and consequently, the exact solution of corresponding SDE is $X = \frac{1}{2}e^{2Y}$ such that its related stochastic equation is:

$$\begin{cases} dY = -Y_t dt + Y_t dW_t^Q, \\ Y(0) = \frac{\ln(2)}{2}. \end{cases}$$

As we know, the exact solution of this linear stochastic differential equation is as follows:

$$Y_t = \frac{\ln(2)}{2} \exp\left(W_t^Q - \frac{3t}{2}\right). \quad (30)$$

Finally, the exact solution of this example is:

$$\mathcal{Z}_t = \ln(X_t) = \ln\left(\frac{1}{2}e^{2Y_t}\right)$$
$$= 2Y_t - \ln 2 = \ln 2\left(\exp\left(W_t^Q - \frac{3t}{2}\right) - 1\right). \quad (31)$$

Conclusions and remarks

In this paper, a couple of analytical solutions of some determined set of stochastic differential equations was indicated via making the Martingale process from a stochastic process. Converting stochastic differential equations to ordinary ones as another suitable method was posed. Indeed, it is tried to omit diffusion part of stochastic equation by applying Martingale processes. In addition, change of variable method on SDEs related to Martingale processeswas discussed. Last of all with some examples, we analyzed and obtained its exact solutions and in some cases their solutions compared with other numerical methods.

References

1. Øksendal, B. K.: Stochastic Differential Equations: An Introduction with Applications, 4th edn. Springer, Berlin (1995)
2. Kloeden, P.E., Platen, E.: Numerical Solution of Stochastic Differential Equations. Springer, Berlin (1999)
3. Higham, D.J.: An algorithmic introduction to numerical simulation of stochastic differential equations. SIAM Rev. **43**, 525546 (2001)
4. Pascucci, A.: PDE and Martingale Methods in Option Pricing. Bocconi & Springer Series, vol. 2. Springer, Milan (2011)
5. Lamperti, J.: A simple construction of certain diffusion processes. J. Math.Kyoto Univ. **4**, 161–170 (1964)
6. Skiadas, C.H.: Exact solutions of stochastic differential equations: Gompertz, generalized logistic and revised exponential. Methodol Comput. Appl. Probab. **12**, 261–270 (2010)
7. Kouritsin, M.A., DeliOn, L.: explicit solutions to stochastic differential equations. Stoch. Anal. Appl. **18**(4), 571–580 (2000)
8. Udriste, C., Damian, V., Matei, L., Tevy, I.: Multitime differentiable stochastic process, diffusion PDEs, Tzitzeica hypersurfaces. UPB Sci. Bull. Ser. A **74**(1), 3–10 (2012)
9. Evans, L.C.: An Introduction to Stochastic Differential Equations. American Mathematical Society, Providence (2013)
10. McKean, H.: Stochastic Integrals. Academic Press, New York (1969)

26

Algorithms for solving reachability problems in 2-link planar arms using Gröbner bases

Zahra Nilforoushan · Keivan Borna

Abstract It is a crucial problem to study the reachability of planar arms inside convex obtuse polygons. In this paper, we studied the reachability problem for 2-link planar arms inside a circle, a general polygon with (without) some holes in it and presented several algorithms for them. Furthermore, we proposed some algorithms for a special case where the shoulder of an arm moves along a given segment or passes through a certain path. It is essential to mention that our approach is based on the Gröbner bases technique.

Keywords 2-link arm · Gröbner bases · Reachability

Mathematics Subject Classification 13P10 · 13P15 · 12Y05 · 14Q99 · 68T40 · 93C10

Introduction

A *linkage* is a collection of rigid rods called *links*. The endpoints of various links are connected by joints, each joint connecting two or more links. The links are free to rotate around the joints.

An *arm* is a simple type of linkage consisting of a sequence of links joined together consecutively with the location of one end fixed which is called *shoulder* and the other mobile end called *hand*; see Fig. 1.

Z. Nilforoushan
Faculty of Engineering, Kharazmi University, Tehran, Iran
e-mail: nilforoushan@khu.ac.ir

K. Borna (✉)
Faculty of Mathematics and Computer Science, Kharazmi University, Tehran, Iran
e-mail: borna@khu.ac.ir

For an arm Γ constrained to lie in a confining region R, given a point $p \in R$ and an initial configuration of Γ inside R, the *reachability problem* for Γ is to determine whether Γ can be moved in R so that the hand of Γ reaches p.

Reachability problem has been studied independently by several researchers. In Pei [11], has studied the reachability problem for a planar chain within two type of confining regions: convex obtuse polygons and circles, and the reachability problem for a planar arms inside a convex obtuse polygon has to be investigated by readers.

The method of *Gröbner bases* is meanwhile well established as one of the fundamental algorithmic tools in polynomial algebra and algebraic geometry; see [5] and [6].

Gröbner bases enable us to solve problems about polynomial ideals in an algorithmic or computational fashion. In its basic version, it deals with multivariate polynomials over a field k. It assigns to any finite set $F \subseteq R = k[x_1, \ldots, x_n]$ another such set G (a Gröbner bases) such that G and F generate the same ideal in R, and such that many problems related to F and its zeros in an algebraically closed extension field $\bar{k}$ can be solved algorithmically using G. Thus the computation of G serves as a preprocessing step (of potentially high complexity) in the solution of these problems. The construction of G depends on F and on a *term-order* $<$ on R; see [12]. If G is taken minimal in a suitable sense (i.e., G is a *reduced Gröbner bases*) then G is even uniquely determined by the ideal generated by F and order lessthan. For example, the Gröbner basis of the following system of polynomials:

$$\begin{cases} x^2 + y^2 + z^2 = 1, \\ x^2 + z^2 = y, \\ x = z \end{cases}$$

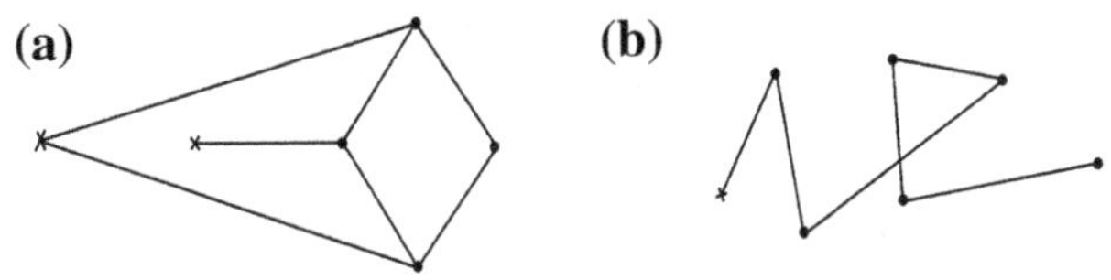

Fig. 1 **a** A planar linkage. **b** An arm in the plane

with respect to the lexical order is:

$$\begin{cases} x - z = 0, \\ -y + 2z^2 = 0, \\ z^4 + \frac{1}{2}z^2 - \frac{1}{4} = 0. \end{cases}$$

The produced Gröbner bases has the remarkable and useful property that it is "triangularized", that is, its third equation depends only on z and by substituting the values of z into the second and first equations, they can be solved uniquely for y and x, respectively. Thus by [Proposition 9, 5], we have found all solutions of the original equations. Of course, in this simple example, the solution could also be derived by a reasonably skillful analysis. However, the Gröbner bases algorithm works in all situations and always results in a "triangularized" system [7].

Buchberger in [7] and Cox et al. in [9] considered a recent application of concepts and techniques from algebraic geometry in areas of computer science. It has developed a systematic approach that uses algebraic varieties to describe the space of possible configuration of mechanical linkages such as robot arms and used this approach to solve the forward and inverse kinematic problems of robotics for certain types of robots. In other words, the inverse kinematic problem for the planar robot arm is defined as the reachability problem for that arm in computational geometry.

Although the computation of Gröbner bases is known to be EXPSPACE-complete problem [2], in this paper we use Gröbner bases to effectively solve several reachability problems only for 2-link planar arms. In the course of computing Gröbner bases by *Buchberger's algorithm*, one has to compute and reduce numerous so called *S-polynomials*. Since these reductions are costly and the number of S-polynomials may be quite large and may, in fact, increase in the course of the algorithm, the overall computation time may be quite high. In fact, in [3] and [4], some criteria are given by which the number of S-polynomials to be considered in the course of the algorithm is reduced drastically. All practical implementations of Buchberger's algorithm make heavy use of these criteria. To be precise let f and g be two polynomials and m be the lcm of their leading monomials, then the S-polynomial of f and g is denoted by $S(f, g)$ and defined as $S(f, g) := \frac{m}{LT(f)}f - \frac{m}{LT(g)}g$. In the univariate case, $S(f, g)$ corresponds to the first step in the Euclidean division of f by g.

Buchberger's Algorithm:

Initialization: $B := \{\}, S := \{f_1, \ldots, f_s\}$.
While $S \neq \{\}$ **do**
    -pick $f \in S, S := S \setminus \{f\}$;
    -reduce f w.r.t. B
    -call g the resulting polynomial;
    -if $g \neq 0$ **then** put $S := S \cup_{b \in B} S(g, b)$; **add** g to B
Return B.

Buchberger proved in his Ph.D thesis (in 1965) that this algorithm terminates and produces Gröbner bases. One of the main difficulties with an actual implementation is that the reduction steps often produce 0 and a lot of time is wasted during these useless reductions. Fortunately, there are many strategies to pick elements in S and predict useless reductions in advance.

Throughout this paper we use the following notations: $s_i = \sin\theta_i, c_i = \cos\theta_i, s = \sin\theta, c = \cos\theta$ and $t = \tan\theta$. The procedures we include in this work are encoded in the Computer Algebra package *CoCoA* [8].

The organization of this paper is as follows. In the following section, we present a general view of our algorithm for the reachability problem for the 2-link planar arm and call appropriate partial algorithms. Then we study our partial algorithms more briefly followed by the reachability problem for 2-link planar arms without confining region. The next sections are devoted to the reachability problem for 2-link planar arms inside the circle, the reachability problem for 2-link planar arms when its shoulder moves and the reachability problem for planar arms inside a general polygon. The final section consists of conclusions and future works.

Our algorithm for reachability problem for 2-link planar arms

In this section we present an algorithm for the reachability problem for 2-link planar arms. First we detect the desired partial algorithm and then the appropriate routine is called:

Main Algorithm:

If there is no confining region for the 2-link planar arms, **then** use Partial Algorithm 1.

If the 2-link planar arms is inside the circle, **then** use Partial Algorithm 2.

If the shoulder of the 2-link planar arms moves, **then** use Partial Algorithm 3.

If the confining region for the 2-link planar arms is a polygon, **then** use Partial Algorithm 4.

Reachability problem for 2-link planar arms without confining region

Consider a general planar robot arm with two joints (see Fig. 2). Using a standard rectangular coordinate system, by (x_1, y_1), we mean the origin of the coordinate system which is placed at the shoulder of the robot arm. In addition call three local rectangular coordinate systems $(x_2, y_2), (x_3, y_3)$ and (x_4, y_4) at shoulder, joint 2, and hand respectively. This coordinate system changes as the position of the arm varies. Take the positive x_2-axis to be the direction of link 1 and the positive x_3-axis and x_4-axis to be the direction of link 2. Then the positive y_2-axis, y_3-axis and y_4-axis are determined to form normal right-handed rectangular coordinate systems. Note that the (x_2, y_2) coordinate of joint 2 is $(l_1, 0)$, where l_1 is the length of link 1 and the (x_3, y_3) and (x_4, y_4) coordinates of hand is $(l_2, 0)$ and $(0, 0)$, where l_2 is the length of link 2.

Partial Algorithm 1:
Task: Reachability for 2-link planar arms without confining region.
Steps:

1 **Find** the (x_1, y_1) coordinate of hand which is a function of c_i and s_i subject to the constraints $c_i^2 + s_i^2 - 1 = 0,\ i = 1, 2.$

2 **Find** all the values of θ_1 and θ_2 that cause the hand of arm to reach the given point $p = (a, b)$: For this we need to know if p can reach to $(l_2(c_1c_2 - s_1s_2) + l_1c_1, l_2(s_1c_2 + c_1s_2) + l_1s_1)$. This is done by solving the system of equations in the next step.

3 Apply **Groebner basis** to solve the following system of equations:

$$\begin{cases} l_2(c_1c_2 - s_1s_2) + l_1c_1 = a, \\ l_2(s_1c_2 + c_1s_2) + l_1s_1 = b, \\ c_1^2 + s_1^2 - 1 = 0, \\ c_2^2 + s_2^2 - 1 = 0. \end{cases} \quad (1)$$

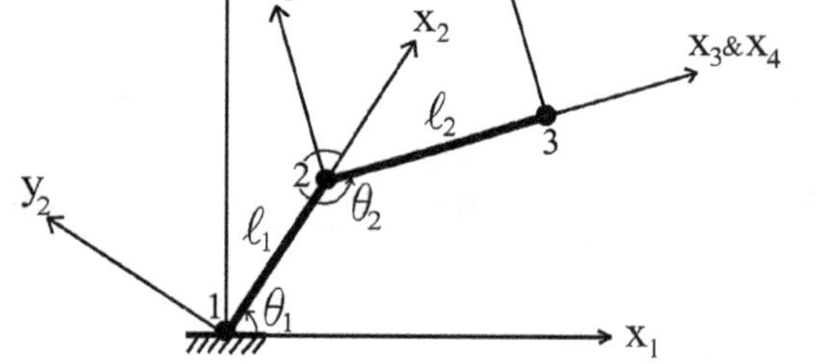

Fig. 2 A planar 2-link arm with it's global and local coordinate systems

Example 1 For the sake of simplicity in $(1-1)$ let $l_1 = l_2 = 1$ and $(a, b) = (1, 1)$. Hence

$$\begin{cases} c_1c_2 - s_1s_2 + c_1 - 1 = 0, \\ s_1c_2 + c_1s_2 + s_1 - 1 = 0, \\ c_1^2 + s_1^2 - 1 = 0, \\ c_2^2 + s_2^2 - 1 = 0. \end{cases}$$

Now use CoCoA to calculate its Gröbner bases:

$$\begin{cases} s_1 + s_2 - c_1 = 0, \\ 2s_1 + s_2 - 1 = 0, \\ s_2^2 + c_2 - 1 = 0, \\ \frac{1}{4}c_2 = 0. \end{cases}$$

Thus:

$c_2 = 0$ and

$(s_1 = 0,\ s_2 = 1$ and $c_1 = 1)$ or $(s_1 = 1,$ and $s_2 = -1$ and $c_1 = 0)$.

As a result $(\theta_1 = 90^\circ$ and $\theta_2 = 270^\circ)$ or $(\theta_1 = 0^\circ$ and $\theta_2 = 90^\circ)$.

Reachability problem for 2-link planar arms inside the circle

We study this problem in two subcases when the shoulder is the origin of the circle and when it is not.

Reachability when the shoulder is the origin of the circle

To reach a point in the circle, joint 2 and hand (Fig. 2) must be inside the circle. Thus for the given circle with radius $m > 0$, there are two certain constraints:

$$\begin{cases} l_1 \leq m, \\ (l_2(c_1c_2 - s_1s_2) + l_1c_1)^2 + (l_2(s_1c_2 + c_1s_2)^2 \leq m^2. \end{cases}$$

Thus we have to add the following equation to (1) with $t \geq 0$, to find all the values of θ_1 and θ_2 (if there exist any).
$l_1^2 + l_2^2 + 2l_1l_2(s_1(s_1c_2 + c_1s_2) + c_1(c_1c_2 - s_1s_2)) - m^2 + t = 0$

with $t \geq 0$ to find all the values of θ_1 and θ_2.

Reachability when the shoulder is not the origin of the circle

Let the vector $\vec{v}$ be the translation vector from the point 0 (the origin of the circle) to point 1 (the shoulder of the arm), we use the above-mentioned argument stated in 1 to the new coordinate system (x, y) with the origin 0 (see Fig. 3). We summarize this observation in the following:

Partial Algorithm 2:

Task: Reachability when the shoulder is (or is not) the origin of a circle with radius $m > 0$.

Steps:

1 **If** the 2-link planar arms is inside the circle and the shoulder is the origin of the circle, **then** apply **Groebner basis** to solve the following system of equations:

$$\begin{cases} l_2(c_1c_2 - s_1s_2) + l_1c_1 = a, \\ l_2(s_1c_2 + c_1s_2) + l_1s_1 = b, \\ c_1^2 + s_1^2 - 1 = 0, \\ c_2^2 + s_2^2 - 1 = 0, \\ l_1^2 + l_2^2 + 2l_1l_2(s_1(s_1c_2 + c_1s_2) \\ \quad +c_1(c_1c_2 - s_1s_2)) - m^2 + t = 0. \end{cases} \tag{2}$$

2 **If** the 2-link planar arms is inside the circle and the shoulder is not the origin of the circle, **then** apply **Groebner basis** to solve the following system of equations:

$$\begin{cases} l_2(c(c_1c_2-s_1s_2)-s(s_1c_2+c_1s_2))+l_1(cc_1-ss_1)=a, \\ l_2(s(c_1c_2-s_1s_2)+c(s_1c_2+c_1s_2)+l_1(sc_1+cs_1)=b, \\ c_1^2+s_1^2-1=0, \\ c_2^2+s_2^2-1=0, \\ l_1^2+2l_1|\vec{v}|c_1+|\vec{v}|^2-m^2+t_1=0, \\ l_1^2+l_2^2+2l_1|\vec{v}|c_1+2l_2|\vec{v}|(c_1c_2-s_1s_2) \\ \quad +2l_1l_2(c_1^2c_2-c_1s_1s_2+cs(s_2-s_1)(c_1c_2-s_1s_2) \\ +(s_1c_2+c_1s_2)(s^2s_1+c^2s_2) \\ \quad +|\vec{v}|^2-m^2+t_2=0. \end{cases} \tag{3}$$

The last condition is obtained after some simplifications for considering the translation vector from the point 0 (the origin of the circle) to point 1. Note that for $t_1, t_2 \geq 0, |\vec{v}|, c$ and s are given and s_1, c_1, s_2 and $c_2 \in \mathbb{R}$ are variables. Using the Gröbner bases, we can find the values of θ_1 and θ_2.

Example 2 Let C be a circle with radius 3 and Γ be a 2-link arm with $l_1 = 1$ and $l_2 = 2$ and $(a,b) = (1,1)$. Then the system to be solved is:

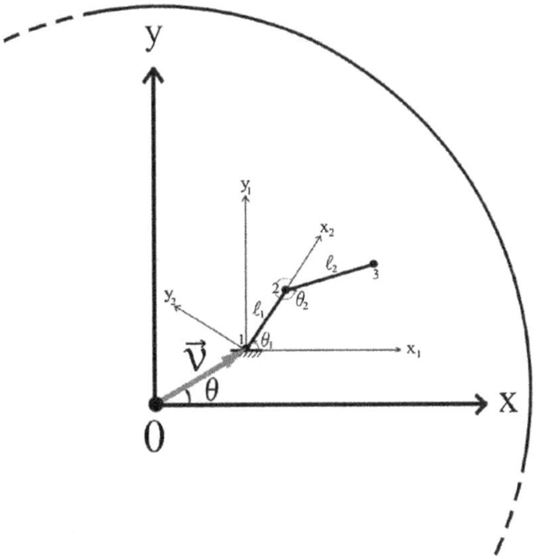

Fig. 3 The shoulder of arm and the origin of circle are not the same

$$\begin{cases} 2(c_1c_2 - s_1s_2) + c_1 - 1 = 0, \\ 2(s_1c_2 + c_1s_2) + s_1 - 1 = 0, \\ c_1^2 + s_1^2 - 1 = 0, \\ c_2^2 + s_2^2 - 1 = 0, \\ 4 + 4(s_1(s_1c_2 + c_1s_2) + c_1(c_1c_2 - s_1s_2)) - 9 + t = 0. \end{cases}$$

CoCoA calculates its Groebner basis as follows:

$$\begin{cases} -\frac{1}{2}c_1 + \frac{1}{2}s_1 + s_2 = 0, \\ s_1 + s_2 + \frac{1}{4} = 0, \\ s_2^2 + \frac{1}{2}c_2 - \frac{1}{16} = 0, \\ \frac{1}{2}c_2 - \frac{3}{8} = 0, \\ -\frac{1}{4}t + 2 = 0. \end{cases}$$

Example 3 In Fig. 3, suppose that $l_1 = 1, l_2 = 2, \theta = 0°, |v| = 0.5, (a,b) = (1,1)$ and $m = 4$:

$$\begin{cases} 2(c_1c_2 - s_1s_2) + c_1 - 3 = 0, \\ 2(s_1c_2 + c_1s_2) + s_1 - 3 = 0, \\ c_1^2 + s_1^2 - 1 = 0, \\ c_2^2 + s_2^2 - 1 = 0, \\ c_1 + t_1 - 14.75 = 0, \\ c_1 + 5 + 2(c_1c_2 - s_1s_2) + 4(c_1^2c_2 - s_1s_2c_1) \\ \quad +s_2(s_1c_2 + c_1s_2) + t_2 - 15.75 = 0. \end{cases}$$

Gröbner bases now gives the following easy system:

$$\begin{cases} c_1 + t_1 - \dfrac{59}{4} = 0, \\ \dfrac{3}{2}s_1 + s_2 + \dfrac{3}{2}t_1 - \dfrac{177}{8} = 0, \\ -\dfrac{9}{4}s_2 - \dfrac{9}{4}c_2 - \dfrac{27}{4}t_1 + \dfrac{1575}{16} = 0, \\ s_2 + 3t_1 - \dfrac{81}{2} = 0, \\ \dfrac{29}{32}t_1 - \dfrac{1}{4}t_2 - \dfrac{1475}{128} = 0, \\ \dfrac{64}{841}t_2^2 - \dfrac{364}{841}t_2 + \dfrac{11289}{6728} = 0. \end{cases}$$

Reachability problem for 2-link planar arms when its shoulder moves

We study three special cases:

The shoulder of the 2-link planar arms moves along a segment.

The shoulder of the 2-link planar arms moves along a parabola.

The shoulder of the 2-link planar arms moves along a circle.

Reachability when the shoulder can move along a segment

Suppose that point 1 (the shoulder) can move along a segment L. Define a global coordinate system at the first point of the segment L. Then as in Fig. 4, θ is constant but distance between 0 and point 1 can vary between 0 and $|L|$ along L, which we call u. Thus u is another variable that must be added to previous variables. Hence the system for solving is the same as (1), just we have to change $|\vec{v}|$ into u and consider u as a variable.

Reachability when the shoulder can move along a known path

In this case it suffices to parameterize the path in the (x, y)-coordinate where the shoulder of the arm has to move along it. On the other hand, if the equation of the path is of the form $y = f(x)$, the coordinate of point 1 (shoulder), point 2, and the end point 3 (hand) of the arm in the (x, y)-coordinate system are respectively as follows:

$$(x, y) = (x, f(x)) = (r\cos\theta, f(r\cos\theta)),$$

$$(x, y) = (l_1\cos(\theta + \theta_1) + r\cos\theta, l_1\sin(\theta + \theta_1) + f(r\cos\theta),$$

$$(x, y) = \left(\sum_{i=1}^{3} l_i \cos\left(\theta + \sum_{j=1}^{i}\theta_j\right)\right.$$
$$\left. + r\cos\theta, \sum_{i=1}^{3} l_i \sin\left(\theta + \sum_{j=1}^{i}\theta_j\right) + f(r\cos\theta)\right).$$

It is clear that by using the parametrization $x = r\cos\theta$ and $y = r\sin\theta$ and $y = f(x)$, one can compute r with respect to θ and then use it in the above equations.

Here we show this in two important cases and the idea for the general case will follow.

Reachability problem for 2-link planar arm when the path is a parabola

For the sake of simplicity take the path $y = x^2$, so by using parametrization, we have $r = \frac{sin\theta}{cos^2\theta}$ and thus the coordinate of point 1 is $(x, y) = (x, x^2) = (\tan\theta, \tan^2\theta)$, the coordinate of point 2 is $(x, y) = (l_1\cos(\theta + \theta_1) + \tan\theta, l_1\sin(\theta + \theta_1) + \tan^2\theta)$ and the coordinate of point 3 is $(x, y) = (l_2\cos(\theta + \theta_1 + \theta_2) + l_1\cos\ (\theta + \theta_1) + \tan\theta, l_2\sin(\theta + \theta_1\theta_2) + l_1\sin(\theta + \theta_1) + \tan^2\theta)$. Therefore $\tan\theta$ and r are new variables that must be substituted with x, y (the previous variables); see Fig. 4.

Reachability problem for 2-link planar arm when the path is a circle

For a 2-link planar arm whose shoulder can move along the circle C with radius $r \neq 0$, the problem would be reduced to the 3-link planar arm that its shoulder is the origin of (x, y)-plane (see Fig. 5). Hence we have the following algorithm.

Partial Algorithm 3:

Task: Reachability when the shoulder can move.

Steps:

1 **If** there is no confining region around the arm, and the joint 1 can move along segment L, **then** we have the following system

$$\begin{cases} l_2(c(c_1c_2 - s_1s_2) - s(s_1c_2 + c_1s_2)) + l_1(cc_1 - ss_1) = a, \\ l_2(s(c_1c_2 - s_1s_2) + c(s_1c_2 + c_1s_2)) + l_1(sc_1 - cs_1) = b, \\ c_1^2 + s_1^2 - 1 = 0, \\ c_2^2 + s_2^2 - 1 = 0. \end{cases} \tag{4}$$

Now we can apply Partial Algorithm 2.

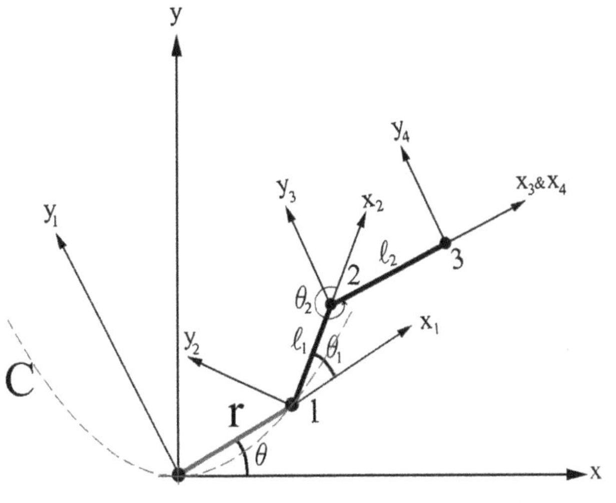

Fig. 4 A 2-link arm when shoulder can move along the parabola $y = x^2$

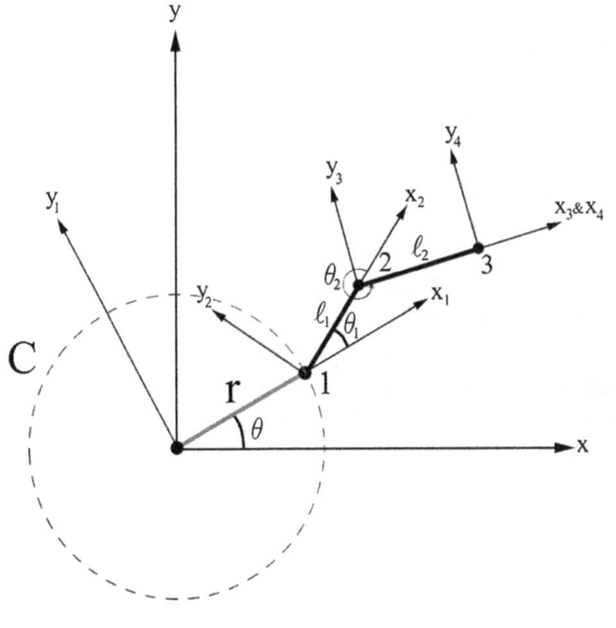

Fig. 5 A 2-link arm when shoulder can move along the circle C

2 **If** the joint can move along a parabola, **then** the new system to solve the reachability of the hand of arm to a given point $p = (a, b)$ is:

$$\begin{cases} l_2(c(c_1c_2 - s_1s_2) - s(s_1c_2 + c_1s_2)) + l_1(cc_1 - ss_1) = a, \\ l_2(s(c_1c_2 - s_1s_2) + c(s_1c_2 + c_1s_2)) + l_1(sc_1 - cs_1) = b, \\ ct - s = 0, \\ t - rc = 0, \\ c_1^2 + s_1^2 - 1 = 0, \\ c_2^2 + s_2^2 - 1 = 0. \end{cases} \tag{5}$$

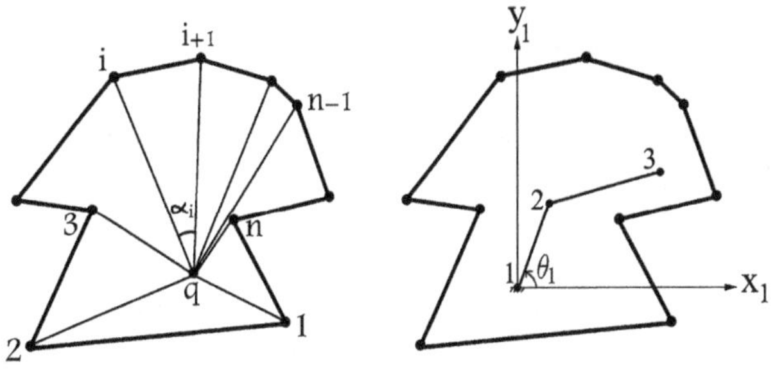

Fig. 6 P with a given point q inside (*in the left*) and a 2-link arm inside P (*in the right*)

3 **If** the joint can move along a circle, **then** the new system to be solved is:

$$\begin{cases} l_2(c(c_1c_2 - s_1s_2) - s(s_1c_2 + c_1s_2)) + l_1(cc_1 - ss_1) = a, \\ l_2(s(c_1c_2 - s_1s_2) + c(s_1c_2 + c_1s_2)) + l_1(sc_1 - cs_1) = b, \\ c^2 + s^2 - 1 = 0, \\ c_1^2 + s_1^2 - 1 = 0, \\ c_2^2 + s_2^2 - 1 = 0. \end{cases} \tag{6}$$

Example 4 For a 2-link arm with $l_1 = 1$ and $l_2 = 2$ which moves along the line $x = 0$ in order to reach the point $(a, b) = (2, 2)$ we have to solve the following sysytem:

$$\begin{cases} 2(c_1c_2 - s_1s_2) + c_1 - 2 = 0, \\ -2(s_1c_2 + c_1s_2) - s_1 - 2 = 0, \\ c_1^2 + s_1^2 - 1 = 0, \\ c_2^2 + s_2^2 - 1 = 0. \end{cases}$$

Using CoCoA we calculate its Gröbner bases as follows:

$$\begin{cases} s_1 + s_1 + c_1 = 0, \\ 2s_1 + s_2 + \frac{5}{4} = 0, \\ \frac{1}{2}s_2^2 + c_2 - \frac{31}{32} = 0, \\ -\frac{1}{2}c_2 + \frac{3}{8} = 0. \end{cases}$$

Reachability problem for planar arms inside a general polygon

We study this problem in two steps:

1. The confining region for the 2-link planar arms is a polygon without hole.
2. The confining region for the 2-link planar arms is a polygon with holes.

Reachability when the confining region is a polygon without any hole

Suppose that the confining region is a polygon P, the 2-link arm (Fig. 2), the global coordinate system (x_1, y_1), the local coordinate systems $(x_2, y_2), (x_3, y_3)$ and (x_4, y_4) are defined as in section "Reachability problem for 2-link planar arms without confining region". Further assume that $p = (a, b)$ in (x_1, y_1) system is given. Let q be a point somewhere inside P and V be the set of vertices of P which are visible from q, as usual its cardinal is denoted by $|V|$. We choose an order on V by rotative plane sweep. Define D_i to be the segment between q and the i-th element of the ordered set V, let m_i be the inclination of D_i and also define α_i to be the angle between D_i and D_{i+1} in q, i.e. $\alpha_i = \tan^{-1}\left(\frac{m_i - m_{i+1}}{1+m_i m_{i+1}}\right)$. Then $\sum_{i=1}^{|V|} \alpha_i = 2\pi$, which we call it *Visibility Condition* in the following. One can independently write this conditions for joint 2 and hand and do the same as before (see Fig. 6). Hence we have the following algorithm:

Partial Algorithm 4:
Task: Reachability for 2-link planar arms inside a polygon without holes.
Steps:

1 **Find** V_1, the set of vertices that are visible from joint 2.
2 **Find** V_2, the set of vertices that are visible from hand (joint 3).
3 **For each** $i = 1, \ldots, |V_1|$, **do**
4 **Compute** m_i, the inclination of D_i (the segment between joint 2 and the i-th element of V_1).
5 **For each** $i = 1, \ldots, |V_2|$, **do**
6 **Compute** m'_i, the inclination of D'_i (the segment between hand and the i-th element of V_2).
7 **Find** the (x_1, y_1) coordinate of hand which is a function of c_i and s_i subject to the constraints $c_i^2 + s_i^2 - 1 = 0;\ i = 1, 2.$
8 **Add** the visibility conditions to the system of equations in (1).
9 Apply **Groebner basis** to solve the following system of equations:

$$\begin{cases} l_2(c_1c_2 - s_1s_2) + l_1c_1 = a, \\ l_2(s_1c_2 + c_1s_2) + l_1s_1 = b, \\ c_1^2 + s_1^2 - 1 = 0, \\ c_2^2 + s_2^2 - 1 = 0, \\ \sum_{i=1}^{|V|} \tan^{-1}\left(\dfrac{m_i - m_{i+1}}{1 + m_i m_{i+1}}\right) = 2\pi, \\ \sum_{i=1}^{|V|} \tan^{-1}\left(\dfrac{m'_i - m'_{i+1}}{1 + m'_i m'_{i+1}}\right) = 2\pi. \end{cases} \tag{7}$$

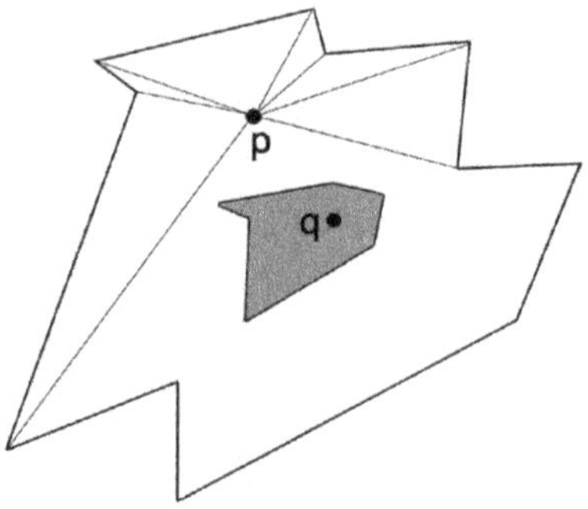

Fig. 7 A polygon P with a hole, p is inside P and q is in the hole

Reachability when the confining region is a polygon with some holes

We explain the problem for polygon P with only one hole. It is easy to check that this method can be used for polygons with more than one hole. Suppose that the confining region is a polygon P with a hole. For any point p in P, we define V to be the set of all vertices in P that are visible from p and are not on the boundary of the hole. It is obvious that if p is in the hole (like q), then it can not see any vertex of the boundary of P except the hole boundary, so in this case $V = \{\}$ and we do not meet the condition of section "Reachability when the shoulder can move along a segment" (see Fig. 7). Otherwise we are able to find all the arm joints that are in P. Hence in this case Partial Algorithm 4 can be applied directly except that V_1 (resp. V_2) are defined as the set of all vertices in P that are visible from joint 2 (resp. hand) and are not on the boundary of the hole.

Conclusions and future works

We studied and presented several algorithms for the reachability problem inside the circle, a general polygon with/without some holes for 2-link planar robot arms. Also we considered the case that the shoulder of arm can move along a given segment, circle, parabola or a certain path. For an unknown path it should be work out more seriously. Furthermore the method could be extended to a 3-dimensional space when joints are spherical by using Euler matrix [1].

Acknowledgments The authors are grateful to the referee for his/her comments and suggestions. The second author is also thankful to the Elite National Foundation of Iran for partial financial support.

References

1. Arfken, G., Weber, H.J.: Mathematical Methods for Physicists. Academic Press, Dublin (1995)
2. Bardet, M.: On the complexity of a Gröbner bases algorithm, algorithms seminar 2002–2004, INRIA, pp. 85–92 (2005). http://algo.inria.fr/seminars/
3. Buchberger, B.: A criterion for detecting unnecessary reductions in the construction of Gröbner bases. In: Ng, E.W. (ed.) Proceedings of the EUROSAM 79 symposium on symbolic
4. Buchberger, B.: Algebraic Manipulation, Marseille, June 26–28, 1979, Lecture Notes in Computer Science, vol. 72. Springer, Berlin (1979)
5. Buchberger, B.: Applications of Gröbner bases in non-linear computational geometry. In: Janssen, R. (ed.) Trends in Computer Algebra, LNCS vol. 296, pp. 25–80. Spiringer, Berlin (1987)
6. Buchberger, B.: Gröbner bases: an algorithmic method in polynomial ideal theory. In: Bose, N.K. (ed.) Multidimensional System Theory, pp. 184–232. D. Reided Publishing Company, Dordrecht (1985)
7. Buchberger, B.: Gröbner bases and Applications. Cambridge University Press, Cambridge (1998)
8. Capani, A., Niesi, G., Robbiano, L.: CoCoA: computations in commutative algebra. http://cocoa.dima.unige.it/
9. Cox, D., Little, J., O'shea, D.: Ideals, Varieties, and Algorithms. Springer, New York (1992)
10. Hopcroft, J., Joseph, D., Whitesides, S.: Movement problems for 2-dimensional linkages. SIAM J. Comput. **13**, 610–629 (1984)
11. Pei, N.: On the reconfiguration and reachability of chains. Ph.D. Thesis, School of Computer Science, McGill University (November 1990)
12. Robbiano, L.: Term orderings on the polynomial ring, in EUROCAL '85. In: Caviness, B.F.: (ed.) Springer LNCS vol. 204, pp. 595–630 (1985)
13. Suzuki, I., Yamashita, M.: Designing multi-link robot arms in a convex polygon. Int J Comput Geom Appl **6**(4), 461–486 (1996)

27

Some applications of a hypergeometric identity

M. R. Eslahchi[1] **· Mohammad Masjed-Jamei**[2]

Abstract In this paper, we use a general identity for generalized hypergeometric series to obtain some new applications. The first application is a hypergeometric-type decomposition formula for elementary special functions and the second one is a generalization of the well-known Euler identity $e^{ix} = \cos x + i \sin x$ and an extension of hyperbolic functions in the sequel. Applying the mentioned identity on classical hypergeometric orthogonal polynomials and deriving summation formulae for some classical summation theorems are two further applications of this identity.

Keywords Hypergeometric functions · Euler identity · Hyperbolic functions · Classical hypergeometric orthogonal polynomials · Classical summation theorems

Mathematical Subject Classification 33C20 · 33C05 · 33C15

Introduction

One of the main reasons for introducing and developing the generalized hypergeometric series is that many special functions [2, 15, 16] can be represented in terms of them and therefore their initial properties can be directly found via the initial properties of hypergeometric functions. Also, they appear as solutions of many important ordinary differential equations [7, 11, 20]. Hence, finding any property of them may be valuable [1, 5, 17]. The generalized hypergeometric function

$$ {}_pF_q\left(\begin{matrix} a_1, & a_2, & \dots & a_p \\ b_1, & b_2, & \dots & b_q \end{matrix} \middle| z\right) = \sum_{k=0}^{\infty} \frac{(a_1)_k (a_2)_k \dots (a_p)_k}{(b_1)_k (b_2)_k \dots (b_q)_k} \frac{z^k}{k!}, \tag{1} $$

in which $(r)_k = \prod_{j=0}^{k-1} (r+j) = \Gamma(r+k)/\Gamma(r)$ denotes the Pochhammer symbol [2, 8] and z may be a complex variable is indeed a Taylor series expansion for a function, say f, as $\sum_{k=0}^{\infty} c_k^* z^k$ with $c_k^* = f^{(k)}(0)/k!$ for which the ratio of successive terms can be written as

$$ \frac{c_{k+1}^*}{c_k^*} = \frac{(k+a_1)(k+a_2)\dots(k+a_p)}{(k+b_1)(k+b_2)\dots(k+b_q)(k+1)}. $$

According to the ratio test [6], the series (1) is convergent for any $p \le q+1$. In fact, it converges in $|z|<1$ for $p=q+1$, converges everywhere for $p<q+1$ and converges nowhere ($z \neq 0$) for $p > q+1$. Moreover, for $p = q+1$ it absolutely converges on $|z| = 1$ if the condition

$$ A^* = \mathrm{Re}\left(\sum_{j=1}^{q} b_j - \sum_{j=1}^{q+1} a_j\right) > 0, $$

holds and is conditionally convergent for $|z| = 1$ and $z \neq 1$ if $-1 < A^* \le 0$ and is finally divergent for $|z| = 1$ and $z \neq 1$ if $A^* \le -1$.

This paper is organized as follows: in the next section, we use a general identity for any arbitrary hypergeometric series of type (1) to obtain some new applications. The first application is a hypergeometric-type decomposition

✉ M. R. Eslahchi
eslahchi@modares.ac.ir

Mohammad Masjed-Jamei
mmjamei@kntu.ac.ir; mmjamei@yahoo.com

1 Department of Applied Mathematics, Faculty of Mathematical Sciences, Tarbiat Modares University, P. O. Box 14115-134, Tehran, Iran

2 Department of Mathematics, K.N.Toosi University of Technology, P. O. Box 18315-1818, Tehran, Iran

formula for elementary special functions. The second application is a generalization of the well-known Euler identity $e^{ix} = \cos x + i \sin x$ and then an extension of hyperbolic functions. Applying the mentioned identity to classical hypergeometric orthogonal polynomials and deriving summation formulae for some classical summation theorems are two further applications of this identity.

Some applications of a general identity for hypergeometric series

The following result is given in [18, p. 441, formula 43] (see also [6]).

Corollary. *For any natural number m we have*

$$
{}_pF_q\left(\begin{matrix} a_1, a_2, \cdots & a_p \\ b_1, b_2, \ldots & b_q \end{matrix}\,\middle|\, z\right) = \sum_{k=0}^{m-1} \frac{(a_1)_k (a_2)_k \cdots (a_p)_k}{(b_1)_k (b_2)_k \cdots (b_q)_k} \frac{z^k}{k!} \, {}_{(mp+1)}F_{(mq+m)} \left(\begin{matrix} \vec{A}_{1,k}, \vec{A}_{2,k}, \ldots & \vec{A}_{p,k}, 1 \\ \vec{B}_{1,k}, \vec{B}_{2,k}, \ldots & \vec{B}_{q,k}, \vec{I}_{1,k} \end{matrix}\,\middle|\, m^{(p-q-1)m} z^m\right), \qquad (2)
$$

where

$$
\vec{A}_{j,k} = \left(\frac{a_j+k}{m}, \frac{a_j+1+k}{m}, \ldots, \frac{a_j+m-1+k}{m}\right) \quad (j = 1, 2, \ldots, p),
$$

$$
\vec{B}_{j,k} = \left(\frac{b_j+k}{m}, \frac{b_j+1+k}{m}, \ldots, \frac{b_j+m-1+k}{m}\right) \quad (j = 1, 2, \ldots, q),
$$

and

$$
\vec{I}_{1,k} = \left(\frac{1+k}{m}, \frac{2+k}{m}, \ldots, \frac{m+k}{m}\right).
$$

Application 1. Identities (2) can be interpreted as a decomposition formula for many elementary special functions whose hypergeometric representations are known.

Example 1. Take $(p, q) = (0, 0)$ and $m = 2, 3, 4$ in (2) to, respectively, obtain

$$
e^z = {}_0F_1\left(\begin{matrix} - \\ 1/2 \end{matrix}\,\middle|\, \frac{1}{4}z^2\right) + z\, {}_0F_1\left(\begin{matrix} - \\ 3/2 \end{matrix}\,\middle|\, \frac{1}{4}z^2\right) = \cosh z + \sinh z,
$$

$$
e^z = {}_0F_2\left(\begin{matrix} - \\ 1/3,\ 2/3 \end{matrix}\,\middle|\, \frac{1}{27}z^3\right) + z\, {}_0F_2\left(\begin{matrix} - \\ 2/3,\ 4/3 \end{matrix}\,\middle|\, \frac{1}{27}z^3\right) + \frac{z^2}{2}\, {}_0F_2\left(\begin{matrix} - \\ 4/3,\ 5/3 \end{matrix}\,\middle|\, \frac{1}{27}z^3\right),
$$

and

$$
e^z = {}_0F_3\left(\begin{matrix} - \\ 1/4,\ 1/2,\ 3/4 \end{matrix}\,\middle|\, \frac{1}{256}z^4\right) + z\, {}_0F_3\left(\begin{matrix} - \\ 1/2,\ 3/4,\ 5/4 \end{matrix}\,\middle|\, \frac{1}{256}z^4\right) + \frac{z^2}{2}\, {}_0F_3\left(\begin{matrix} - \\ 3/4,\ 5/4,\ 3/2 \end{matrix}\,\middle|\, \frac{1}{256}z^4\right) + \frac{z^3}{6}\, {}_0F_3\left(\begin{matrix} - \\ 5/4,\ 3/2,\ 7/4 \end{matrix}\,\middle|\, \frac{1}{256}z^4\right).
$$

Note that the first representation corresponds to the decomposition of the exponential function in the even and odd parts. Moreover, the general case of the three above relations is given as

$$
e^z = \sum_{k=0}^{m-1} \frac{z^k}{k!}\, {}_1F_m\left(\begin{matrix} 1 \\ \frac{1+k}{m}, \frac{2+k}{m}, \ldots, \frac{m+k}{m} \end{matrix}\,\middle|\, \frac{1}{m^m} z^m\right).
$$

Example 2. Since ${}_1F_0\left(\begin{matrix} a \\ - \end{matrix}\,\middle|\, z\right) = (1-z)^{-a}$, identity (2) for $m = 2, 3, 4$, respectively, reads as

$$
(1-z)^{-a} = {}_2F_1\left(\begin{matrix} a/2,\ (a+1)/2 \\ 1/2 \end{matrix}\,\middle|\, z^2\right) + az\, {}_2F_1\left(\begin{matrix} (a+1)/2,\ (a+2)/2 \\ 3/2 \end{matrix}\,\middle|\, z^2\right),
$$

$$
(1-z)^{-a} = {}_3F_2\left(\begin{matrix} a/3,\ (a+1)/3,\ (a+2)/3 \\ 1/3,\ 2/3 \end{matrix}\,\middle|\, z^3\right) + a\frac{z}{1!}\, {}_3F_2\left(\begin{matrix} (a+1)/3,\ (a+2)/3,\ (a+3)/3 \\ 2/3,\ 4/3 \end{matrix}\,\middle|\, z^3\right) + a(a+1)\frac{z^2}{2}\, {}_3F_2\left(\begin{matrix} (a+2)/3,\ (a+3)/3,\ (a+4)/3 \\ 4/3,\ 5/3 \end{matrix}\,\middle|\, z^3\right),
$$

and

$$
(1-z)^{-a} = {}_4F_3\left(\begin{matrix} a/4,\ (a+1)/4,\ (a+2)/4,\ (a+3)/4 \\ 1/4,\ 1/2,\ 3/4 \end{matrix}\,\middle|\, z^4\right) + az\, {}_4F_3\left(\begin{matrix} (a+1)/4,\ (a+2)/4,\ (a+3)/4,\ (a+4)/4 \\ 1/2,\ 3/4,\ 5/4 \end{matrix}\,\middle|\, z^4\right) + a(a+1)\frac{z^2}{2}\, {}_4F_3\left(\begin{matrix} (a+2)/4,\ (a+3)/4,\ (a+4)/4,\ (a+5)/4 \\ 3/4,\ 5/4,\ 3/2 \end{matrix}\,\middle|\, z^4\right) + a(a+1)(a+2)\frac{z^3}{6}\, {}_4F_3\left(\begin{matrix} (a+3)/4,\ (a+4)/4,\ (a+5)/4,\ (a+6)/4 \\ 5/4,\ 3/2,\ 7/4 \end{matrix}\,\middle|\, z^4\right).
$$

Example 3. First, replacing $(p, q) = (0, 1)$ in (2) for $m = 2$ yields

$$
{}_0F_1\left(\begin{matrix} - \\ b \end{matrix}\middle| z\right) = {}_0F_3\left(\begin{matrix} - \\ b/2,\ (b+1)/2,\ 1/2 \end{matrix}\middle| \frac{1}{16}z^2\right) + \frac{z}{b}{}_0F_3\left(\begin{matrix} - \\ (b+1)/2,\ (b+2)/2,\ 3/2 \end{matrix}\middle| \frac{1}{16}z^2\right). \quad (3)
$$

and

$$
\sin z = z{}_0F_3\left(\begin{matrix} - \\ 1/2,\ 3/4,\ 5/4 \end{matrix}\middle| \frac{1}{256}z^4\right) - \frac{z^3}{6}{}_0F_3\left(\begin{matrix} - \\ 5/4,\ 3/2,\ 7/4 \end{matrix}\middle| \frac{1}{256}z^4\right).
$$

Application 2. A generalization of Euler's identity. If $m=2$ is replaced in (2), then we have

$$
{}_pF_q\left(\begin{matrix} a_1,\ a_2,\ \dots\ a_p \\ b_1,\ b_2,\ \dots\ b_q \end{matrix}\middle| z\right) = {}_{2p}F_{2q+1}\left(\begin{matrix} a_1/2,\ (1+a_1)/2,\ \dots\ a_p/2,\ (1+a_p)/2 \\ b_1/2,\ (1+b_1)/2,\ \dots\ b_q/2,\ (1+b_q)/2,\ 1/2 \end{matrix}\middle| 4^{p-q-1}z^2\right) + \frac{a_1a_2\dots a_p}{b_1b_2\dots b_q}z\,{}_{2p}F_{2q+1}\left(\begin{matrix} (a_1+1)/2,\ (a_1+2)/2,\ \dots\ (a_p+1)/2,\ (a_p+2)/2 \\ (b_1+1)/2,\ (b_1+2)/2,\ \dots\ (b_q+1)/2,\ (b_q+2)/2,\ 3/2 \end{matrix}\middle| 4^{p-q-1}z^2\right). \quad (4)
$$

By noting that z may be a complex variable, $z=ix$ in (4) gives

$$
{}_pF_q\left(\begin{matrix} a_1,\ a_2,\ \dots\ a_p \\ b_1,\ b_2,\ \dots\ b_q \end{matrix}\middle| ix\right) = {}_{2p}F_{2q+1}\left(\begin{matrix} a_1/2,\ (1+a_1)/2,\ \dots\ a_p/2,\ (1+a_p)/2 \\ b_1/2,\ (1+b_1)/2,\ \dots\ b_q/2,\ (1+b_q)/2,\ 1/2 \end{matrix}\middle| -4^{p-q-1}x^2\right) + i\frac{a_1a_2\dots a_p}{b_1b_2\dots b_q}x\,{}_{2p}F_{2q+1}\left(\begin{matrix} (a_1+1)/2,\ (a_1+2)/2,\ \dots\ (a_p+1)/2,\ (a_p+2)/2 \\ (b_1+1)/2,\ (b_1+2)/2,\ \dots\ (b_q+1)/2,\ (b_q+2)/2,\ 3/2 \end{matrix}\middle| -4^{p-q-1}x^2\right), \quad (5)
$$

Now since the cosine and sine functions can be written as

$$
\cos z = {}_0F_1\left(\begin{matrix} - \\ 1/2 \end{matrix}\middle| -\frac{1}{4}z^2\right) \text{ and } \sin z = z{}_0F_1\left(\begin{matrix} - \\ 3/2 \end{matrix}\middle| -\frac{1}{4}z^2\right),
$$

relation (3), respectively, yields

$$
\cos z = {}_0F_3\left(\begin{matrix} - \\ 1/4,\ 1/2,\ 3/4 \end{matrix}\middle| \frac{1}{256}z^4\right) - \frac{z^2}{2}{}_0F_3\left(\begin{matrix} - \\ 3/4,\ 5/4,\ 3/2 \end{matrix}\middle| \frac{1}{256}z^4\right),
$$

which is a generalization of Euler's identity for $p=q=0$, because

$$
{}_0F_0\left(\begin{matrix} - \\ - \end{matrix}\middle| ix\right) = {}_0F_1\left(\begin{matrix} - \\ 1/2 \end{matrix}\middle| -x^2/4\right) + ix{}_0F_1\left(\begin{matrix} - \\ 3/2 \end{matrix}\middle| -x^2/4\right) \Leftrightarrow e^{ix} = \cos x + i\sin x.
$$

Subsequently, the well-known hyperbolic functions [2, 16] can be generalized via (4) and (5) and be defined, respectively, as

$$
{}_pCh_q\left(\begin{matrix} a_1,\ a_2,\ \dots\ a_p \\ b_1,\ b_2,\ \dots\ b_q \end{matrix}\middle| x\right) = \frac{1}{2}\left({}_pF_q\left(\begin{matrix} a_1,\ a_2,\ \dots\ a_p \\ b_1,\ b_2,\ \dots\ b_q \end{matrix}\middle| x\right) + {}_pF_q\left(\begin{matrix} a_1,\ a_2,\ \dots\ a_p \\ b_1,\ b_2,\ \dots\ b_q \end{matrix}\middle| -x\right)\right) = {}_{2p}F_{2q+1}\left(\begin{matrix} a_1/2,\ (1+a_1)/2,\ \dots\ a_p/2,\ (1+a_p)/2 \\ b_1/2,\ (1+b_1)/2,\ \dots\ b_q/2,\ (1+b_q)/2,\ 1/2 \end{matrix}\middle| 4^{p-q-1}x^2\right), \quad (6)
$$

Table 1 Characteristics of ten sequences of orthogonal polynomials

Symbol	Weight function	Kind, interval and parameters constraint
$P_n^{(u,v)}(x)$	$W\left(\begin{matrix}-u-v, & -u+v \\ -1, & 0, & 1\end{matrix}\middle\| x\right)=(1-x)^u(1+x)^v$	Infinite, $[-1,1]$, $\forall n,\ u>-1, v>-1$
$L_n^{(u)}(x)$	$W\left(\begin{matrix}-1, & u \\ 0, & 1, & 0\end{matrix}\middle\| x\right)=x^u\exp(-x)$	Infinite, $[0,\infty)$, $\forall n,\ u>-1$
$H_n(x)$	$W\left(\begin{matrix}-2, & 0 \\ 0, & 0, & 1\end{matrix}\middle\| x\right)=\exp(-x^2)$	Infinite, $(-\infty,\infty)$
$J_n^{(u,v)}(x)$	$W\left(\begin{matrix}-2u, & v \\ 1, & 0, & 1\end{matrix}\middle\| x\right)=(1+x^2)^{-u}e^{v\arctan x}$	Finite, $(-\infty,\infty)$, $\max n<u-1/2,\ v\in\mathbb{R}$
$M_n^{(u,v)}(x)$	$W\left(\begin{matrix}-u, & v \\ 1, & 1, & 0\end{matrix}\middle\| x\right)=x^v(x+1)^{-(u+v)}$	Finite, $[0,\infty)$, $\max n<(u-1)/2,\ v>-1$
$N_n^{(u)}(x)$	$W\left(\begin{matrix}-u, & 1 \\ 1, & 0, & 0\end{matrix}\middle\| x\right)=x^{-u}\exp(-1/x)$	Finite, $[0,\infty)$, $\max n<(u-1)/2$
$S_n\left(\begin{matrix}-2u-2v-2, & 2u \\ -1, & 1\end{matrix}\middle\| x\right)$	$W^*\left(\begin{matrix}-2u-2v, & 2u \\ -1, & 1\end{matrix}\middle\| x\right)=x^{2u}(1-x^2)^v$	Infinite, $[-1,1]$, $u>-1/2,\ v>-1$
$S_n\left(\begin{matrix}-2, & 2u \\ 0, & 1\end{matrix}\middle\| x\right)$	$W^*\left(\begin{matrix}-2, & 2u \\ 0, & 1\end{matrix}\middle\| x\right)=x^{2u}\exp(-x^2)$	Infinite, $(-\infty,\infty)$, $u>-1/2$
$S_n\left(\begin{matrix}-2u-2v+2, & -2u \\ 1, & 1\end{matrix}\middle\| x\right)$	$W^*\left(\begin{matrix}-2u-2v, & -2u \\ 1, & 1\end{matrix}\middle\| x\right)=x^{-2u}(1+x^2)^{-v}$	Finite, $(-\infty,\infty)$, $\max n<u+v-1/2,\ u<1/2,\ v>0$
$S_n\left(\begin{matrix}-2u+2, & 2 \\ 1, & 0\end{matrix}\middle\| x\right)$	$W^*\left(\begin{matrix}-2a, & 2 \\ 1, & 0\end{matrix}\middle\| x\right)=x^{-2a}exp(-1/x^2)$	Finite, $(-\infty,\infty)$, $\max n<u-1/2$

and

$$
\begin{aligned}
&{}_pSh_q\left(\begin{matrix}a_1, & a_2, & \ldots & a_p \\ b_1, & b_2, & \ldots & b_q\end{matrix}\middle| x\right)=\frac{1}{2}\left({}_pF_q\left(\begin{matrix}a_1, & a_2, & \ldots & a_p \\ b_1, & b_2, & \ldots & b_q\end{matrix}\middle| x\right)-{}_pF_q\left(\begin{matrix}a_1, & a_2, & \ldots & a_p \\ b_1, & b_2, & \ldots & b_q\end{matrix}\middle| -x\right)\right)\\
&=\frac{a_1a_2\ldots a_p}{b_1b_2\ldots b_q}x\,{}_{2p}F_{2q+1}\left(\begin{matrix}(a_1+1)/2, & (a_1+2)/2, & \ldots & (a_p+1)/2, & (a_p+2)/2 \\ (b_1+1)/2, & (b_1+2)/2, & \ldots & (b_p+1)/2, & (b_p+1)/2, & 3/2\end{matrix}\middle| 4^{p-q-1}x^2\right).
\end{aligned}
\tag{7}
$$

It is clear that for $(p,q)=(0,0)$, relations (6) and (7) reduce to

$$
\begin{aligned}
{}_0Ch_0\left(\begin{matrix}- \\ -\end{matrix}\middle| x\right)&=\cosh x=\frac{1}{2}(e^x+e^{-x})\text{ and }{}_0Sh_0\left(\begin{matrix}- \\ -\end{matrix}\middle| x\right)\\
&=\sinh x=\frac{1}{2}(e^x-e^{-x}).
\end{aligned}
$$

Application 3. A decomposition formula for classical hypergeometric orthogonal polynomials: there are ten sequences of hypergeometric polynomials [7, 12–14] that are orthogonal with respect to the Pearson distribution family

$$W\left(\begin{matrix}d, & e \\ a, & b, & c\end{matrix}\middle| x\right)=\exp\left(\int\frac{dx+e}{ax^2+bx+c}dx\right)\quad(a,b,c,d,e\in\mathbb{R}),$$

and its symmetric analog [11]

$$W^*\left(\begin{matrix}r, & s \\ p, & q\end{matrix}\middle| x\right)=\exp\left(\int\frac{rx^2+s}{x(px^2+q)}dx\right)\qquad(p,q,r,s\in\mathbb{R}).$$

Five of them are infinitely orthogonal with respect to special cases of the two above-mentioned weight functions and five other ones are finitely orthogonal [12–14] which are limited to some parametric constraints. The following Table 1 shows their main characteristics.
where the sequence

$$
\begin{aligned}
\Phi_n(x)=S_n\left(\begin{matrix}r, & s \\ p, & q\end{matrix}\middle| x\right)&=\sum_{k=0}^{[n/2]}\binom{[n/2]}{k}\\
&\left(\prod_{i=0}^{[n/2]-(k+1)}\frac{\left(2i+(-1)^{n+1}+2\,[n/2]\right)p+r}{\left(2i+(-1)^{n+1}+2\right)q+s}\right)x^{n-2k},
\end{aligned}
$$

is a basic class of symmetric orthogonal polynomials [12] satisfying the equation

$$
\begin{aligned}
&x^2(px^2+q)\,\Phi_n''(x)+x(rx^2+s)\,\Phi_n'(x)-\big(n(r+(n-1)p)x^2\\
&\quad+(1-(-1)^n)\,s/2\big)\,\Phi_n(x)=0.
\end{aligned}
$$

For example, consider the shifted Jacobi polynomials [20]

$$P_{n,+}^{(\alpha,\beta)}(x) = {}_2F_1\left(\begin{matrix} -n, & n+\alpha+\beta+1 \\ & \alpha+1 \end{matrix} \middle| x\right) \quad (\alpha,\beta > -1), \tag{8}$$

which are orthogonal with respect to the shifted beta distribution on $[0,1]$ as

$$\int_0^1 x^\alpha (1-x)^\beta P_{n,+}^{(\alpha,\beta)}(x) P_{m,+}^{(\alpha,\beta)}(x) dx = \frac{n!\,\Gamma^2(\alpha+1)\Gamma(n+\beta+1)}{(2n+\alpha+\beta+1)\Gamma(n+\alpha+\beta+1)\Gamma(n+\alpha+1)} \delta_{n,m}.$$

If (8) is replaced in (2) for, e.g., $m = 2$, then one gets (see also [10] in this regard)

$$P_{n,+}^{(\alpha,\beta)}(x) = {}_4F_3\left(\begin{matrix} -n/2, & (1-n)/2, & (n+\alpha+\beta+1)/2, & (n+\alpha+\beta+2)/2 \\ & (\alpha+1)/2, & (\alpha+2)/2, & 1/2 \end{matrix} \middle| x^2\right) - \frac{n(n+\alpha+\beta+1)}{\alpha+1} x\, {}_4F_3\left(\begin{matrix} (1-n)/2, & (2-n)/2, & (n+\alpha+\beta+2)/2, & (n+\alpha+\beta+3)/2 \\ & (\alpha+2)/2, & (\alpha+3)/2, & 3/2 \end{matrix} \middle| x^2\right). \tag{9}$$

Hence, two sequences of polynomials can be defined by (9) as

$$\frac{P_{n,+}^{(\alpha,\beta)}(\sqrt{x}) + P_{n,+}^{(\alpha,\beta)}(-\sqrt{x})}{2} = {}_4F_3\left(\begin{matrix} -n/2, & (1-n)/2, & (n+\alpha+\beta+1)/2, & (n+\alpha+\beta+2)/2 \\ & (\alpha+1)/2, & (\alpha+2)/2, & 1/2 \end{matrix} \middle| x\right),$$

and

$$\frac{P_{n,+}^{(\alpha,\beta)}(\sqrt{x}) - P_{n,+}^{(\alpha,\beta)}(-\sqrt{x})}{2\sqrt{x}} = -\frac{n(n+\alpha+\beta+1)}{\alpha+1} \times {}_4F_3\left(\begin{matrix} (1-n)/2, & (2-n)/2, & (n+\alpha+\beta+2)/2, & (n+\alpha+\beta+3)/2 \\ & (\alpha+2)/2, & (\alpha+3)/2, & 3/2 \end{matrix} \middle| x\right).$$

Another classical example are the Laguerre polynomials

$$y = \widehat{L}_n^{(\alpha)}(x) = {}_1F_1\left(\begin{matrix} -n \\ \alpha+1 \end{matrix} \middle| x\right) \quad (\alpha > -1), \tag{10}$$

that satisfy the differential equation [7]

$$xy'' + (\alpha+1-x)y' + ny = 0,$$

and the orthogonality relation [7]

$$\int_0^\infty x^\alpha e^{-x} \widehat{L}_n^{(\alpha)}(x) \widehat{L}_m^{(\alpha)}(x)\, \mathrm{d}x = \frac{n!\Gamma(\alpha+1)}{(\alpha+1)_n} \delta_{n,m}.$$

If (10) is replaced in (2) for e.g. $m=3$, then one gets

$$\widehat{L}_n^{(\alpha)}(x) = {}_3F_5\left(\begin{matrix} -n/3,\ (1-n)/3,\ (2-n)/3 \\ (\alpha+1)/3,\ (\alpha+2)/3,\ (\alpha+3)/3,\ 1/3,\ 2/3 \end{matrix}\,\middle|\, \frac{x^3}{27}\right)$$
$$-\frac{n}{\alpha+1}x\,{}_3F_5\left(\begin{matrix} (1-n)/3,\ (2-n)/3,\ (3-n)/3 \\ (\alpha+2)/3,\ (\alpha+3)/3,\ (\alpha+4)/3,\ 2/3,\ 4/3 \end{matrix}\,\middle|\, \frac{x^3}{27}\right)$$
$$+\frac{n(n-1)}{2(\alpha+1)(\alpha+2)}x^2\,{}_3F_5\left(\begin{matrix} (2-n)/3,\ (3-n)/3,\ (4-n)/3 \\ (\alpha+3)/3,\ (\alpha+4)/3,\ (\alpha+5)/3,\ 4/3,\ 5/3 \end{matrix}\,\middle|\, \frac{x^3}{27}\right).$$

See also [9] in the sense of incomplete symmetric orthogonal polynomials of Laguerre type. Some other works related to classical hypergeometric orthogonal polynomials can be found in [3, 4, 19].

Application 4. The classical summation theorems of hypergeometric series (such as the Gauss, Kummer and Bailey theorems for ${}_2F_1$ and the Watson and Dixon theorems for ${}_3F_2$) play an important role in evaluating hypergeometric series at specific points [18]. In this section, by expressing the aforesaid theorems we employ identity (2) for them.

- *Gauss's theorem:*

$${}_2F_1\left(\begin{matrix} a,\ b \\ c \end{matrix}\,\middle|\, 1\right) = \frac{\Gamma(c)\Gamma(c-a-b)}{\Gamma(c-a)\Gamma(c-b)},$$

provided that $\mathrm{Re}(c-a-b)>0$.

- *Gauss's second theorem:*

$${}_2F_1\left(\begin{matrix} a,\ b \\ (a+b+1)/2 \end{matrix}\,\middle|\, \frac{1}{2}\right) = \frac{\sqrt{\pi}\,\Gamma\left(\frac{a+b+1}{2}\right)}{\Gamma\left(\frac{a+1}{2}\right)\Gamma\left(\frac{b+1}{2}\right)},$$

provided that $\mathrm{Re}(a) > -1$ and $\mathrm{Re}(b) > -1$.

- *Kummer's theorem:*

$${}_2F_1\left(\begin{matrix} a,\ b \\ 1+a-b \end{matrix}\,\middle|\, -1\right) = \frac{\Gamma(1+a-b)\Gamma\left(1+\frac{a}{2}\right)}{\Gamma(1+a)\,\Gamma\left(1+\frac{a}{2}-b\right)}.$$

provided that $\mathrm{Re}(a) > -1$, $\mathrm{Re}(b)<1$ and $\mathrm{Re}(a-b) > -1$.

- *Bailey's theorem* [18]:

$${}_2F_1\left(\begin{matrix} a,\ 1-a \\ c \end{matrix}\,\middle|\, \frac{1}{2}\right) = \frac{\Gamma\left(\frac{c}{2}\right)\Gamma\left(\frac{c+1}{2}\right)}{\Gamma\left(\frac{c+a}{2}\right)\Gamma\left(\frac{c-a+1}{2}\right)}.$$

provided that $\mathrm{Re}(c) > \mathrm{Re}(a) > 0$.

- *Watson's theorem* [18]:

$${}_3F_2\left(\begin{matrix} a,\ b,\ c \\ (a+b+1)/2,\ 2c \end{matrix}\,\middle|\, 1\right) = \frac{\sqrt{\pi}\,\Gamma\left(c+\frac{1}{2}\right)\Gamma\left(\frac{a+b+1}{2}\right)\Gamma\left(c-\frac{a+b-1}{2}\right)}{\Gamma\left(\frac{a+1}{2}\right)\Gamma\left(\frac{b+1}{2}\right)\Gamma\left(c-\frac{a-1}{2}\right)\Gamma\left(c-\frac{b-1}{2}\right)}.$$

provided that $\mathrm{Re}(2c-a-b) > -1$.

- *Dixon's theorem* [18]:

$${}_3F_2\left(\begin{matrix} a,\ b,\ c \\ 1+a-b,\ 1+a-c \end{matrix}\,\middle|\, 1\right) = \frac{\Gamma\left(1+\frac{a}{2}\right)\Gamma(1+a-b)\,\Gamma(1+a-c)\Gamma\left(1+\frac{a}{2}-b-c\right)}{\Gamma(1+a)\,\Gamma\left(1+\frac{a}{2}-b\right)\Gamma\left(1+\frac{a}{2}-c\right)\Gamma(1+a-b-c)}.$$

provided that $\mathrm{Re}(a-2b-2c) > -2$.

Now, if we substitute Gauss's theorem into (8) for e.g. $m=2,3$, then we, respectively, obtain

$${}_4F_3\left(\begin{matrix} a/2,\ (a+1)/2,\ b/2,\ (b+1)/2 \\ c/2,\ (c+1)/2,\ 1/2 \end{matrix}\,\middle|\, 1\right) + \frac{ab}{c}\,{}_4F_3\left(\begin{matrix} (a+1)/2,\ (a+2)/2,\ (b+1)/2,\ (b+2)/2 \\ (c+1)/2,\ (c+2)/2,\ 3/2 \end{matrix}\,\middle|\, 1\right) = \frac{\Gamma(c)\Gamma(c-a-b)}{\Gamma(c-a)\Gamma(c-b)},$$

and

$${}_6F_5\left(\begin{matrix} a/3,\ (a+1)/3,\ (a+2)/3,\ b/3,\ (b+1)/3,\ (b+2)/3 \\ c/3,\ (c+1)/3,\ (c+2)/3,\ 1/3,\ 2/3 \end{matrix}\,\middle|\, 1\right)$$
$$+\frac{ab}{c}\,{}_6F_5\left(\begin{matrix} (a+1)/3,\ (a+2)/3,\ (a+3)/3,\ (b+1)/3,\ (b+2)/3,\ (b+3)/3 \\ (c+1)/3,\ (c+2)/3,\ (c+3)/3,\ 2/3,\ 4/3 \end{matrix}\,\middle|\, 1\right)$$
$$+\frac{a(a+1)b(b+1)}{2c(c+1)}\,{}_6F_5\left(\begin{matrix} (a+2)/3,\ (a+3)/3,\ (a+4)/3,\ (b+2)/3,\ (b+3)/3,\ (b+4)/3 \\ (c+2)/3,\ (c+3)/3,\ (c+4)/3,\ 4/3,\ 5/3 \end{matrix}\,\middle|\, 1\right)$$
$$= \frac{\Gamma(c)\Gamma(c-a-b)}{\Gamma(c-a)\Gamma(c-b)}.$$

The general case of the two above identities for any natural m is as

$$\sum_{k=0}^{m-1}\frac{(a)_k(b)_k}{(c)_k k!}\,{}_{2m+1}F_{2m}\left(\begin{matrix}\frac{a+k}{m},\ldots,\frac{a+m-1+k}{m},\frac{b+k}{m},\ldots,\frac{b+m-1+k}{m},1\\ \frac{c+k}{m},\ldots,\frac{c+m-1+k}{m},\frac{1+k}{m},\ldots,\frac{m+k}{m}\end{matrix}\middle|\,1\right)=\frac{\Gamma(c)\Gamma(c-a-b)}{\Gamma(c-a)\Gamma(c-b)}.$$

And the general case for Gauss's second theorem takes the form

$$\sum_{k=0}^{m-1}\frac{(a)_k(b)_k\,2^{-k}}{\left(\frac{a+b+1}{2}\right)_k k!}\,{}_{2m+1}F_{2m}\left(\begin{matrix}\frac{a+k}{m},\ldots,\frac{a+m-1+k}{m},\frac{b+k}{m},\ldots,\frac{b+m-1+k}{m},1\\ \frac{k+(a+b+1)/2}{m},\ldots,\frac{m-1+k+(a+b+1)/2}{m},\frac{1+k}{m},\ldots,\frac{m+k}{m}\end{matrix}\middle|\,2^{-m}\right)$$
$$=\frac{\sqrt{\pi}\,\Gamma\left(\frac{a+b+1}{2}\right)}{\Gamma\left(\frac{a+1}{2}\right)\Gamma\left(\frac{b+1}{2}\right)},$$

and for Kummer theorem takes the form

$$\sum_{k=0}^{m-1}\frac{(a)_k(b)_k(-1)^k}{(1+a-b)_k k!}\,{}_{2m+1}F_{2m}\left(\begin{matrix}\frac{a+k}{m},\ldots,\frac{a+m-1+k}{m},\frac{b+k}{m},\ldots,\frac{b+m-1+k}{m},1\\ \frac{1+a-b+k}{m},\ldots,\frac{a-b+m+k}{m},\frac{1+k}{m},\ldots,\frac{m+k}{m}\end{matrix}\middle|\,(-1)^m\right)$$
$$=\frac{\Gamma(1+a-b)\Gamma\left(1+\frac{a}{2}\right)}{\Gamma(1+a)\,\Gamma\left(1+\frac{a}{2}-b\right)},$$

and for Bailey theorem takes the form

$$\sum_{k=0}^{m-1}\frac{(a)_k(1-a)_k(2)^{-k}}{(c)_k k!}\,{}_{2m+1}F_{2m}\left(\begin{matrix}\frac{a+k}{m},\ldots,\frac{a+m-1+k}{m},\frac{1-a+k}{m},\ldots,\frac{-a+m+k}{m},1\\ \frac{c+k}{m},\ldots,\frac{c+m-1+k}{m},\frac{1+k}{m},\ldots,\frac{m+k}{m}\end{matrix}\middle|\,2^{-m}\right)$$
$$=\frac{\Gamma\left(\frac{c}{2}\right)\Gamma\left(\frac{c+1}{2}\right)}{\Gamma\left(\frac{c+a}{2}\right)\Gamma\left(\frac{c-a+1}{2}\right)},$$

and for Watson theorem takes the form

$$\sum_{k=0}^{m-1}\frac{(a)_k(b)_k(c)_k}{((a+b+1)/2)_k(2c)_k k!}\times$$
$$_{3m+1}F_{3m}\left(\begin{array}{c}\frac{a+k}{m},\ldots,\frac{a+m-1+k}{m},\frac{b+k}{m},\ldots,\frac{b+m-1+k}{m},\frac{c+k}{m},\ldots,\frac{c+m-1+k}{m},1\\ \frac{k+(a+b+1)/2}{m},\ldots,\frac{m-1+k+(a+b+1)/2}{m},\frac{2c+k}{m},\ldots,\frac{2c+m-1+k}{m},\frac{1+k}{m},\ldots,\frac{m+k}{m}\end{array}\middle|\,1\right)$$
$$=\frac{\sqrt{\pi}\,\Gamma(c+\frac{1}{2})\Gamma(\frac{a+b+1}{2})\,\Gamma(c-\frac{a+b-1}{2})}{\Gamma(\frac{a+1}{2})\,\Gamma(\frac{b+1}{2})\,\Gamma(c-\frac{a-1}{2})\,\Gamma(c-\frac{b-1}{2})},$$

and finally for Dixon's theorem takes the form

$$\sum_{k=0}^{m-1}\frac{(a)_k(b)_k(c)_k}{(1+a-b)_k(1+a-c)_k k!}\times$$
$$_{3m+1}F_{3m}\left(\begin{array}{c}\frac{a+k}{m},\ldots,\frac{a+m-1+k}{m},\frac{b+k}{m},\ldots,\frac{b+m-1+k}{m},\frac{c+k}{m},\ldots,\frac{c+m-1+k}{m},1\\ \frac{1+a-b+k}{m},\ldots,\frac{a-b+m+k}{m},\frac{1+a-c+k}{m},\ldots,\frac{a-c+m+k}{m},\frac{1+k}{m},\ldots,\frac{m+k}{m}\end{array}\middle|\,1\right)$$
$$=\frac{\Gamma(1+\frac{a}{2})\Gamma(1+a-b)\,\Gamma(1+a-c)\Gamma(1+\frac{a}{2}-b-c)}{\Gamma(1+a)\,\Gamma(1+\frac{a}{2}-b)\,\Gamma(1+\frac{a}{2}-c)\Gamma(1+a-b-c)}.$$

Acknowledgments This work is supported by a Grant from "Iran National Science Foundation" No. 92026373.

References

1. Ancarani, L.U., Gasaneo, G.: Derivatives of any order of the hypergeometric function $_pF_q(a_1, \ldots, a_p; b_1, \ldots, b_q; z)$ with respect to the parameters *ai* and *bi*. J. Phys. A: Math. Theor. **43**, 085210 (2010)
2. Andrews, G.E., Askey, R., Roy, R.: Special Functions. Cambridge University Press, Cambridge (2000)
3. Cheikh, Y.B.: Decomposition of Laguerre polynomials with respect to the cyclic group of order n. Appl. Math. Inform. **81**, 51–64 (1997)
4. Cheikh, Y.B.: Decomposition of the Bessel functions with respect to the cycle group of order n. Matematiche (Catania) **52**, 365–378 (1997)
5. Chaudhry, M.A., Qadir, A., Srivastava, H.M., Paris, R.B.: Extended hypergeometric and confluent hypergeometric functions. Appl. Math. Comput. **159**, 589–602 (2004)
6. Gradshteyn, I.S., Ryzhik, I.M.: Table of Integrals, Series, and Products, 7th edn. Academic Press, Elsevier Inc (2007)
7. Koekoek, R., Lesky, P.A., Swarttouw, R.F.: Hypergeometric Orthogonal Polynomials and Their q-Analogues, Springer Monographs in Mathematics. Springer, Heidelberg (2010)
8. Koepf, W.: Power series in computer algebra. J. Symb. Comput. **11**, 581–603 (1992)
9. Masjed-Jamei, M., Koepf, W.: On incomplete symmetric orthogonal polynomials of Laguerre type. Appl. Anal. **90**, 769–775 (2011)
10. Masjed-Jamei, M., Koepf, W.: On incomplete symmetric orthogonal polynomials of Jacobi type. Integral Transforms Spec. Funct. **21**, 655–662 (2010)
11. Masjed-Jamei, M.: On relationships between classical Pearson distributions and Gauss hypergeometric function. Acta Applicandae Mathematicae **109**, 401–411 (2010)
12. Masjed-Jamei, M.: A basic class of symmetric orthogonal polynomials using the extended Sturm-Liouville theorem for symmetric functions. J. Math. Anal. Appl. **325**, 753–775 (2007)
13. Masjed-Jamei, M.: Classical orthogonal polynomials with weight function: $((ax+b)^2+(cx+d)^2)^{-p}$ exp (q arctan $ax+b/cx+d$) on $(-\infty, \infty)$ and a generalization of T and F distributions. Integral Transforms Spec. Funct. **15**, 137–153 (2004)

14. Masjed-Jamei, M.: Three finite classes of hypergeometric orthogonal polynomials and their application in functions approximation. Integral Transforms Spec. Funct. **13**, 169–190 (2002)
15. Mathai, A.M., Saxena, R.K.: Generalized Hypergeometric Functions with Applications in Statistics and Physical Sciences, Lecture Notes in Mathematics, vol. 348. Springer, Berlin, Heidelberg, New York (1973)
16. Nikiforov, A.F., Uvarov, V.B.: Special Functions of Mathematical Physics. Birkhauser, Boston (1988)
17. Ozergin, E., Ozarslan, M.A., Altin, A.: Extension of gamma, beta and hypergeometric functions. J. Comput. Appl. Math. **235**, 4601–4610 (2011)
18. Prudnikov, A.P., Brychkov, Y.A., Marichev, O.I.: Integrals and Series, Volume 3: More Special Functions. Gordon and Breach Science Publishers, Philadelphia (1990)
19. Ronveaux, A., Zarzo, A., Area, I., Godoy, E.: Decomposition of polynomials with respect to the cyclic group of order m. J. Symb. Comput. **28**, 755–765 (1999)
20. Szegö, G.: Orthogonal Polynomials. AMS, Providence (1978)

Permissions

All chapters in this book were first published in MS, by Springer International Publishing AG.; hereby published with permission under the Creative Commons Attribution License or equivalent. Every chapter published in this book has been scrutinized by our experts. Their significance has been extensively debated. The topics covered herein carry significant findings which will fuel the growth of the discipline. They may even be implemented as practical applications or may be referred to as a beginning point for another development.

The contributors of this book come from diverse backgrounds, making this book a truly international effort. This book will bring forth new frontiers with its revolutionizing research information and detailed analysis of the nascent developments around the world.

We would like to thank all the contributing authors for lending their expertise to make the book truly unique. They have played a crucial role in the development of this book. Without their invaluable contributions this book wouldn't have been possible. They have made vital efforts to compile up to date information on the varied aspects of this subject to make this book a valuable addition to the collection of many professionals and students.

This book was conceptualized with the vision of imparting up-to-date information and advanced data in this field. To ensure the same, a matchless editorial board was set up. Every individual on the board went through rigorous rounds of assessment to prove their worth. After which they invested a large part of their time researching and compiling the most relevant data for our readers.

The editorial board has been involved in producing this book since its inception. They have spent rigorous hours researching and exploring the diverse topics which have resulted in the successful publishing of this book. They have passed on their knowledge of decades through this book. To expedite this challenging task, the publisher supported the team at every step. A small team of assistant editors was also appointed to further simplify the editing procedure and attain best results for the readers.

Apart from the editorial board, the designing team has also invested a significant amount of their time in understanding the subject and creating the most relevant covers. They scrutinized every image to scout for the most suitable representation of the subject and create an appropriate cover for the book.

The publishing team has been an ardent support to the editorial, designing and production team. Their endless efforts to recruit the best for this project, has resulted in the accomplishment of this book. They are a veteran in the field of academics and their pool of knowledge is as vast as their experience in printing. Their expertise and guidance has proved useful at every step. Their uncompromising quality standards have made this book an exceptional effort. Their encouragement from time to time has been an inspiration for everyone.

The publisher and the editorial board hope that this book will prove to be a valuable piece of knowledge for researchers, students, practitioners and scholars across the globe.

List of Contributors

Yuji Liu and Shengping Chen
Department of Mathematics, Guangdong University of Finance and Economics, Guangzhou 510000, People's Republic of China

M. Adabitabar Firozja and F. Rezai Balf
Department of Mathematics Qaemshar Branch, Islamic Azad University, Qaemshahr, Iran

S. Firouzian
Department of Mathematics, Payame Noor University (PNU), Tehran, Iran

Aynur Şahin and Metin Başarır
Department of Mathematics, Sakarya University, 54050 Adapazarı, Sakarya, Turkey

M. Mubeen and V. Narayanan
Department of Mathematics, National Institute of Technology Calicut, Calicut, India

Sumit Chandok
Department of Mathematics, Khalsa College of Engineering and Technology, Punjab Technical University, Amritsar 143001, India

Serkan Araci
Department of Economics, Faculty of Economics, Administrative and Social Science, Hasan Kalyoncu University, 27410 Gaziantep, Turkey

Mehmet Acikgoz and Erkan Ağyüz
Department of Mathematics, Faculty of Arts and Science, University of Gaziantep, 27310 Gaziantep, Turkey

Amir Fallahzadeh and Mohammad Ali Fariborzi Araghi1
Department of Mathematics, Islamic Azad University, Central Tehran Branch, P.O. Box 13185.768, Tehran, Iran

A. Acikgoz and N. A. Tas
Department of Mathematics, Balikesir University, 10145 Balikesir, Turkey

M. S. Sarsak
Department of Mathematics, The Hashemite University, P.O. Box 150459, Zarqa 13115, Jordan

Luca Meacci
Dipartimento di Matematica, "Ulisse Dini", Universita` degli Studi di Firenze, Viale Morgagni, 67/A, 50134 Florence, Italy

H. Azadi Kenary and H. Keshavarz
Department of Mathematics, College of Sciences, Yasouj University, Yasouj 75914-353, Iran

Rouzimaimaiti Mahemuti, Ahmadjan Muhammadhaji and Zhidong Teng
College of Mathematics and System Sciences, Xinjiang University, Urumqi 830046, People's Republic of China

Rami Ahmad El-Nabulsi
College of Mathematics and Information Science, Neijiang Normal University, Neijiang 641112, Sichuan, China

A. Amiraslani
STEM Department, University of Hawaii-Maui College, 310 W Kaahumanu Ave, Kahului, HI 96732, USA

Hidetoshi Murakami
Department of Mathematics, National Defense Academy, 1-10-20 Hashirimizu, Yokosuka, Kanagawa 239-8686, Japan

Qiang Luo
Department of Physics, Renmin University of China, Zhongguancun Street, Beijing 100872, China

Zhidan Wang
School of Mathematical Science, Yangzhou University, Siwangting Road, Yangzhou 225002, China

Maasoomah Sadaf and Ghazala Akram
Department of Mathematics, University of the Punjab, Lahore 54590, Pakistan

Shunyou Xia
School of Mathematics and Computer Science, Guizhou Education University, Guiyang 550018, China
College of Computer Science and Technology, Guizhou University, Guiyang 550025, China

Shuwen Xiang and Yanlong Yang
College of Computer Science and Technology, Guizhou University, Guiyang 550025, China

Deping Xu
College of Information Management, Chengdu University of Technology, Chengdu 610059, China

Alina Gavriluţ
Faculty of Mathematics, "Al.I. Cuza" University, Carol I Bd. 11, Iasi 700506, Romania

Maricel Agop
Department of Physics, Gheorghe Asachi Technical University of Ias¸i, Iasi, Romania

H. Murakami
Department of Mathematical Information Science, Tokyo University of Science, Tokyo, Japan

H. Sadeghi
Department of Mathematics, Fereydan Branch, Islamic Azad University, Isfahan, Iran

M. Lashkarizadeh Bami
Department of Mathematics, Faculty of Science, University of Isfahan, P.O. Box 81745-163, Isfahan, Iran

Asghar Arzhang
Department of Mathematics, Karaj Branch, Islamic Azad University, Karaj, Iran

Birol Gunduz
Department of Mathematics, Faculty of Science and Art, Erzincan University, 24000 Erzincan, Turkey

Sezgin Akbulut
Department of Mathematics, Faculty of Science, Ataturk University, 25240 Erzurum, Turkey

Jamal Rezaei Roshan and Shaban Sedghi
Department of Mathematics, Qaemshahr Branch, Islamic Azad University, Qaemshahr, Iran

Nawab Hussain
Department of Mathematics, King Abdulaziz University, P.O. Box 80203, Jeddah 21589, Saudi Arabia

Nabi Shobkolaei
Department of Mathematics, Babol Branch, Islamic Azad University, Babol, Iran

Dae San Kim
Department of Mathematics, Sogang University, Seoul 121-742, Republic of Korea

Taekyun Kim
Department of Mathematics, Kwangwoon University, Seoul 139-701, Republic of Korea

Rahman Farnoosh, Hamidreza Rezazadeh and Amirhossein Sobhani
School of Mathematics, Iran University of Science and Technology, 16844 Narmak, Tehran, Iran

Maryam Behboudi
Department of Statistics, Science and Research Branch, Islamic Azad University, Tehran, Iran

Zahra Nilforoushan
Faculty of Engineering, Kharazmi University, Tehran, Iran

Keivan Borna
Faculty of Mathematics and Computer Science, Kharazmi University, Tehran, Iran

M. R. Eslahchi
Department of Applied Mathematics, Faculty of Mathematical Sciences, Tarbiat Modares University, P. O. Box 14115-134, Tehran, Iran

Mohammad Masjed-Jamei
Department of Mathematics, K.N.Toosi University of Technology, P. O. Box 18315-1818, Tehran, Iran

Index

www.ingramcontent.com/pod-product-compliance
Lightning Source LLC
Chambersburg PA
CBHW082027190326
41458CB00010B/3297
* 9 7 8 1 6 3 2 3 8 6 2 6 7 *